ELECTRICAL MOTOR CONTROLS
Automated Industrial Systems

Third Edition

AMERICAN TECHNICAL PUBLISHERS, INC.
HOMEWOOD, ILLINOIS 60430

Gary Rockis
Glen Mazur

3 4 5 6 7 8 9 – 92 – 9 8 7 6 5 4 3

Printed in the United States of America

Library of Congress Cataloging-in-Publication Data

Rockis, Gary.
    Electrical motor controls : automated industrial systems / Gary
Rockis, Glen Mazur. -- 3rd ed.
    p.   cm.
    Includes index.
    ISBN 0-8269-1666-X
    1. Electric controllers. 2. Electric motors--Electronic control.
    I. Mazur, Glen. II. Title
TK2851 . R63 1993
621 . 46--dc20

General Electric Company
P.O. Box 2913
Bloomington, Illinois 60525

Gould Inc.
Programmable Control Division
35129 Curtis Boulevard
Eastlake, Ohio 44095

Grayhill Inc.
561 Hillgrove Avenue
La Grange, Illinois 60525

Greenlee Tool Division
Ex-Cell-O Corporation
2136 12th Street
Rockford, Illinois 61101

The Hand Tool Institute
707 Westchester Avenue
White Plains, New York 10604

Harvey Hubbell Incorporated
Ensign Electric Division
P.O. Box 7758
Huntington, West Virginia 25778

Ideal Industries, Inc.
1006 Becker Place
Sycamore, Illinois 60178

IDRECO
Route 56
North Oxford, Massachusetts 06062

Kennedy Manufacturing Company
520 East Sycamore Street
Van Wert, Ohio 45891

Klein Tools, Inc.
7200 McCormick Boulevard.
P.O. Box 599033
Chicago, Illinois 60645

Magnecraft Electric Company
1910 Techny Road
Northbrook, Illinois 60062

MICRO SWITCH
A Division of Honeywell
11 West Spring Street
Freeport, Illinois 61032

Midco Components
9065 South Octavia
Bridgeview, Illinois 60455

Milwaukee Electric Tool Corporation
13135 West Lisbon Road
Brookfield, Wisconsin 53005

NAMCO CONTROLS
7567 Tyler Boulevard
Mentor, Ohio 44060

National Association of Fire Equipment
Distributors
111 East Wacker Drive
Chicago, Illinois 60601

PANDUIT CORP.
17303 Ridgeland Avenue
Tinley Park, Illinois 60477

Poloroid Corporation
2020 Swift Drive
Oak Brook, Illinois 60521

Reliance Electric
P.O. Box 17438
Cleveland, Ohio 44117

The Ridge Tool Company
Subsidary of Emerson Electric Co.
400 Clark Street
Elyra, Ohio 44036

Rockwell International Corp., Inc.
Power Tool Division
401 North Lexington Avenue
Pittsburgh, Pennsylvania 15208

Siemens Energy and Automation, Inc.
P.O. Box 89000
Atlanta, Georgia 30356

Acknowledgments

The author and publisher are grateful to the following companies and organizations for providing technical information and assistance.

The ARO Corp.
One Aro Center
Bryan, Ohio 43506

Action Leathercraft Inc.
Route 3
Box 526ER
San Antonio, Texas 78218

Allen-Bradley Company
1201 South Second Street
Milwaukee, Wisconsin 53204

Amprobe Instrument
Division of Core Industries, Inc.
630 Merrick Road
Lynbrook, New York 11563

Associated Research, Inc.
905 Carriage Park Avenue
Lake Bluff, Illinois 60044

Biddle Instruments
510 Township Line Road
Blue Bell, Pennsylvania 19422

Bryant Wiring Device Division
Westinghouse Electric Corporation
500 Sylvan Avenue
Bridgeport, Connecticut 06606

Bussmann Div. Cooper Industries
P.O. Box 14460
St. Louis, Missouri 63178

Channellock, Inc.
1306 South Main Street
Meadville, Pennsylvania 16335

The Cleveland Twist Drill Co.
1242 East 49th Street
Cleveland, Ohio 44114

Eaton Corp. Cutler-Hammer Products
4201 North 27th Street
Milwaukee, Wisconsin 53216

Dennison Manufacturing Co.
300 Howard Street
Framingham, Massachusetts 01701

DESA Industries
25000 South Western Avenue
Park Forest, Illinois 60466

Dynascan Corporation
6460 West Cortland Street
Chicago, Illinois 60635

Eagle Signal Industrial Controls
A Division of Gulf-Western
Manufacturing Company
736 Federal Street
Davenport, Iowa 52803

Electromatic Controls Corp.
2495 Pembroke Avenue
Hoffman Estates, Illinois 60195

Energy Concepts, Inc.
7440 North Long Avenue
Skokie, Illinois 60077

Fendall Co.
5 East College Drive
Arlington Heights, Illinois 60004

Furnas Electric Company
1000 McKee Street
Batavia, Illinois 60510

Simpson Electric Company
853 Dundee Avenue
Elgin, Illinois 60120

Square D Company
P.O. Box 04549
Milwaukee, Wisconsin 53204

3M Center
Electro-Products Division
St. Paul, Minnesota 55144

Time Mark Corporation
11440 East Pine Street
Tulsa, Oklahoma 74116

Union Carbide Corporation
Battery Products Division
270 Park Avenue
New York, New York 10017

Vaco Products
Div. of Klein Tools Inc.
72 North McCormick Boulevard
P.O. Box 599033
Chicago, Illinois 60649

Waterloo Industries
LUMIDOR
P.O. Box 2095
Waterloo, Iowa 50704

Westinghouse Electric Corporation
Industry Products Company
Medium Motor & Gearing Division
Buffalo, New York 14240

Introduction

Electrical Motor Controls covers the control devices used in modern industrial electrical systems. The devices, applications, and systems are chosen to represent the broad range of uses for electricity found in different industries. The chapters are organized so the content is presented in a logical order. Each new concept builds on the information learned in the previous chapter. The book starts with basics such as tools, symbols, diagrams, and manual controls. The most common control devices and circuits from motor starters to programmable controllers are covered.

Electrical Motor Controls Workbook provides worksheets for each major concept presented in the textbook. The worksheets provide an opportunity for problem solving and circuit design. Typical applications and standard circuits provide the background required to work in the electrical field. For review, a tech-chek follows each chapter.

Electrical Motor Controls Manual provides applications and activities that show the installation and use of electrical control devices. Technical data is provided to show the proper use, sizing, and connection of control devices. Activities provide preparation for proper ordering, installation, maintenance, and troubleshooting of control devices and circuits. Manufacturing data is presented as it appears in service manuals used by industrial electricians.

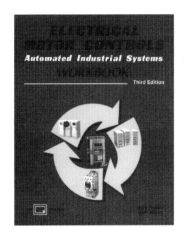

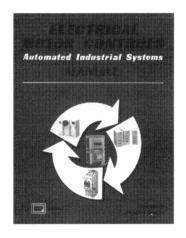

Contents

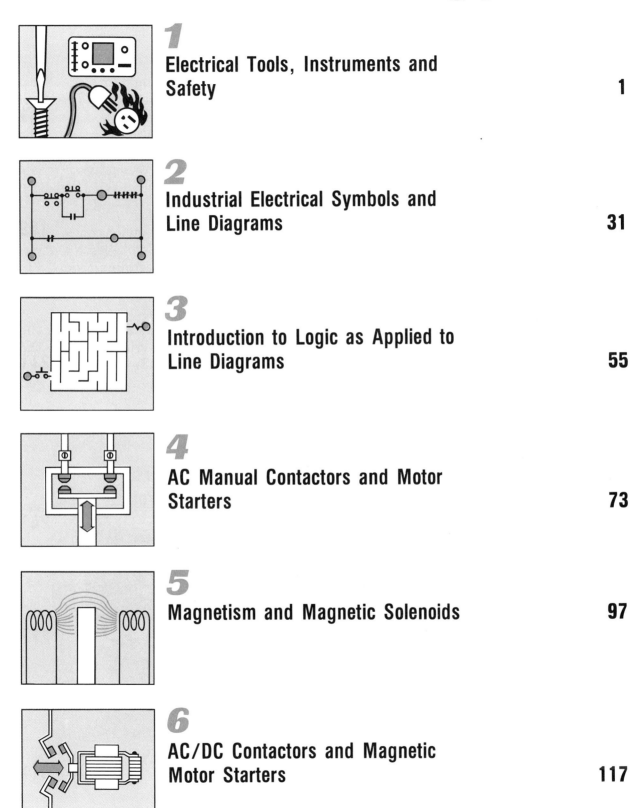

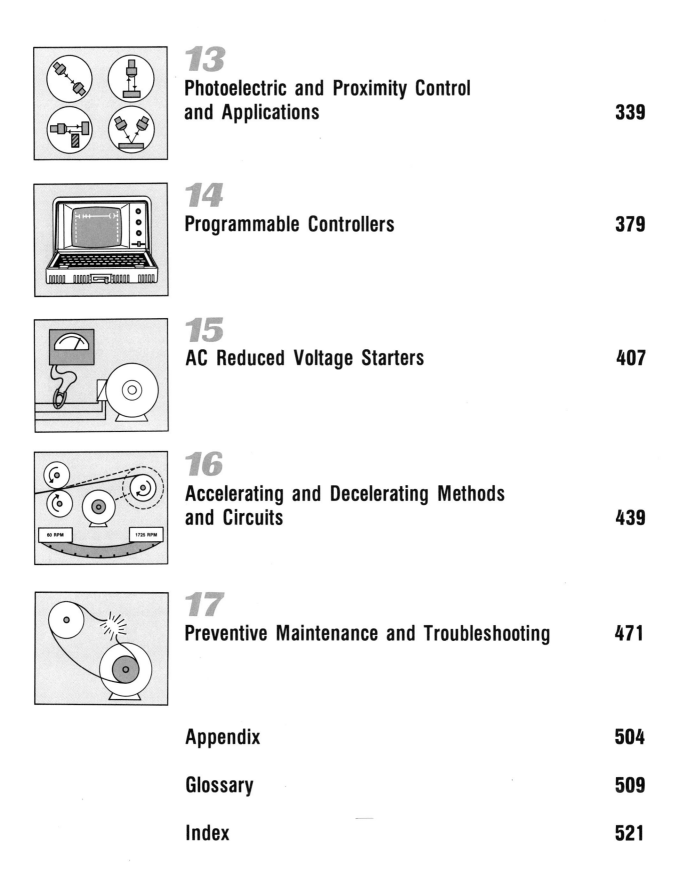

1 Electrical Tools, Instruments and Safety

Electrical tools and electrical instruments are a valuable investment and should be treated accordingly. The proper tools must be selected for each specific job. Tools must be organized and readily available for use. Before using a new tool or instrument the electrician should consult the operator's manual for correct operation.

Electrical power can be dangerous. The electrician must be aware of the dangers associated with electrical power and the potential hazards on the job. Safe work habits and proper procedures will minimize the possibility of an accident.

AN ORGANIZED TOOL SYSTEM

Since tools are expensive to replace, you need to protect your investment. Tools should be marked so that they can be easily identified as belonging to an individual or to a department in the company.

To be effective, tools must be available when needed; they must not be damaged by daily abuse. An organized tool system will provide both a central location and a means of protection for your tools.

Electrical tools can be organized in several ways, depending upon where and how frequently they are used. If the tools are used at a repair bench, a pegboard may be appropriate. If the tools are used only at the construction site, an electrician's leather pouch may be used. When the tools will be used at a bench and on the job, a portable tool box is usually best.

Pegboard. Pegboard is available in 4 feet × 8 feet sheets at most lumber yards. Usually a heavy duty tempered board is best for tools. Once the pegboard is mounted, outlines of the tools can be made to maintain an inventory. These outlines may be painted on the board or cut out of self-adhesive vinyl paper.

Electrician's Pouch. An electrician's pouch is usually made of heavy-duty leather (Figure 1-1). Pouches vary in design and size. Be sure to choose one that meets your specific needs and fits you comfortably. Some hold only a few tools, whereas others hold a wide selection. The type of pouch you need depends on the type of work you are planning to do. It is always wise to choose one that allows extra storage space for tools you may wish to add later.

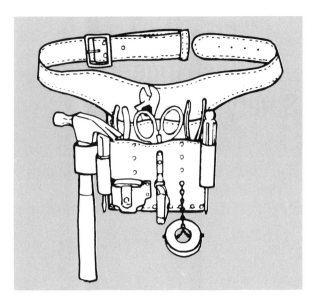

Figure 1-1 Multi-purpose electrician's pouch. (Action Leathercraft)

Figure 1-2 Portable tool box used by electrician. (Waterloo Industries, Inc., "Lumidor")

Portable Tool Box. Many electricians prefer to store their tools in a good portable box (Figure 1-2). A well-designed box can be locked and will keep your tools clean and dry. In addition, the tool box provides a fixed place where all the tools can be collected. To insure a complete inventory after each job, keep a list of all your tools in the box.

Whichever system you choose, organization is necessary. An organized tool system insures that you will always find clean, dry tools when and where you need them. (*Note:* Several manufacturers produce chemicals which can be placed in a tool box to keep the tools and box from rusting.)

TOOL SAFETY

Hand Tools. Use the right tools for the job. Tool safety requires tool knowledge. Use not only the proper tool but also the correct size. Use good quality tools and use them for the job they were designed to accomplish.

Learn how to use the tools properly. Study your tools—learn the safe way of working with each tool. Don't force a tool or use tools beyond their capacity. Don't be afraid to ask questions on the proper and safe use of a tool. It is often tempting to use a screwdriver for a chisel or a pair of pliers for a wrench. Remember, however, that the right tool will do the job faster and safer. The cost of the tool and the time it takes to buy it will prove far less costly than a serious accident.

Keep tools in good condition. Periodic checks on your tools will help to keep them in good condi-

tion. Always inspect a tool before using it. Do not use a tool which is in poor or faulty condition. Use only safe tools. Tool handles should be free of cracks and splinters and should be fastened securely to the working part. Damaged tools are not only dangerous but are also less productive than those in good working condition. When inspection shows a dangerous condition, repair or replace the tool immediately.

Cutting tools should be sharp and clean. Dull tools are dangerous. The extra force exerted in using dull tools often results in losing control of the tool. Dirt or oil on a tool may cause it to slip on the work and thus cause injury.

Keep tools in a safe place. Even good tools can be dangerous when left in the wrong place. Many accidents are caused by tools falling off ladders, shelves and scaffolds that are being moved.

Each tool should have a designated place in the tool box. Do not carry tools in your pockets unless the pocket is designed for that tool. Keep pencils in the pocket designed for them. Do *not* place pencils behind your ear or under your hat or cap.

Keep sharp-edged tools away from the edge of a bench or work area. Brushing against the tool may cause it to fall and injure a leg or foot. When carrying edged and sharply pointed tools, carry with the cutting edge or the point down and outward from your body. Be sure the place you choose to set down a tool is safe.

Power Tools. Do not attempt to use any power tools without knowing their principles of operation, methods of use, and safety precautions.

Obtain authorization from your job supervisor before you use power tools.

Grounding. All power tools should be grounded (unless they are approved double-insulated). Power tools must have a three-wire conductor cord. A three-prong plug connects into a grounded outlet (receptacle). See Figures 1-3 and 1-4 for approved receptacles and consult OSHA (Occupational Safety Health Act) and local codes for proper grounding specifications. It is very dangerous to use an adapter to plug a three-prong plug into a two-hole conductor outlet unless a separate ground wire or strap is connected to an approved type of ground. The ground insures that any short will trip the circuit breaker or blow the fuse. WARNING: An ungrounded power tool can kill you.

Double-insulated tools have two prongs and will have a notation on the specification plate that they are double-insulated. They are safe but are not often used on the job. Electrical parts in the motor of a double-insulated tool are surrounded by extra insulation to help prevent shock; therefore, the tool does not have to be grounded. Both the interior and exterior should be kept clean of grease and dirt that might conduct electricity.

Safety Rules. Study the following rules carefully and keep them in mind when you handle power tools.

1. You must be smarter than the tool you are operating.

2. Know and understand all of the manufacturer's safety recommendations.

3. Be sure that all safety guards are properly in place and in working order.

4. Wear safety goggles and a dust mask when the work requires it.

5. Be sure that the material to be worked is free of obstructions and securely clamped.

6. Before connecting a tool to a power source, be sure that the switch is in the OFF position.

7. Keep your attention focused on the work.

8. A change in sound during tool operation normally indicates trouble. Investigate immediately.

9. Power tools should be inspected and serviced by a qualified repair person at regular intervals as specified by the manufacturer or by OSHA.

10. Inspect electrical cords to see that they are in good condition.

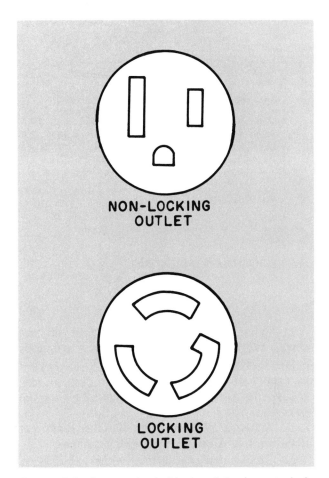

Figure 1-3 Approved electrical outlets (receptacles) commonly used for 110-volt tools and equipment. *Non-locking outlet* for indoor use only in remodeling work in existing structures. *Locking outlet* for exterior use and new construction.

Figure 1-4 Approved electrical outlets (receptacles) commonly used for 220-volt tools and equipment. Locking outlets are required for exterior job use and for new construction.

Red	Fire protection equipment and apparatus
	Portable containers of flammable liquids
	Emergency stop buttons and switches
Yellow	Caution and for marking physical hazards
	Waste containers for explosive or combustible materials
	Caution against starting, using, or moving equipment under repair
	Identification of the starting point or power source of machinery
Orange	Dangerous parts of machines
	Safety starter buttons
	The exposed parts (edges) of pulleys, gears, rollers, cutting devices, power jaws
Purple	Radiation hazards
Green	Safety
	Location of first aid equipment (other than fire fighting equipment)

Figure 1-5 OSHA safety color codes.

11. When work is completed, shut off the power. Wait until all movement of the tool stops before leaving a stationary tool or laying down a portable tool.

12. After use, clean and lubricate tools for the next day's use.

13. When a power tool is defective, remove it from service. Alert others to the situation.

14. Take extra precautions when working on damp or wet surfaces. If necessary, use additional insulation to prevent any part of your body from coming into contact with the wet or damp surface.

15. Whenever working conditions are hazardous, at least two people should work together.

Electrical Safety

OSHA and state safety laws have helped to advance the cause of safety for electricians. With today's safeguards and recommended work practices, plus an understanding of the principles of electricity, people can work safely on electrical equipment. Remember, if you disregard your own safety, you are also disregarding the safety of others. When in doubt about a procedure, ask your supervisor. Report any unsafe condition, equipment, or work practices as soon as possible.

Fuses. Before removing any fuse from a circuit, be sure the switch for that circuit is open or disconnected. When removing fuses, use an approved type fuse puller and break contact on the hot side of the circuit first. When replacing fuses, install the fuse first into the load side of the fuse clip, then into the line side.

Electrical Shock. Electrical shock occurs when a person comes in contact with two conductors of a circuit or when his body becomes a part of the electrical circuit. In either case, a severe shock can cause the heart and lungs to stop functioning. Also, severe burns may occur where current enters and exits the body.

Prevention is the best medicine for electrical shock. Respect all voltages, have a knowledge of the principles of electricity and follow safe work procedures. Do not take chances. All electricians should be encouraged to take a basic course in Cardiac Pulmonary Resuscitation (CPR) so that they can come to the aid of a co-worker in emergency situations.

When using portable electric tools, always make sure they are in safe operating condition. Make sure there is a third wire on the plug for grounding in case of shorts. Theoretically, if electric power tools are grounded and if an insulation breakdown occurs, the fault current should flow through the third wire to ground instead of through the operator's body to ground.

Out of Service Protection. Before any repair is to be performed on a piece of electrical equipment, be absolutely certain the source of electricity is open and *carded* (out of service). Carding is the process of padlocking in the off position the power source and indicating on an appropriate card the procedure which is taking place.

Whenever you leave your job for any reason, or whenever the job cannot be completed the same day, be sure the source of electricity is still open or disconnected when you return to continue the work.

Safety Color Codes

Federal law (OSHA) has established specific colors to designate certain cautions and dangers. Figure 1-5 shows these accepted usages. Study these colors until you are familiar with all of them.

SAFE WORK CLOTHES ON THE JOB

1. Supervisory forces must see that employees do not wear clothing that will create a hazard on their jobs.

a. Oil-or grease-soaked clothing is prohibited as it may catch fire from a match, cigarette, sparks, or any open flame. Overalls or work clothing should be washed regularly.

b. Approved flame resistant clothing must be worn, as specified, on jobs involving exposure to molten metal hazards.

c. A state of partial undress (such as stripping to the waist) on the job is prohibited.

d. Ragged or flapping shoe soles, worn heels, that might become a slipping, tripping, or stumbling hazard, are prohibited. (Shoes should be kept in good repair.)

e. Sneakers, moccasins, canvas or loafer type street shoes are prohibited in the plant.

f. Safety (hard) hats must be worn in areas or departments specified. Conventional head cover must be worn in other areas within the plant where required by District or Division safety policy.

2. Employees working with or around moving machinery or high places are prohibited from wearing:

a. Loose hanging neckties,

b. Long flowing coats, open sweaters,

c. Dusters or aprons,

d. Long belt ends projecting from buckles,

e. Rings, watch chains, fobs, key chains,

f. Baggy, loose or unbuttoned sleeves, loose trouser bottoms, or torn clothing,

g. Gloves, except where authorized by District or Division Safety Committee.

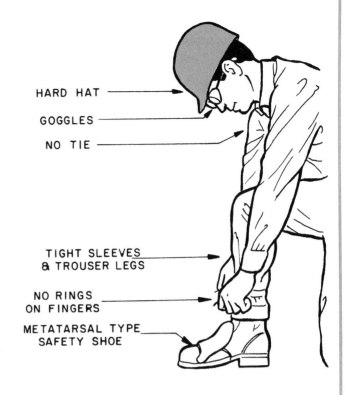

HARD HAT

GOGGLES

NO TIE

TIGHT SLEEVES & TROUSER LEGS

NO RINGS ON FINGERS

METATARSAL TYPE SAFETY SHOE

Clothing should be kept tight as is comfortable and should fit snugly to avoid danger of becoming entangled in moving machinery, or creating a tripping or stumbling hazard.

3. Finger rings shall not be worn by:
ANY employee . . . other than regular office workers whose duties do not require them to operate machines such as multilith, mimeograph machines, etc.

4. Safety shoes are recommended for all supervision and employees. Safety shoes incorporating metatarsal guards afford additional protection to the feet and are particularly recommended.

Clothing and Personal Protective Equipment (Figure 1-6)

1. Wear thick-soled work shoes for protection against sharp objects such as nails. Wear work shoes with safety toes if the job requires. If your shoes are subject to oils and grease, make sure the soles are oil resistant.

2. Wear rubber boots in damp locations.

3. Wear a hat or cap. Wear an approved safety helmet (hard hat) if the job requires. Confine long hair or keep your hair trimmed and avoid placing your head too near rotating machinery.

4. Leave your jewelry in your locker; gold and silver are excellent conductors of electricity.

Figure 1-6 Recommendations for safe working clothes. (Illinois Department of Labor, Division of Safety Inspection and Education)

Fire Safety

The chance of fire is greatly decreased by good housekeeping. Keep rags containing oil, gasoline, alcohol, shellac, paint, varnish, or lacquer in a covered metal container. Keep debris in a designated area away from the building. If a fire occurs, first give an alarm. Alert all workers on the job, and call the fire department. Then make a reasonable effort to contain the fire.

TYPES AVAILABLE BY CLASSIFICATION ▷

| | | | | A | | |

	WATER TYPES		MULTIPURPOSE DRY CHEMICAL		AFFF FOAM	HALON 1211
	STORED PRESSURE	PUMP TANK	STORED PRESSURE	CARTRIDGE OPERATED	STORED PRESSURE	STORED PRESSURE

CLASS A FIRES . . . ordinary combustible materials, such as wood, cloth, paper, rubber, and many plastics.

CLASS B FIRES ✻ . . . flammable liquids/gases, and greases.

CLASS C FIRES . . . energized electrical equipment where the electrical nonconductivity of the extinguishing media is of importance.

CLASS D FIRES . . . combustible metals, such as magnesium, titanium, zirconium, sodium and potassium.

✻ NOTE: O-n-l-y Dry Chemical types effective on *pressurized* flammable gases/liquids; for deep fat fryers, Multipurposes A/B/C dry chemicals NOT acceptablie.

▲ NOTE: Protection required below 40° F.

	▲	▲			▲	
SIZES AVAILABLE	2½ Gal.	2½ and 5-Gal.	2½-30 lb. *ALSO* Wheeled 150-350 lb.	5-30 lb. *ALSO* Wheeled 50-350 lb.	2½ Gal. *ALSO* Wheeled 33 Gal.	9 to 22 lb.
HORIZONTAL RANGE (APPROX.)	30 to 40 ft.	30 to 40 ft.	10-15 ft., (Wheeled-15-45 ft.)	10-20 ft., (Wheeled-15-45 ft.)	20-25 ft., (Wheeled-30 ft.)	14 to 16 ft.
DISCHARGE TIME (APPROX.)	1 Min.	1 to 3 Min.	8-25 Sec., (Wheeled-30-60 Sec.)	8-25 Sec., (Wheeled-20-60 Sec.)	50 sec., (Wheeled-60 sec.)	10 to 18 Sec.

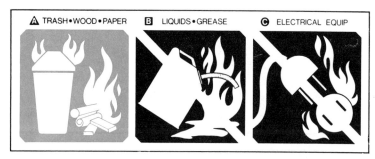

USE ON "A" TYPES

Ⓐ TRASH•WOOD•PAPER   Ⓑ LIQUIDS•GREASE   Ⓒ ELECTRICAL EQUIP

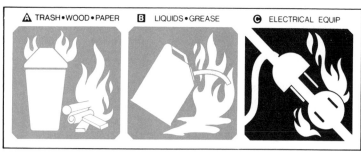

USE ON "A/B" TYPES

Ⓐ TRASH•WOOD•PAPER   Ⓑ LIQUIDS•GREASE   Ⓒ ELECTRICAL EQUIP

Figure 1-7 Common fire extinguishers and their applications. (National Association of Fire Equipment Distributors)

Fire Extinguishers. Always read instructions before using a fire extinguisher. Figure 1-7 illustrates some of the common fire extinguishers and their uses. Careful study of this chart will familiarize you with the types of fires normally encoun-

A/B	B/C				A/B/C			D
AFFF FOAM	**CARBON DIOXIDE**	**DRY CHEMICAL TYPES**		**HALON 1211**	**MULTIPURPOSE DRY CHEMICAL**		**HALON 1211**	**DRY POWDER**
STORED PRESSURE	SELF EXPELLING	STORED PRESSURE	CARTRIDGE OPERATED	STORED PRESSURE	STORED PRESSURE	CARTRIDGE OPERATED	STORED PRESSURE	CARTRIDGE OPERATED
2½ Gal. **ALSO** Wheeled 33 Gal.	5-20 lb. **ALSO** Wheeled 50-100 lb.	2½-30 lb. **ALSO** Wheeled 150-350 lb.	4-30 lb. **ALSO** Wheeled 50-350 lb.	2 to 22 lb.	2½-30 lb. **ALSO** Wheeled 150-350 lb.	5-30 lb. **ALSO** Wheeled 50-350 lb.	9 to 22 lb.	30 lb. **ALSO** Wheeled 150-350 lb.
20-25 ft., (Wheeled-30 ft.)	3-8 ft., (Wheeled-10 ft.)	10-15 ft., (Wheeled-15-45 ft.)	10-20 ft., (Wheeled-15-45 ft.)	10 to 16 ft.	10-15 ft., (Wheeled-15-45 ft.)	10-20 ft., (Wheeled-15-45 ft.)	14 to 16 ft.	5 ft., (Wheeled-15 ft.)
50 sec., (Wheeled-60 sec.)	8-15 Sec., (Wheeled-8-30 Sec.)	8-25 Sec., (Wheeled-30-60 Sec.)	8-25 Sec., (Wheeled-20-60 Sec.)	8 to 18 Sec.	8-25 Sec., (Wheeled-30-60 Sec.)	8-25 Sec., (Wheeled-20-60 Sec.)	10 to 18 Sec.	20 Sec., (Wheeled-150 lb. 70 Sec., 350 lb. 1¾ Min.)

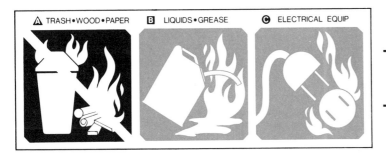

A TRASH•WOOD•PAPER   B LIQUIDS•GREASE   C ELECTRICAL EQUIP

USE ON "B/C" TYPES

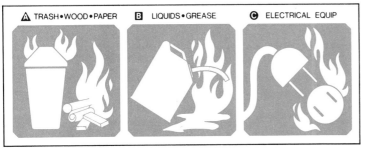

A TRASH•WOOD•PAPER   B LIQUIDS•GREASE   C ELECTRICAL EQUIP

USE ON "A/B/C" TYPES

tered and the best type of extinguisher used to combat that type of fire.

Fire extinguishers are normally red. If they are not red, they should have a red background so they can be easily located.

If firemen are called, be ready to direct them to the fire. Also inform them of any special problems or conditions that exist, such as downed electrical wires or leaks in gas lines.

A final thought: report to your supervisor any

accumulations of rubbish or unsafe conditions that could be fire hazards. Also, if a portable tool bin is used on the job, a good practice is to store a CO_2 extinguisher in it.

In-Plant Training for Selected Personnel. It is highly recommended that a select group of personnel (if not *all* personnel) be acquainted with all extinguisher types and sizes you have on hand in the plant or work area. Such training should include a tour of your facility indicating special fire hazard operations.

In addition, it will be helpful to periodically practice a dry run, discharging each type of extinguisher. Such practice is essential to:

1. learn how to activate each type
2. know the discharge ranges
3. realize which types are affected by winds and drafts
4. be familiar with discharge duration
5. learn of any precautions to take as noted on the name plate.

BASIC RULES

• *Lightly oil all moving parts such as drawers, trays, and hinges at regular intervals.*

• *Use graphite, not oil, on locks and padlocks.*

• *Touch up all rusted spots; pay particular attention to the bottom of tool boxes.*

• *Make sure that the wheels on tool cabinets are turning freely.*

• *Drawers and trays that hold sharp-edged tools such as chisels, screwdrivers, etc., should be lined with cork, felt, or scrap carpeting.*

• *Check the handle; is it firmly attached to the tool box?*

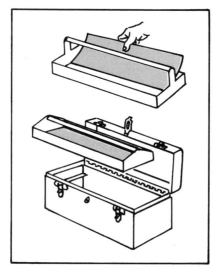

Figure 1-8 Tool box with lift-out tray. Protect tools by lining the bottom of box and tray with felt. (The Hand Tool Institute)

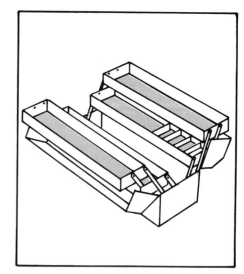

Figure 1-9 Tool box with lever-operated trays that open automatically as the cover is lifted. (The Hand Tool Institute)

Maintenance Tips for Extinguishers.

1. Inspect extinguishers at least once a month. (It is very common to find units that are missing, damaged, or used. Consider contracting for such a service.)

2. Contract for annual maintenance with a *qualified* service agency.

3. *Never* attempt to make repairs to extinguishers. This is the chief cause of dangerous shell ruptures.

TOOLS AND TOOL USAGE

In the remainder of this unit you will find illustrated tips on the proper use of the tools you may need for various jobs (Figures 1-8 through 1-34). Study these tips for proper usage and become familiar with the tools designed for specific jobs. In addition, this unit shows photographs of 80 tools with a brief description of the operation of each (Figures 1-35 through 1-104). This is intended as a reference aid, not as a recommended tool list.

• *Keep your tool box or chest locked when not in use.*

• *Sand, or file down, any sharp edges that may cause damage to clothes or fingers. Such sharp edges are usually caused by dropping the tool box to the floor instead of placing it on the floor.*

• *Always replace your tools in the same tray or drawer that you removed them from. Use this system and you will not waste time hunting for a particular tool that you know is "just there."*

• *Wipe away all grease and moisture from tools before storing them in the tool box, chest, or cabinet.*

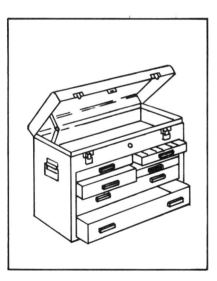

Figure 1-10 Tool chests, more substantial than tool boxes, may have from two to ten drawers. (The Hand Tool Institute)

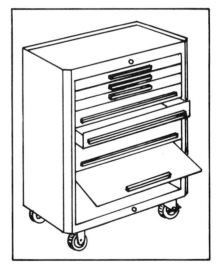

Figure 1-11 Tool cabinets are always mounted on casters; chests can be added to the top of the cabinet. (The Hand Tool Institute)

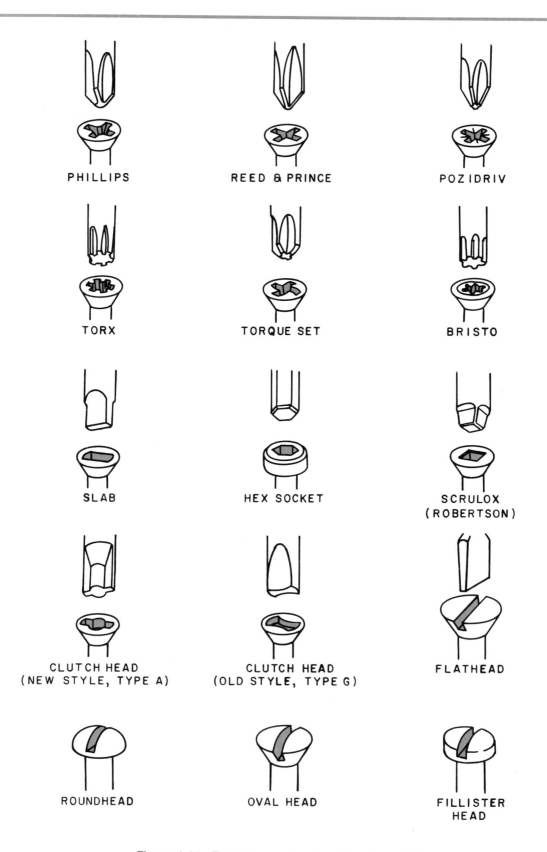

Figure 1-12 Typical screw head configurations. (The Hand Tool Institute)

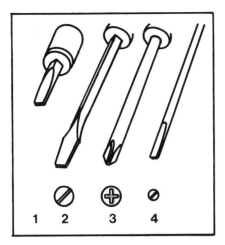

Figure 1-13 Typical screwdriver assemblies. (The Hand Tool Institute)

1. Stubby screwdriver for working in close quarters.

2. Screwdriver with a square shank to which a wrench can be applied to remove stubborn screws.

3. Screwdriver for Phillips screws.

4. Cabinet screwdriver has a thin shank to reach and drive screws in deep, counterbored holes.

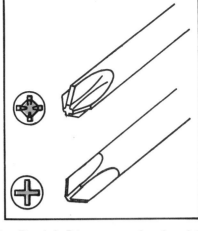

Figure 1-14 The Reed & Prince screw head and its matching screwdriver (bottom) somewhat resembles the Phillips type (top), but there is a difference. The slots in a Phillips screw end in a slight curve at their intersection, unlike the Reed & Prince screw, whose slots meet at an exact right angle. (The Hand Tool Institute)

1. Make sure that the tip fits the slot of the screw not too loose and not too tight.

2. Do not use a screwdriver as a cold chisel or punch.

3. Do not use a screwdriver near live wires (do not use any other tool, for that matter).

4. Do not expose a screwdriver to excessive heat.

5. Redress a worn tip with a file in order to regain a good straight edge.

6. Discard a screwdriver that has a worn or broken handle.

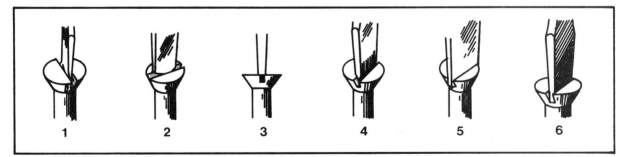

Figure 1-15 Proper and improper use of flathead screwdrivers. (The Hand Tool Institute)

1. This tip is too narrow for the screw slot; it will bend or break under pressure.

2. A rounded or worn tip. Such a tip will ride out of the slot as pressure is applied.

3. This tip is too thick. It will only serve to chew up the slot of the screw.

4. A chisel ground tip will also ride out of the screw slot. Best to discard it.

5. This tip fits, but it is too wide and will tear the wood as the screw is driven home.

6. The right tip. This tip is a snug fit in the slot and does not project beyond the screw head.

Figure 1-16 A ratchet-type offset screwdriver for working in tight spots; it is reversible. (The Hand Tool Institute)

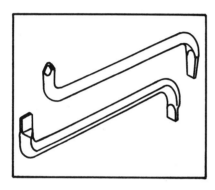

Figure 1-17 Offset screwdrivers for driving screws in awkward places. (The Hand Tool Institute)

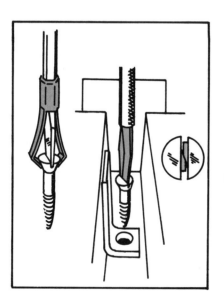

Figure 1-18 The screw-holding screwdriver is a must for working in close quarters as shown. The type shown at the left uses clips to hold the screw. The one at the right has a sliding collar that spreads the split blade of the screwdriver to hold the screw. After the screw has been firmly started, further driving can be done with a conventional screwdriver. (The Hand Tool Institute)

BASIC SAFETY RULES WHICH APPLY TO THE USE OF PLIERS.

1. Pliers should not be used for cutting hardened wire unless specifically manufactured for this service.

*2. **Never** expose pliers to excessive heat. This may draw the temper and ruin the tool.*

*3. Always cut at right angles. **Never** rock from side to side or bend the wire back and forth against the cutting blades.*

4. Do not bend stiff wire with light pliers like needle nose pliers. This can damage the pliers. Use a sturdier tool, like lineman's pliers.

*5. **Never** use pliers as a hammer nor hammer on the handles. They may crack or break, or blades may be nicked by such abuse.*

*6. **Never** extend the length of handles to secure greater leverage. Use a larger pair of pliers or a bolt cutter.*

7. Pliers should not be used on nuts or bolts. A wrench will do the job better and with less risk of damage to the fastener.

8. Oil pliers occasionally. A drop of oil at the hinge will lengthen tool life and assure easy operation.

9. Safety glasses should be worn when cutting wire, etc. to protect eyes from being struck by the end of the object being cut.

*10. **WARNING.** Ordinary plastic dipped handles are designed for comfort—not electrical insulation. Tools having high dielectric insulation are available and are so identified. Do not confuse the two.*

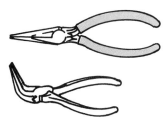

Figure 1-21 Long nose pliers. (The Hand Tool Institute)

Figure 1-22 Slip joint pliers. (The Hand Tool Institute)

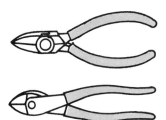

Figure 1-23 Diagonal cutting pliers. (The Hand Tool Institute)

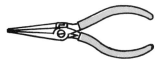

Figure 1-24 Flat nose pliers. (The Hand Tool Institute)

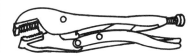

Figure 1-19 Locking pliers. (The Hand Tool Institute)

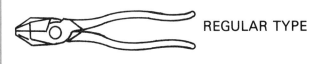

REGULAR TYPE

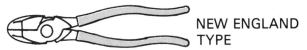

NEW ENGLAND TYPE

Figure 1-20 Lineman's side cutting pliers. (The Hand Tool Institute)

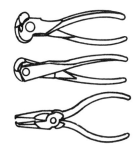

Figure 1-25 End cutting pliers. (The Hand Tool Institute)

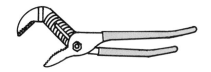

Figure 1-26 Utility pliers. (The Hand Tool Institute)

BASIC SAFETY RULES WHICH APPLY TO THE USE OF WRENCHES

*1. **Never** use a pipe extension or other form of "cheater" to increase the leverage of any wrench.*

2. Select a wrench whose opening exactly fits the nut. Too large an opening can spread the jaws of an open-end wrench and batter the points of a box or socket wrench. Care should be exercised in selecting inch wrenches for inch fasteners and metric for metric fasteners.

*3. If possible, always **pull** on a wrench handle and adjust your stance to prevent a fall if something lets go.*

4. The safest wrench is a box or socket type — particularly on hex nuts or fittings. Always use a straight, rather than an offset handle if conditions permit.

BOX WRENCHES

COMBINATION BOX— AND OPEN-END WRENCHES

SOCKET WRENCHES

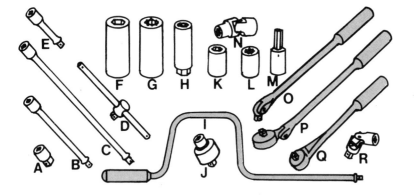

SOCKETS

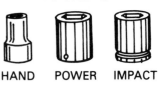

HAND   POWER   IMPACT

A	— ADAPTER	K —	REGULAR 6-POINT SOCKET
B,C,E	— EXTENSION BARS	L —	REGULAR 12-POINT SOCKET
D	— SLIDING T HANDLE	M —	HOLLOW SCREW SOCKET BIT
F	— DEEP 6-POINT SOCKET	N —	UNIVERSAL 12-POINT SOCKET
G	— DEEP 12-POINT SOCKET	O —	FLEX HANDLE
H	— SPARK PLUG SOCKET	P —	FLEX HEAD RATCHET
I	— SPEEDER HANDLE	Q —	REVERSIBLE RATCHET
J	— RATCHET ADAPTER	R —	UNIVERSAL JOINT

Figure 1-27 Various types of wrenches (The Hand Tool Institute)

Proper Uses *The jaw and chain type wrenches are designed for turning and holding pipe fittings, within their capacities, in service where tooth marks are not objectionable. To avoid tooth marks, use the strap wrench. In using the jaw type wrench it is important to maintain a gap between the back of the hook jaw and the pipe. This concentrates the pressure at two points only and produces the maximum gripping action and rotating force. The offset handle pattern is ideal for close quarters where the normal swing of a straight handle pattern is limited. If possible, always pull rather than push on the wrench handle.*

ADJUSTABLE WRENCHES

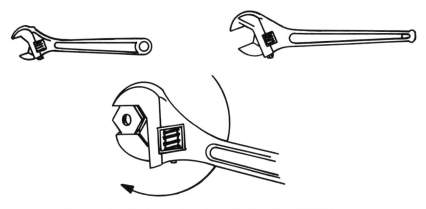

SAFE WAY TO USE ADJUSTABLE WRENCH

Figure 1-28 Adjustable wrenches. (The Hand Tool Institute)

PIPE WRENCHES

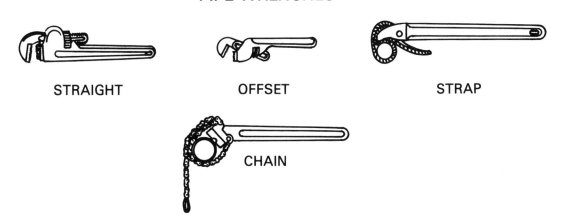

STRAIGHT   OFFSET   STRAP

CHAIN

Figure 1-29 Pipe wrenches. (The Hand Tool Institute)

PIPE VISES

*Pipe vises are especially designed to hold pipe or round stock. They are often mounted on trucks and beams as well as on workbenches. They are available with capacities to hold pipe up to eight inches in diameter. The two main types are the **yoke vise** and the **chain vise**, with the latter especially designed to hold irregular work. Both types are available with tripods and are called tripod vises. A **clamp kit vise** can be mounted without drilling holes for temporary attachment where light-duty work is to be performed.*

Pipe vises are made in a number of different forms, including vises with bolt holes for permanent mounting and portable vises with clamp attachments for temporary mounting on benches, studs, posts, etc.

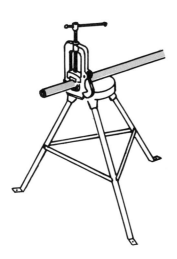 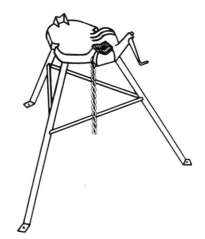

Figure 1-30 The yoke type vise and the chain vise are available in portable workbench models. (The Hand Tool Institute)

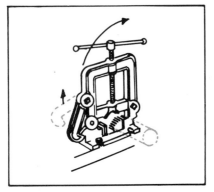

The yoke type pipe vise is bolted to the workbench. Note the hinge at one end and the hook at the other so that the pipe need not be "threaded" through the vise jaws to be worked on.

Figure 1-31 Yoke type pipe vise. (The Hand Tool Institute)

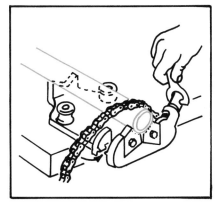

The chain vise is designed to hold pipe as well as irregular work. Work is released from the vise by loosening the nut and then removing the pipe—or other work—from the vise. This allows the pipe to be installed or removed without having to slide its entire length through the vise.

Figure 1-32 Chain vise. (The Hand Tool Institute)

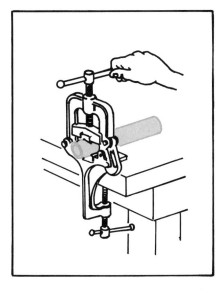

The clamp kit vise can be temporarily mounted without drilling holes, sometimes on the nearest 2x4, for light-duty work.

Figure 1-33 Clamp kit vise. (The Hand Tool Institute)

Nut drivers are screwdriver-type tools which in their simplest form, have a one-piece shank and socket secured in a fixed handle. Socket heads have openings for hex nuts, bolts and screws up to 3/4" nut size. Shafts may be solid, drilled part way or full hollow—plain or magnetic for holding small fasteners.

Handles, both regular and reversible ratcheting, are available to accommodate interchangeable shanks for recessed head screws such as Phillips. Frearson, Hex, Spline and Scrulox and for slotted screws as well as for standard 1/4" drive sockets.

Figure 1-34 Nutdrivers are excellent close-quarters tools. (The Hand Tool Institute)

TOOL CABINET

Figure 1-35 Used to organize and store a variety of tools and materials in one location. (Kennedy Manufacturing Company)

TOOL BOX

Figure 1-36 Used to organize, store, and carry tools. (Courtesy of The Ridge Tool Company)

TOOL POUCH AND ELECTRICIAN'S BELT

Figure 1-37 Used to safely transport and store many small electrical hand tools and instruments. (Klein Tools, Inc.)

FLATHEAD SCREWDRIVER

Figure 1-38 Used to tighten or loosen flathead screws. Be sure slot fits screw head. (Vaco Products Company)

PHILLIPS SCREWDRIVER

Figure 1-39 Used to tighten or loosen Phillips screws. Be sure that the screwdriver fits the screw head. (Vaco Products Company)

PHILLIPS OFFSET SCREWDRIVER

Figure 1-40 Provides means for reaching difficult Phillips head screws. (Vaco Products Company)

PHILLIPS SCREW-HOLDING DRIVER

Figure 1-41 Used to hold Phillips screws in place while working in tight spots. Once started, screw is released and tightened with standard screwdriver. (Vaco Products Company)

FLATHEAD SCREW-HOLDING DRIVER

Figure 1-42 Used to hold flathead screws in place when working in tight spots. Once started, screw is released and tightened with a standard screwdriver. (Vaco Products Company)

FLATHEAD OFFSET SCREWDRIVER

Figure 1-43 Provides means for reaching difficult flathead screws. (Vaco Products Company)

LONG NEEDLE-NOSE PLIERS AND CUTTER

Figure 1-44 Appropriate for bending wire, cutting wire and positioning small components. (Vaco Products Company)

LINEMAN'S SIDE CUTTING PLIERS

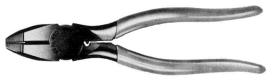

Figure 1-45 Appropriate for cutting cable, removing knockouts, twisting wire and deburring conduit. (Ideal Industries, Inc.)

UTILITY PLIERS

Figure 1-46 Tightens box connectors, lock nuts, and small size conduit couplings. (Channellock, Inc.)

SLIP JOINT PLIERS

Figure 1-47 Used for a wide range of service involving gripping, turning, and bending. (Vaco Products Company)

DIAGONAL SIDE-CUTTING PLIERS

Figure 1-48 Useful for cutting cables and wires too difficult for side cutting pliers. (Courtesy of The Ridge Tool Company)

END-CUTTING PLIERS

Figure 1-49 Used for cutting wire, nails, rivets, etc., close to work (Channellock, Inc.)

ELECTRICIAN'S KNIFE

Figure 1-50 Removes insulation from cables and service conductors. Keep cutting blade sharp. (Klein Tools, Inc.)

SKINNING KNIFE

Figure 1-51 Removes insulation from cables and service conductors. Keep cutting blade sharp. (Klein Tools, Inc.)

CABLE STRIPPER

Figure 1-52 Removes insulation from heavy duty cables. (Ideal Industries, Inc.)

WIRE STRIPPERS

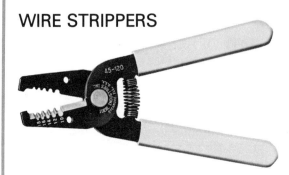

Figure 1-53 Removes insulation from small diameter wire. (Ideal Industries, Inc.)

COMBINATION WIRE STRIPPER, BOLT CUTTER AND CRIMPER

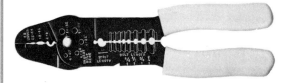

Figure 1-54 Removes insulation from small diameter wire, secures crimp type connectors, and cuts small bolts to given lengths. (Vaco Products Company)

CABLE CUTTER

Figure 1-55 Provides shear-type cable cutting for relatively large diameter cables. Makes a clean, even cut for ease in fitting lugs and terminals. (Klein Tools, Inc.)

ELECTRICIAN'S HAMMER

Figure 1-56 Can be used to mount electrical boxes and drive nails. Can also be used to determine height of receptacle box since most hammers are 12″ in length from head to end of handle, or can be so marked. (Klein Tools, Inc.)

SLEDGE HAMMER

Figure 1-57 Medium sized (5 lb) hammer used for driving stakes and other heavy duty pounding. (Courtesy of The Ridge Tool Company)

BALL PEEN HAMMER

Figure 1-58 Ball peen hammers of the proper size are designed for striking chisels and punches. They also may be used for riveting, shaping, and straightening unhardened metal. (Channellock, Inc.)

FOLDING RULE

Figure 1-59 Works better than tapes for measuring layout work. (Klein Tools, Inc.)

POWER RETURN STEEL TAPE

Figure 1-60 Used for rapid layout in measurements. Should be as wide as possible for easy extension. (Courtesy of The Ridge Tool Company)

POCKET TORPEDO LEVEL

Figure 1-61 Useful in leveling electrical control panels and conduit bends. (Klein Tools, Inc.)

24-INCH LEVEL

Figure 1-62 Useful in leveling long conduit runs and bus bar installations. (Courtesy of The Ridge Tool Company)

FISH TAPE WITH HOLDER

Figure 1-63 Useful to pull wire through conduit and "fish" wires around obstructions in walls. (Ideal Industries, Inc.)

POWER WIRE PULLER

Figure 1-64 Used to pull larger cables and wires into place. (Greenlee Tool Division, Ex-Cell-O Corporation).

REAMING TOOL

Figure 1-65 Used to deburr or remove rough edges from inside cut conduit. (Courtesy of The Ridge Tool Company)

HEAVY DUTY HAMMER DRILL

Figure 1-66 Because of the hammer action of the drill, it is ideal for drilling concrete when fasteners are being installed. (Milwaukee Electric Tool Corporation)

ADJUSTABLE WRENCH

Figure 1-67 Tightens positive sized items such as hex head lag screws, bolts, and larger size conduit couplings. (Vaco Products Company)

CHAIN PIPE WRENCH

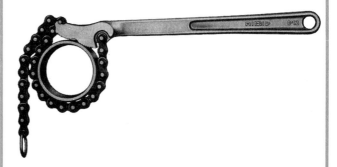

Figure 1-68 Used to tighten and loosen larger pipes and conduit. (Courtesy of The Ridge Tool Company)

HEX KEY WRENCH SET

Figure 1-69 Used for tightening positive sized hex head bolts. (Vaco Products Company)

NUT DRIVER SET

Figure 1-70   Used to tighten positive sized hex head nuts and screws. (Vaco Products Company)

HACKSAW

Figure 1-71   Cuts heavy cable, pipe, and conduit. (Courtesy of The Ridge Tool Company)

SOCKET WRENCH SET

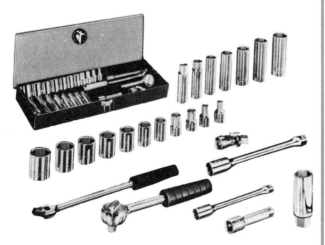

Figure 1-72   Used to tighten a variety of positive sized items such as hex head lag screws, bolts, and various electrical connectors. (Klein Tools, Inc.)

RIGID CONDUIT HICKEY

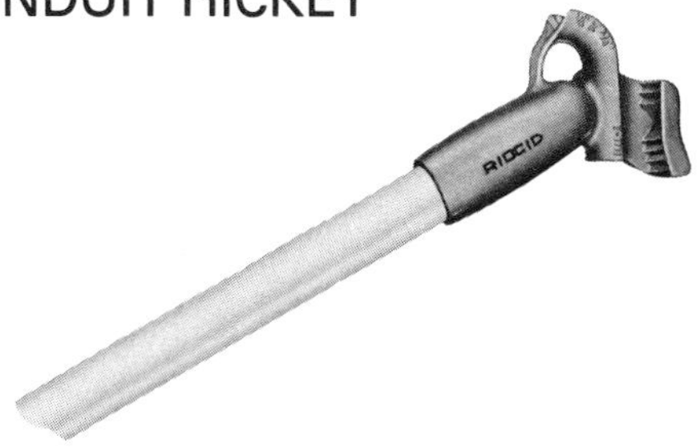

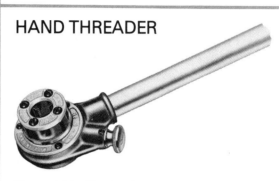

Figure 1-73   Used to bend rigid conduit to a variety of shapes. Available for different size conduits. (Courtesy of The Ridge Tool Company)

HAND THREADER

Figure 1-74   Used to thread rigid conduit pipe at a variety of locations on the job site. (Courtesy of The Ridge Tool Company)

POWER THREADER WITH STAND

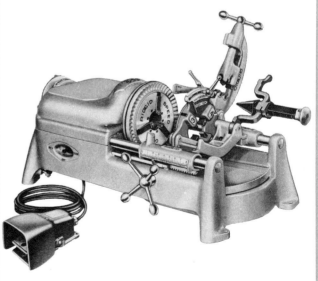

Figure 1-75   Used to thread rigid conduit pipe at a more centralized location. (Courtesy of The Ridge Tool Company)

HYDRAULIC PUNCH SET

Figure 1-76 Used to punch holes in metal enclosures and cabinets through use of hydraulic cylinder. (Greenlee Tool Division, Ex-Cell-O Corporation)

KNOCK OUT PUNCH

Figure 1-77 Used to punch holes in metal enclosures and cabinets through the use of wrenches or sockets. (Greenlee Tool Division, Ex-Cell-O Corporation)

TAP AND DIE SET

Figure 1-78 Used to provide threading for control panel mounting, bolts, nuts, and steel rod. (The Cleveland Twist Drill Company)

POWER STUD GUN

Figure 1-79 Used to firmly secure fasteners in concrete. Extreme care must be exercised when using this device. In certain cases one must be licensed to use it. (DESA Industries)

WIRE MARKERS

Figure 1-80 Used to properly mark and identify wires and matching terminal ends. (Ideal Industries, Inc.)

TIE-RAP GUN

Figure 1-81 Used to firmly secure wires in wire runs and harnesses. (Dennison)

WIRE LUBRICANT

Figure 1-82 When applied to outside of electrical wires, it allows them to be pulled more easily through conduit. (Ideal Industries, Inc.)

ELECTRICIAN'S TAPE

Figure 1-83 Used to provide insulation to electrical connections once repairs or installation have been made. (3M Corporation)

SAFETY GLASSES

Figure 1-84 Provide eye protection from flying objects and debris. Required on all construction sites. (Fendall Company)

SAFETY HARD HAT

Figure 1-85 Usually made of high density polyethylene to protect against impact, chemicals, and high voltage. Required on all construction sites. (Klein Tools, Inc.)

LOCK-OUT TAG

Figure 1-86 Used in conjunction with padlock to warn others that equipment is shut down for repair and that someone is working on it. (General Electric)

PADLOCK

Figure 1-87 Used to provide lock-out protection for electrical equipment. May also be used on a tool box. (Cutler Hammer)

FUSE PULLER

Figure 1-88 Safely removes cartridge type fuses. (Ideal Industries, Inc.)

FLASHLIGHT

Figure 1-89 Used for emergency lighting and inspection. (Union Carbide Corporation).

POLAROID CAMERA

Figure 1-90 When making circuit changes, especially in conduit, use instant photographs as a means of evaluating changes without returning to the job site. (Polaroid Corporation)

POLARIZED RECEPTACLE TESTER

Figure 1-91 Used to verify grounding and proper installation of three-wire outlets. (Ideal Industries, Inc.)

INDUSTRIAL VOM WITH CLAMP-ON AMMETER

Figure 1-92 Used to measure voltage, current, and resistance. Clamp-on ammeter attachment allows AC/DC current to be measured without breaking circuit. (Courtesy of Simpson Electric Company, Elgin, Illinois)

DIGITAL AC / DC AMMETER

Figure 1-93 Used to measure high values of AC/DC current. (Courtesy of Simpson Electric Company, Elgin, Illinois)

OSCILLOSCOPE

Figure 1-94 Used to troubleshoot and verify outputs and inputs of electronic devices. Used with transducers, it may be used in vibration analysis. (Courtesy of Simpson Electric Company, Elgin, Illinois)

DIGITAL LOGIC PROBE

Figure 1-95 Used to troubleshoot and verify outputs and inputs of digital logic circuits. (B & K Dynascan Corporation)

DIGITAL HAND TACHOMETER

Figure 1-96 Used to record the rotational speed of motors in revolutions per minute. (James G. Biddle Company)

VOLTAGE TESTER

Figure 1-97 Used to measure AC/DC voltages up to 600 volts. (Amprobe Instrument, Division of Core Industries, Inc.)

GROUNDED CIRCUIT TESTER

Figure 1-98 Provides method of testing for ground in AC or DC circuits from 100 to 500 volts. (Vaco Products Company)

CONTINUITY TESTER

Figure 1-99 Checks for defective controls and switches, broken leads and lines, and motors. Tests for continuity of circuit with power off. (Vaco Products Company)

TEST LEAD SET

Figure 1-100 Used with test equipment and control circuits in isolation problems. (Vaco Products Company)

MEGOHMMETER

Figure 1-101 Used for testing the insulation properties of motors, transformers, cables, generators, and other related equipment. (Associated Research, Inc.)

STRIP CHART RECORDER

Figure 1-102 Used to monitor voltage and current variations over extended periods of time. (Amprobe Instrument, Division of Core Industries, Inc.)

OHMMETER (VIBROGROUND)

Figure 1-103 Used for testing of resistance of man-made grounds and solid resistivity for corrosion control. (Associated Research, Inc.)

PHASE SEQUENCE INDICATOR

Figure 1-104 Used to determine proper phasing of three-phase systems. (Associated Research, Inc.)

HAZARDOUS LOCATIONS

The use of electrical equipment in areas where explosion hazards are present can lead to an explosion and fire. This danger exists in the form of escaped flammable gases such as naptha, benzine, propane, and others. Coal, grain, and other dust suspended in air can also cause an explosion.

Article 500 of The National Electrical Code® covers hazardous locations. Figure 1-105 outlines an example of part of this classification system. Any hazardous location requires the maximum in safety and adherence to local, state, and federal guidelines and laws as well as in-plant safety rules.

CLASSIFICATION OF HAZARDOUS LOCATIONS		
CLASS	DIVISION	GROUP
I. Flammable gases or vapors are, or may be, present in quantity to produce explosion. II. Flammable dust in suspension in quantity to produce explosion. III. Flammable fibers are present but are not likely to be in suspension to explode.	1. Locations where flammables are openly handled may be present continuously. 2. Locations where flammables are normally contained or are in the form of deposits of dust. Applies also to locations adjacent to Division 1 areas.	A. Acetylene B. Hydrogen, butadiene ethylene oxide, propylene oxide, manufactured gas with more than 30% by volume hydrogen C. Ethyle-ether, ethylene, cyclopropane, hydrogen sulfide, etc. D. Gasoline, hexane, naptha, benzine, etc. E. Metal dust; aluminum, magnesium, their alloys and others F. Carbon black, coal, coke, dust
EXAMPLE CLASS I, DIVISION 1, GROUPS B AND C		G. Flour, starch or grain dust

NOTE: For further details, see National Electrical Code®, Article 500.

Figure 1-105. Article 500 of the National Electrical Code® covers the requirements for electrical equipment in locations where fire or explosion hazards may exist.

1. What are three of the different methods used to organize electrical tools?
2. What are 10 of the basic rules to follow for proper and safe tool usage?
3. Why should all power tools be grounded before working with them?
4. List all safety precautions that should be taken before working on an electrical circuit in order to prevent an electrical shock.
5. What do the following OSHA safety color codes designate? Red; yellow; orange; purple; green.
6. What is a Class "A" fire?
7. What is a Class "B" fire?
8. What is a Class "C" fire?
9. What is a Class "D" fire?
10. Why is it dangerous to use a screwdriver around live electrical wires?
11. Is it safe to use a pipe extension to increase the leverage of a wrench?
12. What tool should be used for bending and cutting small diameter wire?
13. What tool should be used for cutting cable and twisting wire?
14. What tool should be used for removing insulation from small diameter wire?
15. What tool should be used to pull wire through conduit?
16. What tool should be used to bend conduit?
17. What tool should be used to remove cartridge type fuses?
18. What instrument can be used to measure voltage in a circuit?
19. What instrument can be used to measure current in a circuit?

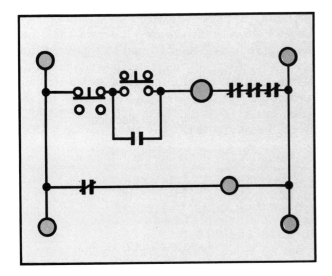

2 Industrial Electrical Symbols and Line Diagrams

Industrial electrical symbols and line (ladder) diagrams provide the information necessary to understand the operation of any electrical control system. Industrial electrical symbols identify electrical devices within a circuit. The operation of a circuit is determined by the location of electrical devices. The line diagram indicates the location of electrical devices within a circuit. Circuit arrangement and modifications to existing circuits is possible using line diagrams. To read and troubleshoot an electrical control system, the electrician must have a working knowledge of line diagrams.

LANGUAGE OF CONTROL

In every trade or profession, there is a certain language which must be understood in order to transfer information and ideas rapidly and efficiently. This language may include words and phrases as well as symbols. The object of this unit is to convey a thorough understanding of how each of these parts is related. Once this language of controls is learned, you will know how to read the stories which are told by line diagrams.

LINE DIAGRAMS (LADDER DIAGRAMS)

The basic means of communicating the language of control is through the use of the line diagram or ladder diagrams (Figure 2–1). The line diagram consists of a series of symbols interconnected by lines to indicate the flow of current through the various devices. The line diagram tells us in a remarkably short time a series of relationships that would take many words to explain. The line diagram shows us basically two things: (1) The power source (shown as the heavier black line) and (2) how current flows through the various parts of the circuit, such as pushbuttons, contacts, coils, and overloads (shown by thinner control circuit lines).

The line diagram is intended to show only the circuitry which is necessary for the basic operation of the controller. It is not intended to show the physical relationship of the various devices in the controller. Rather, it leans toward simplicity, emphasizing only the operation of the control circuit. (Another diagram, called the wiring diagram, will be covered in Unit 4.)

MANUAL CONTROL CIRCUITS

Figure 2-2, top, illustrates a very basic manual control circuit in line diagram form. The line diagram shown is a pushbutton controlling light. This circuit is considered to be manual because someone (a person) must initiate an action for the circuit to operate. Figure 2-2, bottom, illustrates

pictorially what is included in the circuit of Figure 2-2. Note from the line diagram that the heavy dark lines labeled L1 and L2 represent the power source or power circuit. The voltage of the power circuit is usually indicated somewhere on the circuit near these lines. In this case, it is 115 volts, but may be 230, 460, 575 or 2300 VAC. When the voltage is DC, it may be marked with a

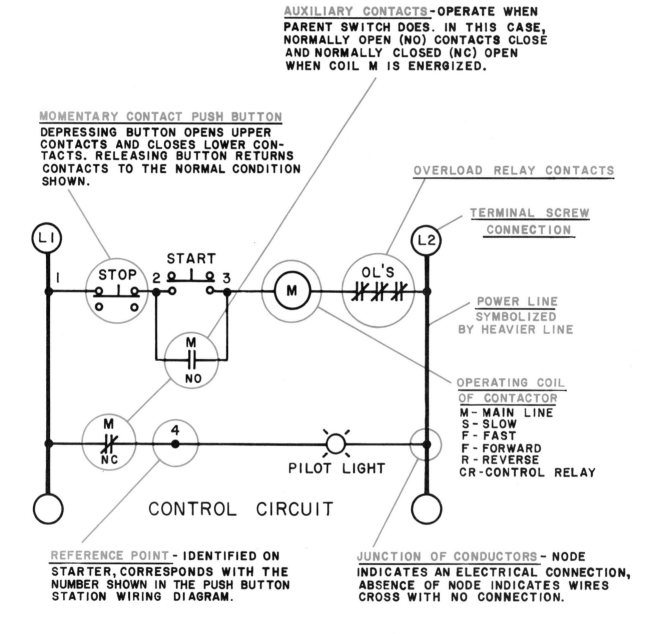

AUXILIARY CONTACTS - OPERATE WHEN PARENT SWITCH DOES. IN THIS CASE, NORMALLY OPEN (NO) CONTACTS CLOSE AND NORMALLY CLOSED (NC) OPEN WHEN COIL M IS ENERGIZED.

MOMENTARY CONTACT PUSH BUTTON
DEPRESSING BUTTON OPENS UPPER CONTACTS AND CLOSES LOWER CONTACTS. RELEASING BUTTON RETURNS CONTACTS TO THE NORMAL CONDITION SHOWN.

OVERLOAD RELAY CONTACTS

TERMINAL SCREW CONNECTION

POWER LINE SYMBOLIZED BY HEAVIER LINE

OPERATING COIL OF CONTACTOR
M - MAIN LINE
S - SLOW
F - FAST
F - FORWARD
R - REVERSE
CR - CONTROL RELAY

PILOT LIGHT

CONTROL CIRCUIT

REFERENCE POINT - IDENTIFIED ON STARTER, CORRESPONDS WITH THE NUMBER SHOWN IN THE PUSH BUTTON STATION WIRING DIAGRAM.

JUNCTION OF CONDUCTORS - NODE INDICATES AN ELECTRICAL CONNECTION, ABSENCE OF NODE INDICATES WIRES CROSS WITH NO CONNECTION.

Figure 2-1 The line diagram shown consists of a series of symbols interconnected by lines to indicate the flow of current through the various devices.

polarity of negative or positive (− or +) and may be voltages of 90, 120, 180, 240, 500 or 550 VDC.

The dark black nodes on a circuit indicate that an electrical connection is made. If the node is *not present,* the wires merely cross each other and are not electrically connected.

The line diagram is always read from left (L1) to right (L2). In the case of Figure 2-2, the circuit would be read in the following manner: Depressing PB1 (Pushbutton 1) allows current to pass across the closed contacts of PB1 through the pilot light (PL1) and on to L2, forming a complete circuit which activates the pilot light (PL1). Releasing PB1 opens contacts PB1, stopping the current flow to the pilot light, making the pilot light go out.

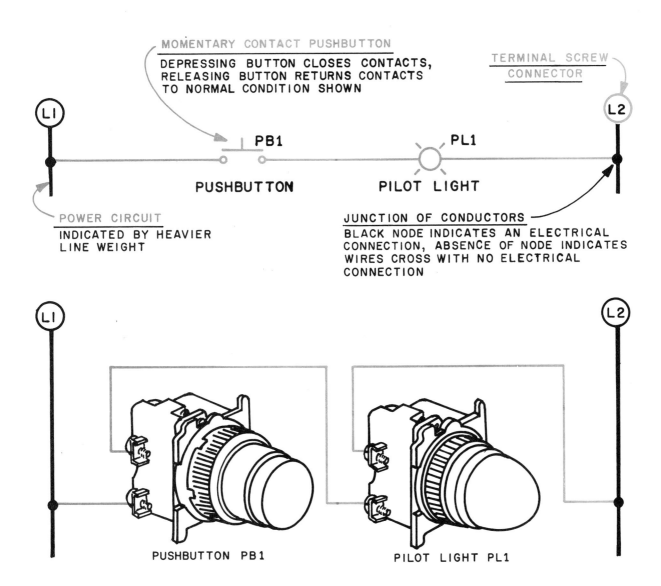

Figure 2-2 Line diagram and pictorial diagram of pushbutton switch controlling a pilot light. Circuit Operation: Depressing pushbutton (PB1) allows current to flow from L1 through PB1, through pilot light (PL1) and on to L2, causing the light to glow. Releasing PB1 opens contacts PB1, stopping the current flow to the light. The light then goes out.

Figure 2-3, top, illustrates another simple manual control circuit. In this circuit, we are using a simple manual motor starter with overload protection to control and protect a single phase motor. Figure 2-3, bottom, indicates pictorially how the device would actually look. It should be apparent from the pictorial diagrams that line diagrams are much easier to draw than pictorial diagrams.

The manual starter switch in Figure 2-3 is represented in the line diagram by the set of normally open (NO or N.O.) contacts S1 and by the overload contacts OL1. Recall that the line diagram is set up to be read easily and does not necessarily indicate where the devises are physically located. This is why the overloads are shown

between the motor and line L2 in the line diagram but are physically located together in the actual manual starter.

Figure 2-3 would be read in the following manner: When the switch portion of the starter (normally open contacts S1) is closed by hand, current passes through the contacts S1 through the motor and through the overloads to L2, starting the motor. The motor will continue to run unless contacts S1 are opened, a power failure exists or until the motor experiences an overload. In the case of an overload, the OL1 contacts would open and the motor would stop. The motor could not be restarted until the overload is removed and the overload contacts reset to their normally closed (NC or N.C.) position. (*Note:* In a line diagram the

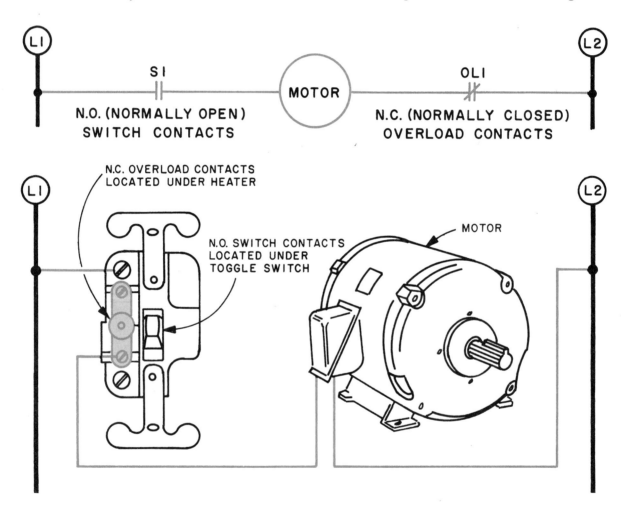

Figure 2-3 Line diagram and pictorial diagram of a manual switch with overload protection controlling a motor. Circuit Operation: When the switch contacts S1 are closed by hand through the toggle switch, current passes from L1 through contacts S1, through the motor and through the overloads to L2, starting the motor. Opening contacts S1 stops the current flow to the motor and the motor stops. Operation of overload OL1 may also stop the operation of the motor.

overload comes after the motor, but in the actual circuit the overload comes before the motor. Since the overload is a series device and will open the motor control circuit in either position, it does not matter that it is shown after the motor in the line diagram.)

AUTOMATIC CONTROL

With the advent of the industrial revolution, automatically controlled devices replaced many jobs that were once done manually. As a part of automation, control circuits had to be designed to replace manual devices.

Figure 2-4, top, illustrates how the manual cir-

cuit of Figure 2-3 could be converted to automatic operation. By adding one more device, such as the float switch, the electric motor on a sump pump can be turned on and off automatically. (see Figure 2-4, bottom).

Reading from left to right on the line diagram, we understand that the circuit would respond in the following manner. With switch contacts S1 in a closed position, float switch contacts FS1 will determine if current passes through the circuit. When the float switch contacts FS1 are closed, current passes through contacts S1 through contacts FS1 and on to the motor and its overload to L2 so that the pump motor is started. The pump will continue to pump water until the water level drops enough to cause the contacts FS1 to open

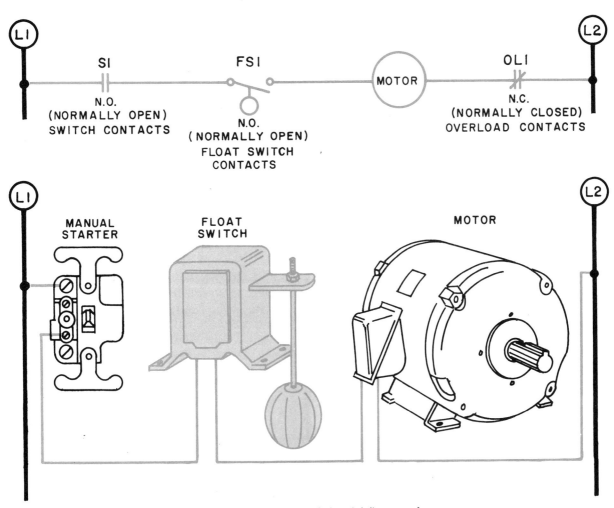

Figure 2-4 Line diagram and pictorial diagram of a manual switch with overload protection and a float switch controlling a pump motor for "SUMP" operation.

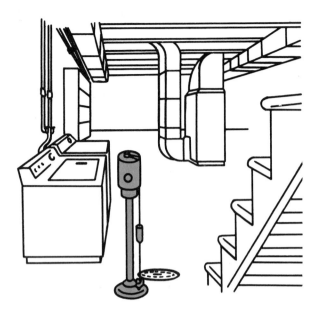

Figure 2-5 A sump pump control responds automatically to prevent flooding in basements. Through the use of a float switch (NO), water is automatically removed when the water level begins to rise.

and shut off the pump motor. Only an overload in the motor, a power failure or the manual opening of contacts S1 can stop the pump motor from automatically pumping water when the level is too high.

This type of control circuit is used in basements to control a sump pump in order to prevent flooding (see Figure 2-5). When the water level begins to rise, the float switch senses the change and

automatically starts the pump, removing the water again.

Float switches generally are designed with two sets of contacts: (1) normally open (NO) and (2) normally closed (NC) (Figure 2-6). Since the pair of normally open (NO) and normally closed (NC) contacts of the float switch can either *close* or *open* contacts with changes in liquid level, variations in the application of the control device are possible. For example, when the normally closed contacts are used as in Figure 2-6, the float switch can be used to keep the water at a predetermined level.

Reading from left to right on the line diagram (Figure 2-6), we see that the circuit would perform in the following manner. With contacts S1 in a closed position, float switch contacts FS1 allow current to pass across them through the motor and through the overloads to L2 so that the pump motor is started. The pump motor will continue to pump water until the water level rises high enough to cause contacts FS1 to open and shut off the pump. Only an overload in the motor, a power failure or the manual opening of contacts S1 can stop the pump from automatically filling the tank to a predetermined level.

This type of circuit could be used to maintain a certain level of water in a cattle tank similar to that shown in Figure 2-7. When the water level drops because of evaporation or periodic drinking, the normally closed (NC) contacts will close and start the pump motor. When the water level rises to the predetermined level, the float will cause the normally closed (NC) contacts to open and the pump motor will shut off. In certain circuits, such as the one shown for liquid level, you will find overloads which automatically reset themselves as part of the motor. In certain situations, such as

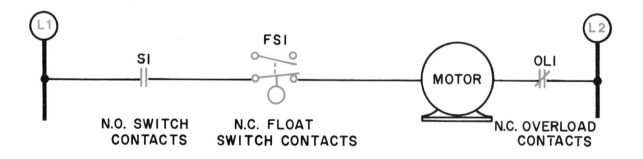

Figure 2-6 Line diagram of a manual switch with overload protection and a float switch controlling a pump motor for "PUMP" operation.

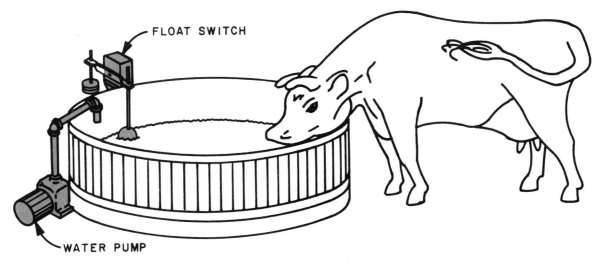

Figure 2-7 Pictorial diagram of a manual switch and float switch controlling a pump motor. Through the use of a float switch (NC), water is automatically replaced when the water level begins to fall.

flooding, it is considered less serious to let the motor burn out than it is to shut the operation down.

MAGNETIC CONTROL

Although manual controls are compact and sometimes less expensive than magnetic controls, industrial and commercial installations will often require that certain electrical control equipment be located in one area while the load device must be located in another area. To accomplish this remote control function, simple electromagnetic devices called solenoids, contactors and magnetic motor starters can be used.

Solenoids

A solenoid (Figure 2-8) consists of a frame, plunger and coil. When the coil is energized (by an electric current passing through it), a magnetic field is set up in the frame. This magnetic field causes the plunger to move into the frame. The result is a straight line force, usually as a push or pull action. In the picture shown, the action is a push operation.

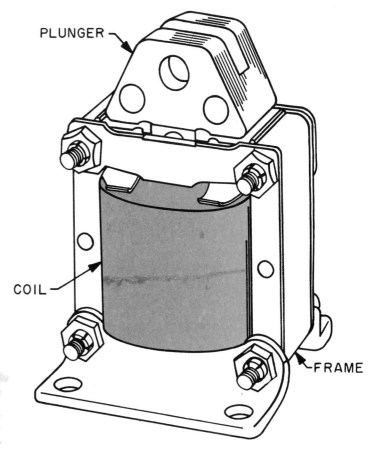

Figure 2-8 Pictorial diagram of a solenoid consisting of a frame, plunger and coil.

Figure 2-9, top, illustrates symbolically in a line diagram how the solenoid circuit operates. The circuit would be read in the following manner: Depressing the NO (normally opened) contacts of pushbutton PB1 allows an electric current to flow through the solenoid, creating a magnetic field. The solenoid, depending on construction, will cause the plunger to push or pull (Figure 2-9, bottom).

This type of circuit could be used to control a door lock or holding clamp which would be opened only when the pushbutton is depressed. In the case of a door lock (Figure 2-10), as long as the pushbutton is depressed the door may open. When the button is released, however, the door is locked.

Use of this circuit would provide security access to a building. With the addition of a holding clamp, a part may be held in position or released, based on the decision of the pushbutton operator.

Contactors

A contactor (Figure 2-11) is constructed like and operates much like a solenoid. The contactor has a frame, plunger and coil like that of the solenoid. However, the action of the plunger is directed to close sets of contacts as shown in Figure 2-12. The closing of these contacts within the contactor allows electrical devices to be controlled from remote locations.

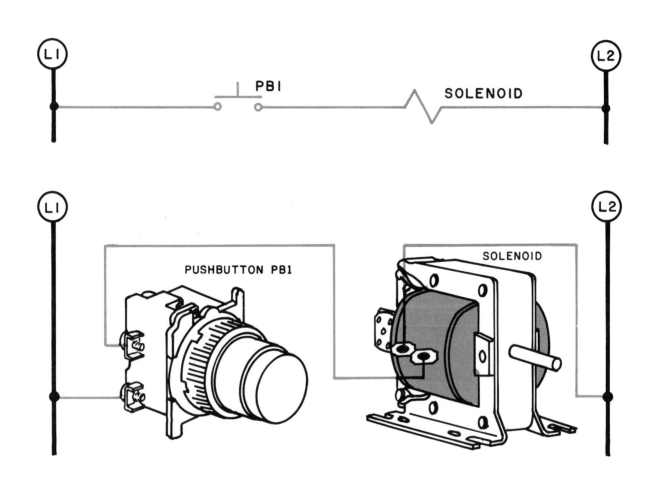

Figure 2-9 Line diagram and pictorial diagram of a push-button switch controlling a solenoid. Circuit Operation: Depressing PB1 allows current to flow from L1 through PB1 through the solenoid and on to L2, causing the solenoid to be activated. Releasing PB1 opens contacts PB1, stopping current flow and deactivating the solenoid.

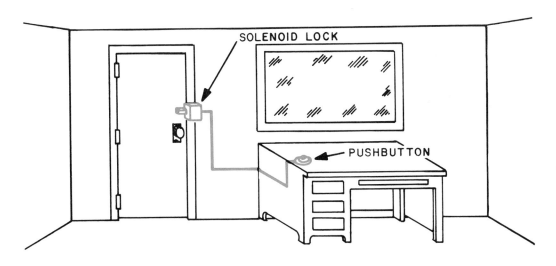

Figure 2-10 Pictorial diagram of a pushbutton switch operating a solenoid door lock. When the button is depressed, the lock is opened; when the button is released, the lock closes.

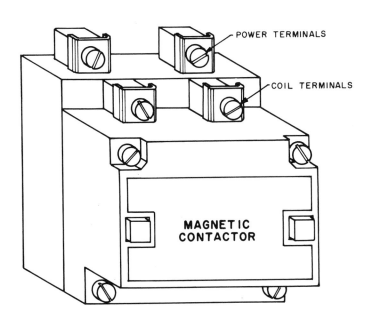

Figure 2-11 The contactor shown is constructed internally much like a solenoid. The contactor has a frame, plunger and coil like that of a solenoid. However, the action of the plunger in this case is directed to close sets of contacts.

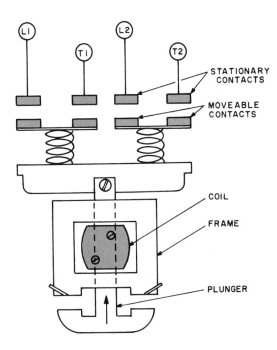

Figure 2-12 The internal construction of the solenoid shown indicates how the frame, plunger and coil mechanism operate to close two sets of electrical contacts.

Figure 2-13, top, illustrates in a line diagram how the contactor operates electrically. Figure 2-13, bottom, illustrates pictorially how the circuit would be constructed. It should be noted immediately that the line diagram does not show the power contacts in the control circuit. It is important to remember this point, since the line diagram is designed to illustrate the control circuit only and not the power circuit. The line diagram is planned this way to keep the line diagram simple. The power contacts will be discussed later in the unit.

The circuit of Figure 2-13 would be read in the following manner: Depressing pushbutton PB1 allows current to pass through contacts PB1 through the coil of contactor C1 through to line 2, (L2) causing coil C1 to be energized. The activation of coil C1 in turn causes the power contacts of the contactor to close, as indicated in Figure 2-14. Figure 2-14 is used at this time only to point out that there are power contacts in the contactor and that they close each time the control circuit is activated. It is up to the electrician to know that

these contacts exist, but that they are usually never shown in the line diagram. More information on the power circuit will be provided in the future units on contactors, magnetic motor starters and overloads.

Releasing PB1 stops the flow of current to the contactor coil C1 and causes the coil to de-energize. When the coil de-energizes, the power contacts return to their normally open condition, causing the lights, heaters or other types of loads connected to them to be shut off (see Fig 2-13).

This circuit works well in turning on and off various loads remotely. However, in its present design it does require someone to hold the contacts closed if they wish to have the coil continuously energized. To eliminate the necessity of someone holding the pushbutton continuously, devices called auxiliary contacts (Figure 2-15) can be added to a contactor to form an *electrical holding circuit*. These auxiliary contacts are attached physically to the side of the contactor and will be opened and closed with the power contacts, as the coil is energized or de-energized. These contacts

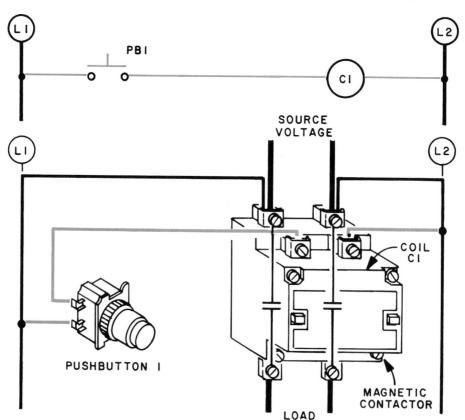

Figure 2-13 Line diagram and pictorial diagram of a pushbutton switch controlling the coil in a magnetic contactor. Circuit operation: Depressing PB1 allows current to pass from

L1 through closed contacts of PB1, through coil C1 and on to L2, causing coil C1 to be energized. Releasing PB1 stops current to coil C1, de-energizing C1.

will be shown on the line diagram because they are part of the control circuit.

Figure 2-16, top, shows the line diagram of the electrical holding circuit and Figure 2-16 bottom shows pictorially how this circuit would be wired.

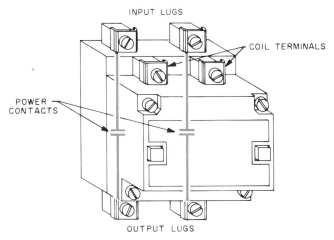

Figure 2-14 Although power contacts are usually not shown in the line diagram, they do exist in the contactor and close each time the contactor is energized. It is up to the electrician to know that these contacts exist and what effect they have on the power circuit.

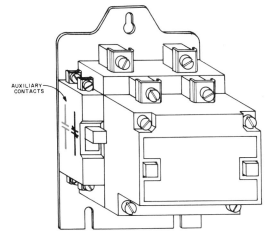

Figure 2-15 Auxiliary contacts can be added to a contactor to form an electrical holding circuit. These contacts can be incorporated into the control circuit to eliminate the necessity of someone holding the start pushbutton once it has been energized.

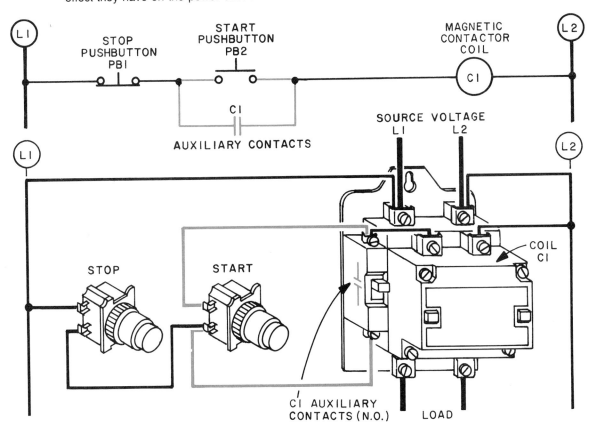

Figure 2-16 Line diagram and pictorial diagram of two pushbuttons and a set of auxiliary contacts controlling the coil in a magnetic contactor.

Auxiliary contacts C1 form the electrical holding circuit in this control diagram.

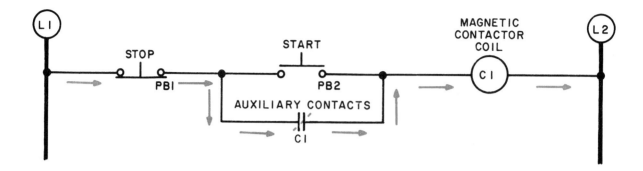

Figure 2-17 Line diagram indicating path of current flow when auxiliary contacts C1 are energized by coil C1. Circuit Operation: Depressing PB2 allows current to pass from L1 through closed contacts PB1, through contacts PB2, through coil C1 and on to L2, causing coil C1 to be energized. With coil C1 energized, auxiliary contacts C1 close and remain closed as long as coil C1 is energized. Auxiliary contact C1 forms a continuous electrical path around contacts PB2, allowing the coil C1 to remain energized even if contacts PB2 open. Depressing PB1 stops current to coil C1, de-energizing coil C1 and causing auxiliary contacts C1 to return to their normally open position.

Reading from left to right (Figure 2-17), we see that the circuit would respond in the following manner: Depressing pushbutton PB2 would allow current to pass through the closed contacts of the stop button PB1 and through the closed contacts of PB2 through to the coil C1 and on to L2, causing the coil C1 to be energized. With coil C1 energized, auxiliary contacts C1 would close and remain closed as long as the coil is energized. Note from the illustration of Figure 2-17 that this forms a continuous electrical path around the pushbutton PB2, so that even if pushbutton PB2 is released, the circuit will remain energized since the coil is still energized.

The only way the circuit may be de-energized is through a power failure or by depressing the normally closed (NC) stop button PB1. In either of these cases, the current flow to the coil C1 would stop and the coil would de-energize, causing the auxiliary contacts C1 to return to their normally open position. To re-energize this circuit, pushbutton PB2 would again have to be depressed so that the procedure previously explained would repeat itself.

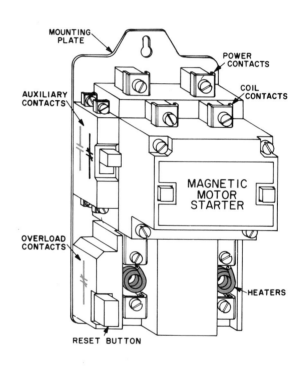

Figure 2-18 Pictorial diagram of a magnetic motor starter with overloads and auxiliary contacts.

Magnetic Motor Starters

Magnetic motor starters are used to start and stop motors. In addition, magnetic motor starters are required to provide protection for the motor in case of an overload. Magnetic starters are identical to contactors except that they have another section, called the overloads, attached to them as shown in Figure 2-18.

Overloads will be discussed only briefly at this point because they have a definite effect on the operation of the control circuit. How overloads operate in detail will be covered in a future unit. For now it will be enough to know that the overload device has heaters (in the power circuit) which sense excessive current going to the motor and cause the normally closed (NC) contacts (in the control circuit) to open when the overload becomes dangerous to the motor.

Figure 2-19, top, illustrates a line diagram using a magnetic starter with overloads and Figure 2-19, bottom, illustrates the physical wiring. Note that the only difference in this line diagram and that of the contactor used in Figure 2-16, is the use of an M coil, indicating motor starter coil.

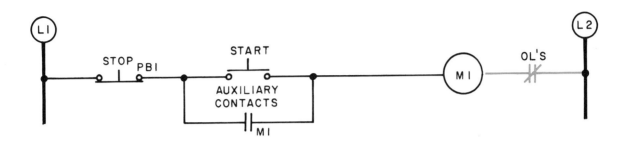

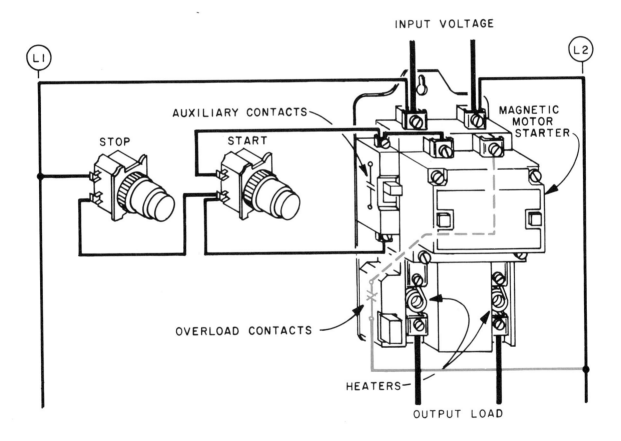

Figure 2-19 Line diagram and pictorial diagram of two pushbuttons and a set of auxiliary contacts controlling the coil in a magnetic starter with overload protection.

Figure 2-20 shows that electrically the circuit would operate in the following manner: Depressing the START pushbutton allows current to pass through coil M1 and through overload OL1, causing coil M1 to energize. With coil M1 energized, auxiliary contacts M1 close, and the circuit will remain energized even if the START pushbutton is released.

The circuit can be de-energized only if STOP is depressed, a power failure occurs, or any one of the overloads sense a problem in the power circuit. If one of these situations results, coil M1 will de-energize, causing auxiliary contacts M1 to return to their normally open (NO) condition. When the motor stops due to an overload, the overload must be removed, the overload device reset, and pushbutton START would again have to be depressed in order to repeat the procedure in restarting the motor.

In this unit, you have covered some of the most basic control devices in industrial control. Each of these circuits was treated as it would operate in the line diagram. In further units, you will apply more of these devices to control additional circuits. As an aid in helping you interpret these diagrams, you are provided with a pictorial description and schematic symbol of other common control devices at the end of this unit (Pages 45 through 52). Study these devices carefully and refer to them as you cover more complex line diagrams in future units.

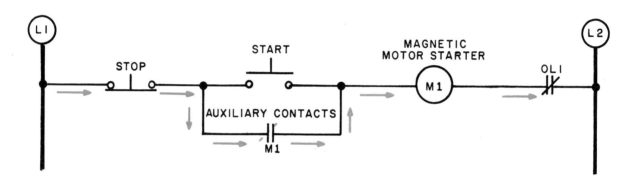

Figure 2-20 Line diagram indicating path of current flow when auxiliary contacts M1 are energized by coil M1. Overload contacts OL1 remain normally closed and part of the continuous electrical circuit unless the motor experiences an overload, causing these contacts to open and de-energize coil M.

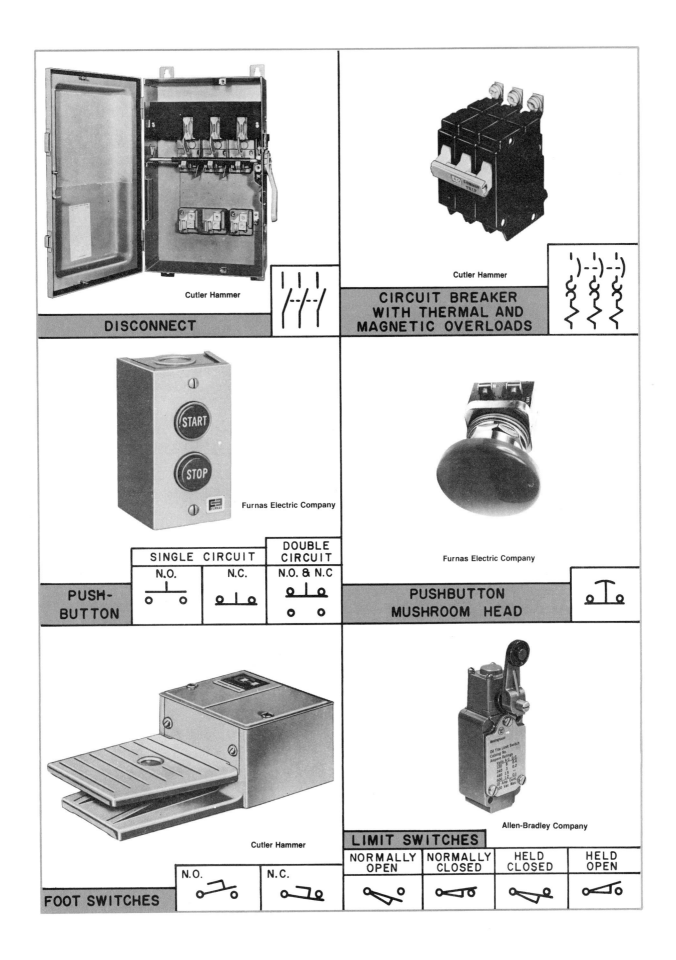

Cutler Hammer

DISCONNECT

Cutler Hammer

CIRCUIT BREAKER WITH THERMAL AND MAGNETIC OVERLOADS

Furnas Electric Company

SINGLE CIRCUIT		DOUBLE CIRCUIT
N.O.	N.C.	N.O. & N.C

PUSH-BUTTON

Furnas Electric Company

PUSHBUTTON MUSHROOM HEAD

Cutler Hammer

Allen-Bradley Company

LIMIT SWITCHES

NORMALLY OPEN	NORMALLY CLOSED	HELD CLOSED	HELD OPEN

FOOT SWITCHES

N.O.	N.C.

Allen-Bradley Company

Transamerica Delaval

LIQUID LEVEL SWITCH	N.O.	N.C.

FLOW SWITCH (AIR, WATER, ETC)	N.O.	N.C.

Allen-Bradley Company

Square D Company

TEMPERATURE ACTUATED SWITCH	N.O.	N.C.

PRESSURE & VACUUM SWITCHES	N.O.	N.C.

Allen-Bradley Company

Furnas Electric Company

SELECTOR SWITCH 2 POSITION

J   K

O—A1
O   O A2

	J	K
A1	X	
A2		X

X-CONTACT CLOSED

SELECTOR SWITCH 3 POSITION

J K L

O   O A1
O   O A2

	J	K	L
A1	X		
A2			X

X-CONTACT CLOSED

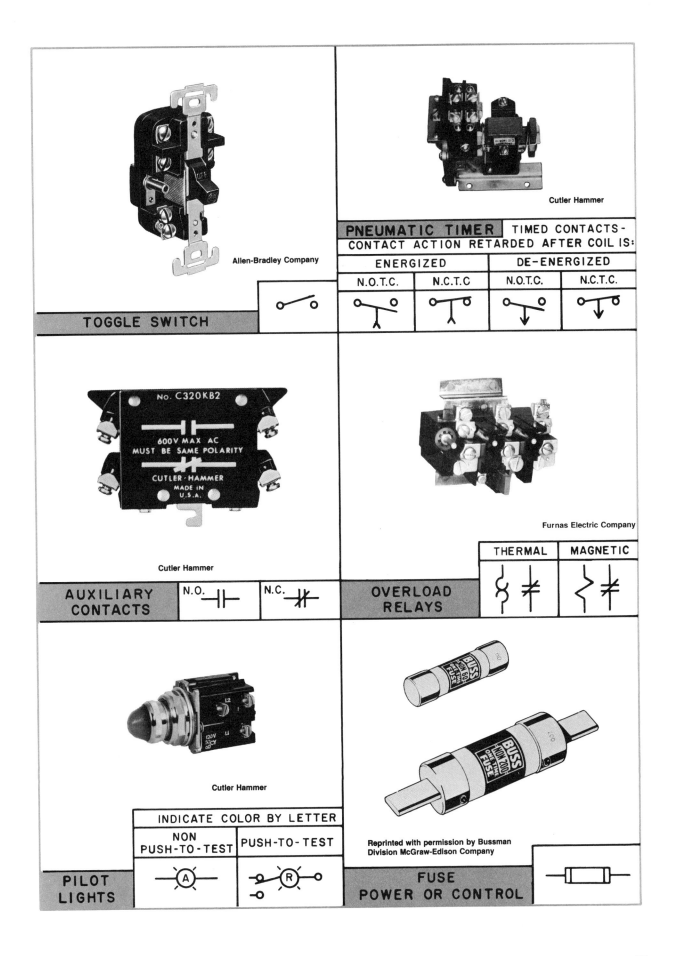

Allen-Bradley Company

Cutler Hammer

TOGGLE SWITCH		PNEUMATIC TIMER	TIMED CONTACTS- CONTACT ACTION RETARDED AFTER COIL IS:			
			ENERGIZED		DE-ENERGIZED	
			N.O.T.C.	N.C.T.C	N.O.T.C.	N.C.T.C.

No. C320KB2

600V MAX AC
MUST BE SAME POLARITY

CUTLER·HAMMER
MADE IN
U.S.A.

Cutler Hammer

Furnas Electric Company

AUXILIARY CONTACTS	N.O.	N.C.	OVERLOAD RELAYS	THERMAL	MAGNETIC

Cutler Hammer

PILOT LIGHTS	INDICATE COLOR BY LETTER		FUSE POWER OR CONTROL	
	NON PUSH-TO-TEST	PUSH-TO-TEST		

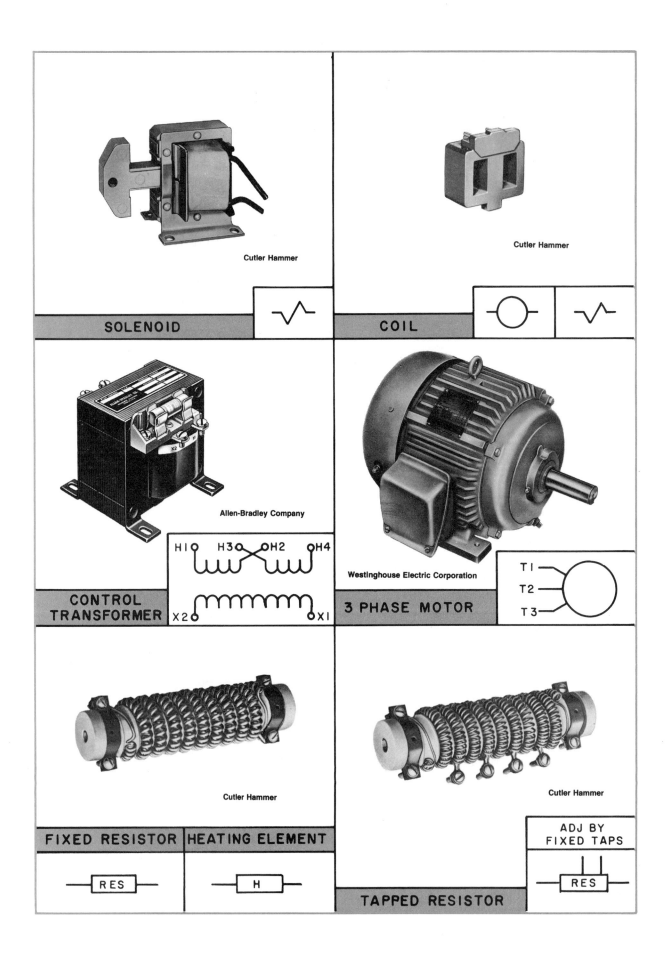

Cutler Hammer

Cutler Hammer

SOLENOID

COIL

Allen-Bradley Company

Westinghouse Electric Corporation

CONTROL TRANSFORMER

H1   H3   H2   H4

X2   X1

3 PHASE MOTOR

T 1
T 2
T 3

Cutler Hammer

Cutler Hammer

FIXED RESISTOR   HEATING ELEMENT

RES

H

ADJ BY FIXED TAPS

RES

TAPPED RESISTOR

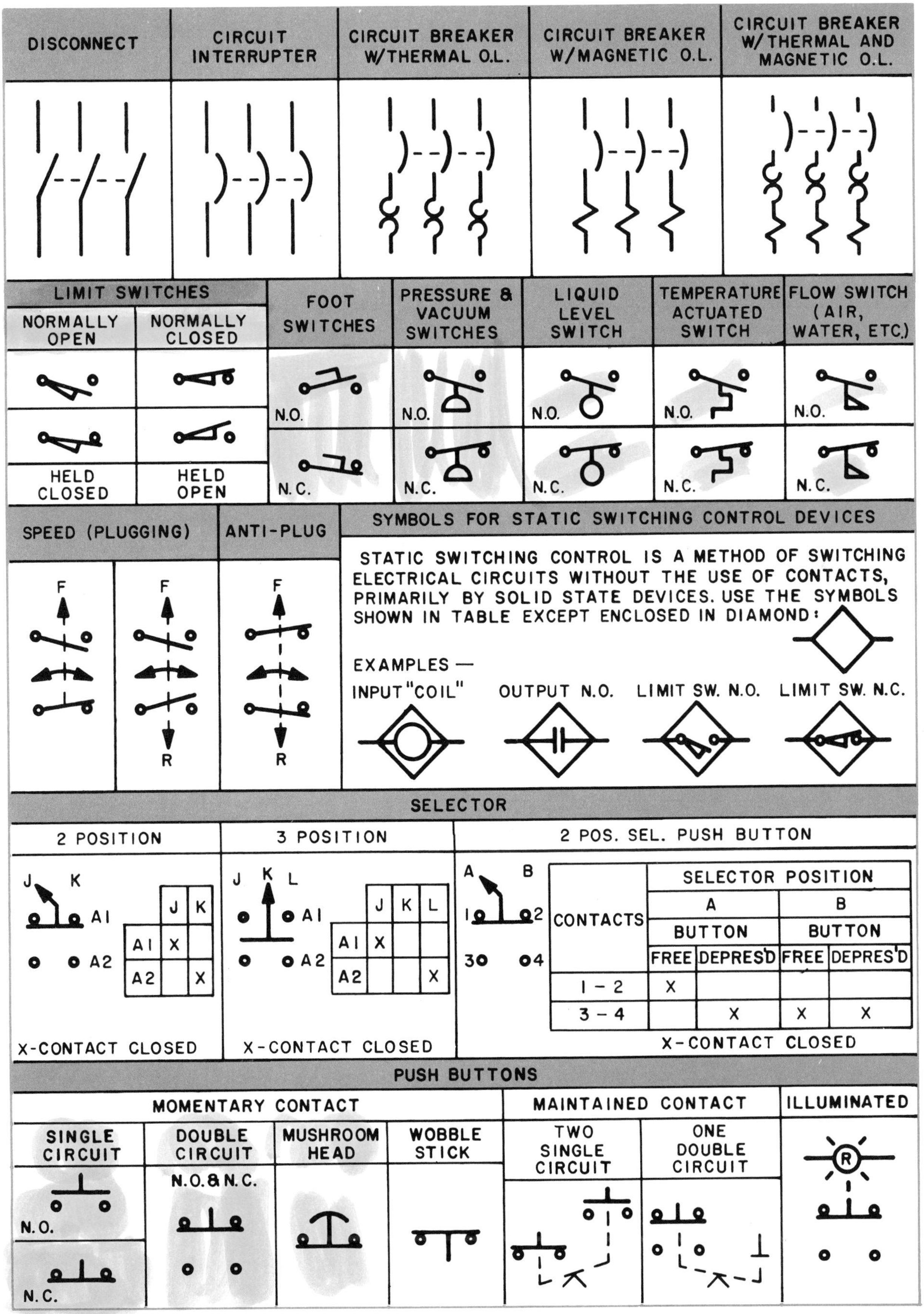

DISCONNECT	CIRCUIT INTERRUPTER	CIRCUIT BREAKER W/THERMAL O.L.	CIRCUIT BREAKER W/MAGNETIC O.L.	CIRCUIT BREAKER W/THERMAL AND MAGNETIC O.L.

LIMIT SWITCHES		FOOT SWITCHES	PRESSURE & VACUUM SWITCHES	LIQUID LEVEL SWITCH	TEMPERATURE ACTUATED SWITCH	FLOW SWITCH (AIR, WATER, ETC.)
NORMALLY OPEN	NORMALLY CLOSED					
		N.O.	N.O.	N.O.	N.O.	N.O.
		N.C.	N.C.	N.C.	N.C.	N.C.
HELD CLOSED	HELD OPEN					

SPEED (PLUGGING)	ANTI-PLUG	SYMBOLS FOR STATIC SWITCHING CONTROL DEVICES

STATIC SWITCHING CONTROL IS A METHOD OF SWITCHING ELECTRICAL CIRCUITS WITHOUT THE USE OF CONTACTS, PRIMARILY BY SOLID STATE DEVICES. USE THE SYMBOLS SHOWN IN TABLE EXCEPT ENCLOSED IN DIAMOND:

EXAMPLES — INPUT "COIL"   OUTPUT N.O.   LIMIT SW. N.O.   LIMIT SW. N.C.

SELECTOR

2 POSITION	3 POSITION	2 POS. SEL. PUSH BUTTON

2 POSITION

	J	K
A1	X	
A2		X

X-CONTACT CLOSED

3 POSITION

	J	K	L
A1	X		
A2			X

X-CONTACT CLOSED

CONTACTS	SELECTOR POSITION				
	A		B	
	BUTTON		BUTTON	
	FREE	DEPRES'D	FREE	DEPRES'D
1 – 2	X			
3 – 4		X	X	X

X-CONTACT CLOSED

PUSH BUTTONS

MOMENTARY CONTACT				MAINTAINED CONTACT		ILLUMINATED
SINGLE CIRCUIT	DOUBLE CIRCUIT	MUSHROOM HEAD	WOBBLE STICK	TWO SINGLE CIRCUIT	ONE DOUBLE CIRCUIT	
N.O.	N.O. & N.C.					
N.C.						

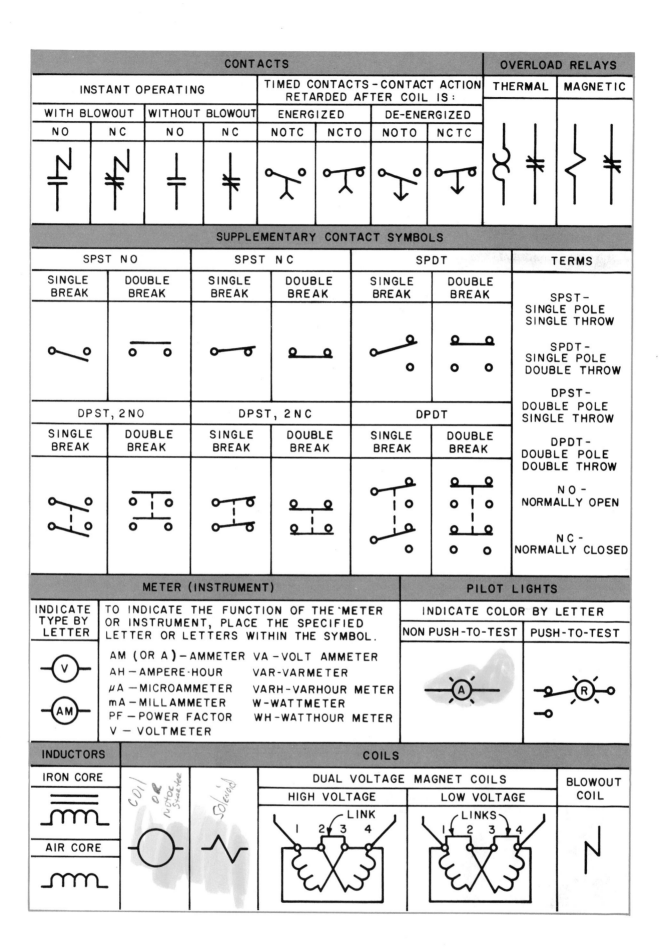

CONTACTS

INSTANT OPERATING				TIMED CONTACTS – CONTACT ACTION RETARDED AFTER COIL IS:				OVERLOAD RELAYS	
WITH BLOWOUT		WITHOUT BLOWOUT		ENERGIZED		DE-ENERGIZED		THERMAL	MAGNETIC
NO	NC	NO	NC	NOTC	NCTO	NOTO	NCTC		

SUPPLEMENTARY CONTACT SYMBOLS

SPST NO		SPST NC		SPDT		TERMS
SINGLE BREAK	DOUBLE BREAK	SINGLE BREAK	DOUBLE BREAK	SINGLE BREAK	DOUBLE BREAK	SPST – SINGLE POLE SINGLE THROW

DPST, 2NO		DPST, 2NC		DPDT		
SINGLE BREAK	DOUBLE BREAK	SINGLE BREAK	DOUBLE BREAK	SINGLE BREAK	DOUBLE BREAK	

TERMS:

SPST – SINGLE POLE SINGLE THROW

SPDT – SINGLE POLE DOUBLE THROW

DPST – DOUBLE POLE SINGLE THROW

DPDT – DOUBLE POLE DOUBLE THROW

NO – NORMALLY OPEN

NC – NORMALLY CLOSED

METER (INSTRUMENT)

INDICATE TYPE BY LETTER	TO INDICATE THE FUNCTION OF THE METER OR INSTRUMENT, PLACE THE SPECIFIED LETTER OR LETTERS WITHIN THE SYMBOL.

AM (OR A) – AMMETER   VA – VOLT AMMETER
AH – AMPERE-HOUR   VAR – VARMETER
μA – MICROAMMETER   VARH – VARHOUR METER
mA – MILLAMMETER   W – WATTMETER
PF – POWER FACTOR   WH – WATTHOUR METER
V – VOLTMETER

PILOT LIGHTS

INDICATE COLOR BY LETTER

NON PUSH-TO-TEST	PUSH-TO-TEST

INDUCTORS

IRON CORE

AIR CORE

COILS

DUAL VOLTAGE MAGNET COILS

HIGH VOLTAGE	LOW VOLTAGE	BLOWOUT COIL

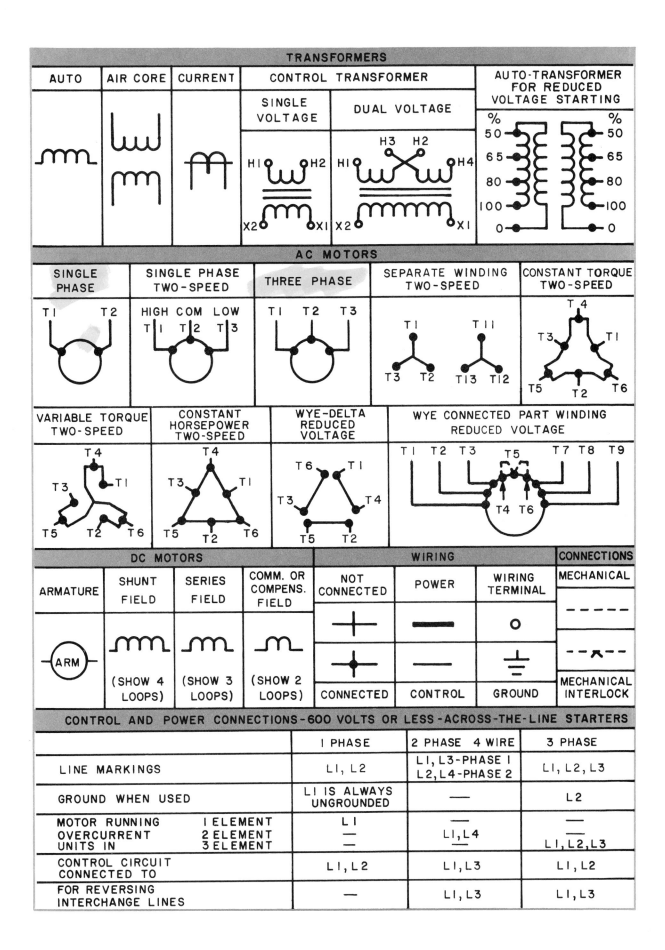

TRANSFORMERS

AUTO	AIR CORE	CURRENT	CONTROL TRANSFORMER		AUTO-TRANSFORMER FOR REDUCED VOLTAGE STARTING
			SINGLE VOLTAGE	DUAL VOLTAGE	

AC MOTORS

SINGLE PHASE	SINGLE PHASE TWO-SPEED	THREE PHASE	SEPARATE WINDING TWO-SPEED	CONSTANT TORQUE TWO-SPEED

VARIABLE TORQUE TWO-SPEED	CONSTANT HORSEPOWER TWO-SPEED	WYE-DELTA REDUCED VOLTAGE	WYE CONNECTED PART WINDING REDUCED VOLTAGE

DC MOTORS / WIRING / CONNECTIONS

ARMATURE	SHUNT FIELD	SERIES FIELD	COMM. OR COMPENS. FIELD	NOT CONNECTED	POWER	WIRING TERMINAL	MECHANICAL
	(SHOW 4 LOOPS)	(SHOW 3 LOOPS)	(SHOW 2 LOOPS)	CONNECTED	CONTROL	GROUND	MECHANICAL INTERLOCK

CONTROL AND POWER CONNECTIONS - 600 VOLTS OR LESS - ACROSS-THE-LINE STARTERS

		I PHASE	2 PHASE 4 WIRE	3 PHASE
LINE MARKINGS		LI, L2	LI, L3 - PHASE I L2, L4 - PHASE 2	LI, L2, L3
GROUND WHEN USED		LI IS ALWAYS UNGROUNDED	—	L2
MOTOR RUNNING OVERCURRENT UNITS IN	I ELEMENT	LI	—	—
	2 ELEMENT	—	LI, L4	—
	3 ELEMENT	—		LI, L2, L3
CONTROL CIRCUIT CONNECTED TO		LI, L2	LI, L3	LI, L2
FOR REVERSING INTERCHANGE LINES		—	LI, L3	LI, L3

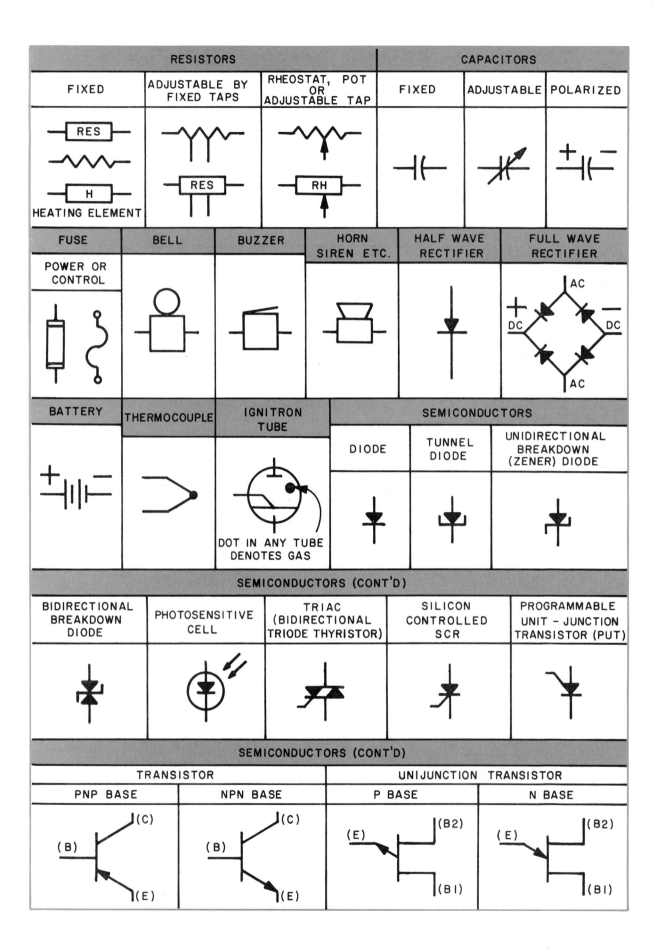

RESISTORS			CAPACITORS		
FIXED	ADJUSTABLE BY FIXED TAPS	RHEOSTAT, POT OR ADJUSTABLE TAP	FIXED	ADJUSTABLE	POLARIZED
RES / HEATING ELEMENT H	RES	RH			+ −

FUSE	BELL	BUZZER	HORN SIREN ETC.	HALF WAVE RECTIFIER	FULL WAVE RECTIFIER
POWER OR CONTROL					AC + DC − DC AC

BATTERY	THERMOCOUPLE	IGNITRON TUBE	SEMICONDUCTORS		
+ −		DOT IN ANY TUBE DENOTES GAS	DIODE	TUNNEL DIODE	UNIDIRECTIONAL BREAKDOWN (ZENER) DIODE

SEMICONDUCTORS (CONT'D)

BIDIRECTIONAL BREAKDOWN DIODE	PHOTOSENSITIVE CELL	TRIAC (BIDIRECTIONAL TRIODE THYRISTOR)	SILICON CONTROLLED SCR	PROGRAMMABLE UNIT – JUNCTION TRANSISTOR (PUT)

SEMICONDUCTORS (CONT'D)

TRANSISTOR		UNIJUNCTION TRANSISTOR	
PNP BASE	NPN BASE	P BASE	N BASE
(C) (B) (E)	(C) (B) (E)	(B2) (E) (B1)	(B2) (E) (B1)

Common Abbreviations for Electrical Terms and Devices

AC	Alternating current		MB	Magnetic brake
ALM	Alarm		MCS	Motor circuit switch
AM	Ammeter		MEM	Memory
ARM	Armature		MTR	Motor
AU	Automatic		MN	Manual
BAT	Battery (electrical)		NEG	Negative
BR	Brake relay		NEUT	Neutral
CAP	Capacitor		NC	Normally closed
CB	Circuit breaker		NO	Normally open
CEMF	Counter electromotive force		OHM	Ohmmeter
CKT	Circuit		OL	Overload relay
CONT	Control		PB	Pushbutton
CR	Control relay		PH	Phase
CRM	Control relay master		PLS	Plugging switch
CT	Current transformer		POS	Positive
D	Down		PRI	Primary switch
DB	Dynamic braking contactor or relay		PS	Pressure switch
DC	Direct current		R	Reverse
DIO	Diode		REC	Rectifier
DISC	Disconnect switch		RES	Resistor
DP	Double pole		RH	Rheostat
DPDT	Double pole, double throw		S	Switch
DPST	Double pole, single throw		SCR	Semiconductor-controlled rectifier
DS	Drum switch		SEC	Secondary
DT	Double throw		1PH	Single phase
EMF	Electromotive force		SOC	Socket
F	Forward		SOL	Solenoid
FLS	Flow switch		SP	Single pole
FREQ	Frequency		SPDT	Single pole, double throw
FS	Float switch		SPST	Single pole, single throw
FTS	Foot switch		SS	Selector switch
FU	Fuse		SSW	Safety switch
GEN	Generator		T	Transformer
GRD	Ground		TB	Terminal board
IC	Integrated circuit		3PH	Three phase
INTLK	Interlock		TD	Time delay
IOL	Instantaneous overload		THS	Thermostat switch
JB	Junction box		TR	Time delay relay
LS	Limit switch		U	Up
LT	Lamp		UCL	Unclamp
M	Motor starter		UV	Under voltage

1. What is the function of the line diagram?
2. How are wires that are electrically connected illustrated on the line diagram?
3. Where are the overload contacts drawn in a line diagram?
4. When a float switch is used to maintain a predetermined level, are normally open (NO) or normally closed (NC) contacts used?
5. How is a solenoid illustrated in the line diagram?
6. How is a contactor illustrated in the line diagram?
7. What are auxiliary contacts and how are they illustrated in the line diagram?
8. When using auxiliary contacts to maintain an electrical holding circuit, are normally closed (NC) or normally open (NO) contacts used?
9. What is the difference between a contactor and a magnetic motor starter?
10. In a line diagram that uses auxiliary contacts to form a holding circuit around the start pushbutton, how can the power be removed from the magnetic motor starter coil after the start button is pressed?

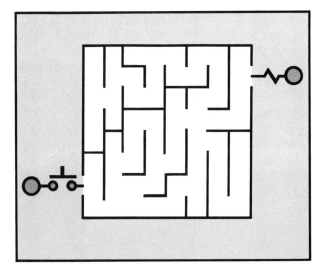

3 Introduction to Logic as Applied to Line Diagrams

All electrical circuits can be grouped into three basic sections: signal section, decision section, and action section. The signal section starts or stops the flow of current through the control device. The decision section of the circuit determines the type and sequence of work to occur. The action section may be direct or indirect. For example, a direct action section will turn on a motor, whereas an indirect section will activate a starter to turn on a motor.

Within each electrical circuit section certain logic functions exist. The six most common logic functions are AND, OR, NOT, MEMORY, NOR, and NAND. Combinations of these circuits can be used to develop a complex automated decision making circuit.

BASIC RULES OF LINE DIAGRAMS

When reading and interpreting line diagrams, it is important for the electrician to master the basic rules of line diagrams. Although there may be many ways to connect a circuit, the electrical industry has established a universal set of symbols and rules on how line diagrams (circuits) should be laid out. In Unit 2 you were introduced to accepted electrical symbols; in this unit you will learn how to use these symbols in line diagrams.

It is important for electricians to master the art of reading, interpreting, and applying the basic rules of line diagrams. By applying these standards, the electrician will have established a working practice common to all electricians.

One Load per Line

Figure 3-1 illustrates the proper way to connect a pilot light into a circuit with a single pole switch. In this circuit the power lines (heavier lines) are drawn vertically on the left and right sides of the drawing. The lines are marked line one (L1) and line two (L2). The space between L1 and L2 represents the voltage of the control circuit. When the switch (S1) is closed, this voltage will appear across the pilot light (PL1). Since the voltage between L1 and L2 is the proper voltage for the pilot light, current will flow through the switch and pilot light. The pilot lamp will then glow.

It is important to note that no more than one load should be placed in any one circuit line between L1 and L2. Figure 3-2 illustrates two loads (a pilot light and a solenoid) connected improperly in one line. The two loads in this circuit are connected in series. When the switch (S1) is closed, the voltage between L1 and L2 must divide across both loads, the result being that neither device will receive the entire 120 volts necessary for proper operation.

When more than one load must be connnected in the line diagram, the loads must be connected in parallel. Figure 3-3 illustrates the proper way of connecting two loads—a pilot light and sole-

noid—into a circuit. In this circuit there is only one load for each line between L1 and L2, even though there are two loads in the circuit. The voltage from L1 and L2 will appear across each load for proper operating of the pilot light and solenoid. This circuit then has two lines, one for the pilot light and one for the solenoid.

Loads Are Connected to L2

Loads in General. In the line diagram all loads have one side connected to L2. The load is the electrical device in the line diagram that uses

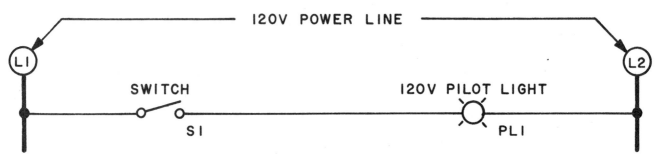

Figure 3-1 Line diagram of a single-pole switch controlling a pilot light.

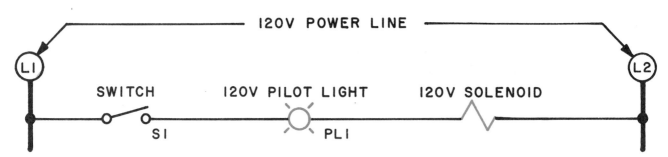

Figure 3-2 Line diagram of two loads (a pilot light and solenoid) improperly connected in one line of a line diagram.

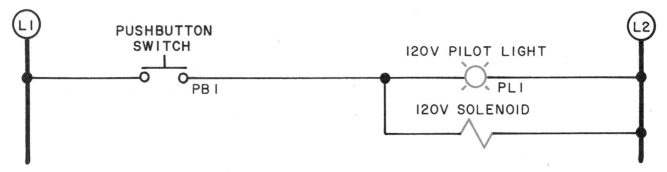

Figure 3-3 Line diagram of two loads (a pilot light and solenoid) properly connected in *two* separate lines of a line diagram.

the electrical power from L1 to L2. Control relays, solenoids, and pilot lights are examples of loads that are connected directly or indirectly to L2. Figure 3-4 illustrates the proper connection for each of these loads.

Magnetic Motor Starters. Motor starting coils are connected to L2 indirectly through the normally closed overload contacts. Figure 3-5 illustrates the proper way to connect a magnetic motor starter and its overloads to L2. The overload contact is normally closed and will open only if an overload condition exists in the motor. The number of normally closed overload contacts between

the starting coil and L2 will depend on the type of starter and power that is used in the circuit. There will be anywhere from one to three normally closed contacts shown between the starter and L2 for all line diagrams. To avoid confusion it is common practice to draw one set of normally closed contacts (see fourth line, Figure 3-5) and mark these contacts "All OL'S." An overload marked this way indicates that the circuit will be correct for any motor or starter used. The electrician should then know to connect all the normally closed overload contacts that the starter is designed for in series.

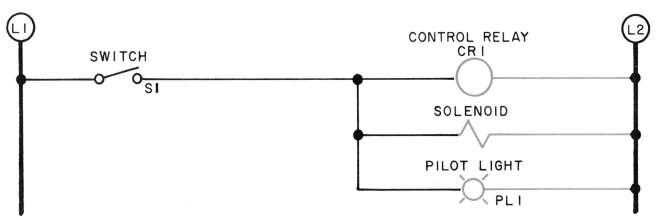

Figure 3-4 Line diagram of loads (control relay, solenoid and pilot light) connected directly to L2.

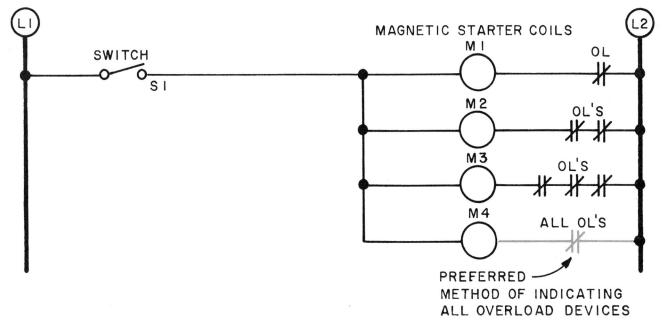

Figure 3-5 Line diagram of a magnetic motor starter and its overloads properly connected to L2. Note

on line four the *preferred* method of indicating overloads in a line diagram.

Control Devices Are Connected between L1 and the Operating Coil

Operating coils of contactors and starters are activated by control devices such as pushbuttons, limit switches, and pressure switches. Examples of each of these control devices are illustrated in Figure 3-6. Each will include at least one control device. If no control device were included in a line, then the operating coil would be on all the time. A circuit may contain as many control devices as is required to make the operating coil function as specified. These control devices may be connected in series or parallel with each other when controlling an operating coil.

Figure 3-7 illustrates how two control devices (a flow switch and a temperature switch) are connected in a series to control a coil in a magnetic motor starter. Both the flow switch and the temperature switch must close to allow current to pass from L1 through the control device, the mag-

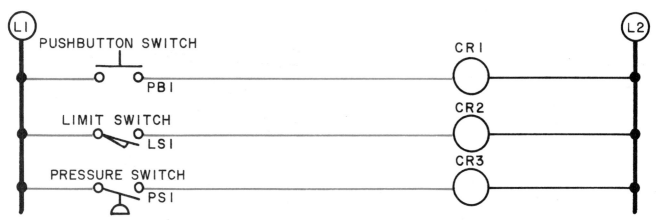

Figure 3-6 Line diagram of control devices properly connected between L1 and the load.

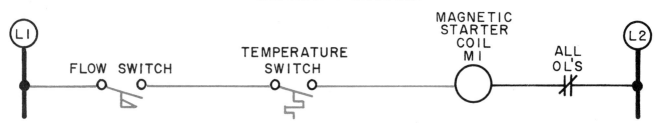

Figure 3-7 Line diagram of two control devices (flow switch and temperature switch) wired in series to control the operation of a coil in a magnetic motor starter.

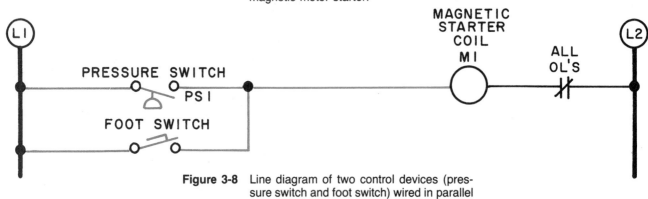

Figure 3-8 Line diagram of two control devices (pressure switch and foot switch) wired in parallel so that each may control the operation of a coil in a magnetic motor starter.

netic coil and the overloads on to L2. Figure 3-8 illustrates how two control devices (a pressure switch and a foot switch) are connected in parallel to control a coil in a magnetic motor starter. Either the pressure switch or the foot switch can be closed to allow current to pass from L1 through the control device, the magnetic coil and the overloads on to L2. Regardless of how the control devices are arranged in a circuit, they must be connected between L1 and the operating coil. The contacts of the control device may be either normally open or normally closed. The contacts used and the way the control devices are connected into a circuit (series or parallel) will determine the function of the circuit.

Each Line Is Numbered

Each line in a line diagram should be marked (line one, two, three, etc.), starting with the top line and reading down. A line can be defined as a complete path from L1 to L2 that contains a load. Figure 3-9 illustrates the marking of each line in a line diagram with three separate lines. Line one connects a push button to the solenoid to complete the path from L1 to L2. Line two connects a pressure switch to the solenoid to complete the path from L1 to L2. Since either the push button or the pressure switch will complete the path from L1 to L2, they are marked as two separate lines even though they control the same load. Line three

illustrates a foot switch and a temperature switch to complete the path through to the pilot light. Since it takes both the foot switch and the temperature switch to complete the path to the pilot light, they both appear in the same line—line three.

Numbering each line is important for understanding the function of a circuit. Apply this rule to all line diagrams. As circuits become more complex and additional lines are added, the importance of this numbering system will become clear. For this reason, we will number all line diagrams, no matter how simple they may seem.

LINE DIAGRAMS: SIGNALS, DECISIONS AND ACTION

In the opening paragraphs of this section we have set down some accepted rules which should help you simplify the reading and designing of control circuits in line diagram form. It should be apparent that the importance of these rules lies in the fact that all electricians can follow them to interpret consistently how a device or series of devices is controlled.

The entire concept of control is to accomplish specific work in a predetermined manner. In other words, we want the circuit to respond as we have designed it, without any changes or surprises. To accomplish this consistency, the electrician must

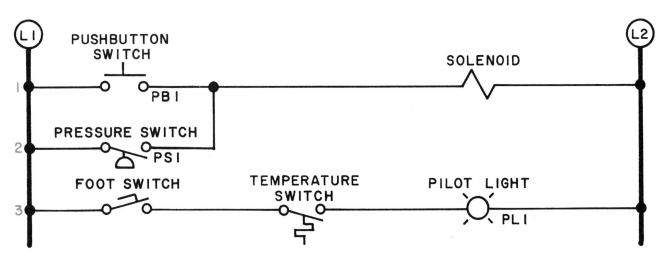

Figure 3-9 Numbering each line in a line diagram makes it easier to understand more complex diagrams. Starting with the top line and reading down, each line can be defined as a complete path from L1 to L2 that contains a load.

realize that all control circuits are composed of three basic sections as illustrated in Figure 3-10: the signal(s), the decision, and the action. Each part—the signal(s), the decision and the action—will be treated in depth in this unit. Complete understanding of these concepts will pave the way for understanding any of the existing industrial control circuits as well as those which are being created by industry as it becomes more mechanized and automated.

Signals: Manual, Mechanical and Automatic

A signal starts or stops the flow of current by closing or opening the control device's contacts. If the contacts are closed, current is allowed to flow through the control device. If the contacts are opened, current is not allowed to flow through the control device. In the circuits presented so far, the following devices have served as the signal portion of the control circuit: pushbuttons, limit switches, flow switches, foot switches, temperature switches and pressure switches.

All signals are dependent upon some condition that must take place. This condition can be manual, mechanical or automatic. A manual condition is any input into the circuit by a person. Foot switches and pushbuttons are examples of control devices that would respond to a manual input. A mechanical condition is any input into the circuit by some mechanically moving part. A limit switch is a good example of a control device that would respond to a mechanical input. When a moving object, such as a box, hits a limit switch, the limit switch usually has a lever, roller, ball or plunger type actuator to cause a set of contacts to open or close without further assistance. An automatic condition is one which will respond to changes in a system. Flow switches, temperature switches and pressure switches are good examples of automatic control. When a change in the flow of a liquid is created, a change in temperature is sensed, or pressure varies, these devices can automatically open and close sets of contacts. Remember—the signal accomplishes no work by itself; it merely starts or stops the flow of current in that part of the circuit.

Decisions

The decision part of the circuit determines what work is to be done and in what order the work is to occur. The decision part of the circuit is the part that adds, subtracts, sorts out, selects, and redirects the signals from the control devices to the load. In order for the decision part of the circuit to perform a definite sequence, it must perform in a logical manner. The way the control devices are connected into the circuit gives the circuit some "logic" function. Six of these "logic" functions that you will learn in this unit are AND, OR, NOT, MEMORY, NOR, and NAND logic. The decision part of the circuit accepts informational inputs (signals), makes logical decisions based on the way the control devices are connected into the circuit, and provides the output signal that controls the load, such as an operating coil.

Action

Once a signal is generated and the decision has been made within a circuit, some type of action should result. In most cases it is the operating coil in the circuit which is responsible for initiating this type of action. This action is direct when devices such as motors, lights and heating elements are turned on as a direct result of the signal and the decision. This action is indirect when the coils in solenoids, magnetic starters and relays

Figure 3-10 All control circuits are composed of 3 basic sections: the signals, the decision and the action.

are energized. The action is indirect because the coil energized by the signal and the decision may in turn energize a magnetic motor starter which actually starts the motor. Regardless of how this action takes place, the load causes some action (direct or indirect) in the circuit and for this reason is called the action part of the circuit.

LOGIC FUNCTIONS

In order to fully understand an industrial electrical circuit, the electrician must understand the "logic" of the circuit. Control devices such as pushbuttons, limit switches and pressure switches have no intelligence of their own. The electrician must connect these control devices into a circuit so that the circuit can be caused to function in a predetermined manner. By connecting control devices to perform logical functions, the electrician has given the circuit intelligence. All control circuits are merely basic logic functions or combinations of logic functions. The logic functions that you will learn in this unit are common to all areas of industry. This includes electricity, electronics, hydraulics, pneumatics, math and many of the routine activities you perform every day.

AND Function

The first logic function you will learn is called the AND function. A simple example of AND logic takes place whenever you start an automobile that has an automatic transmission. The ignition switch must be turned to the start position *and* the transmission selector must be in the park position before the starter is energized. Before the action (load on) in this automobile circuit can take place, the control signals (manual), must be performed in a logical manner (decision).

An example of AND logic used in industry is illustrated in Figure 3-11. In this circuit two push buttons connected in series are controlling a solenoid. Before the solenoid is energized, both push button one *and* two must be pressed. The logic function that makes up the decision part of this circuit is AND logic. There is a reason for using logic in a circuit, just as there is a reason for using it in the automobile circuit. The reason illustrated in Figure 3-11 could be to build in safety for the operator of this circuit. If the solenoid were operating a punch press or shear, we could space the pushbuttons far enough apart so that the operator would have to use both hands to make the machine operate, thus making sure they were not near the machine when it was activated. AND logic can be simplified by saying the load will be on only if all the control signal contacts are closed. As with any logic function, the signals may be manually, mechanically, or automatically controlled. Any control device such as limit switches, pressure switches, etc., with normally open contacts can be used in developing AND logic. The normally open contacts of each control device must be connected in series for AND logic.

OR Function

The second logic function you will learn is called the OR function. A simple example of OR logic is in a home that has two doorbells (pushbuttons) controlling one bell in the house. The bell (load) may be energized by pressing (signal on) either the front *or* the back pushbutton (control device). Here, as in the automobile circuit, the control devices are connected to respond in a logical manner. The logic of these two circuits is not the same, yet each has a specific purpose.

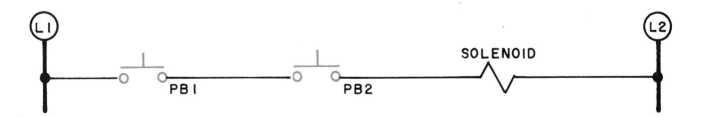

Figure 3-11 Line diagram of two control devices (pushbuttons) connected in series for AND logic. Both PB1 *and* PB2 must be activated for the coil to be energized.

An example of OR logic used in industry is illustrated in Figure 3-12. In this circuit a pushbutton and a temperature switch are connected in parallel. The load is a heating element that is controlled by either of two control devices. The logic of this circuit is OR logic because the push button *or* the temperature switch will energize the load. In this circuit the temperature switch is an example of an automatic control device that will automatically turn the heating element on and off to maintain the temperature setting for which the temperature switch is set. The manually controlled pushbutton switch could be used to test or turn on the heating element when the temperature switch contacts are open.

OR logic can be defined by saying the load will be on if any one of the control signal's contacts is closed. The control devices are all connected in parallel, in this case using their normally open contacts for OR logic. It is important at this point that you do not start associating all series circuits with AND logic and all parallel circuits with OR logic. Series and parallel refer to the physical relationship of each control device to other control devices or components in the circuit. This series and parallel relationship is only part of what determines the logic function of any circuit. Normally open and normally closed contacts are not always perfect examples of pure logic and we must consider their other possibilities as we develop a total logic system.

AND Logic Combined with OR Logic

The decision part of any circuit may contain one or more logic functions. An example of a circuit that contains both logic functions learned so far is illustrated in Figure 3-13. In this circuit both pressure *and* flow must be present in addition to the pushbutton *or* the foot switch being engaged to energize the load. This provides the circuit with the advantage of both AND logic and OR logic. The machine is protected because both pressure and flow must exist before it is started and there is a choice between using a pushbutton or a foot switch for final operation. The action taking place in this circuit is energizing a coil in a magnetic motor starter. The signal inputs for this circuit would have to be two automatic and at least one manual.

As you can see from the circuit of Figure 3-13, each control device will respond to its own input signal and has its own decision-making capability. When multiple control devices are used in combination with other control devices making their own decisions, a more complex decision can be made through the combination of all control devices used in the circuit. All industrial control circuits consist of control devices capable of making decisions in accordance with the input signals received. It is important for the electrician to recognize simple logic functions and combinations of logic functions in order to sort out and understand how the circuits are applied to industry.

NOT Function

The logic functions you have learned so far were for control devices that had normally open contacts. This means that when there was no input signal to the control device the contacts were open in the circuit. If normally closed contacts were used instead of normally open contacts, the logic function would then change. For example, if the normally open contacts on the pushbutton in Figure 3-14 were changed to normally closed contacts, the solenoid and pilot light would be energized without pressing the pushbutton. Pressing the pushbutton in this circuit would de-energize the loads. If the loads are to remain energized, then there must *not* be a signal. This logic function is called NOT logic. NOT logic means that there is an output if the control signal is off. NOT logic can be simplified by saying the output will remain only if the control signal contacts remain closed.

An example of NOT logic is the courtesy light in a refrigerator. The light is on if the control signal is off. The control signal would be the door of the refrigerator. Any time it is open (signal off), the load (courtesy light) is on. Remember: the condition that controls the signal can be anything that is actuating the control device, whether it is manual, mechanical or automatic. With the refrigerator door the condition was mechanical.

MEMORY Function

Many of today's industrial circuits require their control circuits not only to make logic decision, such as AND, OR, NOT, but also to be capable of storing, memorizing or retaining the signal inputs to keep the load energized even after the signals are removed. A switch in your home that controls a light from only one location is an example of memory circuit. When it is ON, it remains ON until it is turned OFF, and remains OFF until it is turned ON. It performs a MEMORY function because the output corresponds to the last input information until new input information is received to change it. In the case of the home light switch, the memory circuit was accomplished by a switch that would mechanically stay in one position or another. In industrial control circuits, it is more common to find pushbutton switches with return spring contacts than those which mechanically stay held in one position. To give circuits

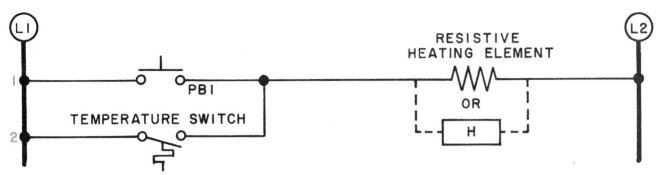

Figure 3-12 Line of two control devices (pushbutton and temperature switch) connected in parallel for OR logic. Either PB1 or the temperature switch may be used to activate the heating element.

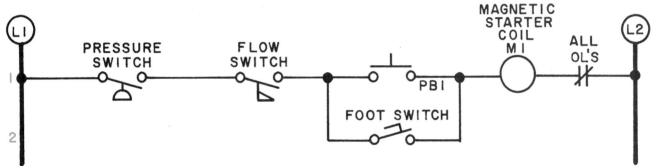

Figure 3-13 Line diagram of combination circuit showing AND logic with OR logic. Both pressure *and* flow in addition to the pushbutton *or* the foot switch will energize the coil in the magnetic motor starter.

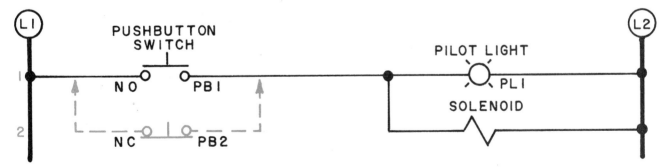

Figure 3-14 Line diagram of how a circuit may be modified to produce NOT logic. When the NC (normally closed) contacts are substituted for the NO (normally open) contacts, the pilot light and solenoid would be energized without pressing the pushbutton. If the loads are to remain energized, then there must *not* be a signal. This logic function is called NOT logic.

with pushbutton switches MEMORY, an additional set of contacts must be added to this circuit, as in Figure 3-15. Note that this MEMORY circuit is merely the auxiliary holding contacts we discussed in Unit 2. Once coil M of the magnetic motor starter is energized, coil M will cause coil contacts M1 to close and remain closed (MEMORY) until the coil is de-energized. To de-energize the coil and make the circuit more useful, NOT logic can be added to MEMORY logic, as in Figure 3-16, to provide a useful and common START/STOP control circuit. When STOP button PB1 is activated, current to the coil stops and contacts M1 open, returning the circuit to its original condition.

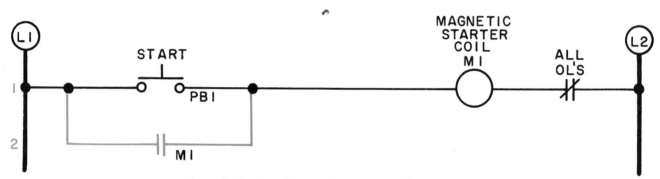

Figure 3-15 By adding auxiliary contacts M1 to a circuit, MEMORY can be added to the circuit of pushbutton PB1 when it is released.

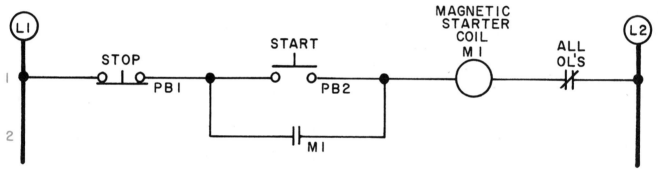

Figure 3-16 By adding the NOT logic of stop button PB1 to the MEMORY logic of contacts M1, a useful and common start/stop control circuit is created.

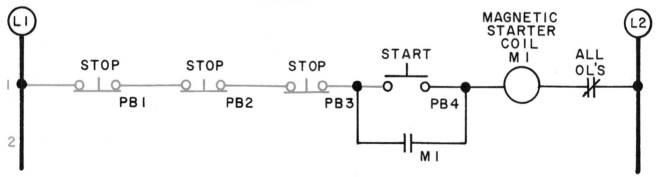

Figure 3-17 Line diagram of a circuit designed to provide NOR logic. Pressing any stop pushbutton (PB1, PB2, or PB3) in this circuit will de-energize the load (coil M). In other words, each not *or* the other not (signal off) will de-energize the load.

NOR Function

The fourth logic element to be discussed in this unit is the NOR functions. NOR logic is an extension of NOT logic in that two or more normally closed contacts in series are used to control a load. Figure 3-17 is a good example of NOR logic. In figure 3-17 we have provided additional safety for the operator of this circuit by adding several emergency stop buttons (NOT logic) to the control circuit. Pressing any emergency stop button in this circuit will de-energize the load (coil M). In other words each NOT *or* the other NOT (NOR) (SIGNAL OFF) will de-energize the load. With assembly lines so common in industry, the need for stopping all machines that are interconnected can be of prime importance. By incorporating NOR logic, each machine may be controlled by one operator, but any operator or supervisor will have the capability of turning off all the machines on the assembly line to protect individual operators or the entire system.

It should be pointed out that this addition of emergency stops to existing equipment is commonplace in industry and the ability to do this is expected from the electrician. With the knowledge of NOR logic, the electrician can readily add additional stops by merely wiring them in series to perform their necessary function.

NAND Function

The last logic element to be discussed in this unit is the NAND function. NAND logic is also an extension of NOT logic, except that two or more normally closed contacts are connected in parallel to control a load. A good example of a NAND circuit is the courtesy light in your automobile. In an automobile the light is on if the control signal is off. This circuit is different from a refrigerator door in that an automobile may have two or more doors which must not be open for the load to be off. In other words, each NOT *and* every NOT (SIGNAL OFF) will de-energize the load. Figure 3-18 is an example of NAND logic applied to an industrial circuit. With this circuit we wish to fill two interconnecting tanks with a liquid. When pushbutton PB3 is depressed, Coil M will be energized and auxiliary contacts M1 will close and remain closed (MEMORY) until both tanks are filled. The circuit will cause both tanks to fill to a predetermined level because the float switch in tank 1 *and* tank 2 must NOT open until both tanks are full. In other words each NOT *and* every NOT must be open (signal off) to stop the filling process based on the input of the float switches. NOR logic is also present in this circuit because the emergency stops (NOT) at tank 1 *or* tank 2 may stop the process if an operator at either of the tanks see a problem.

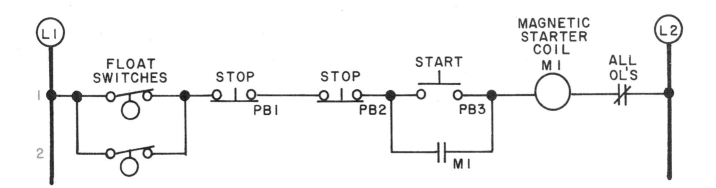

Figure 3-18 Line diagram of a circuit designed to provide NAND logic through the use of float switches (FL1 and FL2) and NOR logic through the use of emergency stops (PB1 and PB2). MEMORY is provided in the circuit through the use of auxiliary contacts M1. The NAND logic in this circuit causes both tanks to fill to a predetermined level because the float switch in Tank 1 (NOT) *and* Tank 2 (NOT) must be open (signal off) to stop the filling process.

COMMON CONTROL CIRCUITS

In the remainder of this unit we are going to describe several control circuits commonly in use for commercial and industrial electrical circuits. Each circuit will be described in depth regarding the entire circuit operation. In each case the circuit will be described in accordance with the accepted rules of line diagrams. To reinforce your knowledge of basic logic functions, know how these basic logic functions and combination of logic functions are used to accomplish the overall intelligence or function of the circuit. Although an electrician needs knowledge of each logic element, he or she must also understand quickly what the entire circuit is designed to do in order to be able to begin wiring or troubleshooting the circuit.

Two or More Start-Stop Stations Used to Control a Magnetic Starter

Often it is necessary to start and stop a load from more than one location. The circuits of Figure 3-19 allow us to start or stop the magnetic starter from two locations. If additional stops were to be connected into this circuit, they would be connected in series (NOR logic) with existing stops. If additional starts were to be connected into this circuit, they would be connected in parallel (OR

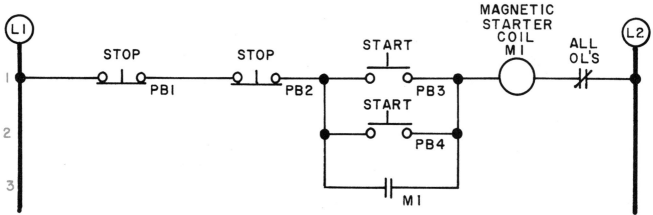

Figure 3-19 Line diagram of a circuit designed for two or more start/stop stations used to control a magnetic starter. (For detailed operation of this circuit, refer to text.)

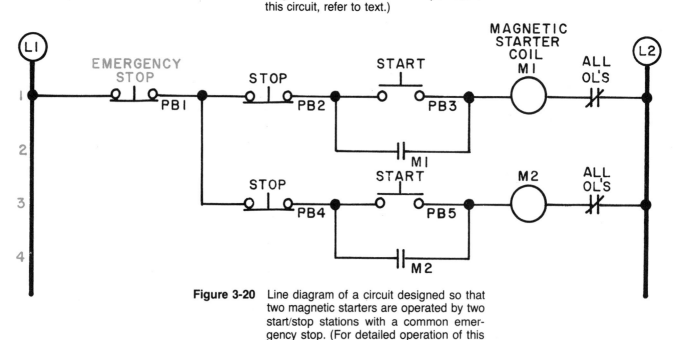

Figure 3-20 Line diagram of a circuit designed so that two magnetic starters are operated by two start/stop stations with a common emergency stop. (For detailed operation of this circuit, refer to text.)

logic) with the existing starts. The circuit of Figure 3-19 would be read as follows: Pushing any one of the start buttons (PB3, or PB4) will cause coil M to energize, causing auxiliary contacts M1 to close (MEMORY) and remain closed until coil M is de-energized. Coil M can be de-energized by depressing stop buttons PB1 or PB2, by an overload which would activate the OL's, or by a loss of voltage to the circuit. In the case of an overload, the overload would have to be removed and the circuit overload devices reset before the circuit would be returned to normal starting condition.

Two Magnetic Starters Operated by Two Start-Stop Stations with Common Emergency Stop

In almost all electrical systems, several devices can be found running off a common supply voltage. Magnetic motor starters are no exception to this practice. Figure 3-20 illustrates how two start-stop stations are used to control two separate magnetic motor starter coils with a common emergency stop protecting the entire system. The circuit of Figure 3-20 would be read as follows: Pushing the start button PB3 will cause coil M1 to energize and "seal in" auxiliary contacts M1. Pushing the start button PB5 will cause coil M2 to energize and "seal in" auxiliary contacts M2. Once the entire circuit is operational, emergency stop button PB1 can shut down the entire circuit or the individual stop buttons PB2 or PB4 can stop or de-energize the coils in their respective circuits. Note also that each individual circuit is

overload protected and will not affect the other when one magnetic motor starter experiences a problem.

One Start Stop Station Used to Control Two or More Magnetic Starters

Steel mills, paper mills, bottling plants and canning plants are typical examples of industries which require simultaneous operation of two or more motors. In each of these industries, one will find products or materials spread out over great lengths which must be started together so as not to separate the products or stretch the materials. Figure 3-21 illustrates a circuit where two motors can be started almost simultaneously from one location. The circuit of Figures 3-21 would be read as follows: Pushing the start button energizes coil M1 and "seals in" *both* sets of auxiliary contacts M1. (*Note:* It is perfectly acceptable to have more than one set of auxiliary contacts controlled by one coil). When *both* sets of contacts close, the first set of M1 contacts (line 2) provide MEMORY for start pushbutton PB2 and complete the circuit to energize coil M1. The second set of M1 contacts (line 3) complete the circuit to coil M2, energizing coil M2. Since both coils energize almost simultaneously, the motors associated with these magnetic motor starters also start almost simultaneously. Pushing the stop button will break the circuit (line 1), de-energizing coil M1. When coil M1 drops out, both sets of M auxiliary contacts will be activated, M1 will open and coil M2 will de-energize. Since both coils de-energize almost

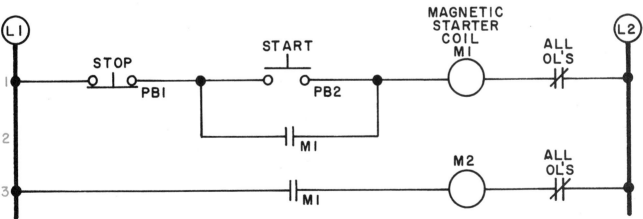

Figure 3-21 Line diagram of a circuit designed so that one start/stop station is used to control two or more magnetic starters. (For detailed operation of this circuit, refer to text.)

simultaneously, the motors associated with these magnetic motor starters also stop almost simultaneously. Note that an overload in magnetic motor starter M2 will affect only the operation of coil M2. If, however, an overload exists in motor starter M1, the entire circuit will be shut down. The entire circuit will stop because de-energizing coil M1 also affects both sets of auxiliary contacts M1. This type of protection might be used where a machine such as an industrial drill would be damaged if the cooling líquid pump shut off while the drill was still operating. If the drill continued without cooling liquid, it would heat up and ruin the drill bit.

Pressure Switch with Pilot Light to Indicate When a Device Is Activated

Indicator lights are manufactured in a variety of colors, shapes, and sizes to meet the needs of a heavily automated industry. In future units, we will look at the various types of indicating devices. For now it is important only to know that the illumination of these indicators serves notice to an operator that any one of a sequence of events may be taking place. Figure 3-22 and Figure 3-23 illustrate circuits in which a pilot light is used to indicate when a device is activated. The circuit of Figure 3-22 would be read as follows: When toggle switch S1 is closed, pressure switch S2 has automatic control over the circuit. When the pressure to switch S2 drops, the pressure switch S2 will close and activate coil M controlling the magnetic starter of the compressor motor, starting the compressor. At the same time, contacts M1 will close and the indicator light PL1 will turn on. The compressor will continue to run and the pilot light will stay on as long as the motor runs. When pressure builds sufficiently to open pressure switch S2, coil M will de-energize and the magnetic motor starter will drop out, stopping the compressor motor. In addition, the pilot light will go out because contact M1 controlled by coil M will open. In other words, the pilot light will be on *only* when the compressor motor is running. This type of cir-

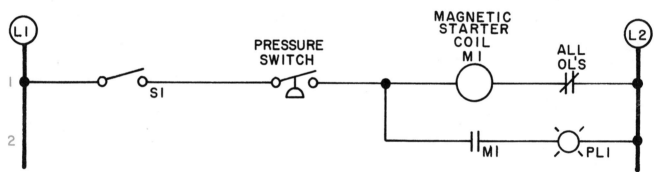

Figure 3-22 Line diagram of a circuit designed with a pressure switch and pilot light to indicate when a device is activated. (For detailed operation of this circuit, refer to text.)

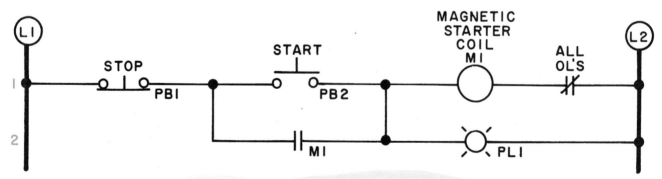

Figure 3-23 Line diagram of a circuit designed with a start/stop station and pilot light to indicate when a device is activated. (For detailed operation of this circuit, refer to text.)

cuit might be used in a garage to let the owner know when the air compressor is on or off.

Figure 3-23 would be read as follows: Pressing start button PB2 energizes coil M, causing auxiliary contacts M1 to close. Closing contacts M1 not only provides MEMORY for start button PB2, but also maintains an electrical path for the pilot light. In this circuit, as long as coil M is energized, the pilot light will stay on. Pressing stop button PB1 de-energizes coil M, opening contacts M1 and turns off the pilot light. An overload in this circuit would also de-energize coil M, opening contacts M1 and turning off the pilot light. A circuit like this can be used as a positive indicator that some process is taking place, although it may be in a remote place such as in a pump well or in another building.

Start-Stop Station with Pilot Light to Indicate a Device Is Not Running

Indicator lights can be used to show when an operation is stopped as well as when it is started. To use an indicator light to show that an operation has stopped, NOT logic must be introduced into the circuit. You will recall that NOT logic can be established by placing one set of NC (normally closed) contacts in series with a device. Since auxiliary contacts may be normally open or normally closed, we need use only normally closed auxiliary contacts to create NOT logic. Figure 3-24 illustrates a circuit using NOT logic to indicate when a device is NOT operating. Figure 3-24 would be read as follows: Pressing start button PB2 ener-

gizes coil M, causing *both* sets of auxiliary contacts M1 to energize. Normally open contacts M1 (line 2) close, providing MEMORY for PB 2, and normally closed contacts M1 (line 3) open, disconnecting pilot light PL1 from the line voltage, causing the light to go out. Pressing stop button PB1 de-energizes coil M, causing *both* sets of contacts to return to their normal positions. Normally open contacts M1 (line 2) return to their normally open position and normally closed contacts M1 (line 3) return to their normally closed position, causing the pilot light to be reconnected to the line voltage and causing it to go on. In other words, the pilot light will be *on* only when the coil to the magnetic motor starter is *off*. Although we used a pilot light in this circuit, a bell or siren could be substituted for the pilot light to serve as a warning device. A circuit like this would be used to monitor critical operating procedures such as a cooling pump for a nuclear reactor. When the cooling pump stopped, the pilot light, bell or siren would immediately call attention to the fact that the process had been stopped.

Pushbuttons Arranged for a Sequence Control of Devices

When working with machines like conveyor systems, it is often necessary to have one conveyor system feeding boxes or other types of materials onto another conveyor system. Since material could pile up on the second conveyor from the first conveyor if the second conveyor stopped, some type of circuit is needed to prevent the pile-up on

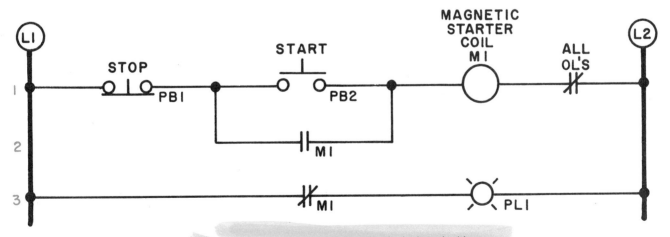

Figure 3-24 Line diagram of a circuit designed with a start/stop station and pilot to indicate when a device is not activated. (For detailed operation of this circuit, refer to text.)

the second conveyor. Figure 3-25 illustrates a circuit which will not let the first conveyor operate unless the second conveyor has started and is running. The circuit of Figure 3-25 would read as follows: Pressing start button PB2 will energize coil M1 and cause auxiliary contacts M1 to close. With auxiliary contacts M1 closed, PB2 has MEMORY and has provided an electrical path to allow coil M2 to be energized when start button PB4 is pressed. With start button PB4 pressed, coil M2 energizes and closes contacts M2, providing MEMORY for start button PB4 so that both conveyors are now running. Note, however, from the circuit that conveyor one (coil M2) cannot be started unless conveyor two (coil M1) is energized. Note also that if an overload occurs in the circuit with coil M1 or if the stop button PB1 is pressed, both conveyors will shut down. Note also that if conveyor one (coil M2) experiences an overload, only conveyor one will shut down. Since conveyor one is feeding conveyor two, a problem in conveyor one will not have any serious consequences with conveyor two.

Jogging with a Selector Switch

The last circuit that we will describe in this unit involves two new concepts: namely, jogging and the use of simple selector switches. Jogging is a term which describes the frequent starting and stopping of a motor for short periods of time. Jog-

ging is used to position materials by moving the materials small distances each time the motor starts. In Figure 3-26 we will use a selector switch to provide a jog/run circuit common to industry. The selector switch (two-position switch) will be used to manually open or close a portion of the electrical circuit. It is often called a selector switch because it will allow us to choose which way we want the circuit to operate. In this case, the selector switch will determine if the circuit is a jog circuit or run circuit, hence the name jog/run circuit. The circuit of Figure 3-26 would read as follows: With selector switch S1 in the open position (jog), pressing start button PB2 energizes coil M, causing the magnetic motor starter to operate. Releasing start button PB2 de-energizes coil M, causing the magnetic motor starter to stop. With selector Switch S1 in the closed position (run), pressing start button PB2 energizes coil M, closing auxiliary contacts M1 (MEMORY) so that the magnetic starter operates and continues to operate until stop button PB1 is pressed.

When stop button PB1 is pressed, coil M de-energizes and all circuit components return to their original position. In this circuit, as in others discussed, the overloads may also open the circuit and must be reset after the overload is removed in order to return the circuit to normal operation. A circuit of this type may be found where an operator may run a machine continuously for produc-

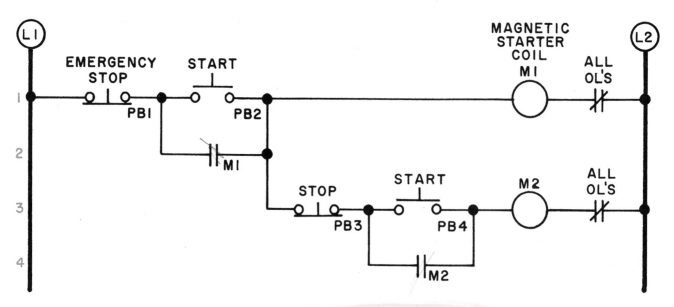

Figure 3-25 Line diagram of a circuit designed with pushbuttons arranged for a sequence control of devices. (For detailed operation of this circuit, refer to text).

tion, but may stop it at any time for small adjustments or repositioning.

Although many of the sample circuits shown in this introductory unit illustrated many applications for pushbuttons, there is no reason why pressure, temperature, flow, foot and limit switches could not be substituted to meet special needs. In future units, we will inject these and many more devices to expand your knowledge of more complex and automated systems.

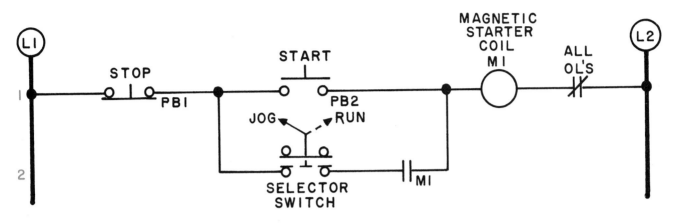

Figure 3-26 Line diagram of a circuit designed for jogging with a selector switch. (For detailed operation of this circuit, refer to text).

QUESTIONS ON UNIT 3

1. How many electrical loads can be placed in the control circuit of any one line between L1 and L2? *1*
2. If more than one load must be connected in the line diagram, how are the loads connected? *Parallel*
3. All loads are connected directly or indirectly to which power line? *both; L2*
4. Where are control devices connected in a line diagram? *L1*
5. How is each line in a line diagram marked to distinguish that line from all other lines? *Numbers*
6. What are the three basic sections that all control circuits are composed of? *Sig, dec, action*
7. What is the signal(s) part of the control circuit? *switch*
8. What is the decision part of the control circuit? *Logic*
9. What is the action part of the control circuit? *Load*
10. What is meant by AND logic as applied to control circuits? *series + normally open.*
11. What is meant by OR logic as applied to control circuits? *parallel normally open.*
12. What is meant by NOT logic as applied to control circuits? *normally closed + don't have a signal*
13. What is meant by MEMORY logic as applied to control circuits? *contact in parallel with switch*
14. What is meant by NOR logic as applied to control circuits? *Nor closed*
15. What is meant by NAND logic as applied to control circuits? *Parallel closed series*
16. When additional "stops" are added to a control circuit, should they be connected in series or parallel? *Series*
17. When additional "starts" are added to a control circuit, should they be connected in series or parallel? *parallel*
18. What is meant by "jogging"?

Logic Symbols Cross-reference Chart

LOGIC ELEMENT	AND	OR	NOT	NAND	NOR
LOGIC ELEMENT FUNCTION	OUTPUT IF ALL CONTROL INPUT SIGNALS ARE ON	OUTPUT IF ANY ONE OF THE CONTROL INPUTS IS ON	OUTPUT IF SINGLE CONTROL INPUT SIGNAL IS OFF	OUTPUT IF ALL CONTROL INPUT SIGNALS ARE ON	OUTPUT IF ALL CONTROL INPUT SIGNALS ARE OFF
N.F.P.A. STANDARD			SUPPLY	SUPPLY	SUPPLY
BOOLEAN ALGEBRA SYMBOL	$(\ )\bullet(\ )$	$(\ )+(\ )$	$\overline{(\ )}$	$\overline{(\ )\bullet(\ )}$	$\overline{(\ )+(\ )}$
ARO PNEUMATIC LOGIC SYMBOL					
MIL-STD-806B AND ELECTRONIC LOGIC SYMBOL					
NEMA LOGIC SYMBOL					
ELECTRICAL RELAY LOGIC SYMBOL					
ELECTRICAL SWITCH LOGIC SYMBOL					
A.S.A.(J.I.C.) VALVING SYMBOL					
FLUIDIC DEVICE TURBULENCE AMPLIFIER					
N.F.P.A. STANDARD T3.7.68.2					

Source: The ARO Corporation

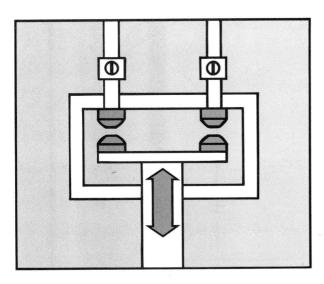

4 AC Manual Contactors and Motor Starters

AC manual contactors and manual motor starters have been modified and improved over the years. A manual contactor is a device that must be physically opened and closed to control any electrical circuit. Manual motor starters are contactors with overload protectors used only to control motors. Overload protection is used in a circuit to allow a motor to start but will open if the motor overloads during operation. Manual starters are available in a variety of sizes, poles, voltages, and phases. Most manual motor starters are mounted in an enclosure to provide protection from water, dust, and chemicals.

Manual Switching

When electric motors were first introduced in the late 1800's for industrial use, the starting and stopping of motors was done through the use of a simple knife switch similar to the one shown in Figure 4-1. This type of knife switch remained popular for quite some time but was eventually discontinued as a means of connecting and disconnecting line voltage directly to motor terminals for three basic reasons.

First, the open knife switch had exposed (live) parts which presented an extreme electrical hazard to the operator. In addition, any applications where dirt or moisture were present made this open concept too vulnerable to problems.

Second, the speed of opening and closing contacts was determined solely by the operator. If the operator did not open or close the switch quickly, considerable arcing and pitting of the contacts

soon lead to rapid wear and eventual replacement.

The third problem regarding knife switches was the material from which they were made. Most knife switches were made of soft copper, which, after repeated arcing, heat generation, and mechanical fatigue soon had to be replaced.

Mechanical Improvements

As industry demanded more electric motors at the turn of the century, certain improvements had to be made upon the knife switch to make it more acceptable as a controller.

First, the knife switch was enclosed in a steel case to protect the switch, and an insulated external handle was added to protect the operator (Figure 4-2).

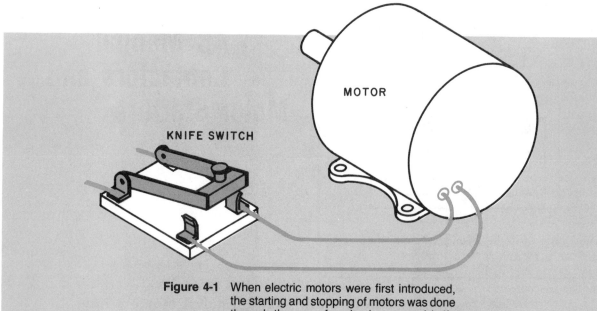

Figure 4-1 When electric motors were first introduced, the starting and stopping of motors was done through the use of a simple exposed knife switch.

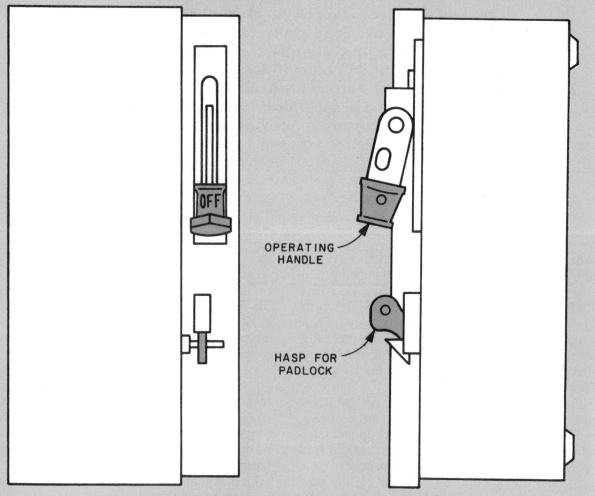

Figure 4-2 Enclosing the knife switch in a steel enclosure protected the switch, and the insulated handle protected the operator.

Second, an operating spring was attached to the handle of the enclosure to assure quick opening and closing of the knife blade (Figure 4-3). The switch handle was designed so that once the handle was moved a certain distance, the tension on the spring forced the contacts to open or close at the same continuous speed each time it was operated.

Even with these improvements, the knife switch had one serious flaw: the blade and jaw mechanism of the knife switch had a short mechanical life when the knife switch was used as a direct controller. Because of this persistent problem, the knife blade mechanism was discontinued as a means of direct control for motors; however, it is still maintained for use as a disconnect. The disconnect is a device used only periodically to remove electrical circuits from their source of supply. Since the disconnect is used less frequently in this situation, the mechanical life of the blade mechanism is not of major concern.

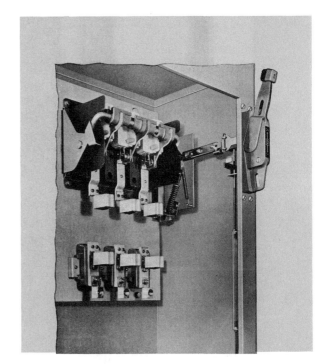

Figure 4-3 An operating spring is attached to the handle of a disconnect enclosure to assure quick opening and closing of the knife blade mechanism each time it is operated. (Square D Company)

MANUAL CONTACTORS

Double-Break Contacts. Another major reason for the discontinued use of the knife blade as a direct controller was the development of double-break butt contacts similar to those shown in Figure 4-4.

With double-break contacts, a device could be designed which would have a higher contact rating (current rating) in a smaller space than devices designed with single-break contacts. One such device was the manual contactor. Figure 4-5 illustrates how the double-break contacts operate in a contactor. When a set of NO (normally open)

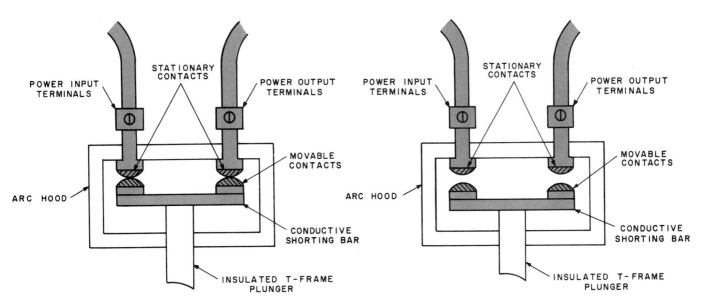

Figure 4-4 Double-break contacts allow for a higher contact rating (current rating) in a smaller space than those designed with single-break contacts.

Figure 4-5 When a set of NO double-break contacts are energized, the movable contacts are forced against the two stationary contacts to complete the electrical circuit.

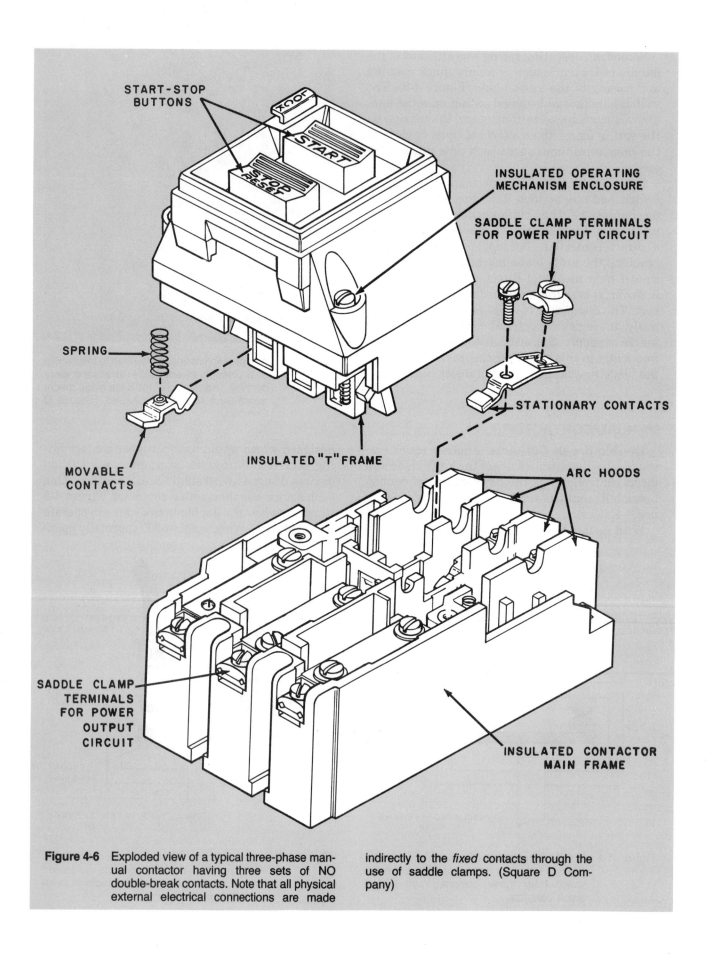

Figure 4-6 Exploded view of a typical three-phase manual contactor having three sets of NO double-break contacts. Note that all physical external electrical connections are made indirectly to the *fixed* contacts through the use of saddle clamps. (Square D Company)

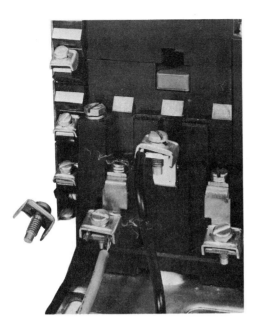

Figure 4-7 Saddle clamps provide a convenient means for attaching circuit wires to the input and output terminal of a contactor. (Cutler Hammer)

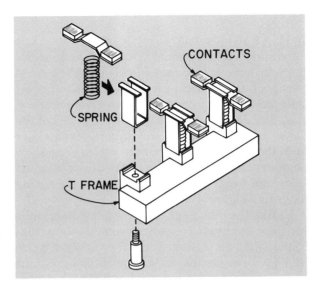

Figure 4-8 One advantage of silver alloy contacts is the fact that the oxide (rust) which forms on these materials is an excellent conductor of electricity. Even when the contacts appear dull or tarnished, they are still capable of operating normally.

double-break contacts are energized, the movable contacts are forced against the two stationary contacts to complete the electrical circuit. For all practical purposes, this movable contact is a shorting bar between the two stationary contacts completing the electrical circuit. When the manual contactor is de-energized, the movable contacts are forced away from the stationary contacts and the circuit is again open. When NC (normally closed) double-break contacts are used, the procedure is reversed.

Figure 4-6 illustrates an exploded view of a typical three-phase manual contactor having three sets of NO double-break contacts. One set of NO double-break contacts is used to open and close each phase in the circuit. Note in this case that the movable contacts are all located on an insulated main T-frame and are provided with springs to soften their impact. The main T-frame can be set into action by the pushbutton mechanism shown or by a toggle switch mechanism. When activated, the mechanical linkage consistently and quickly makes or breaks the circuits similar to that of a disconnect.

It should be noted also from Figure 4-6 that the movable contacts have no physical electrical connection to any external electrical wires. When the movable contacts make or break the circuit, they move into the arc hoods and merely bridge electri-

cally the gap between the set of fixed contacts. All physical electrical connections are made indirectly to the fixed contacts through saddle clamps such as those shown in Figure 4-7. The arc hoods mentioned earlier not only insulate each set of contacts from the others, but also help contain and quench the arcs drawn across these contacts as they open and close.

Contact Construction. The second major breakthrough which helped contactors become popular was the introduction of new metal alloys. As you will recall from an earlier discussion, one of the major problems with the knife switch was the fact that it was constructed from soft copper.

Most contacts today are made of a low-resistance silver alloy. Silver is alloyed (mixed) with cadmium or cadmium oxide to make an exceptionally arc-resistant material which still has good conductivity. In addition, the silver alloy has good mechanical strength, making it able to endure the continual wear encountered by many openings and closings. One other advantage of silver alloy contacts is the fact that the oxide (rust) which forms on these materials is an excellent conductor of electricity. Even when the contacts appear dull or tarnished, they are still capable of operating normally (Figure 4-8).

In Unit 3 on industrial electrical symbols and line diagrams, we emphasized the use of the line diagram in determining very quickly how the control circuit of an electrical system would perform. At that point we only briefly discussed the *power circuit*. Since manual contactors directly control the power circuit, we must now take a much closer look at how the power circuit is wired. For a thorough understanding of this relationship, we must introduce a new type of diagram called the wiring diagram (Figure 4-9).

Unlike the line diagram, the wiring diagram is intended to show as closely as possible the actual connection and placement of *all* component parts in a device or circuit, including the power circuit wiring. An understanding of the wiring diagram is extremely important because the electrician may be called upon to make changes in the power circuit as well as in the control circuit.

Figure 4-9 illustrates a typical wiring diagram for a single-pole manual contactor and a pilot light. It should be immediately apparent that this type of diagram not only shows the power contacts, but also shows how they would be connected to the load. Like the line diagram, the power circuit is indicated through heavy dark lines and the control pilot circuit is indicated by a thinner line. The circuit would be read as follows: When power contacts in L1 and L2 close, current passes from L1 through the pilot light on to L2, causing the pilot lamp to glow. At the same time, current

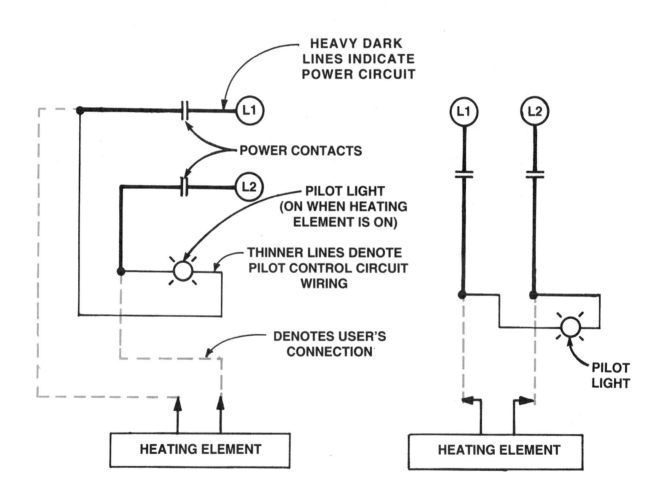

Figure 4-9 Wiring diagram for a single-pole manual contactor with pilot light. Note that a wiring diagram shows the power contacts and how the load is attached to them.

passes from L1 through the heating element and on to L2, causing the heating element to be activated. Note that the pilot light and heater element are in parallel with each other.

It would appear from the circuit just covered in Figure 4-9 that wiring diagrams are as easy to understand as line diagrams. Actually, wiring diagrams can get very complex, as shown by the dual-element heating load in Figure 4-10. This wiring diagram must be studied very carefully to determine the operation of this circuit. The circuit would be read as follows: When the first heating element is to be operated, the power contacts in L1 and L2 must be closed so that connection is made to the low and common terminals of the heater,

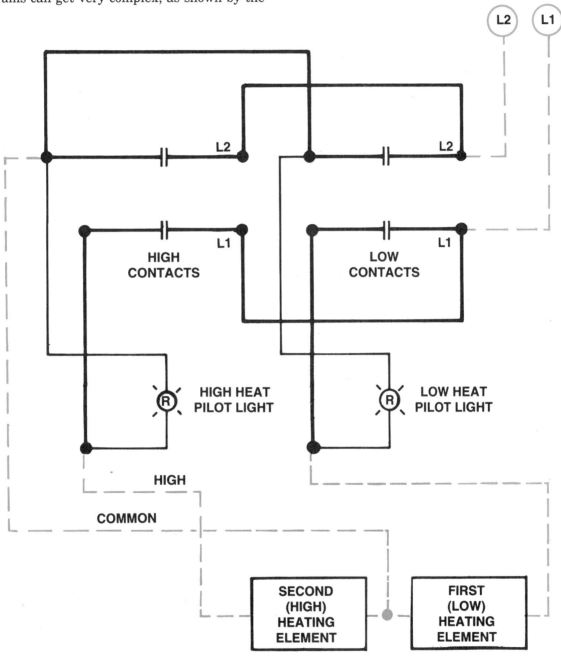

Figure 4-10 Wiring diagram of a dual-element heater with pilot lights. Note that the power circuit is indicated by the heavier dark lines and the pilot control circuit by the thinner lines.

thus allowing the "low heat" element to be energized (Figure 4-11). When the heater is to be operated at high heat, the high heat power contacts in L1 and L2 must be closed so that connection is made to the "high" and common terminals of the heater, thus allowing the "high element" to be energized (Figure 4-12). Since a pilot light is in parallel with each appropriate heater, a "low" light and "high" light would turn on to indicate each condition.

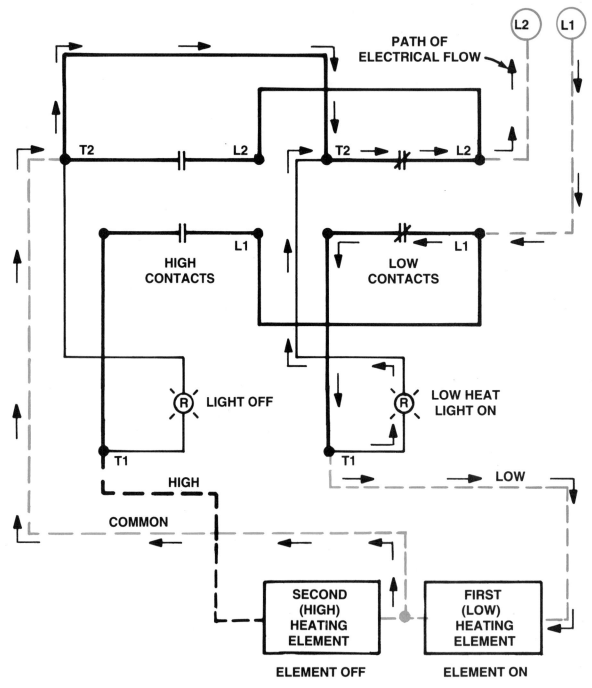

Figure 4-11 Circuit operation–low heat: When operated at "Low Heat", the low heat power contacts L1 and L2 must be closed so that connection is made to the LOW and COMMON terminals of the load, thus allowing the "low heat" element to be energized.

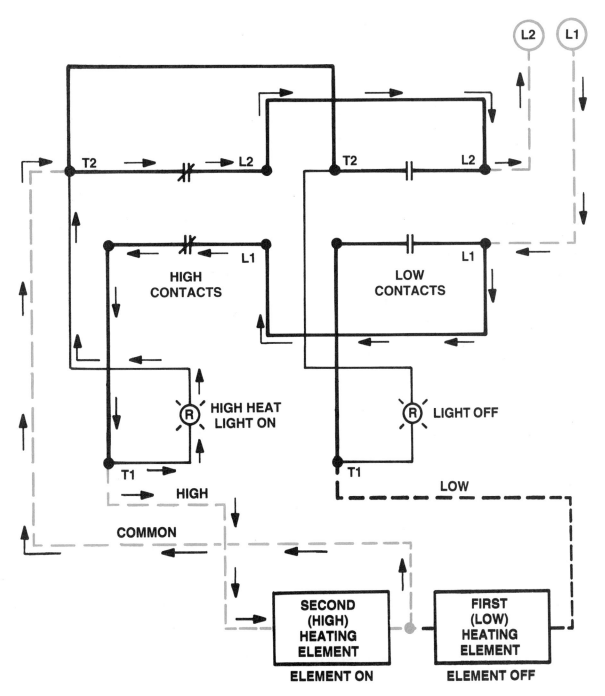

Figure 4-12 Circuit operation–high heat: When operated at "High Heat", the high heat power contacts L1 and L2 must be closed so that connection is made to the HIGH and COMMON terminals of the load, thus allowing the "high heat" element to be energized.

One problem which could arise with a dual element start is the possibility that someone would try to energize both sets of elements at the same time, causing serious damage to the heater. To prevent this from happening, most manual contactors of this type are equipped with a mechanical interlock. Mechanical interlocking means that the contacts are set up in such a way that both sets of contacts cannot be closed at the same time. Mechanical interlocking can be established by a mechanism which forces open one set of contacts if the other contacts are being closed. Another method is to provide a blocking bar or holding mechanism which will not allow the first set of contacts to close until the second set of contacts is open. The electrician can determine if a device is mechanically interlocked by consulting the wiring diagram information provided by the manu-

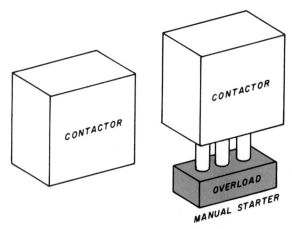

Figure 4-13 The primary difference between a contactor and a motor starter is the addition of an overload device. A motor starter is the combination of a contactor and an overload device.

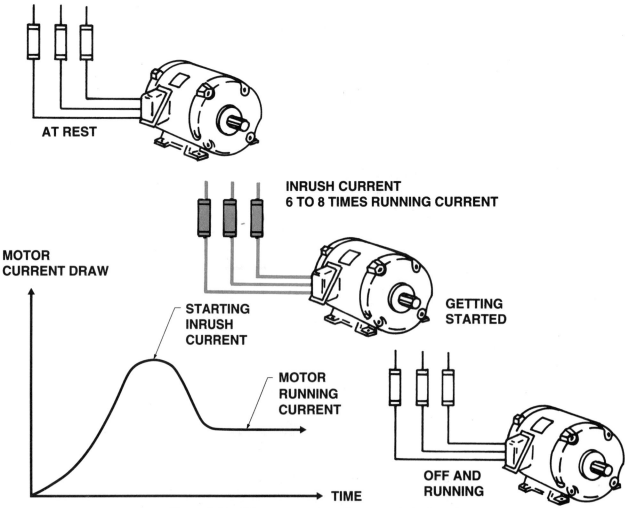

Figure 4-14 When a motor is getting started, the inrush current may be six to eight times more than the running current.

facturer. This information is usually packaged with the equipment when it is delivered or is glued to the inside of the enclosure.

MANUAL CONTACTORS VS. MANUAL STARTERS

It should be pointed out quite clearly that there is a difference between manual contactors and manual starters. A manual contactor is a device which manually opens and closes contacts in *any* electrical circuit. Manual starters, however, are used only in electrical motor circuits.

The primary difference between a contactor and a starter is the addition of a second component called the overload protection (Figure 4-13). The starter consists of a contactor *and* an overload protective device. The overload protection must be added because the *National Electrical Code®* requires that the control device must not only turn a motor on and off, but must also protect the motor from destroying itself under an overloaded situation such as a locked rotor. A locked rotor exists when a motor is loaded so heavily that the motor shaft cannot turn. When the rotor is jammed in this manner, it will draw excessive current and burn up if not disconnected from the line voltage. It is the job of the overload device to sense this and open the circuit.

Need for Overload Protection

You may wonder why a special overload relay instead of a circuit breaker or fuse must protect the motor. To understand this concept, we must first understand something about the electric motor itself. Figure 4-14 illustrates the three stages that a motor must go through in normal operation: resting, starting, operating under load.

When a motor is at rest, obviously it requires no current because the circuit is open. When the circuit is closed, however, the motor draws a tremendous inrush current to start (usually six to eight times the running current). To avoid shutting down the circuit immediately by this large inrush current, the fuse or circuit breaker must have a sufficiently high ampere rating to avoid opening the circuit when the motor is starting.

Since the fuse or circuit breaker is rated at six to eight times the running current, it is possible for the motor to encounter a load while running which does not draw enough current to blow the fuse or trip a breaker but is still large enough to produce sufficient heat to burn up the motor. (*Note:* The intensive heat concentration generated by excessive current in the windings is what causes the insulation to fail and burn up the motor. It has been estimated that for every 1°C (Celsius) rise over normal ambient temperatures, ratings for insulation can reduce the life expectancy of a motor almost a year per degree. Ambient temperature is the temperature of the air surrounding the motor. The normal rating for many motors is about 40°C or 104° Fahrenheit.)

A dilemma then occurs in that we must keep the fuse or breaker to protect the circuit against very high currents of a short circuit or a ground while at the same time finding a device that: (1) will not open the circuit while the motor is getting started; (2) will open the circuit if the motor gets overloaded when the fuse won't blow (Figure 4-15). What is needed is a device specifically designed to protect against overloads: an overload relay.

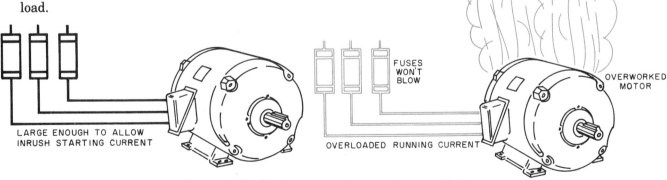

Figure 4-15 Overload protection must be provided in such a way that the circuit will not open while the motor is getting started, but will open if the motor gets overloaded when the fuses won't blow.

To meet these motor protective needs, overload relays were designed to: (1) have a time delay to allow harmless temporary overloads without disrupting the circuit; (2) have a trip capability to open the circuit if mildly dangerous currents which could result in motor damage continue over a period of time; (3) have some means of resetting the circuit once the overload is removed.

Melting Alloy Overloads

One of the most popular methods of providing overload protection is through the melting alloy overload relay. As mentioned previously, heat is the final end product which destroys a motor. To be effective, the overload relay must measure the temperature of the motor. Since the overload relay is normally some distance from the motor, the overload relay must indirectly monitor the temperature conditions of the motor.

To effectively monitor the heat generated by excessive current and the heat created through ambient temperature rise, a device called a heater coil, in combination with a sensing means, was devised.

Many different types of heaters are available today (Figure 4-16). However, the operating principle of each is the same. The heater coil converts the excess current drawn by the motor into additional heat which the sensing device uses to determine whether the motor is in danger.

Most manufacturers today rely on a unique eutectic alloy in conjunction with some type of mechanism to activate a tripping device when an overload occurs. The nature of the eutectic alloy is such that it will respond to the rise of a definite temperature level. It has a fixed temperature at which it will flow from a solid state to a liquid state without going through a mushy condition. This temperature never changes and is not affected by repeated melting and resetting.

Although each manufacturer designs overloads somewhat differently, most use some type of ratchet wheel and eutectic alloy combination to activate a trip mechanism in the event of an overload.

Figure 4-17 illustrates how one manufacturer has constructed the overload relay device. The heart of the overload device is the melting solder eutectic alloy tube. The tube consists of two parts: the outer spindle and an inner spindle with a ratchet wheel held firmly into the tube by means of a locking ring and solid eutectic alloy solder film.

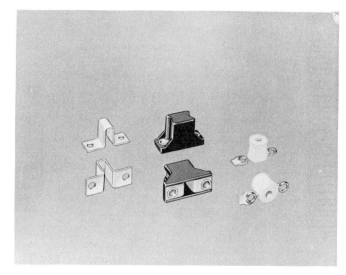

Figure 4-16 Many different types of heaters are available; however, the operating principle of each is the same. The heater coil converts the excess current drawn by the motor into additional heat which the sensing device uses to determine whether the motor is in danger. (Allen-Bradley Company)

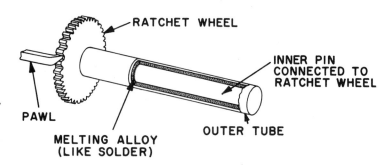

Figure 4-17 The heart of the overload device is the melting solder eutectic alloy tube. The tube consists of two parts: the outer spindle and an inner spindle with a ratchet wheel which is held firmly into the tube by means of a locking ring and solid eutectic alloy solder film.

Looking at a side view of the device (Figure 4-18) you can see that if the alloy is cool, the pin and ratchet wheel are "frozen" into position so that the wheel cannot be turned.

If, however, we apply excessive current to the heater coil (Figure 4-19), the heater coil will act like a toaster element, causing the alloy to melt from the extra heat generated from the excessive current. The direct result of the melting is that the ratchet wheel is free to turn.

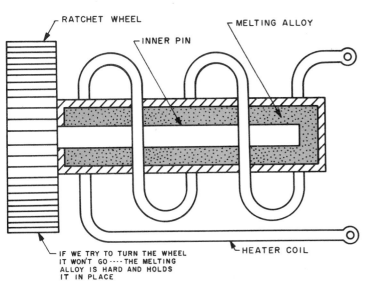

Figure 4-18 Looking at a side view of the alloy tube, you can see that if the alloy is cool the pin and ratchet wheel are "frozen" into position so that the wheel cannot be turned.

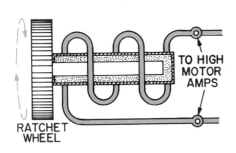

Figure 4-19 When excessive current passes through the heater coil, it will act like a toaster element, causing the alloy to melt. The direct result of this melting is that the ratchet wheel is free to turn.

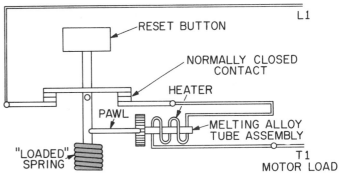

Figure 4-20 With the melting alloy tube assembly in place, an overload device would appear as shown when in a normal condition. Note that the loaded spring tries to push the normally closed contacts up and open; however, the pawl (locking mechanism) is caught in the ratchet wheel and will not let the spring push up.

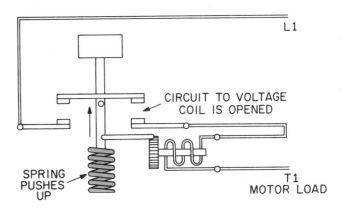

Figure 4-21 In the event of an overload, the following events take place: (1) The heater coil heats the melting alloy tube. (2) The alloy melts. (3) The ratchet wheel is free to turn. (4) The spring pushes the reset button up. (5) The contact to the voltage coil of the contactor opens. (6) The contactor opens the circuit to the motor. (7) The motor stops; no current *now* flows through the heater coil. (8) The heater coil cools, and the melted alloy, along with the ratchet wheel, is again firm.

If we place this melting alloy tube assembly into an overload device, we will have a device like that pictured in Figure 4-20. In this illustration the motor current conditions are normal. The loaded spring tries to push the normally closed contacts up and open; however, the pawl (locking mechanism) is caught in the ratchet wheel and will not let the spring push up.

In the event of an overload (Figure 4-21), the following events take place. (1) The heater coil heats the melting alloy tube. (2) The alloy melts. (3) The ratchet wheel is free to turn. (4) The spring pushes the reset button up. (5) The contact to the voltage coil of the contactor opens. (6) The contactor opens the circuit to the motor. (7) The motor stops; no current *now* flows through the heater coil. (8) The heater coil cools, and the melted alloy, along with the ratchet wheel, is again firm.

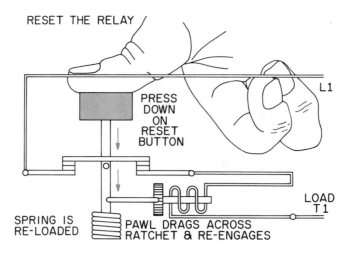

RESET THE RELAY

PRESS DOWN ON RESET BUTTON

L1

LOAD T1

SPRING IS RE-LOADED

PAWL DRAGS ACROSS RATCHET & RE-ENGAGES

Figure 4-22   Once the overload is removed, the overload device may be reset using the reset button to force the pawl (latching mechanism) across the ratchet wheel until the contacts are closed and the spring and ratchet wheel are returned to their original condition.

SMALL HORSEPOWER MOTOR

SMALL HEATER COIL

LARGE HORSEPOWER MOTOR

LARGE HEATER COIL

Figure 4-23   Basically, the same overload device is used with all sizes of motors; only the heater coil size is changed. For small horsepower motors, you use a small heater coil and for larger horsepower motors you use a larger heater coil. Consult the *NEC* for appropriate selection of overload heater sizes.

Resetting the Overload Device

Before resetting the overload device, it is important to find the cause of the overload. If the overload is not removed, the device will again trip on resetting. Once the overload is removed, the device can be reset as shown in Figure 4-22. The reset button is pushed, forcing the pawl (latching mechanism) across the ratchet wheel until the contacts are closed and the spring and ratchet wheel are returned to their original condition. The start button can then be pressed and the motor will start again.

It should be apparent that there is nothing to replace or repair when an overload device trips. You merely remove the cause of the overload and press the reset button.

Basically, the same overload device is used with all sizes of motors; only the heater coil size is changed. For small horsepower motors, you use a small heater coil and for larger horsepower motors you use a larger heater coil (Figure 4-23). Check the *National Electrical Code®* section 430 for appropriate selection of overload heater sizes.

SELECTING AC MANUAL STARTERS

Electricians may be called upon to select manual starters for new installations or replace ones which have been severely damaged due to an electrical fire or explosion. In either case, the electrician must specify certain characteristics of the starter in order to get a proper replacement. Figure 4-24 has been designed to aid in the proper selection of manual starters. Note the important points covered in this illustration, namely phasing, number of poles, voltage consideration, starter size and enclosure type.

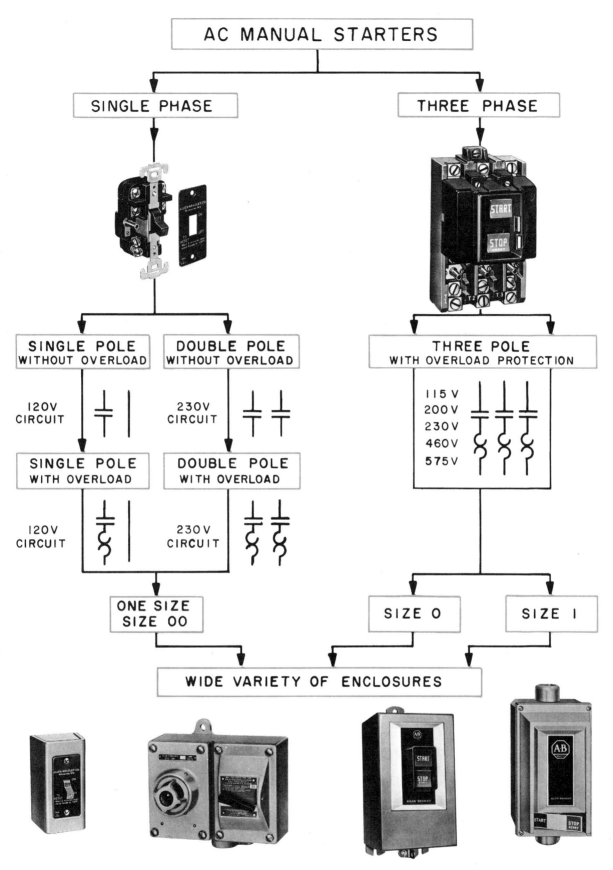

Figure 4-24 This has been designed to aid in the proper selection of manual starters. Note the important points covered in this illustration, namely phasing, number of poles, voltage consideration, starter size, and enclosure type. (Allen-Bradley Company)

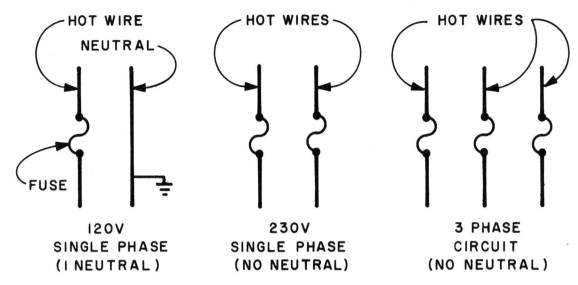

HOT WIRE
NEUTRAL
FUSE

120V
SINGLE PHASE
(1 NEUTRAL)

HOT WIRES

230V
SINGLE PHASE
(NO NEUTRAL)

HOT WIRES

3 PHASE
CIRCUIT
(NO NEUTRAL)

Figure 4-25 A 120V single-phase source has one hot wire (ungrounded conductor) and one neutral wire (grounded conductor) while a 230V single-phase source has two hot wires (ungrounded conductors) and no neutral. A three-phase source has three hot wires and no neutral.

Phasing

Categorically, AC manual contactors can be divided into two groups: single-phase and three-phase (Figure 4-25). In the unit on power distribution, the concept of what single-phase and three-phase is will be treated in depth, but for now it will be important only to know that 120V single-phase source has one hot wire (ungrounded conductor) and one neutral wire (grounded conductor) while a 230V single-phase source has two hot wires (ungrounded conductors) and no neutral. A three-phase source has three hot wires and no neutral.

Single-Phase Starters. Since the *National Electrical Code®* requires that each ungrounded conductor (hot wire) be open when disconnecting a device, single-phase starters are available as single-pole and double-pole devices (Figure 4-24). The single-pole device is used on 120V circuits while the double-pole is used for 230V circuits.

Since single-phase manual starters are physically small (Figure 4-24), they have been limited in horsepower rating. They may be used as starters for motors of one horsepower maximum. Single-phase starters come in only one size for all motors rated at one horsepower or less. The size established for single-phase starters is size 00.

Three-Phase Starters. Three-phase manual starters are physically larger than single-phase devices and may be used on motors up to ten horsepower. Three-phase contactors are usually push-button operated instead of toggle operated as in single-phase.

Although contactors can be used in certain applications without overload devices, a majority of the motor circuits require a manual starter which has overloads.

Since three-phase devices have three hot wires which must be disconnected, three-phase devices are designed with three-pole switching (Figure 4-24). Three-phase, like single-phase, use butt type contacts and have quick-make and quick-break mechanisms.

Three-phase contactors and starters are designed to be used on circuits from 115V up to and including 575V.

General Applications of Manual Contactors and Starters

Single Phase

1. Single-phase motors one horsepower and under where low voltage protection is not needed.

2. Single-phase motors that do not require a high frequency of operation.

Three Phase

1. Three-phase motors 7.5 horsepower and under operating at 208/230V or 10 horsepower and under operating at 380/575V.

2. Three-phase motors where low voltage protection is not needed.

3. Three-phase motors that do not require a high frequency of operation.

4. Three-phase motors that do not need remote operation by push buttons or limit switches.

ENCLOSURES

Enclosures provide mechanical and electrical protection for the operator and the starter (Figure 4-26). Although the enclosures are designed to provide protection in a variety of situations (water, dust, oil and hazardous location), the internal electrical wiring and physical construction of the starter remain the same. Consult the

NEMA TYPE 1
SURFACE MOUNTING

Type 1 enclosures are intended for indoor use primarily to provide a degree of protection against contact with the enclosed equipment in locations where unusual service conditions do not exist. The enclosures are designed to meet the rod entry and rust-resistance design tests. Enclosure is sheet steel, treated to resist corrosion.

NEMA TYPE 3R

Type 3R enclosures are intended for outdoor use primarily to provide a degree of protection against falling rain, sleet, and external ice formation. They are designed to meet rod entry, rain, external icing, and rust-resistance design tests. They are not intended to provide protection against conditions such as dust, internal condensation, or internal icing. Entire enclosure is finished in a weather resistant gray baked enamel.

NEMA TYPE 1
FLUSH MOUNTING

Flush mounted enclosures for installation in machine frames and plaster wall. These enclosures are for similar applications and are designed to meet the same tests as NEMA Type 1 surface mounting.

NEMA TYPE 4

Type 4 enclosures are intended for indoor or outdoor use primarily to provide a degree of protection against windblown dust and rain, splashing water, and hose-directed water. They are designed to meet hosedown, dust, external icing, and rust-resistance design tests. They are not intended to provide protection against conditions such as internal condensation or internal icing. Enclosures are made of heavy gauge stainless steel, cast aluminum or heavy gauge sheet steel, depending on the type of unit and size. Cover has a synthetic rubber gasket.

Figure 4-26 Enclosures provide mechanical and electrical protection for the operator and the starter. NEMA means National Electrical Manufacturers Association. (Allen-Bradley Company)

**NEMA TYPE 3R, 7 & 9
UNILOCK ENCLOSURE
FOR HAZARDOUS
LOCATIONS**

This enclosure is of cast aluminum alloy with a baked paint coating on the outside for extra corrosion resistance. The V-Band permits easy removal of the cover for inspection and for making field modifications. This enclosure is for similar applications and is designed to meet the same tests as NEMA Types 3R, 7 and 9. For NEMA Type 3R application, it is necessary that a drain or breather-drain combination be added.

NOTE: Enclosures do not normally protect devices against conditions such as condensation, icing, corro- *sion or contamination which may occur within the enclosure or enter via the conduit or unsealed openings. Users must make adequate provisions to safeguard against such conditions, and satisfy themselves that the equipment is properly protected. For additional information of definitions, descriptions and test criteria, see National Electrical Manufacturers Association (NEMA) Standards Publication No. 250-1979.*

**NEMA TYPE 4X
NON-METALLIC,
CORROSION-RESISTANT
GLASS POLYESTER**

Type 4X enclosures are intended for indoor or outdoor use primarily to provide a degree of protection against corrosion, windblown dust and rain, splashing water, and hose-directed water. They are designed to meet the hosedown, dust, external icing, and corrosion-resistance design tests. They are not intended to provide protection against conditions such as internal condensation or internal icing. Enclosure is glass reinforced polyester with a synthetic rubber gasket between cover and base. Ideal for such industries as chemical plants and paper mills.

NEMA TYPE 6P

Type 6P enclosures are intended for indoor or outdoor use primarily to provide a degree of protection against the entry of water during prolonged submersion at a limited depth. They are designed to meet air pressure, external icing, and corrosion-resistance design tests. They are not intended to provide protection against conditions such as internal condensation or internal icing.

NEMA TYPE 7
FOR HAZARDOUS
GAS LOCATIONS
BOLTED ENCLOSURE

Type 7 enclosures are for indoor use in locations classified as Class 1, Groups C or D, as defined in the National Electrical Code. Type 7 enclosures are designed to be capable of withstanding the pressures resulting from an internal explosion of specified gases, and contain such an explosion sufficiently that an explosive gas-air mixture existing in the atmosphere surrounding the enclosure will not be ignited. Enclosed heat generating devices are designed not to cause external surfaces to reach temperatures capable of igniting explosive gas-air mixtures in the surrounding atmosphere. Enclosures are designed to meet explosion, hydrostatic, and temperature design tests. Finish is a special corrosion-resistant, gray enamel.

NEMA TYPE 12

Type 12 enclosures are intended for indoor use primarily to provide a degree of protection against dust, falling dirt, and dripping noncorrosive liquids. They are designed to meet drip, dust, and rust-resistance tests. They are not intended to provide protection against conditions such as internal condensation.

NEMA TYPE 9
FOR HAZARDOUS
DUST LOCATIONS

Type 9 enclosures are intended for indoor use in locations classified as Class II, Groups E, F, or G, as defined in the National Electrical Code. Type 9 enclosures are designed to be capable of preventing the entrance of dust. Enclosed heat generating devices are designed not to cause external surfaces to reach temperatures capable of igniting or discoloring dust on the enclosure or igniting dust-air mixtures in the surrounding atmosphere. Enclosures are designed to meet dust penetration and temperature design tests, and aging of gaskets. The outside finish is a special corrosion-resistant gray enamel.

NEMA TYPE 13

Type 13 enclosures are intended for indoor use primarily to provide a degree of protection against dust, spraying of water, oil, and noncorrosive coolant. They are designed to meet oil exclusion and rust-resistance design tests. They are not intended to provide protection against conditions such as internal condensation.

STEP 1: REMOVE COVER

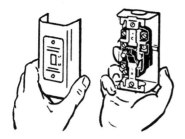

REMOVE TWO SCREWS AND PULL THE METAL COVER OFF.

STEP 3: PULL IN WIRES AND CONNECT

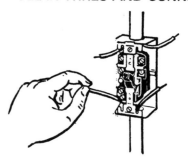

WIRES CAN BE PULLED INTO THE BOX AND CONNECTED TO THE TERMINALS WITHOUT REMOVING THE SWITCH FROM THE ENCLOSURE.

STEP 2: MOUNT UNIT

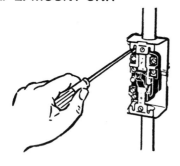

MOUNT SWITCH ON THE WALL WITH TWO SCREWS. CONVENIENT CONDUIT OPENINGS ON TOP, BOTTOM AND BACK PERMIT EASY CONNECTION.

STEP 4: PUT COVER BACK ON

COVER SLIPPED ON AND FASTENED WITH TWO SCREWS. ENTIRE OPERATION IS SIMPLE AND EASY AND SAVES TIME AND MONEY.

National Electrical Code and local codes to determine the proper selection of an enclosure for a particular application. Figure 4-27 illustrates a single-phase manual starter being mounted into a general purpose enclosure.

Figure 4-27 Single-phase manual starter being mounted into a general purpose enclosure. (Allen-Bradley Company)

SPECIFIC APPLICATION FOR MANUAL STARTERS

Figure 4-28 illustrates several specific applications for manual motor starters, namely drill presses, conveyor systems, air compressors, table saws and band saws. In most of these applications the manual starter provides the means of turning on and off the device while providing motor overload protection.

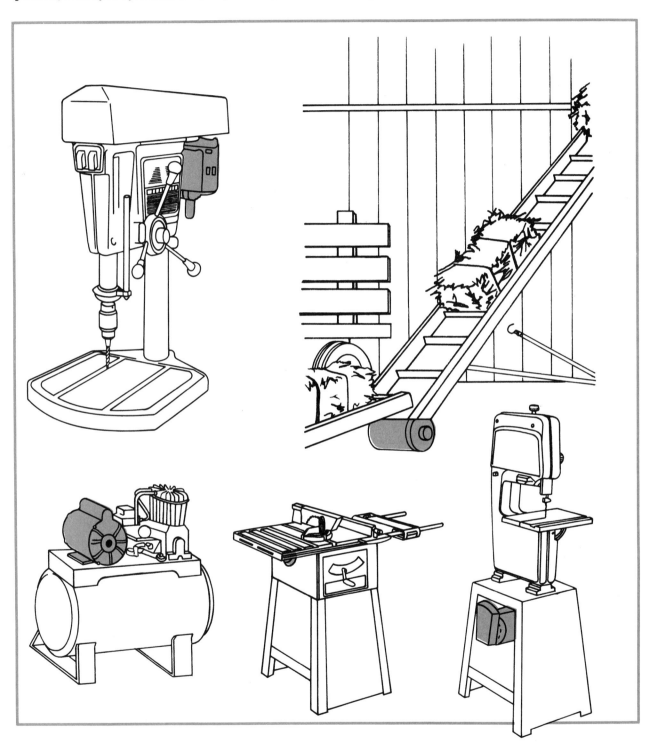

Figure 4-28 Specific applications for manual motor starters, namely drill presses, conveyor systems, air compressors, table saws and band saws.

General Motor Protection Table

(For exact motor protection, refer to NEC)

Three Phase 230 Volt Motors and Circuits.

1 Size Of Motor (NEC Table 430-150) HP	Amp Rating	2 Dual-Element Fuse For Motor Running Overload Protection (430-32) (Also branch circuit protection) FUSETRON or LOW-PEAK Fuses (Amps): Motor Rated Not Over 40°C Or Not Less Than 1.15 S.F. (Max Fuse 125%)	All Other Motors (Max. Fuse 115%)	Switch (115% Min. Or HP Rated) Or Fuse-holder Size	3 †Branch Circuit Protection (430-52) (Short-Circuit Protection Only; Fuses do not give motor running protection): †††Class For Motor Start Inrush And Code Letter	†Time-Delay Fuse (FUSETRON or LOW-PEAK Dual-Element) Fuse Amps	Switch or Fuse-holder Size	†Non-Time Delay Fuse: Fuse Amps	Switch Or Fuse-holder Size	4 Min. Size Of Starter (NEMA)	5 ††Min. Size Of Copper Wire (AWG or MCM) a-TW (60°C) b-THW (75°C) c-THWN (75°C) d-THHN (90°C) e-XHHW (90°C)§	6 Min. Size Of Trade Conduit For Copper Wire (inches)
½	2	2½	2¼	30	Any	4	30	6	30	00	14 (a,b,c,d,e)	½
¾	2.8	3½	3²⁄₁₀	30	Any	4	30	10	30	00	14 (a,b,c,d,e)	½
1	3.6	4½	4	30	Any	6¼	30	15	30	00	14 (a,b,c,d,e)	½
1½	5.2	6¼	5⁶⁄₁₀	30	Any	8	30	15	30	00	14 (a,b,c,d,e)	½
2	6.8	8	7	30	1	10	30	25	30	0	14 (a,b,c,d,e)	½
					2	10	30	20	30			
					3-4	10	30	15	30			
3	9.6	12	10	30	1	15	30	30	30	0	14 (a,b,c,d,e)	½
					2	15	30	25	30			
					3	15	30	20	30			
					4	15	30	15	30			
5	15.2	17½	17½	30	1	25	30	50	60	1	14 (a,b,c,d,e)	½
					2	25	30	40	60			
					3	25	30	35	60			
					4	25	30	25	30			
7½	22	25	25	30	1	35	60	70	100	1		
					2	35	60	60	60		10 (a,b,c)	½
					3	35	60	45	60		12 (d,e)	½
					4	35	60	35	60			
10	28	35	30	60	1	40	60	90	100	2		
					2	40	60	70	100		8 (a)	¾
					3	40	60	60	60		10 (b,c,d,e)	½
					4	40	60	45	60			
15	42	50	45	60	1	60	60	125	200	2	6 (a,b)	1
					2	60	60	110	200		6 (c)	¾
					3	60	60	90	100		8 (d)	½
					4	60	60	70	100		8 (e)	¾
20	54	60	60	100	1	80	100	175	200	3		
					2	80	100	150	200		4 (a,b,c)	1
					3	80	100	110	200		6 (d,e)	¾
					4	80	100	90	100			
25	68	80	70	100	1	100	100	225	400	3		
					2	100	100	175	200		3 (a)	1¼
					3	100	100	150	200		4 (b,c,d,e)	1
					4	100	100	110	200			
30	80	100	90	100	1	125	200	250	400	3		
					2	125	200	200	200		1 (a)	1¼
					3	125	200	175	200		3 (b)	1¼
					4	125	200	125	200		3 (c,d,e)	1
40	104	125	110	200	1	150	200	350	400	4		
					2	150	200	300	400		2/0 (a)	1½
					3	150	200	225	400		1 (b,c)	1¼
					4	150	200	175	200		2 (d,e)	1
50	130	150	150	200	1	200	200	400	400	4		
					2	200	200	350	400		3/0 (a)	2
					3	200	200	300	400		2/0 (b,c,d,e)	1½
					4	200	200	200	200			
60	154	175	175	200	1	250	400	500	600	5	4/0 (a)	2
					2	250	400	400	400		3/0 (b)	2
					3	250	400	350	400		3/0 (c)	1½
					4	250	400	250	400		2/0 (d,e)	1½
75	192	225	200	400	1	300	400	600	600	5	300 (a)	2½
					2	300	400	500	600		250 (b)	2½
					3	300	400	400	400		250 (c)	2
					4	300	400	300	400		4/0 (d,e)	2
100	248	300	250	400	1-2	400	400	▼▼		5	500 (a)	3
					3	400	400	500	600		350 (b,c)	2½
					4	400	400	400	400		300 (d,e)	2
125	312	350	350	400						6	4/0-2/Ø* (a)	2-2*
					1-2-3	450	600	▼▼			3/0-2/Ø* (b)	2-2*
					4	450	600	500	600		3/0-2/Ø* (c)	2-1½*
											500 (d,e)	3
150	360	450	400	600	1-2-3	500	600	▼▼		6	300-2/Ø* (a)	2-2½*
					4	500	600	600	600		4/0-2/Ø* (b,c)	2-2*
											3/0-2/Ø* (d,e)	2-1½*
200	480	600	500	600						6	500-2/Ø* (a)	2-3*
					Any	▼		▼▼			300-2/Ø* (b)	2-2½*
											300-2/Ø* (c,d,e)	2-2*

† If manufacturer's overload relay table states a maximum branch circuit protective device of a lower rating, that lower maximum rating must be used in lieu of above recommendation (430-52).

▼ Use KRP-C Fuses sized at 150% to 250% depending on code letter and starting methods.

▼▼ Use KTU Fuses sized at 150% to 300% depending on code letter and starting methods.

* Indicates two sets of multiple conductors and two runs of conduit.

§ For dry locations only.

†† Subject to interpretation of code enforcing authorities (refer to footnotes of NEC Table 310-16).

†††Class	Motors with Code Letter Marking (430-152)
1	F to V full voltage start or resistor or reactor start.
2	F to V with auto transformer start. B to E full voltage start or resistor or reactor start.
3	B to E with auto transformer start.
4	A—

Source: Bussmann Div. Cooper Industries.

General Motor Protection Table

(For exact motor protection, refer to NEC)

Three Phase 460 Volt Motors and Circuits.

1		2			3					4	5	6
Size Of Motor (NEC Table 430-150)		**Dual-Element Fuse For Motor Running Overload Protection** (430-32) (Also branch circuit protection)			**†Branch Circuit Protection** (430-52) (Short-Circuit Protection Only; Fuses do not give motor running protection)					**Min. Size Of Starter** (NEMA)	**††Min. Size Of Copper Wire** (AWG or MCM) **a-TW** (60°C) **b-THW** (75°C) **c-THWN** (75°C) **d-THHN** (90°C) **e-XHHW** (90°C) §	**Min. Size Of Trade Conduit For Copper Wire** (inches)
HP	**Amp Rating**	**FUSETRON or LOW-PEAK Fuses (Amps)** Motor Rated Not Over 40°C Or Not Less Than 1.15 S.F. (Max Fuse 125%)	All Other Motors (Max. Fuse 115%)	**Switch** (115% Min. Or HP Rated) **Or Fuse-holder Size**	**†††Class For Motor Start Inrush And Code Letter**	**†Time-Delay Fuse (FUSETRON or LOW-PEAK Dual-Element)** Fuse Amps	Switch or Fuse-holder Size	**†Non-Time Delay Fuse** Fuse Amps	Switch Or Fuse-holder Size			
½	1	1¼	1⅛	30	Any	2	30	3	30	00	14 (a,b,c,d,e)	½
¾	1.4	1⁶⁄₁₀	1⁶⁄₁₀	30	Any	2½	30	6	30	00	14 (a,b,c,d,e)	½
1	1.8	2¼	2	30	Any	3²⁄₁₀	30	6	30	00	14 (a,b,c,d,e)	½
1½	2.6	3²⁄₁₀	2⁸⁄₁₀	30	Any	4	30	10	30	00	14 (a,b,c,d,e)	½
2	3.4	4	3½	30	Any	5	30	15	30	00	14 (a,b,c,d,e)	½
3	4.8	5⁶⁄₁₀	5	30	Any	8	30	15	30	0	14 (a,b,c,d,e)	½
					1	15	30	25	30			
5	7.6	9	8	30	2	15	30	20	30	0	14 (a,b,c,d,e)	½
					3-4	15	30	15	30			
					1	20	30	35	60			
7½	11	12	12	30	2	20	30	30	30	1	14 (a,b,c,d,e)	½
					3	20	30	25	30			
					4	20	30	20	30			
					1	20	30	45	60			
10	14	17½	15	30	2	20	30	35	60	1	14 (a,b,c,d,e)	½
					3	20	30	30	30			
					4	20	30	25	30			
					1	30	30	70	100			
15	21	25	20	30	2	30	30	60	60	2	10 (a,b,c)	½
					3	30	30	45	60		12 (d,e)	½
					4	30	30	35	60			
					1	40	60	90	100			
20	27	30	30	60	2	40	60	70	100	2	8 (a)	¾
					3	40	60	60	60		10 (b,c,d,e)	½
					4	40	60	45	60			
					1	50	60	110	200		6 (a)	1
25	34	40	35	60	2	50	60	90	100	2	8 (b,e)	¾
					3	50	60	70	100		8 (c,d)	½
					4	50	60	60	60			
					1	60	60	125	200		6 (a,b)	1
30	40	50	45	60	2	60	60	100	100	3	8 (d)	½
					3	60	60	80	100		8 (c,e)	¾
					4	60	60	60	60			
					1	80	100	175	200		4 (a)	1
40	52	60	60	100	2	80	100	150	200	3	6 (b)	1
					3	80	100	110	200		6 (c,d,e)	¾
					4	80	100	80	100			
					1	100	100	200	200			
50	65	80	70	100	2	100	100	175	200	3	3 (a)	1¼
					3	100	100	150	200		4 (b,c,d,e)	1
					4	100	100	100	100			
					1	125	200	250	400		1 (a)	1¼
60	77	90	80	100	2	125	200	200	200	4	3 (b)	1¼
					3	125	200	175	200		3 (c,d,e)	1
					4	125	200	125	200			
					1	150	200	300	400		1/0 (a)	1½
75	96	110	110	200	2	150	200	250	400	4	1 (b,c)	1¼
					3	150	200	200	200		2 (d,e)	1
					4	150	200	150	200			
					1	200	200	400	400		3/0 (a)	2
100	124	150	125	200	2	200	200	350	400	4	2/0 (b,c)	1½
					3	200	200	250	400		1/0 (d,e)	1¼
					4	200	200	200	200			
					1	250	400	500	600		4/0 (a)	2
125	156	175	175	200	2	250	400	400	400	5	3/0 (b)	2
					3	250	400	350	400		3/0 (c)	1½
					4	250	400	250	400		2/0 (d,e)	1½
					1	300	400	600	600		300 (a)	2½
150	180	225	200	400	2	300	400	450	600	5	4/0 (b,c)	2
					3	300	400	400	400		3/0 (d,e)	1½
					4	300	400	300	400			
					1	400	400	▼▼			500 (a)	3
200	240	300	250	400	2	400	400	600	600	5	350 (b,c)	2½
					3	400	400	500	600		300 (d,e)	2
					4	400	400	400	400			

† If manufacturer's overload relay table states a maximum branch circuit protective device of a lower rating, that lower maximum rating must be used in lieu of above recommendation (430-52).

▼▼ Use KTU Fuses sized at 150% to 300% depending on code letter and starting methods.

§ For dry locations only.

†† Subject to interpretation of code enforcing authorities (refer to footnotes of NEC Table 310-16).

†††Class	Motors with Code Letter Marking (430-152)
1	F to V full voltage start or resistor or reactor start.
2	F to V with auto transformer start. B to E full voltage start or resistor or reactor start.
3	B to E with auto transformer start.
4	A—

Source: Bussmann Div. Cooper Industries.

QUESTIONS ON UNIT 4

1. What were the disadvantages of using a knife switch for starting and stopping electric motors?
2. Where are knife switches used today?
3. What is meant by a NO double-break contact?
4. What is the function of the arc hoods?
5. Why is a silver alloy used on the switching contacts?
6. How does a wiring diagram differ from a line diagram?
7. What do the heavy dark lines indicate on a wiring diagram?
8. What is a mechanical interlock used for?
9. How does a manual starter differ from a manual contactor?
10. Why does a motor have to be protected by both an overload relay and a fuse or breaker?
11. What is meant by ambient temperature?
12. What are the basic requirements of an overload relay?
13. How does the basic overload relay determine an overload?
14. How is the overload relay reset?
15. How many contacts are required to switch a 120 volt circuit?
16. How many contacts are required to switch a 240 volt circuit?
17. How many contacts are required to switch a three-phase circuit?
18. Why are contactors and starters placed in an enclosure?
19. What is meant by an NEMA type 12 enclosure?
20. What type of enclosure would be used in a hazardous dust location?

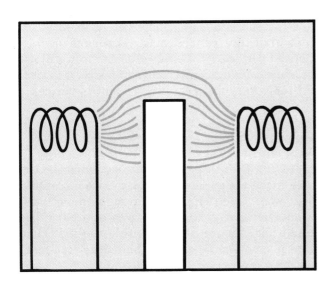

5 Magnetism and Magnetic Solenoids

There are two types of magnets: permanent magnets and temporary magnets. Permanent magnets retain their magnetism after the magnetizing force has been removed. Temporary magnets, or electromagnets, most often are used with electrical control devices. Electromagnets, called solenoids, can be used to activate levers to turn or move parts back and forth. Solenoids are used to control devices such as valves, relays, aircraft parts, and other industrial machinery.

MAGNETISM

Kinds of Magnets

Categorically there are only two types of magnets: permanent magnets and temporary magnets. The natural magnet, magnetite (Figure 5-1), and the manufactured magnets of Figure 5-2 are examples of permanent magnets. Figure 5-3, by contrast, shows an example of a manufactured temporary magnet.

Permanent magnets are magnets which can retain their magnetism after a magnetizing force has been removed. Permanent magnets are said to have a high retentivity because they can retain the residual magnetism (leftover magnetism) once a magnetizing force has been removed.

Temporary magnets are magnets which have extreme difficulty in retaining any magnetism after the magnetizing force has been removed. Because of this they are said to have a low reten-

tivity. In other words, very little residual magnetism (leftover magnetism) remains once the magnetizing force has been removed.

Molecular Theory of Magnetism

The idea that certain materials could hold their magnetism while other materials could not interested scientists for centuries. It was not until the development of the atomic theory and molecular action that there was any satisfactory theory or idea presented as to why this happened. Basically the molecular theory of magnetism states that all substances are made up of an infinite number of molecular magnets that can be arranged in two ways: organized or disorganized (Figure 5-4). If the material has disorganized molecular magnets (left), then the material is considered to be unmagnetized. If, however, the molecular magnets are organized (right), that material is considered to be magnetized.

The molecular theory of magnetism is important to understand because it explains a great deal about how certain materials react to magnetic fields when used as control devices. For example, it explains why hard steel is used for permanent magnets while soft iron is used on temporary magnets found in control devices.

If hard steel is magnetized (its molecular structure organized), the dense molecular structure will not very easily disorganize once a magnetizing force has been removed. Thus hard steel is hard to magnetize; but it is also hard to demagnetize, making it a good permanent magnet.

When soft iron is used, however, its loose molecular structure can be magnetized and demagnetized very easily. Because the soft iron will not retain residual magnetism very easily, it is ideal for temporary magnets used in control devices.

Relationship between Electricity and Magnetism

In 1819 a Danish physicist named Hans C. Oersted made a significant discovery. He discovered that when electricity flows through an electrical conductor, a magnetic field is created around that conductor (Figure 5-5). This was the first time that magnetism other than that of the rock magnetite was observed.

Figure 5-1 Permanent natural magnet, magnetite.

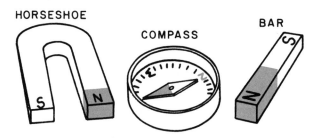

Figure 5-2 Permanent manufactured magnets: horseshoe, compass, and bar magnets.

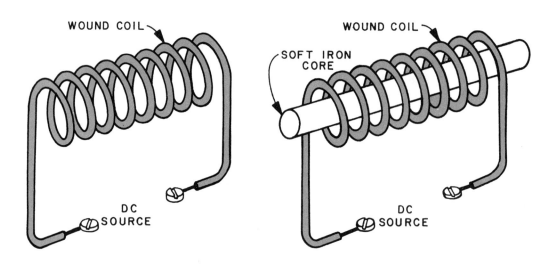

Figure 5-3 Temporary man-made magnets.

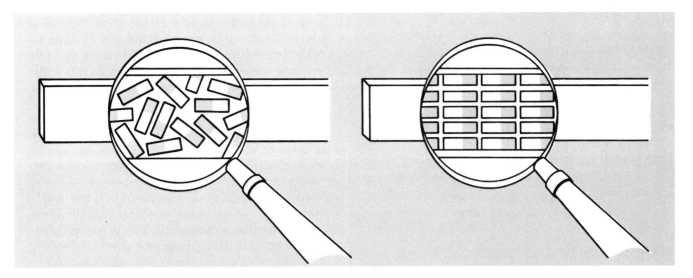

Figure 5-4 The molecular theory of magnetism states that all substances are made up of an infinite number of molecular magnets that can be arranged in two ways: organized or disorganized. If the substance has disorganized molecular magnets (left), it is considered to be unmagnetized. If, however, the magnets are organized (right), the substance is considered to be magnetized.

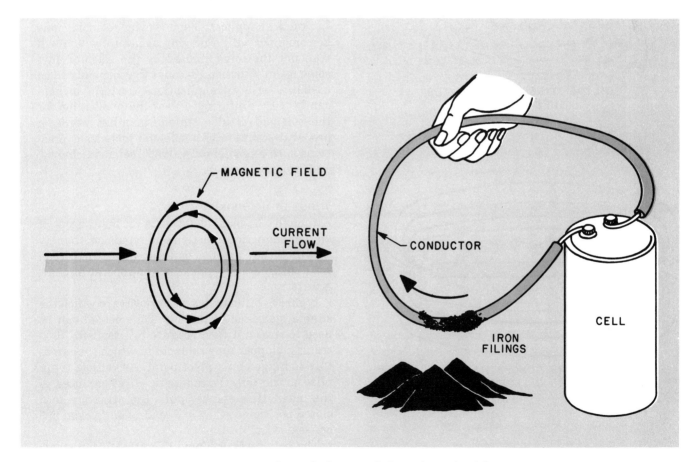

Figure 5-5 Oersted discovered that when electricity flows through an electrical conductor, a magnetic field is created around that conductor.

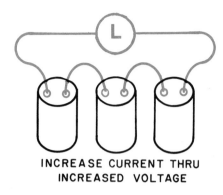

INCREASE CURRENT THRU
INCREASED VOLTAGE

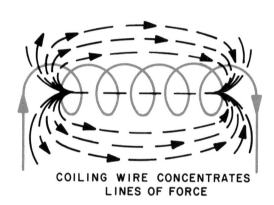

COILING WIRE CONCENTRATES
LINES OF FORCE

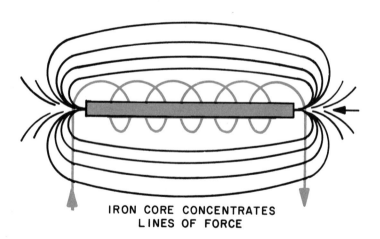

IRON CORE CONCENTRATES
LINES OF FORCE

Figure 5-6 Oersted discovered that there were three ways to increase the strength of a magnet: increase the current; increase the number of turns in the coil; add an iron core through the center of the coil.

Since the magnetic effects created in Oersted's first experiments were quite small, Oersted tried several experiments to increase the strength of his magnets. After many attempts, Oersted finally decided that there were only three ways to increase the amount of magnetism that a coil can produce. Figure 5-6 illustrates the three ways to increase the magnetic effect of a coil. First, you can increase the amount of current by stepping up the voltage. Second, you can increase the number of turns of wire in the coil. Third, you can insert an iron core through the coil to concentrate the lines of force.

The outcome of these experiments is the development, in modern times, of a huge control industry which relies on magnetic coils to convert electric energy into usable magnetic energy. One such magnetic coil device to be explained in this unit is the solenoid.

SOLENOID ACTION

A solenoid technically is a simple electromagnet consisting of a coil of wire and a source of voltage. In practical applications, however, solenoids may be combined with a moving armature which will transmit the force created by the solenoid into some useful function. Because the solenoid can be used in a variety of applications, the term solenoid can be very confusing unless the application for the solenoid is also stated. In other words, a proper description of a solenoid includes expressions such as "solenoid valve," "solenoid clutch," and "solenoid switch."

Types of Solenoids

Although all solenoids operate on the concept of electromagnetic attraction and repulsion, the way in which they are mechanically constructed can result in different applications and operating characteristics.

Figure 5-7 illustrates several ways in which the simple magnetic attraction of a solenoid can be used to transmit force. Figure 5-7, top left, illustrates a clapper type solenoid in which the armature is hinged on a pivot point. As voltage is applied to the coil, the magnetic effect produced in the magnetic assembly pulls the armature to a closed position so that it is said to be "picked up" or "sealed in."

The bell-crank solenoid (Figure 5-7, top right) uses a lever mechanism attached to the armature to transform the vertical action of the armature into a horizontal motion. The use of this lever mechanism allows the shock of the armature pick

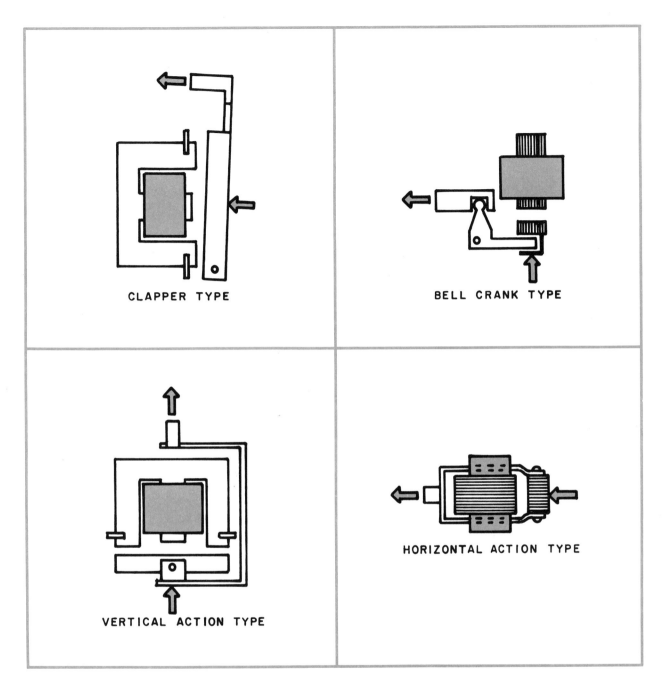

Figure 5-7 Solenoids are constructed in four basic configurations: the clapper, the bell-crank, the vertical action and the horizontal action type. Each has its own unique operating characteristics.

up to be absorbed by the lever mechanism and not transmitted to the end of the lever. This can be very beneficial when a soft but firm motion is required in certain controls.

The vertical action solenoid (Figure 5-7, bottom left) also uses a mechanical assembly but transmits the vertical action of the armature in a straight line motion as the armature is picked up.

The horizontal solenoid (Figure 5-7, bottom right) is a direct action device. The movement of the armature moves the resultant force in a

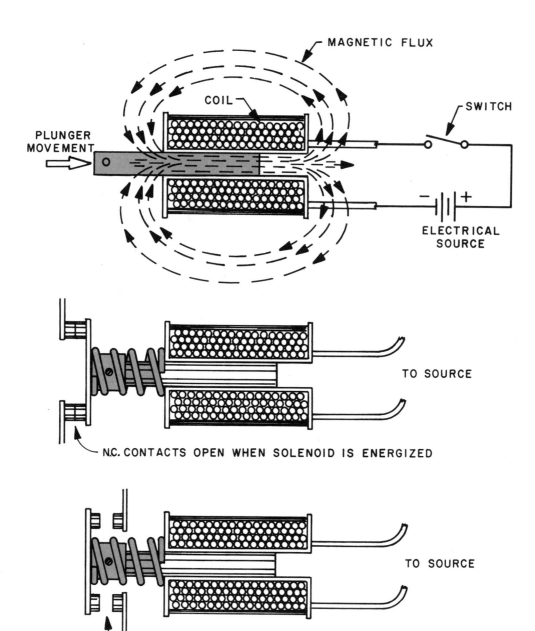

MAGNETIC FLUX

COIL

SWITCH

PLUNGER
MOVEMENT

ELECTRICAL
SOURCE

TO SOURCE

N.C. CONTACTS OPEN WHEN SOLENOID IS ENERGIZED

TO SOURCE

N.O. CONTACTS CLOSE WHEN SOLENOID IS ENERGIZED

Figure 5-8 Plunger Solenoid: If an iron rod that is free to move is placed within an electrical coil, the rod will tend to equalize, or align itself within the coil, when current is passing through the coil. If a spring is attached to the rod and the current controlled, the rod can be made to do useful work.

straight line. Horizontal action solenoids are one of the most common types.

Still another type of solenoid is one employing only a moving iron cylinder (Figure 5-8) as is illustrated. If an iron rod that is free to move is placed within an electrical coil, the rod will tend

to equalize, or align itself within the coil, when current is passing through the coil. If the rod and solenoid are of equal length, for example, a current will cause the rod to center itself so that the rod ends line up with the ends of the solenoid. In such a device, a spring is used to move the rod a

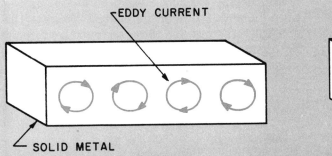

EDDY CURRENT

SOLID METAL

LAMINATION   RIVETS

Figure 5-9 Left: solid iron allows eddy currents within the solid metal. Right: Laminated thin sheets of metal are used on AC solenoids to reduce eddy currents created by the changing magnetic field.

short distance from its center place in the coil. Then, when the current is turned on, the rod moves against the spring tension to re-center itself. When the current is turned off, the spring returns the rod to its off-center position. The motion of the iron rod can therefore be used to operate any number of mechanical devices such as those found on pinball machines or solenoid valves.

Solenoid Construction

Eddy Currents. Although the construction of a solenoid appears to be quite simple, certain points in the construction are worth noting. One of the first things to note is that in AC solenoids, the magnetic assembly and armature are not solid pieces of metal but rather a number of thin pieces laminated together. The reason these thin sheets are used in AC solenoids is to reduce eddy currents produced by transformed action in the metal. If solid cores were used, as in Figure 5-9, left, eddy currents would be generated in the solid core when used with AC and create heat build up. Through the use of laminated thin sheets of metal in the core, eddy currents are confined to each lamination, thus reducing the intensity of the magnetic effect and subsequent heat build up. (Figure 5-9, right.) For DC solenoids a solid core is acceptable, since the current is in one direction and continuous.

Armature Air Gap. When solenoids are designed, it is important that the armature be attracted to its sealed-in position in such a way that it completes the magnetic circuit as com-

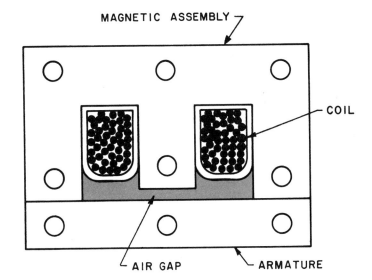

MAGNETIC ASSEMBLY

COIL

AIR GAP   ARMATURE

Figure 5-10 A small air gap is always left in the magnetic assembly to break the magnetic field and allow the armature to drop away freely after being de-energized.

pletely as possible in order to avoid chattering. To assure this, both the faces on the magnetic assembly and those on the armature must be machined flat to a very close tolerance.

These closely machined surfaces, however, could lead to a problem when the solenoid armature is sealed in. As the coil is de-energized, some magnetic flux (residual magnetism) always remains and could be enough to hold the armature in the sealed position. To eliminate this possibility, a small air gap is always left in the iron circuit to break the magnetic field and allow the armature to drop away freely after being de-energized (Figure 5-10).

Shading Coil. A shading coil (Figure 5-11) is a single turn of conducting material (generally copper or aluminum) mounted on the face of the magnetic assembly or armature. The purpose of the shading coil is to set up an auxiliary magnetic attraction which is out of phase (out of time) from the main coil attraction such that it will help hold in the armature as the main magnetic coil attraction drops to zero in an AC circuit. Figure 5-12 illustrates the benefit of the shading coil. With AC current the magnetic flux periodically drops to zero and the armature would have a tendency to drop out, or chatter. The out of phase attraction of the shading coil adds enough pull to the unit to keep the armature firmly seated. Without the shading coil, excessive noise, wear and heat would build up on the armature faces, reducing the armature's life expectancy.

SOLENOID CHARACTERISTICS

Coils

As described earlier, magnetic coils are generally constructed of many turns of insulated copper wire wound on a spool. The mechanical life of most coils today is improved by encapsulating them in an epoxy resin or glass-reinforced alkyd material (Figure 5-13). In addition to increasing mechanical strength, these materials greatly increase the moisture resistance of the magnetic coil. Since these devices are encapsulated and cannot be repaired when they go bad, it is more practical to replace coils than to repair them.

Coil Inrush and Sealed Currents. Solenoid coils, like motors, tend to draw more current when first energized than is required to keep them running. (Figure 5-14.) In a solenoid coil, the inrush current will be approximately six to ten times the sealed current. After the solenoid has been energized for some time, the coil will become hot, causing the coil current to fall again and stabilize at approximately 80% of its value when cold. The reason for such a high inrush current is the fact that the basic opposition to current flow when a solenoid is energized is only the resistance of the copper coil. Upon energizing, however, the armature begins to move iron into the core of the coil.

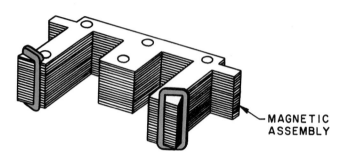

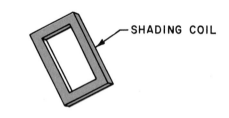

Figure 5-11 A shading coil is mounted on the face of the magnetic assembly to set up an auxiliary magnetic attraction out of phase with the main field to help hold the armature in place when AC current passes through.

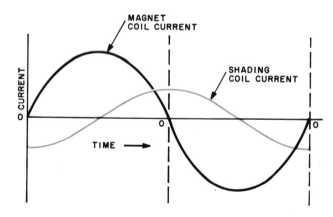

Figure 5-12 The current and magnetic field of the shading coil is out of time (phase) with the magnet coil current and magnet coil magnetic field.

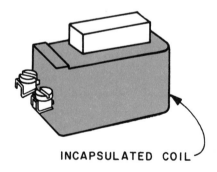

Figure 5-13 The mechanical life of most coils is improved by incapsulating them in an epoxy resin or glass-reinforced alkyd material. In addition to increasing mechanical strength, these materials greatly increase moisture resistance of the magnetic coil.

This large amount of iron in the magnetic circuit greatly increases the impedance (alternating current opposition to current flow) of the coil and thus decreases the current through the coil. The heating effect of the coil further reduces current flow because copper wire when heated increases in resistance, limiting some current flow. This high inrush current is worth noting, especially when several solenoids may be energized at the same time on a common power supply.

Coil Inrush and Sealed Current Ratings. Magnetic coil data is usually given in volt amperes (volt times amperes) (VA). For example, given a solenoid whose coil is rated at 600VA inrush and 60VA sealed, the inrush current of the 120V coil is 600/120, or 5 amps, and the sealed current is 60/120, or 0.5 amperes. The same solenoid with a 480V coil will only draw 600/480, or 1.25 ampere inrush, and 60/480 or 0.125 amperes sealed. The VA rating can be very helpful in determining the starting and energized current load drawn from the supply line.

Coil Voltage Characteristics. Since the voltage applied to a coil determines what the operating current will be, certain voltage terms are used to explain the way in which the coil interacts with the armature. Pick-up voltage is minimum control voltage which will cause the armature to start to move. The seal-in voltage is the minimum control voltage required to cause the armature to seat against the pole faces of the magnet. The drop-out voltage is the voltage which exists when the voltage has reduced sufficiently enough to allow the solenoid to open. Although most solenoids will have a seal-in voltage that is less than the pick-up voltage, there are some exceptions. The bell-crank armature solenoid and magnetic assembly is one such exception. The bell-crank armature assembly has characteristics which allow its design to have a lower seal-in voltage than pick-up voltage. This can be a decided advantage in certain situations, since the device will definitely seal in if it has enough voltage to be picked up.

Effects of Voltage Variations

High Voltage. If the voltage applied to the coil is too high, the coil will draw more than its desired or rated current. Excessive heat will then be produced and will cause early failure of the coil insulation. The magnetic pull will also be too high and will cause the armature to slam in with excessive force, causing the magnetic faces to wear rapidly and reducing the expected life of the solenoid.

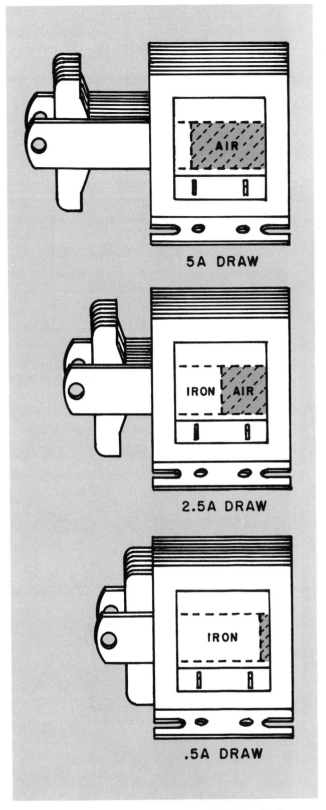

5A DRAW

2.5A DRAW

.5A DRAW

Figure 5-14 Inrush current to a solenoid is 6 to 10 times the sealed current. The movement of the steel armature into the solenoid greatly increases the impedance (AC opposition to current flow) of the solenoid and substantially reduces the amount of current drain.

ORDERING INFORMATION 1/2 INCH PULL

	Quiet Force Seated (Lbs)		Plunger Weight (Lbs)	Shipping Weight (Lbs)	Volt-Amps 100% Voltage Seated	1/2 INCH MAXIMUM STROKE		
	85% Voltage	100% Voltage				Force in Lbs Horizontal 85% Voltage	Volt-Amps 100% Voltage	Duty Cycle 50% Time On (Ops/min)
A 100	7	9	.2	1.3	40	2.7	230	240
A 101	9	12	.3	1.5	50	4.0	322	190
A 102	11	15	.3	1.7	50	4.7	420	180
B 100	11	15	.4	2.3	60	6.2	520	200
B 101	13	18	.5	2.6	70	9.6	790	109

Figure 5-15 Manufacturers provide specification charts like the one shown to help determine the amount of force created by each type of solenoid.

FORCE AND CURRENT CURVES

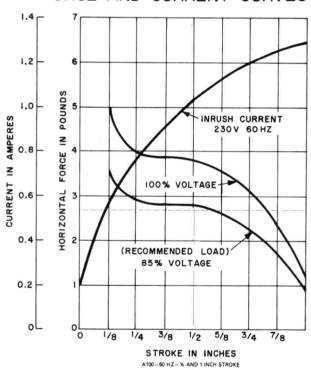

STROKE IN INCHES

A100—60 HZ—½ AND 1 INCH STROKE

Figure 5-16 Manufacturers provide specification curves like the one shown to help determine the overall operating characteristics of a solenoid.

Low Voltage. Low voltage on the coil will produce low coil currents and reduce magnetic pull. When the voltage is greater than the pick-up voltage but less than the seal-in voltage, the solenoid may pick up but will not seal. As the coil is not designed to carry continuously, the greater pick-up current (remember that pick-up current is six to ten times sealed current) will quickly get hot and burn out. The armature will also chatter—this would, in addition to creating a great deal of noise, increase the wear of the magnetic faces.

SELECTING THE PROPER SOLENOID

Basic Rules of Solenoid Application

In replacement or modification of a system using solenoids, you should always strive to get the best results from solenoid application. To do so you must consider the loading conditions and select the solenoid which gives the optimum performance. As a guide, the following four rules of good solenoid application should be considered.

1. Obtain complete data on load requirements. Both the ultimate life of the solenoid and the life of its linkage depend upon the loading of the solenoid. If overloaded, the solenoid may not seal correctly, resulting in coil noise, overheating and eventual burn out. Therefore, an accurate estimate of the required force at specific inching strokes (pounds load *vs.* inches travelled) should be made.

2. Allow for possible low voltage conditions of the power supply. Since the pull of the solenoid varies as the square of the voltage (4 pounds at 10 volts, 16 pounds at 20 volts) some allowance must be made for low voltage condition of the power supply. It is recommended that the solenoid be applied in accordance with recommended load. This rating is based on the amount of force the solenoid can develop with 85% of the rated voltage applied to the coil.

3. Use shortest possible stroke. Length of stroke is very important to maximize solenoid performance. First, shorter strokes mean faster operating rates. Shorter strokes also require less power, and this decreases coil heating. Any decrease in heating will increase the life expectancy of the coil. Second, more force is available at shorter strokes. This allows a smaller, lower rated, lower cost solenoid to be used. Third, less destructive mechanical energy is usually available from shorter strokes. This decrease in destructive energy, or impact force will help to reduce solenoid wear.

4. Don't use an oversized solenoid. Use of an oversized solenoid is inefficient, resulting in higher initial cost, a physically larger unit and greater power consumption. Since energy produced by a solenoid is constant regardless of the load, any energy not expended in useful work must be absorbed by the solenoid in the form of impact force. This results in reduced mechanical life and subjects the linkage mechanism to unnecessary strain.

Method of Determining Proper Solenoid

Push or Pull. A solenoid may push or pull, depending upon application. In the case of the door latch, the unit must *pull*. In a clamping jig, the unit must *push*.

Length of Stroke. After you have determined whether the solenoid must push or pull, you need to determine the length of the stroke. In the case of a door latch, let us assume that one-half inch would be the maximum stroke length.

Required Force. Next, the force required to perform this operation must be determined. As-

suming the force has been calculated to be 2.7 pounds, we can refer to manufacturer's specifications similar to those of Figure 5-15 to find the proper solenoid force or one close to it. From the column headed "Force in lbs." at 85% voltage (received load), find the assumed force or the nearest higher acceptable force. In this case for 2.7 pounds force in a horizontal position, the A100 solenoid is acceptable.

Uniform Force Curve. Next, you must check the force curve for the solenoid (Figure 5-16) to see that the solenoid meets the load throughout its length of travel. In this case we check our solenoid against the force curve entitled A100 60 Hertz (Hz) 1/2 and 1-inch stroke. Since the solenoid will give us at least 2.7 pounds throughout the specified stroke, it is acceptable.

Duty Cycle. Check the duty cycle requirements of your application against the duty cycle information given for your solenoid in the table under duty cycle (Figure 5-15). In this case, the duty cycle is 240 operations per minute, which is sufficient for this application.

Type of Mounting. Specify the type of mounting required (Figure 5-17). For the door latch, we want an end-mounting solenoid, so we would use sample A.

Voltage Rating. Assume we specify the 60 Hertz (Hz) coil voltage required from Figure 5-18. For a 115V coil we would select sample 2A. The

```
A - END MOUNTING
B - RIGHT SIDE MOUNTING
C - THROAT MOUNTING
D - NO MOUNTING (FOR THRU-BOLTS)
E - LEFT SIDE MOUNTING
F - BOTH SIDE MOUNTING
```

Figure 5-17 Manufacturers provide letter or number codes to indicate the type of mounting the solenoid has.

NO.	VOLTS
2 A	115
3 A	230
4 A	460
5 A	575

Figure 5-18 Manufacturers provide letter or number codes to indicate what voltages are available for a given solenoid.

```
LINEAR SOLENOID DESIGN DATA SHEET

(1) TYPE: PUSH ___   PULL ___

(2) STROKE: _____ INCHES

(3) FORCE: _____ LBS AT START _____ LBS AT END OF STROKE (FOR HOLD UNIT)

(4) DUTY CYCLE: CONTINUOUS ___   INTERMITTENT ___

(5) TYPE OF MOUNTING: HORIZONTAL ___   VERTICAL ___

(6) CYCLES PER SECOND: _____

(7) VOLTAGE: D.C. ___ MIN ___ MAX                    SKETCH

            A.C. ___ MIN ___ MAX

(8) AMBIENT TEMPERATURE: _____

(9) BODY SIZE: LENGTH _____

             DIAMETER _____



(10) TYPE OF ELECTRICAL CONNECTION:

    _____

(11) ENVIRONMENTAL CONDITIONS

    DUST ___   WATER ___   OIL ____

    OTHER _____


DESCRIPTION OF SOLENOID APPLICATION CONSIDERED _____
_____
_____
_____
```

Figure 5-19 Most manufacturers will, upon request, provide a specification ordering sheet like that shown. The more complete this sheet is, the better the chance of receiving the proper solenoid for a given application.

final solenoid selection, then, based on performance, would be an A100A2A solenoid.

Other Considerations. Refer to Figure 5-19 for other additional background information which may be helpful in obtaining the proper solenoid in other situations.

SOLENOID APPLICATIONS

In the next few paragraphs, we intend to show some of the most typical uses of solenoids in commercial and industrial use. It should be apparent just from these few examples what enormous possibilities solenoids have for control circuit applications.

Hydraulics/Pneumatics

Although hydraulic and pneumatic equipment can be switched through the use of fluids or air, it is often more convenient for it to be done by electrical solenoids. Figure 5-20 depicts the action of a two-position, double solenoid valve controlling a standard hydraulic cylinder. When pushbutton PB1 is momentarily depressed, the valve spool in the solenoid valve is shifted so that the fluid is directed to the back side of the cylinder piston, causing it to move forward. The piston will continue to travel to its full extension or until it stalls against the work it is performing.

When PB2 is momentarily depressed, the valve spool in the solenoid valve shifts to the opposite direction so that the fluid is directed to the front side of the cylinder, causing the piston to move backward. (*Note:* Coils and buttons should be wired so that they are interlocked. Failure to do so is a common error, resulting in coil burn out.)

With this type of circuit, reversal can be made to take place at any point in the piston cycle with the piston still moving. The unit cannot, however, be stopped in either direction part-way through a stroke and held there.

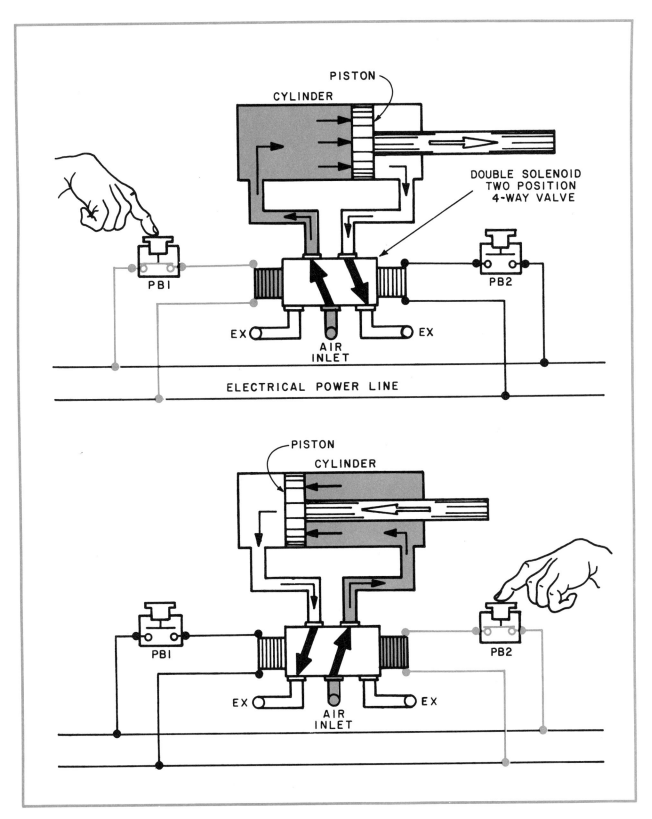

Figure 5-20 Two-position double solenoid valves can be used to electrically control a standard hydraulic or pneumatic cylinder. As shown at top of this figure, cylinder extends when pushbutton at left is pressed. As shown at bottom, cylinder retracts when pushbutton at right is pressed.

Refrigeration

In refrigeration equipment, it is common to find direct acting two-way valves like the one in Figure 5-21. Two-way (shut-off) valves have one inlet and one outlet pipe connection. These units may be constructed normally closed (NC), where the valve would be closed when de-energized and open when energized; or they may be constructed normally open (NO), where the valve would be open when de-energized and closed when energized.

Figure 5-22 illustrates the number and type of solenoids in a typical refrigeration system. In the case of the liquid stop and suction gas solenoid applications, the solenoid valves could be operated by two-wire or three-wire thermostats. In the case of the hot gas defrost solenoid valve, it remains closed until the defrost cycle and then feeds the evaporator with hot gases for the defrosting operation.

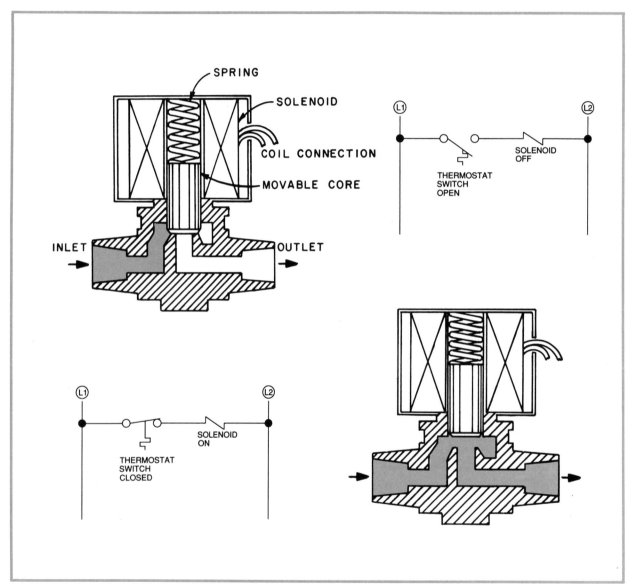

Figure 5-21 In refrigeration equipment, it is common to find direct acting two-way valves like the one shown. These units can be used to electrically open and close lines in a refrigeration system through the use of thermostats and other control devices.

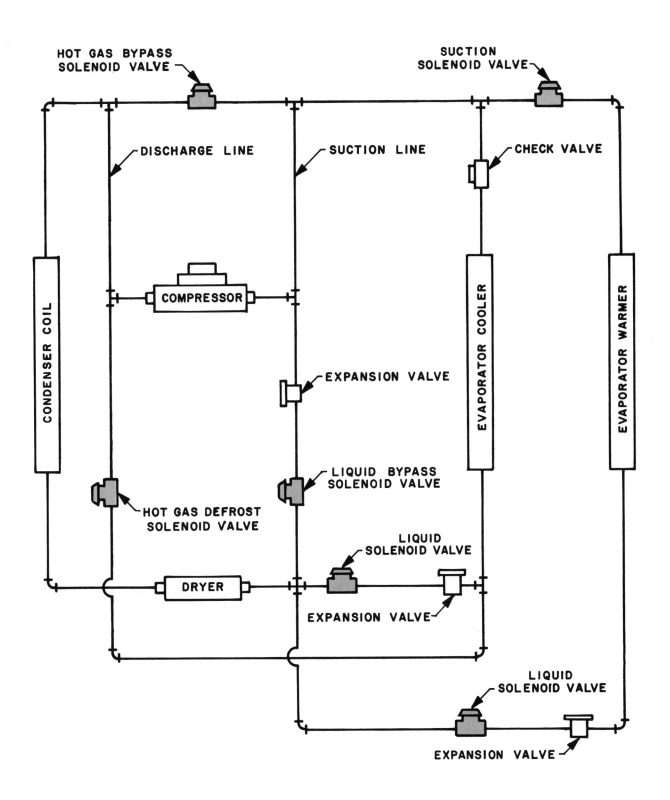

HOT GAS BYPASS
SOLENOID VALVE

SUCTION
SOLENOID VALVE

DISCHARGE LINE

SUCTION LINE

CHECK VALVE

CONDENSER COIL

COMPRESSOR

EVAPORATOR COOLER

EVAPORATOR WARMER

EXPANSION VALVE

LIQUID BYPASS
SOLENOID VALVE

HOT GAS DEFROST
SOLENOID VALVE

LIQUID
SOLENOID VALVE

DRYER

EXPANSION VALVE

LIQUID
SOLENOID VALVE

EXPANSION VALVE

Figure 5-22 Typical refrigeration system illustrating the multiple use of solenoid valves.

Combustion

Figure 5-23 illustrates the number and type of solenoids in an oil-fired singleburner system. Note that solenoids are crucial in the start up and normal operating functions of this system.

Safety

Figure 5-24 illustrates how a solenoid can be attached to a manual starter to provide low voltage protection. When small machine safety is critical, this type of device is the logical answer. Low voltage protection is accomplished by a continuous duty solenoid which is energized whenever the line-side voltage is present. When the line voltage is lost, the solenoid de-energizes, opening the starter contacts. The contacts will not automatically close when the line voltage is restored because of a latching mechanism. To close the contacts, the device must be manually reset.

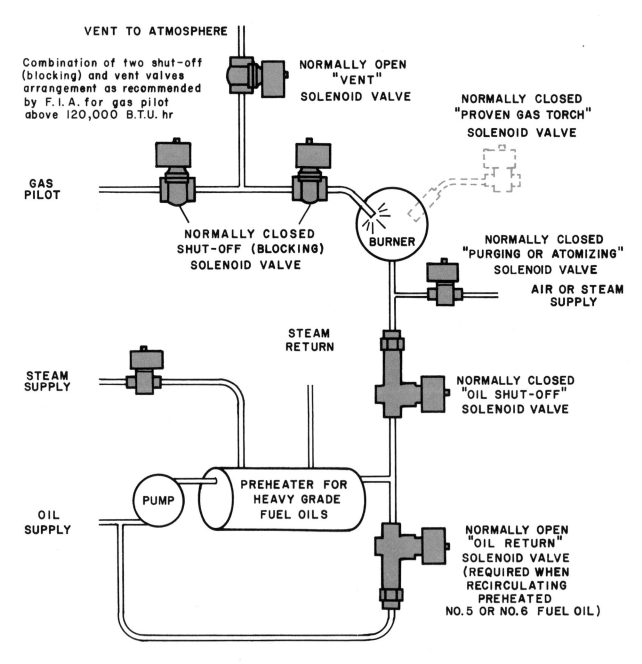

Figure 5-23 Oil-fired combustion system illustrating the multiple use of solenoid valves.

General Purpose

In addition to commercial and industrial use, you will find solenoids for general purpose application like those depicted in Figure 5-25; namely, business machines, photography and aircraft. You need only extend your imagination to determine how many and what type of solenoids are used in these devices.

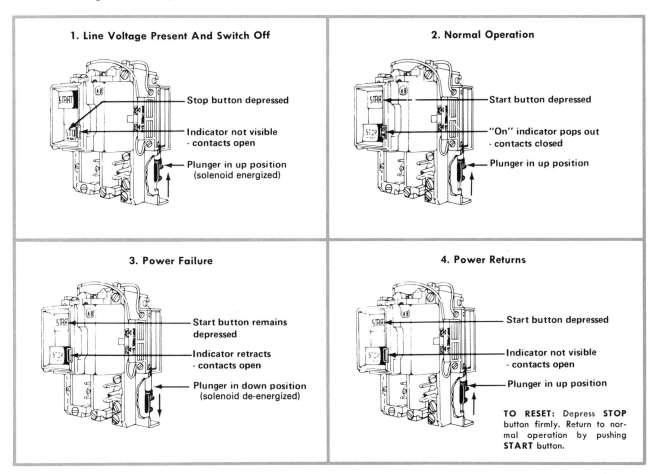

1. Line Voltage Present And Switch Off

Stop button depressed

Indicator not visible - contacts open

Plunger in up position (solenoid energized)

2. Normal Operation

Start button depressed

"On" indicator pops out - contacts closed

Plunger in up position

3. Power Failure

Start button remains depressed

Indicator retracts - contacts open

Plunger in down position (solenoid de-energized)

4. Power Returns

Start button depressed

Indicator not visible - contacts open

Plunger in up position

TO RESET: Depress **STOP** button firmly. Return to normal operation by pushing **START** button.

Figure 5-24 Sequence of operation illustrating how a solenoid can be added to a manual starter to provide low voltage protection.

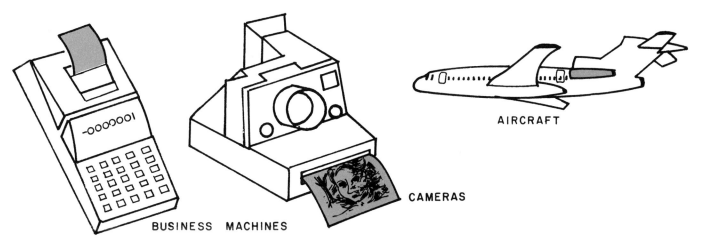

BUSINESS MACHINES

CAMERAS

AIRCRAFT

Figure 5-25 Solenoids have found enormous applications as general purpose solenoids in business machines, photography and aircraft systems.

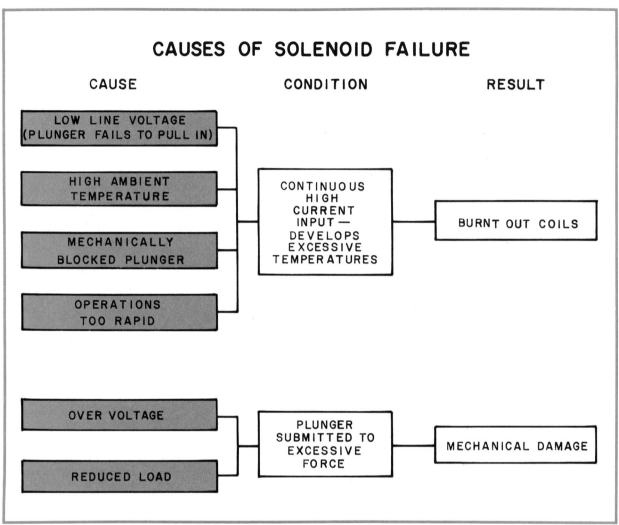

Figure 5-26 Block diagram for troubleshooting solenoids when they become defective or noisy.

TROUBLESHOOTING MAGNETIC SOLENOIDS

General Troubleshooting

Troubleshooting magnetic solenoids is relatively easy because the two major problems which are possible revolve around the coil and its mechanical linkage. Figure 5-26 illustrates in block diagram a logical technique for isolating most of these problems.

AC Hum

All AC devices which incorporate a magnetic coil produce a characteristic hum. This hum or noise is due mainly to the changing magnetic field, inducing mechanical vibration. Solenoids could become excessively noisy as a result of some of the following operating conditions:

1. Broken shading coil
2. Operating voltage too low
3. Wrong coil
4. Misalignment between the armature and magnetic assembly so that the armature is then unable to seat properly.

1. What are the two main types of magnets?
2. Why is soft iron more easily magnetized than hard steel?
3. How can the strength of an electromagnet be increased?
4. What are the different types of solenoids used to transmit force?
5. Why is the armature of the solenoid made from a number of thin laminated pieces instead of a solid piece?
6. What is the purpose of having a small air gap on the armature face?
7. What function does the shading coil perform?
8. Why does a solenoid coil have a much higher inrush current than sealed current?
9. What is the effect of a higher-than-rated voltage applied to the solenoid coil?
10. What is the effect of a lower-than-rated voltage applied to the solenoid coil?
11. What are the four basic rules to follow when selecting the proper solenoid?
12. What are some of the requirements that must be determined when selecting the proper solenoid?
13. What are some common applications of solenoids?
14. What are some of the problems that may develop in a solenoid?

Typical Application of Solenoids Used in an Industrial Process

FLOW SWITCH MIX TANK PUMP

FLOW SWITCH

FILL TANK PUMP M2

1ST CONVEYOR

CAN

PROXIMITY SWITCH TO ACTIVATE FILL CYLINDER

SOLENOID ACTIVATED TO PUSH CAN OFF WHEEL

WHEEL ROTATES TO FILL CAN

CHAMBER FOR CAN

SOLENOID

CAN

FILL CYLINDER

PROXIMITY SWITCH

METAL TO ACTIVATE

CAN SPACER

CAN LIDDER

CAN LID SEALER

2ND CONVEYOR

IN SOLENOID

TURN ON SOLENOID

OUT SOLENOID

PNEUMATIC CIRCUIT FOR CAN FILL CYLINDER

COUNTER

PROXIMITY SWITCH AND COUNTER WITH SOLENOID USED TO INSPECT EVERY 100TH CAN

MIX TANK

2 SPEED MIX MOTOR

OVERFLOW PHOTO ELECTRIC EYE

FILL TANK

UPPER LEVEL PHOTO ELECTRIC EYE

LOWER LEVEL PHOTO ELECTRIC EYE

PAINT MIXING AND CAN FILLING PROCESS

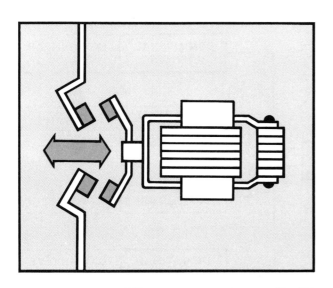

6 AC/DC Contactors and Magnetic Motor Starters

Electromagnets applied to contactors can be used to make magnetic contactors and magnetic motor starters. Magnetic contactors and magnetic motor starters can be electrically controlled from remote places and are available in AC or DC power. When large contactors are utilized, some type of arc suppression is required to keep the contactors in the device from burning up.

All magnetic motor starters require overload protection. In all types of overload protection, a signal is sent back to the magnetic starter to cause it to open and shut down the motor. Magnetic motor starters are available in a variety of sizes, poles, voltages, and phases.

CONTACTORS

Types of Contactors

Contactors have been defined by the National Electrical Manufacturers Association (NEMA) as devices for repeatedly establishing and interrupting an electrical power circuit. Contactors are used to make and break the electrical power circuit to such loads as lights, heaters, transformers and capacitors (Figure 6-1).

Two types of contactors are defined by the National Electrical Manufacturers Association (NEMA): electronic and magnetic. In this unit we will limit our discussion to magnetic contactors which are activated by electromechanical means (solenoid action). A future unit will deal exclusively with electronic solid state control devices.

Contactor Construction

Figure 6-2 illustrates the construction of three commonly used magnetic contactor assemblies. It should be immediately apparent that the solenoid action studied in Unit 5 is the principal operating mechanism for a magnetic contactor. Instead of pushing and pulling levers and valves, however, the linear action of the solenoid is now being used to open and close sets of contacts. The use of solenoid action rather than manual input is a definite advantage of the magnetic contactor over the manual contactor. With magnetic contactors and their associated electrical circuits, remote control and automation can be designed into a system which would be impossible with a manual contactor.

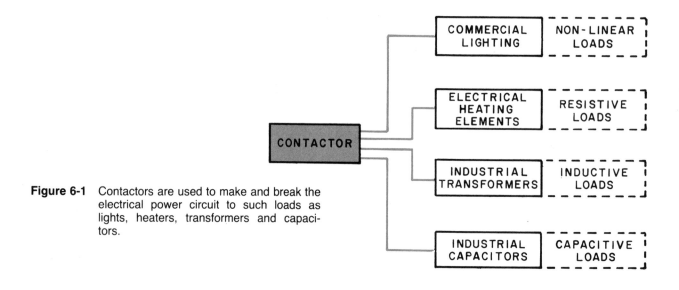

Figure 6-1 Contactors are used to make and break the electrical power circuit to such loads as lights, heaters, transformers and capacitors.

Contactor Wiring

Over a period of years it has become commonplace to refer to certain control circuits by the number of conductors used in the *control circuit;* namely, two-wire and three-wire control. In the next few paragraphs we will describe characteristics which are often attributed to these names but caution against the widespread use of these names to describe other than circuits in their most simple forms. The applied logic function approach covered in Unit 3 is a more accurate description for the broad range of circuitry you will encounter.

Two-Wire Control. In the line diagram of Figure 6-3, two wires connect the control device contacts to the coil of a magnetic contactor, indicating the use of two-wire control. (*Note:* These contacts could be a thermostat float switch or other maintained contact device.) When the contacts of the control device close, they complete the coil circuit of the contactor, causing it to be energized, and connect the load to the line through the power contacts. When the control device contacts open, the contactor coil is de-energized, opening the contacts, thus stopping the load. Wired as illustrated, the contactor will function automatically in response to the condition of the control device without the attention of an operator.

A two-wire control circuit provides low voltage release but not low voltage protection. This means that in the event of a power loss in the control circuit, the contactor will de-energize (low voltage release); but it will also re-energize if the control device is still closed when the circuit has power restored. Low voltage protection is not provided

for in this circuit because there is no way for the operator to be protected from the circuit once it has been re-energized.

Caution must be exercised in the use and service of two-wire control circuits because of this lack of protection. Generally speaking, two-wire control is used only for remote or inaccessible installations such as pumping stations, water or sewage treatment, air conditioning or refrigeration systems and process line pumps where an immediate return to service after a power failure is essential.

Three-Wire Control (Memory). Three-wire control is the MEMORY circuit discussed in Unit 3. The three-wire control circuit illustrated in Figure 6-4 uses a momentary contact stop button wired in series with a momentary contact start button which in turn is parallel wired to a set of contacts which form a holding circuit interlock (memory). When the NO start button is depressed, current flows through the stop button (NC) through the momentarily closed start, through the coil and overloads to L2, causing the magnetic coil to energize. Upon energizing, the auxiliary holding circuit interlock contacts (MEMORY) close, sealing the path through to the coil circuit even if the start button is released.

Pressing the NC stop button will open the circuit to the coil, causing the contactor to de-energize. A power failure will also de-energize the contactor. When the contactor de-energizes, the interlock (MEMORY) contactor re-opens, and both current paths to the coil, through the start button and the interlock, are now open.

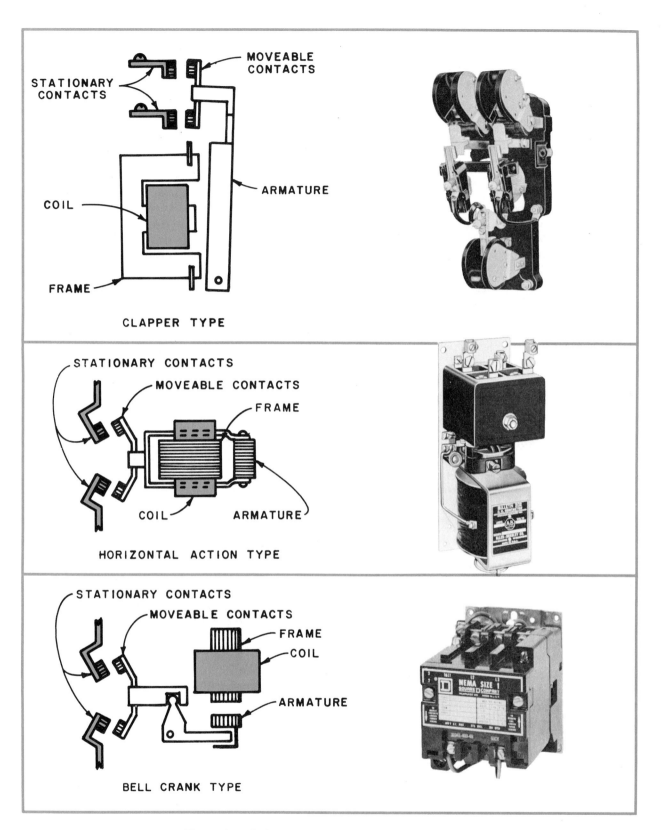

STATIONARY
CONTACTS

MOVEABLE
CONTACTS

ARMATURE

COIL

FRAME

CLAPPER TYPE

STATIONARY CONTACTS

MOVEABLE CONTACTS

FRAME

COIL   ARMATURE

HORIZONTAL ACTION TYPE

STATIONARY CONTACTS

MOVEABLE CONTACTS

FRAME

COIL

ARMATURE

BELL CRANK TYPE

Figure 6-2 Different types of magnetic contactor assemblies are used with a variety of manufacturers. (Cutler Hammer, Allen-Bradley Company, Square D Company)

Three-wire control will provide both low voltage release and low voltage protection. The coil will not only drop out at low or no voltage, but also cannot be reset unless the voltage returns and the operator again depresses the start button. Since three wires from the pushbutton station are connected into the contactor at points L1, L2 and L3, this wiring scheme is commonly referred to as three-wire control.

AC vs DC Contactors

Although AC and DC contactors both operate on the same principle of solenoid action, there are some major differences.

A DC solenoid contact assembly may have only one set of contacts, while an AC contactor may have several sets of contacts (Figure 6-5). In a DC contactor it is necessary to break only one side of the line, while in three-phase AC circuits you would be required to break all three lines, creating the need for several sets of contacts. For multiple contact control, the T-bar assembly of Figure 6-5 allows several sets of contacts to be activated simultaneously.

Another distinguishing difference between DC and AC contactors is their magnetic assembly. In a DC contactor the magnetic assembly is made of solid steel. Laminations are unnecessary in a DC coil since the current is traveling in one direction at a continuous rate and does not create eddy current problems.

The other major differences between AC and DC contactors are the electrical and mechanical requirements necessary for suppressing the arcs created in opening and closing contacts under load.

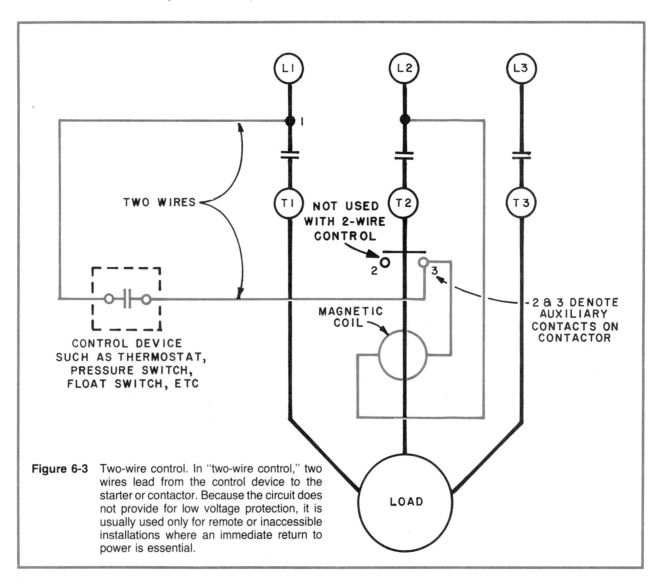

Figure 6-3 Two-wire control. In "two-wire control," two wires lead from the control device to the starter or contactor. Because the circuit does not provide for low voltage protection, it is usually used only for remote or inaccessible installations where an immediate return to power is essential.

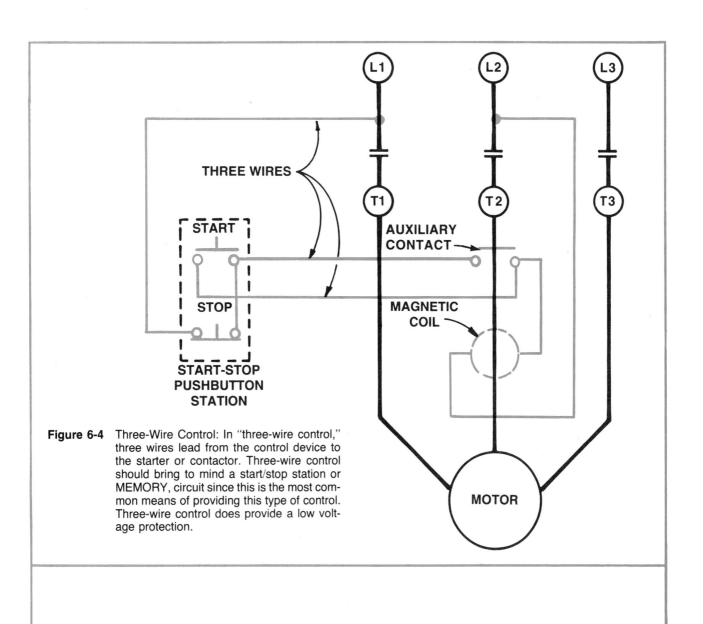

THREE WIRES

L1 L2 L3

T1 T2 T3

START

STOP

START-STOP
PUSHBUTTON
STATION

AUXILIARY
CONTACT

MAGNETIC
COIL

MOTOR

Figure 6-4 Three-Wire Control: In "three-wire control," three wires lead from the control device to the starter or contactor. Three-wire control should bring to mind a start/stop station or MEMORY, circuit since this is the most common means of providing this type of control. Three-wire control does provide a low voltage protection.

Figure 6-5 In a DC contactor assembly, you may find only one set of contacts (left), while AC contactors may have several sets of contacts (right). AC contactor assemblies are made of laminated steel while DC assemblies are solid due to the lack of eddy currents with continuous DC. (Allen-Bradley Company, Cutler Hammer)

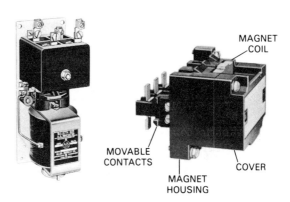

MAGNET
COIL

MOVABLE
CONTACTS

COVER

MAGNET
HOUSING

ARC SUPPRESSION

Opening Contact Arc

When a set of contacts are opened under load (Figure 6-6), there is a short period of time (a few thousandths of a second) during which the contacts are neither fully in touch with each other yet not completely separated. As they continue to separate, the contact surface area becomes less and less, increasing the electrical resistance. With full load current passing through this increasing resistance, a substantial temperature rise is created on the surface of the contacts. This temperature rise is often high enough to cause the contact surfaces to become molten and emit ions of vaporized metal into the gap between the contacts. Since this hot ionized vapor is obviously electrically conductive, it will permit the current to continue to flow in the form of an arc, even though the contacts are completely separated. These arcs in turn produce additional heat which, if continued, can be injurious to the contact surfaces.

Since prolonged arcing is considered harmful to contact surfaces, the sooner we can extinguish the arc, the longer the life expectancy of the contacts. Figure 6-7 illustrates the damage which may occur from prolonged arcing.

DC Arc Suppression

DC arcs are considered the most difficult to extinguish because the continuous DC supply causes current to flow constantly and with great stability across a much wider gap than does an AC supply of equal voltage.

To combat prolonged arcing in DC circuits, the switching mechanism must be such that the contacts will separate rapidly and with enough air gap to extinguish the arc as soon as possible on opening. It is also necessary in closing DC contacts to move the contacts together as quickly as possible to avoid some of the same problems encountered in opening them.

DC contactors, then, will be somewhat larger than AC contactors in order to allow for the additional air gap. They will also tend to have faster acting operating characteristics.

One disadvantage of rapid closing in a DC contactor is that the contacts must be buffered to eliminate contact bounce due to excessive closing forces. This bounce effect can be minimized through the use of certain types of solenoid action and through the use of springs attached under the contacts to absorb some of the shock.

AC Arc Suppression

It has been established that a DC arc will last until the contacts are drawn so far apart that the arc becomes unstable and is extinguished. An AC arc, on the other hand, tends to be self-extinguishing when a set of contacts is being opened. In contrast to a DC supply of constant voltage, the AC supply has a voltage which reverses its polarity every 1/120 of a second (Figure 6-8) when operated on a 60 Hertz (Hz) line frequency. In brief, this alternation means that the arc will have a maximum duration of no more than half a cycle. Furthermore, during any half cycle, the maximum arcing current is reached only once in that half cycle.

Since the AC arc is self-extinguishing, the contacts can be separated more slowly and the gap length may be shortened. This short gap keeps the voltage across the gap low and the arc energy low. With low gap energy, ionizing gases cool more rapidly, extinguishing the arc and making it dif-

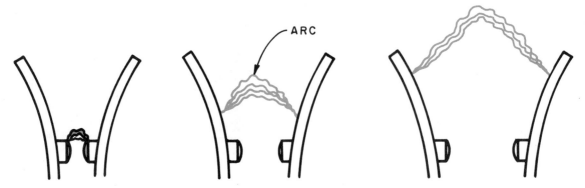

Figure 6-6 When a set of contacts is opened, an electrical arc will be created between the sets of contact tips as they separate.

ficult to restart. The obvious benefit of this is that AC contactors need less room for operation and will run cooler, increasing contact life.

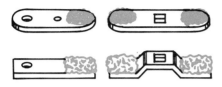

Figure 6-7 Prolonged arcing can result in damage to contact surfaces. The sooner we can extinguish the arc, the longer the life expectancy of the contacts.

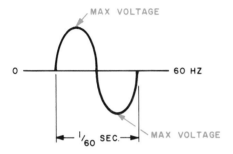

Figure 6-8 In contrast to a DC supply of constant voltage, the AC supply voltage reverses polarity every one-half cycle. Furthermore, during any one-half cycle, the maximum arcing current is reached only once in that half cycle.

Arc at Closing

Arcing may also occur on AC/DC contactors when they are closing. There are two ways in which the arc may develop. The most common arcing occurs when the contacts come close enough together that a voltage breakdown occurs and the arc is able to bridge the open space between the contacts.

The second type occurs if a whisker or rough edge of the contact touches first and melts, causing an ionized path which allows current to flow. In either case, the arc lasts until the contact surfaces are fully closed. Contactor design for both AC and DC devices are quite similar. The contactor should be designed so that the contacts will close as rapidly as possible without bouncing to minimize the arc at each closing.

AUXILIARY ARC SUPPRESSING TECHNIQUES

Arc Chutes

In addition to the mechanical movement of contacts, certain devices can be added to contactors—especially large contactors—to confine, divide and extinguish arcs drawn between contacts opened under load. Figure 6-9, left, illustrates how arc chutes are used to contain large arcs and the gases created by them. The arc chutes employ

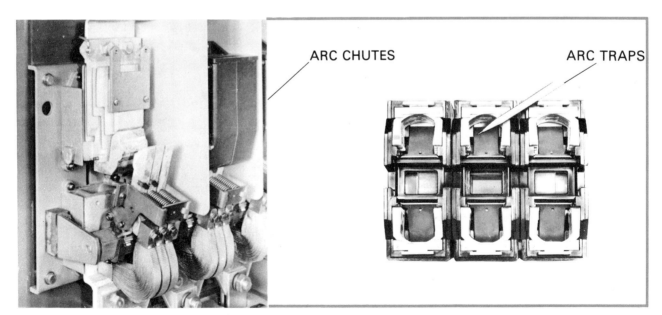

Figure 6-9 In addition to the mechanical movement of contacts, certain devices like those shown above can be added to contactors—especially large contactors—to confine, divide

and extinguish arcs drawn between contacts opened under load. (Cutler Hammer, General Electric)

the "de-ion" principle which confines, divides and extinguishes the arc for each set of contacts. Figure 6-9, right, employs another arc quenching technique by utilizing special arc traps and arc quenching compounds. This proven circuit breaker technique attracts, splits, and quickly cools arcs as well as venting ionized gasses. Vertical barriers between each set of contacts, plus arc covers, confine arcs to separate chambers and quickly quench them.

DC Magnetic Blow Out Coils

When a DC circuit carrying large amounts of current is interrupted, the collapsing magnetic field (flux) of the circuit current may induce a voltage which will help sustain the arc. Since a sustained electrical arc will either melt the contacts, weld them together, or severely damage them, some type of action must be taken to quickly limit the damaging effect of these heavy current arcs.

One way to stop the arc quickly would be to

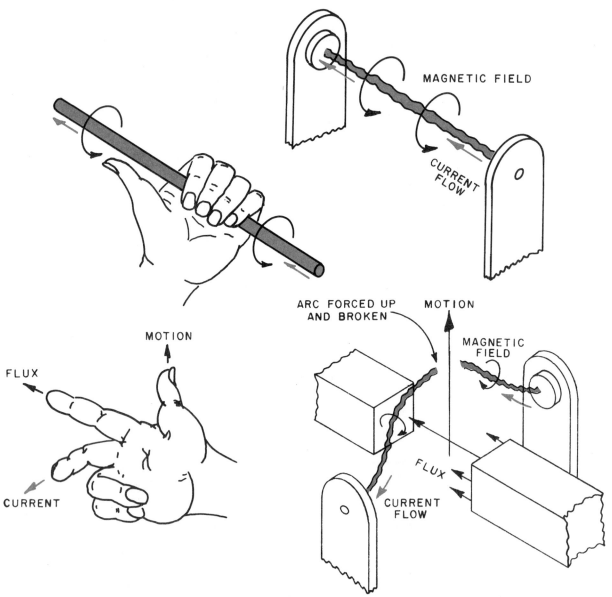

Figure 6-10 Top: The left-hand rule adapted for electron flow states: When the thumb on the left hand points in the direction of the current flow, the wrapping fingers will point in the direction of the resulting magnetic field. Bottom: The right-hand (motor) rule adapted for electron flow further states: When a current-carrying conductor (represented by the middle finger) is placed in a parallel magnetic field (represented by the "flux" finger), the resultant force or movement will be in the direction of the thumb. In this case, the arc is forced up, making it longer and easier to extinguish.

move the contacts some distance from each other as quickly as possible. The problem with this method, however, is that the contactor would have to be quite large to accommodate such a large air gap. To reduce distance and still quench arcs quickly, magnetic blowout coils were developed. Blowout coils provide a magnetic field which blows out the arc in much the same manner as you would blow out a match.

Figure 6-10 and Figure 6-11 illustrate the operating principle behind a magnetic blowout coil. Whenever a current flows through a conductive medium (in this case ionized air), a magnetic field is created around that current flow. The direction of the magnetic field (flux) around the conductor can be readily determined by the left-hand rule (adapted for electron flow), which states: When the thumb on the left hand points in the direction of the current flow, the wrapping fingers will point in the direction of the resulting magnetic flux. Since current is moving toward you in the illustration (Figure 6-10, top), the magnetic flux will be formed in a clockwise direction. The right-hand (motor) rule, adapted for electron flow, fur-

ther states that when a current carrying conductor (represented by the middle finger) is placed in a parallel magnetic field (represented by the "flux" finger), the resulting force or movement will be in the direction of the thumb (Figure 6-10, bottom). The reason for this action is the fact that the magnetic field around the current flow opposes the parallel magnetic field above the current flow, making the magnetic field weaker while at the same time aiding the magnetic field below the current flow, making the magnetic field stronger. The net result of this action is an upward push which very quickly elongates the arc current so that it will break. This electromagnetic blowout coil is sometimes called a "puffer" because of its blowout ability (Figure 6-11).

CONTACT CONSTRUCTION ON HIGH AMPERAGE CONTACTS

How contacts are designed and from what materials they are made depend upon the size, current rating and application of the contactor. When contactors use double-break contacts, they

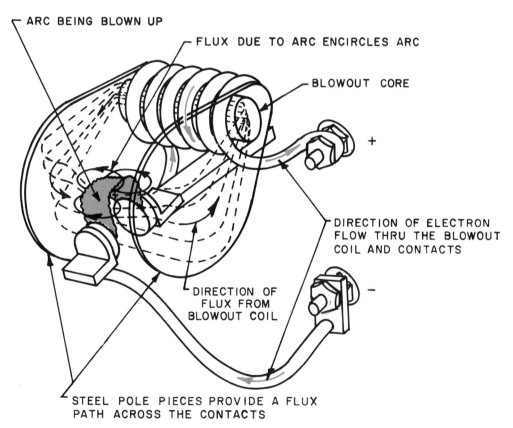

ARC BEING BLOWN UP

FLUX DUE TO ARC ENCIRCLES ARC

BLOWOUT CORE

+

DIRECTION OF ELECTRON FLOW THRU THE BLOWOUT COIL AND CONTACTS

DIRECTION OF FLUX FROM BLOWOUT COIL

−

STEEL POLE PIECES PROVIDE A FLUX PATH ACROSS THE CONTACTS

Figure 6-11 Electromagnetic blowout coils (sometimes called "puffers") can be used to rapidly extinguish a DC arc.

are usually made of a silver cadmium alloy. In large contactors with single break contacts (Figure 6-12), contacts are frequently made of copper because of lower cost.

Because of the nature of copper contacts, the single-break contacts are designed with a wiping-sliding action to remove the copper oxide film which forms on the copper tips. This wiping action is necessary because copper oxide formed on the contact when not in use is a nonconductor and must be eliminated for good circuit conductivity.

In most cases the slight rubbing action and burning that occur during normal operation will keep the contact surfaces clean enough for good service. Copper contacts that seldom open or close, however, or those being replaced, should be cleaned to reduce contact resistance, which is often the cause of serious heating of the contacts.

GENERAL PURPOSE AC/DC CONTACTOR SIZES AND RATINGS

Magnetic contactors, like manual starters, are rated according to the size and type of load by the National Electrical Manufacturers Association (NEMA). The tables of Figures 6-13 and 6-14 indicate the number size designation—00, 1, 2, 9, etc.—for general purpose AC and DC contactors

Figure 6-12 Arc chutes removed to illustrate the use of copper contacts on large contactors. (Cutler Hammer)

and establish the current load carried by *each* contact in the contactor. Note that the rating is for each contact individually, not for the whole contactor. In other words, a three-pole contactor rated at 18 amperes is actually capable and rated for switching three separate 18 ampere loads simultaneously.

		STANDARD NEMA RATINGS OF AC CONTACTORS, 60HZ				
	8-HR OPEN RATING (A)	POWER (HP)				
SIZE		THREE PHASE			SINGLE PHASE	
		200 V	230 V	230/460 V	115 V	230 V
00	9	1½	1½	2	⅓	1
0	18	3	3	5	1	2
1	27	7½	7½	10	2	3
2	45	10	15	25	3	7½
3	90	25	30	50	-	-
4	135	40	50	100	-	-
5	270	75	100	200	-	-
6	540	150	200	400	-	-
7	810	-	300	600	-	-
8	1,215	-	450	900	-	-
9	2,250	-	800	1,600	-	-

Table based on ICS 2-321B-1 (Revised)

Figure 6-13 Size and ratings of AC contactors.

		STANDARD NEMA RATINGS OF DC CONTACTORS		
SIZE	8-HR OPEN RATING (A)	POWER RATING (HP)		
		115 V	230 V	550 V
1	25	3	5	-
2	50	5	10	20
3	100	10	25	50
4	150	20	40	75
5	300	40	75	150
6	600	75	150	300
7	900	110	225	450
8	1,350	175	350	700
9	2,500	300	600	1,200

Table based on ICS 2-211

Figure 6-14 Size and ratings of DC contactors.

To give some perspective of how greatly in size contactors vary from size 00 to size 9, Figures 6-15 illustrates pictorially the different physical sizes involved in going from the smallest to the largest

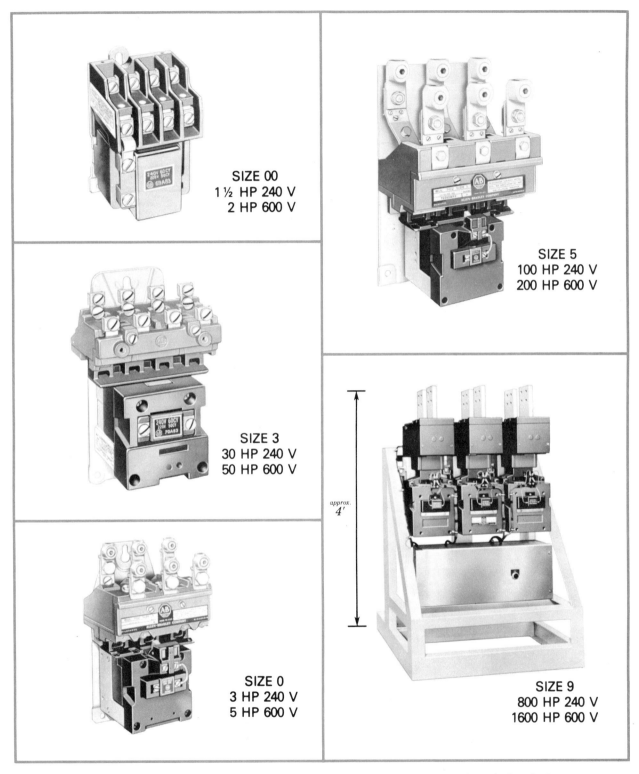

SIZE 00
1 ½ HP 240 V
2 HP 600 V

SIZE 3
30 HP 240 V
50 HP 600 V

SIZE 0
3 HP 240 V
5 HP 600 V

SIZE 5
100 HP 240 V
200 HP 600 V

approx. 4'

SIZE 9
800 HP 240 V
1600 HP 600 V

Figure 6-15 Different physical sizes involved in going from the smallest (size 00), which is approximately six inches, to the largest (size 9), which is approximately four feet, magnetic contactors. (Allen-Bradley Company)

contactor size. You will note that the range is from inches to several feet in length. For selection purposes, Figure 6-16 illustrates another chart with the type, size and voltage available for common magnetic contactors in a wide variety of enclosures.

INTRODUCTION TO MAGNETIC MOTOR STARTERS

Once you have a thorough grasp of the fundamentals of magnetic contactors, understanding magnetic motor starters requires only one more

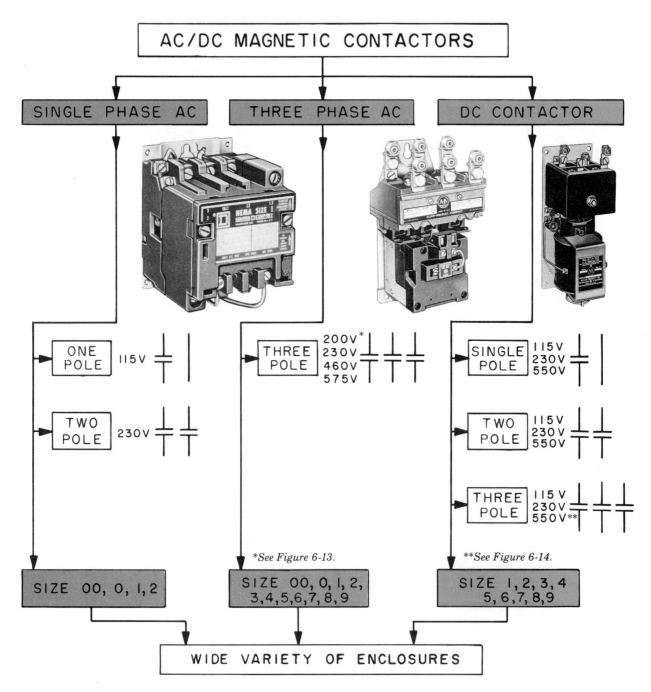

Figure 6-16 Block diagram illustrating the sizes, types and voltages available for common AC/DC contactors in a wide variety of enclosure types. (Allen-Bradley Company)

building block: the overload relay. Figure 6-17 indicates that a magnetic motor starter, like the manual starter, is no more than a contactor with an overload relay physically and electrically attached. Figure 6-18, top, illustrates a free standing overload relay, and Figure 6-18, bottom, illustrates an overload relay attached to a contactor to form a magnetic motor starter. Since overloads are so important to understanding the operation of a magnetic motor starter, we will review the melting alloy overload device mentioned in Unit 4 and introduce some other new overload devices:

Overload Protection

Melting Alloy Overload Relay. In Unit 4, "AC Manual Contactor and Motor Starters," we discussed in depth how a melting alloy overload relay operated. From Figure 6-18, it should be apparent that the sensing device for a manual starter and a magnetic starter are quite similar. The main difference between the two devices is that the manual overload actually opens the power contacts on the starter, while the overload device on a magnetic starter opens a set of contacts only to the magnetic coil, which de-energizes the coil, allowing the power to open. Selection of heater sizes for this device will be treated at the end of this unit.

Magnetic Overload Relay. Magnetic overload relays similar to that of Figure 6-19 provide another means of monitoring the amount of cur-

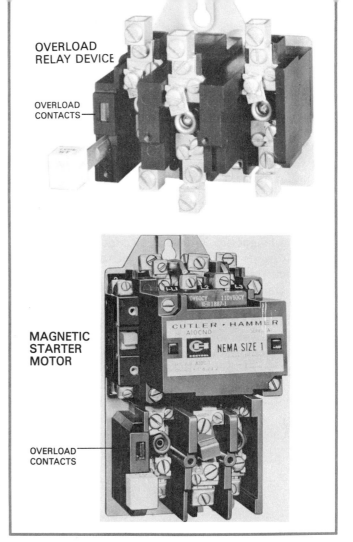

Figure 6-18 An overload device and a magnetic motor starter. (Cutler Hammer)

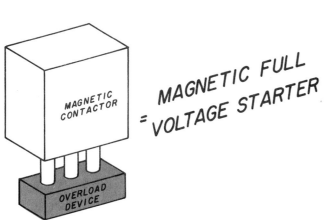

Figure 6-17 A magnetic motor starter is merely a contactor with an overload relay physically and electrically attached.

Figure 6-19 Magnetic overload relays provide another means of monitoring the amount of current drawn by a motor. (General Electric)

rent drawn by a motor. The device operates through the use of a current coil, as shown in Figure 6-20. At a specified overcurrent value, the current coil acts as a solenoid, causing a set of NC contacts to open, causing the circuit to open and thus protect the motor by disconnecting it.

The use of magnetic overload relays is found in special applications such as steel mill processing lines or other heavy duty industrial applications where holding a specified level of motor current is considered essential.

The magnetic overload is also ideal for special applications such as slow acceleration motors, high current inrush motors or any use where normal time current curves of thermal overload relays do not provide satisfactory operation. This flexibility is made possible because the magnetic unit can be set for either instantaneous or inverse time-tripping characteristics. The device may also offer independent adjustable trip time and trip current.

Another advantage of the magnetic overload is that it is extremely quick on reset, since it does not require a cooling off period before it can be reset. One disadvantage of the magnetic overload is that it is much more expensive than the thermal overload units.

Bi-Metallic Automatic Reset Overloads. In certain applications, such as walk-in meat coolers, remote pumping stations and some chemical process equipment, overload relays which can reset automatically to keep the unit operating up to the last possible moment may be required. One such device for this purpose is the bi-metallic overload relay. Bi-metallic overloads operate on the principle of the bi-metallic strip. The bi-metallic strip is made with two pieces of different metal. These dissimilar metals are then permanently joined by lamination. Because metals expand and contract at different rates, heating the bi-metallic strip will cause it to warp or curve. The warping effect of the bi-metallic strip can then be utilized as a trip lever or means for separating contacts (Figure 6-21). Once the tripping action has taken place, the bi-metallic strip will cool and reshape itself. In certain applications,

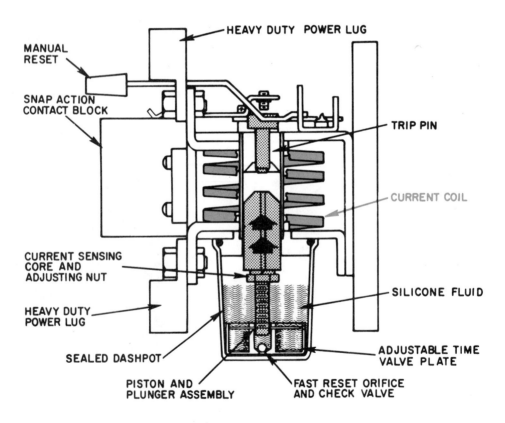

Figure 6-20 The heart of the magnetic overload relay is the current coil. At a specified overcurrent value, the current coil acts as a solenoid, causing a set of NC contacts to open. (General Electric)

such as circuit breakers, a trip lever will need to be reset to make the circuit operate again. In other applications, such as an automatic reset overload relay shown in Figure 6-22, the device will automatically reset the circuit when the bi-metallic strip cools and reshapes itself. With this type of device, the motor will restart even if the overload has not been cleared and will again trip and reset at given intervals. Care must be exercised in the selection of this type of overload, since repeated cycling will eventually burn out the motor. Note from the diagram of Figure 6-22 that the bi-metallic strip may often be shaped in the form of a U. The U-shape provides a more uniform temperature response.

Trip Indicators. Many overload devices like the automatic reset now have a trip indicator built into the unit to indicate to the operator that an overload has taken place within the device. Figure 6-23 indicates one manufacturer's method of illustrating visually the condition of the overload device. When the overload relay has tripped, a red metal indicator appears in a window located above the reset button. This red tag informs the operator or electrician both why the unit is not operating and that it potentially is capable of coming back on line with an automatic reset.

Overload Current Transformers for Large HP Motors. Large horsepower motors have currents that exceed the values of standard overload relays. To make the overload relays any larger would greatly increase their physical size and would become a real space problem in relation to

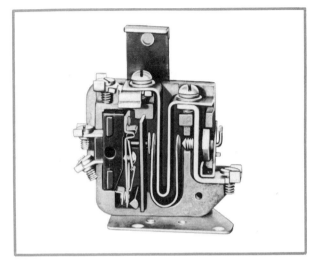

Figure 6-22 Cutaway view showing construction of bi-metallic overload relay. The bi-metallic strip in this case is U-shaped to provide uniform temperature response. (Square D Company)

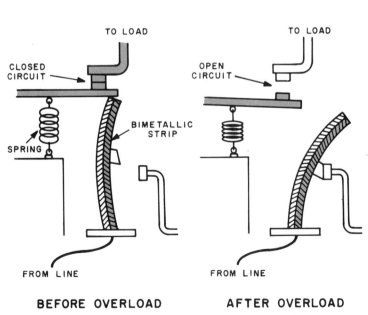

Figure 6-21 The warping effect of a bi-metallic strip can be utilized as a trip lever or means for separating contacts.

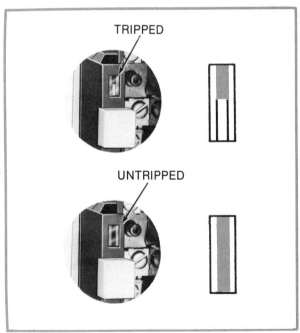

Figure 6-23 One manufacturer's method of indicating the condition of the overload relay. (Cutler Hammer)

the magnetic motor starter. To avoid such a conflict, current transformers similar to those in Figure 6-24 are used to reduce the current in a fixed ratio. In other words, the current transformer determines how much current is going to the motor but reduces that current to a lower value for the overload relay. For example, if 50 amperes were going to the motor, only 5 amperes may be going to the overload relay through the use of the current transformer. Standard current transformers are usually rated, giving both primary and secondary rated current, such as 50/5 or 100/5.

Because the ratio is always the same, in this case 50 to 5, an increase to the motor will also increase the current to the overload relay. If the right current transformer and overload relay combination are selected, the same overload protection can be provided to the motor as if the overload relay were actually in the load circuit. When an excessive current is sensed, the overload relay contacts open and the coil to the magnetic motor starter is de-energized, thereby shutting off the contactor containing the motor. Several different current transformer ratios are available and make this type of overload protection easy to provide.

Selection of Overload Heater Sizes

In installing magnetic motor starters, the selection of the proper heater size is probably one of the most important and most often overlooked parts. In the remainder of this unit we will outline some of the basic procedures used in the proper selection of overload heaters.

Motor Characteristics and Surroundings. When selecting thermal overload heaters, it is extremely important that each motor be sized according to its own unique operating characteristics and applications. Specifically, you must know the full load current rating (FLC), the service factor (SF) and ambient temperature (surrounding air temperature) of the motor when it is operating.

Full Load Current Rating (FLC). Selection of thermal heaters should always be based on the actual full load current shown on the motor nameplate or in the motor manufacturer's specification sheet (See Figure 6-25). These current figures reflect as closely as possible the currents to be expected when the motor is running at specified voltages, speeds and normal torque characteristics. Heater manufacturers have for your conve-

nience prepared current charts indicating which heater should be used with each (FLC) full load current. These charts will be discussed in detail further in the unit.

Service Factor. In practically every motor application, there are times when the motor must produce more than its rated horsepower for a short period of time without damage. The service

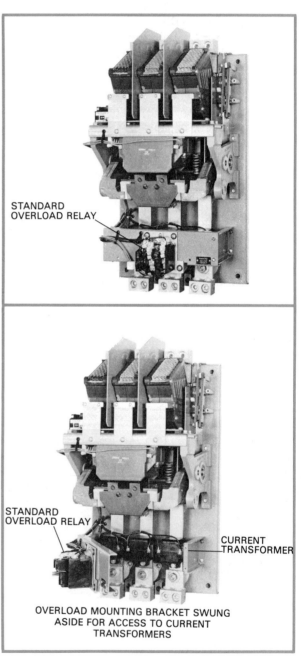

Figure 6-24 By using current transformers with specific reduction ratios, standard overload relays may be used on very large starters. (Cutler Hammer)

factor rating, then, is a number designation to tell the user what percentage of extra demand can be placed on the motor for short intervals without damaging the motor. Typical service factor numbers range from 1.00 to 1.25, indicating that the motor can produce 0% to 25% extra demand over that for which it is normally rated. Obviously a 1.00 service factor means the motor can produce no more power than it was rated for and to do so would result in damage.

In contrast, a service factor of 1.25 means that a motor can produce up to 25% more power than it was rated for within short periods of time.

Excessive Current Limitation. In terms of current, the excessive current which can be safely handled by any given motor for short periods of time is determined by multiplying the service factor (1.15) × actual full load current (FLC) to determine excessive current. For example, if a motor were rated at an FLC of 10 amps with a service factor of 1.15, the excess short term current would be 1.15 × 10 amps, or 11.5 amps. In other words, the motor could handle an additional 1.5 amps for a short period of time.

Ambient Temperature. In Unit 4 we learned that a thermal overload relay operated on the principle of heat. When an overload was taking place, sufficient heat was generated by the excessive current to melt the metal alloy and allow the device to trip. Since this type of relay is sensitive to any type of heat, we must also consider in some cases the temperature surrounding this device (ambient temperature). This concept may not

seem significant at first glance; however, it can definitely be a factor when you consider moving this control device from a refrigerated meat packing plant to a location near a blast furnace.

Overload relay devices are generally rated to trip at a specific current when surrounded by an ambient temperature of 40°C (104°F). This standard ambient temperature will be acceptable for most control applications; however, further in the unit we will show how to compensate for higher and lower ambient temperatures.

Selection Procedure for Typical Overload Heaters

In the following procedure we will use one manufacturer's approach to the selection of overload heater coils. For maximum motor protection and compliance with Article 430-32 of the *National Electrical Code,* select the heater coils from any manufacturer's tables on the basis of motor nameplate full load current (FLC).

Gathering Information. In our first problem we will assume that we have a class 8536 Type SCG-3 Size 1 magnetic starter for controlling a three-phase, 1.15 service factor motor with a full load current of 17.0 amperes with both the motor and controller located in a 40°C (104°F) ambient temperature.

For this example we can determine the class, type and size of the magnetic starter by looking at the nameplate on the face of the magnetic starter (Figure 6-26). The phase, service factor and full load current of the motor can be determined by looking at the nameplate found on the motor (Figure 6-25). Because our application is a typical one, the ambient temperature has been established at 40°C, and we will accept this as fact. When ambient temperature is questionable, it should actually be measured at the job site or determined by some other method.

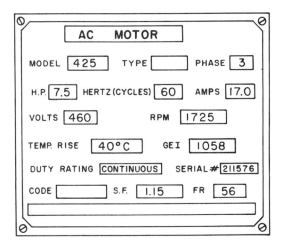

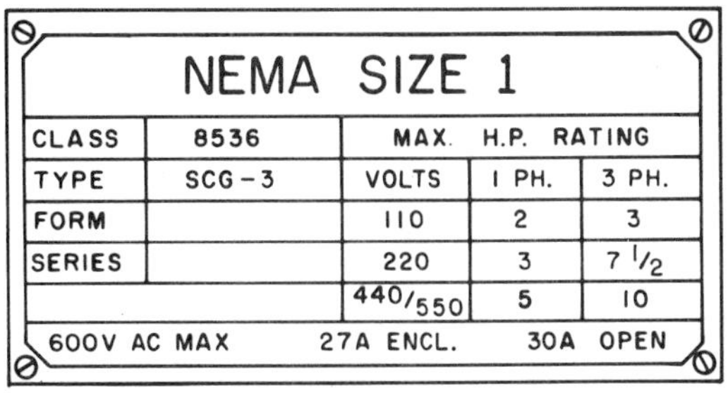

Figure 6-25 Selection of thermal heaters should always be based on the actual full load current shown on the motor nameplate or in the motor manufacturer's specification sheet.

Figure 6-26 The class, type, and size of the magnetic starter can be determined by looking at the nameplate on the magnetic starter.

Even with all of this information known, we should refer to the manufacturer's section on thermal unit selection to see if any restrictions may have been placed on the class of starter we intend to size. Using the information available in Figure 6-27, we find that unless we have a class 8198 starter, we can use 100% of the motor full load current for thermal unit selection with a 1.15 to 1.25 service factor. Since our motor has a 1.15 service factor, we will use 100% of the motor full load current when making our calculations.

Using Manufacturer's Heater Selection Charts. When paging through manufacturers' catalogs, you will generally find a section toward the back of the catalog which provides information for the proper selection of overload heaters. Be sure to look for a chart which is appropriate for the size of starter you are going to use. In our example we want a chart which will properly provide information on a size 1 starter. In this case Figure 6-28 provides the correct information. This information is also found within the enclosure of many motor starters.

Using column three, labeled "3 TU", we can begin our search for the proper current range for our heater. We use column three in this case because we have a three-phase motor and all three phases must have thermal unit protection as indicated.

Since our motor nameplate information indicates a maximum full load current (FLC) of 17 amperes, we must have a heater that will provide protection around this value. Reading down column 3 TU, we discover toward the bottom of the column a heater size which will provide protection within the range of 16.0 to 17.5 amperes. Since this is the only size close to our specified value, it is considered as acceptable. In this case, we would order a B32 heater for proper protection. Manufacturers will obviously have different numbers to relate to their specific heaters, but the selection procedure will be almost identical.

Checking Your Selection. If there is any doubt as to whether the heater you have selected will in fact provide the protection you desire, one further step can be used to check your work. The *National Electrical Code* section 430 says that a motor must be protected up to 125% of its full load current. Since the *minimum* full load of this overload device is 16.0 amperes, the device will trip at 125% of this value, or 1.25 × 16.0, which is 20 amperes. To see if this range is acceptable, we

Motor and controller in *same ambient temperature:*
a. All starter classes, except Class 8198:
   1. For 1.15 to 1.25 service factor motors use 100% of motor full load current for thermal unit selection.
   2. For 1.0 service factor motors use 90% of motor full load current for thermal unit selection.
b. Class 8198 only:
   1. For 1.0 service factor motors use 100% of motor full load current for thermal unit selection.
   2. For 1.15 to 1.25 service factor motors use 110% of motor full load current for thermal unit selection.

Figure 6-27 Manufacturer's instructions for classes of starters. (Square D Company)

Motor Full-Load Current (Amp.)			Thermal Unit Number
1 T. U.	2 T. U.	3 T. U.	
0.29– 0.31	0.29– 0.31	0.28– 0.30	B 0.44
0.32– 0.34	0.32– 0.34	0.31– 0.34	B 0.51
0.35– 0.38	0.35– 0.38	0.35– 0.37	B 0.57
0.39– 0.45	0.39– 0.45	0.38– 0.44	B 0.63
0.46– 0.54	0.46– 0.54	0.45– 0.53	B 0.71
0.55– 0.61	0.55– 0.61	0.54– 0.59	B 0.81
0.62– 0.66	0.62– 0.66	0.60– 0.64	B 0.92
0.67– 0.73	0.67– 0.73	0.65– 0.72	B 1.03
0.74– 0.81	0.74– 0.81	0.73– 0.80	B 1.16
0.82– 0.94	0.82– 0.94	0.81– 0.90	B 1.30
0.95– 1.05	0.95– 1.05	0.91– 1.03	B 1.45
1.06– 1.22	1.06– 1.22	1.04– 1.14	B 1.67
1.23– 1.34	1.23– 1.34	1.15– 1.27	B 1.88
1.35– 1.51	1.35– 1.51	1.28– 1.43	B 2.10
1.52– 1.71	1.52– 1.71	1.44– 1.62	B 2.40
1.72– 1.93	1.72– 1.93	1.63– 1.77	B 2.65
1.94– 2.14	1.94– 2.14	1.78– 1.97	B 3.00
2.15– 2.40	2.15– 2.40	1.98– 2.32	B 3.30
2.41– 2.72	2.41– 2.72	2.33– 2.51	B 3.70
2.73– 3.15	2.73– 3.15	2.52– 2.99	B 4.15
3.16– 3.55	3.16– 3.55	3.00– 3.42	B 4.85
3.56– 4.00	3.56– 4.00	3.43– 3.75	B 5.50
4.01– 4.40	4.01– 4.40	3.76– 3.98	B 6.25
4.41– 4.88	4.41– 4.88	3.99– 4.48	B 6.90
4.89– 5.19	4.89– 5.19	4.49– 4.93	B 7.70
5.20– 5.73	5.20– 5.73	4.94– 5.21	B 8.20
5.74– 6.39	5.74– 6.39	5.22– 5.84	B 9.10
6.40– 7.13	6.40– 7.13	5.85– 6.67	B 10.2
7.14– 7.90	7.14– 7.90	6.68– 7.54	B 11.5
7.91– 8.55	7.91– 8.55	7.55– 8.14	B 12.8
8.56– 9.53	8.56– 9.53	8.15– 8.72	B 14
9.54–10.6	9.54–10.6	8.73– 9.66	B 15.5
10.7 –11.8	10.7 –11.8	9.67–10.5	B 17.5
11.9 –13.2	11.9 –12.0	10.6 –11.3	B 19.5
13.3 –14.9	. . . .	11.4 –12.0	B 22
15.0 –16.6	. . . .	. . . .	B 25
16.7 –18.0	. . . .	. . . .	B 28.0
Following Selections for Size 1 Only			
. . . .	11.9 –13.2	. . . .	B 19.5
. . . .	13.3 –14.9	11.4 –12.7	B 22
. . . .	15.0 –16.6	12.8 –14.1	B 25
16.7 –18.9	16.7 –18.9	14.2 –15.9	B 28.0
19.0 –21.2	19.0 –21.2	16.0 –17.5	B 32
21.3 –23.0	21.3 –23.0	17.6 –19.7	B 36
23.1 –25.5	23.1 –25.5	19.8 –21.9	B 40
25.6 –26.0	25.6 –26.0	22.0 –24.4	B 45
. . . .	. . . .	24.5 –26.0	B 50

Figure 6-28 Heater selection chart for size 1 starters. The columns marked 1 TU, 2 TU and 3 TU, denote the number of thermal overload heaters to be used in the relay. (Square D Company)

divide the minimum trip current (20 amps) by the full load current of our motor (17 amps)—20/17—and multiply it by 100% to equal 118% protection. Since the minimum trip current is below 125%, the heater selection in this example is correct.

Ambient Temperature Compensation. In cases where ambient temperature varies above or below the standard temperature of 40°C, most heater manufacturers will provide special selection tables to compensate or allow for temperature changes in the overload heaters. This compensation provides that the increase or decrease in temperature will not affect the proper protection provided by the overload relay.

Figure 6-29 illustrates an example of a manufacturer's chart used in the selection of thermal units for special application. From this chart it should be apparent that as ambient temperature increases, less current is needed to trip the overload device. In contrast, as ambient temperature decreases, more current is needed to trip the overload device.

Using our previous example of 17.0 as full load current, let us determine what correction factor would be necessary if the ambient temperature increased 10°C (50°F) to a higher value of 50°C (122°F). Since the ambient temperature increased 10°C, the correction factor would be 0.9. Multiplying the correction factor 0.9 × the (FLC) of 17, we find the current rating now would be 15.3 amps. Using our selection chart for heaters, we see that the current range closest to acceptable is 14.2 to 15.9, or a B28 heater, which is one size smaller.

If the ambient temperature decreased 10°C, then the temperature surrounding the device would be 30°C (86°F). For a 10°C *decrease* in ambient temperature, the correction factor would be 1.05. Using an (FLC) of 17 again, the corrected current would be 1.05 × 17, or 17.85 amps. In the selection guide, a range of 17.5 to 19.7 appears acceptable. In this case a B36 heater, which is one size larger, is appropriate.

Although it appears that for each 10°C change in temperature you need move only one heater size above or below that which is standard, we caution you to always consult manufacturers' specifications and tables for proper sizing.

Full Load Current Unknown. In rare instances, such as older installations or severely damaged equipment, it may be impossible to determine from a motor nameplate its full load current. When this situation exists, full load cur-

Class of Controller	Continuous Duty Motor Service Factor†	Melting Alloy and Non-Compensated Bimetallic Relays			Ambient Temperature-Compensated Relays
		Ambient Temperature of Motor			
		Same as Controller Ambient	Constant 10°C (18°F) **Higher** Than Controller Ambient	Constant 10°C (18°F) **Lower** Than Controller Ambient	Constant 40°C (104°F) or less, for **Any** Controller Ambient
		Full Load Current Multiplier			
All Classes, Except 8198	1.15 to 1.25	1.0	0.9	1.05	1.0
	1.0	0.9	0.8	.95	0.9
Class 8198	1.15 to 1.25	1.1	1.0	1.15	1.1
	1.0	1.0	0.9	1.05	1.0

†For intermittent (short time) rated motors consult local Square D Field Office.

Figure 6-29 Most manufacturers will provide special selection tables like the one shown to compensate or allow for temperature changes in the overload heaters. (Square D Company)

rent may be approximated on the basis of average full load currents like those shown in the chart of Figure 6-30.

This technique should be used only as a last resort. To illustrate the fact that this technique is not suggested as a standard procedure, most manufacturers present the following statement with these charts: *"CAUTION: The average rating could be high or low for a specific motor and therefore selection on this basis always involves risk. For fully reliable motor protection, select heat coils on the basis of full load current ratings as shown on the motor nameplate."* If these charts are used, however, the stated full load current of the motor should be used in the selection of the heater, using the same procedure as if it were the motor nameplate information.

In other words, these charts provide approximately the same information that could be found on the motor nameplate but should be used only if motor nameplate information is not available.

INHERENT MOTOR PROTECTORS

Inherent motor protectors are overload devices which are located directly on or in a motor to provide overload protection. Certain inherent motor protection will base the sensing element on the amount of heat generated or the amount of current consumed by the motor. Based on what the device senses, it will directly or indirectly (using contactors) trip a circuit which will disconnect the motor from the power circuit. Bi-metallic thermo discs and solid state thermistors are examples of inherent motor protectors.

Ampere ratings of motors vary somewhat, depending upon the type of motor. The values given below are for drip-proof, Class B insulated (T Frame) where available, 1.15 service factor, NEMA Design B motors. These values represent an average full load motor current which was calculated from the motor performance data published by several motor manufacturers. In the case of high torque squirrel cage motors, the ampere ratings will be at least 10% greater than the values given below.

Caution — These average ratings could be high or low for a specific motor and therefore heater coil selection on this basis always involves risk. For fully reliable motor protection, select heater coils on the basis of full load current rating as shown on the motor nameplate.

AMPERE RATINGS OF THREE PHASE, 60 HERTZ, A-C INDUCTION MOTOR

Hp	Syn. Speed RPM	200 Volts	230 Volts	380 Volts	460 Volts	575 Volts	2200 Volts
¼	1800	1.09	.95	.55	.48	.38	....
	1200	1.61	1.40	.81	.70	.56	....
	900	1.84	1.60	.93	.80	.64	....
⅓	1800	1.37	1.19	.69	.60	.48	....
	1200	1.83	1.59	.92	.80	.64	....
	900	2.07	1.80	1.04	.90	.72	....
½	1800	1.98	1.72	.99	.86	.69	....
	1200	2.47	2.15	1.24	1.08	.86	....
	900	2.74	2.38	1.38	1.19	.95	....
¾	1800	2.83	2.46	1.42	1.23	.98	....
	1200	3.36	2.92	1.69	1.46	1.17	....
	900	3.75	3.26	1.88	1.63	1.30	....
1	3600	3.22	2.80	1.70	1.40	1.12	....
	1800	4.09	3.56	2.06	1.78	1.42	....
	1200	4.32	3.76	2.28	1.88	1.50	....
	900	4.95	4.30	2.60	2.15	1.72	....
1½	3600	5.01	4.36	2.64	2.18	1.74	....
	1800	5.59	4.86	2.94	2.43	1.94	....
	1200	6.07	5.28	3.20	2.64	2.11	....
	900	6.44	5.60	3.39	2.80	2.24	....
2	3600	6.44	5.60	3.39	2.80	2.24	....
	1800	7.36	6.40	3.87	3.20	2.56	....
	1200	7.87	6.84	4.14	3.42	2.74	....
	900	9.09	7.90	4.77	3.95	3.16	....
3	3600	9.59	8.34	5.02	4.17	3.34	....
	1800	10.8	9.40	5.70	4.70	3.76	....
	1200	11.7	10.2	6.20	5.12	4.10	....
	900	13.1	11.4	6.90	5.70	4.55	....
5	3600	15.5	13.5	8.20	6.76	5.41	....
	1800	16.6	14.4	8.74	7.21	5.78	....
	1200	18.2	15.8	9.59	7.91	6.32	....
	900	18.3	15.9	9.60	7.92	6.33	....
7½	3600	22.4	19.5	11.8	9.79	7.81	....
	1800	24.7	21.5	13.0	10.7	8.55	....
	1200	25.1	21.8	13.2	10.9	8.70	....
	900	26.5	23.0	13.9	11.5	9.19	....
10	3600	29.2	25.4	15.4	12.7	10.1	....
	1800	30.8	26.8	16.3	13.4	10.7	....
	1200	32.2	28.0	16.9	14.0	11.2	....
	900	35.1	30.5	18.5	15.2	12.2	....
15	3600	41.9	36.4	22.0	18.2	14.5	....
	1800	45.1	39.2	23.7	19.6	15.7	....
	1200	47.6	41.4	25.0	20.7	16.5	....
	900	51.2	44.5	26.9	22.2	17.8	....
20	3600	58.0	50.4	30.5	25.2	20.1	....
	1800	58.9	51.2	31.0	25.6	20.5	....
	1200	60.7	52.8	31.9	26.4	21.1	....
	900	63.1	54.9	33.2	27.4	21.9	....
25	3600	69.9	60.8	36.8	30.4	24.3	....
	1800	74.5	64.8	39.2	32.4	25.9	....
	1200	75.4	65.6	39.6	32.8	26.2	....
	900	77.4	67.3	40.7	33.7	27.0	....
30	3600	84.8	73.7	44.4	36.8	29.4	....
	1800	86.9	75.6	45.7	37.8	30.2	....
	1200	90.6	78.8	47.6	39.4	31.5	....
	900	94.1	81.8	49.5	40.9	32.7	....
40	3600	111.	96.4	58.2	48.2	38.5	....
	1800	116.	101.	61.0	50.4	40.3	....
	1200	117.	102.	61.2	50.6	40.4	....
	900	121.	105.	63.2	52.2	41.7	....
50	3600	138.	120.	72.9	60.1	48.2	....
	1800	143.	124.	75.2	62.2	49.7	....
	1200	145.	126.	76.2	63.0	50.4	....
	900	150.	130.	78.5	65.0	52.0	....
60	3600	164.	143.	86.8	71.7	57.3	....
	1800	171.	149.	90.0	74.5	59.4	....
	1200	173.	150.	91.0	75.0	60.0	....
	900	177.	154.	93.1	77.0	61.5	....
75	3600	206.	179.	108.	89.6	71.7	....
	1800	210.	183.	111.	91.6	73.2	....
	1200	212.	184.	112.	92.0	73.5	....
	900	222.	193.	117.	96.5	77.5	....
100	3600	266.	231.	140.	115.	92.2	....
	1800	271.	236.	144.	118.	94.8	23.6
	1200	275.	239.	145.	120.	95.6	24.2
	900	290.	252.	153.	126.	101.	24.8
125	3600	....	292.	176.	146.	116.	....
	1800	....	293.	177.	147.	117.	29.2
	1200	....	298.	180.	149.	119.	29.9
	900	....	305.	186.	153.	122.	30.9
150	3600	....	343.	208.	171.	137.	....
	1800	....	348.	210.	174.	139.	34.8
	1200	....	350.	210.	174.	139.	35.5
	900	....	365.	211.	183.	146.	37.0
200	3600	....	458.	277.	229.	184.	....
	1800	....	452.	274.	226.	181.	46.7
	1200	....	460.	266.	230.	184.	47.0
	900	....	482.	279.	241.	193.	49.4
250	3600	....	559.	338.	279.	223.	....
	1800	....	568.	343.	284.	227.	57.5
	1200	....	573.	345.	287.	229.	58.5
	900	....	600.	347.	300.	240.	60.5
300	1800	....	678.	392.	339.	271.	69.0
	1200	....	684.	395.	342.	274.	70.0
400	1800	....	896.	518.	448.	358.	91.8
500	1800	....	1110.	642.	555.	444.	116.

Figure 6-30 Most manufacturers provide charts like the one shown so that full load current may be approximated when motor nameplate information is not available. Charts like these should be used only if motor nameplate information is not available from any other source. (Cutler Hammer)

Bi-Metallic Thermo Disc

The bi-metallic thermo disc (Figure 6-31) operates on the same principle as the bi-metallic strip discussed in the automatic reset overload relay. The main differences between these devices are the shape of the device and its location. The thermo disc has the shape of a miniature dinner plate and is located within the frame of the motor. When the motor is overloaded, the bi-metallic disc warps and opens a circuit. In most cases, bi-metallic thermo discs are used on smaller horsepower motors to directly disconnect the motor from the power circuit. It could, however, be possible to tie it into the control circuit of a magnet contactor coil circuit, where it would be used as an indirect control device.

Thermistor Overload Device

The thermistor based overload is another more sophisticated form of inherent motor protection (Figure 6-32). This type of overload combines a thermistor, transistorized relay and contactor into a custom-built overload protector.

The thermistor is basically a resistor whose resistance changes with the amount of heat

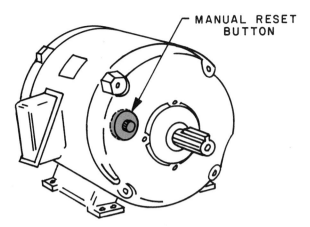

Figure 6-31 Bi-metallic overload protectors like the one shown are usually used on smaller horse-power motors to directly disconnect the motor from the power circuit.

applied to it. As the temperature increases, the resistance of the thermistor decreases, and the amount of current passing through the thermistor increases. Since the thermistor is a low power device (usually in the thousandths of ampere range), the changing signal must be amplified before it can do any work such as triggering a

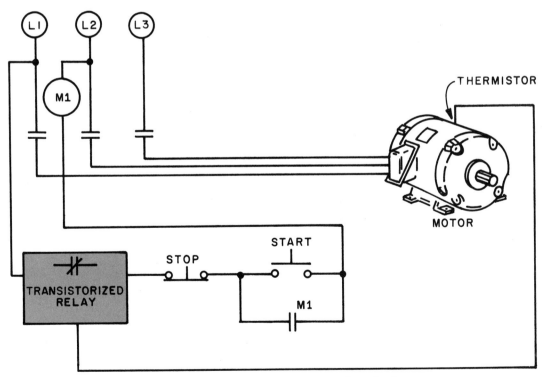

Figure 6-32 Illustrates a thermistor-based overload protection circuit.

relay. With an amplifier, however, it can cause a relay to open a set of contacts in the control circuit of a magnetic contactor, causing the power circuit of the motor to be de-energized.

The major drawback to this system is that it requires a great deal of co-ordination between the user and the manufacturer to customize the design. Custom work costs more than standard off-the-shelf overload protectors and, thus, makes it uneconomical except for the very special and high-priced motor which requires this type of protection.

ELECTRONIC OVERLOAD PROTECTION

Although the most common method used today of protecting a motor is the overload heater, a new method of protecting the motor with an electronic overload is becoming increasingly popular. The electronic overload functions on the same basic principal as the overload heater; namely, monitoring the current in the motor directly by measuring the current in the power lines of the motor.

OPTIONAL MODIFICATION FOR CONTACTORS AND MAGNETIC MOTOR STARTERS

When space is at a premium in certain control panels, certain devices can be added to a basic contactor or magnetic motor starter to expand its capability. Figure 6-33 illustrates several optional devices which are available.

Additional Electrical Contacts

Most contactors and starters have the ability to control several additional electrical contacts simply by adding them to existing auxiliary contacts. In addition to providing electrical interlock (memory), they can be used as auxiliary overload alarm contacts. When a set of NO or NC contacts are used on the side of the contactor or starter, they can be wired into an alarm circuit to provide warning when the device drops out. The NO contacts close, completing an alarm circuit to a pilot light or audible alarm, thereby notifying the operator that something is wrong.

Power Poles

In certain cases, additional power poles (contacts

capable of carrying a load) can be added to a basic contactor. These power poles are available with normally open or normally closed contacts. Usually only one power pole unit with one or two contacts can be added per contactor or starter.

In certain cases with larger size contactors or starters, it may be necessary to replace the coil to handle the additional load created in energizing these additional poles. Most power poles can be factory or field installed.

Pneumatic Timer

For applications requiring the simultaneous operation of a timer and a contactor, a mechanically operated pneumatic timer can be mounted on some sizes of contactors or starters. The use of mechanically operated timers results in considerable savings in panel space over a separately mounted timer. Available in time delay after de-energization (off delay) or time delay after energization (on delay), the timer attachment has an adjustable timing period over a specified range.

Most manufacturers provide units that are field convertible from on-delay to off-delay, or vice-versa, without any additional parts. They are, however, ordered either fixed or variable.

Most timers mount on the side of the basic contactor and can be secured firmly. Electrical contacts, one NO and one NC, are provided.

Transient Suppression Module

Transient suppression modules are designed to be added where the transient voltage, generated when opening the coil circuit, interferes with the power operation of nearby components and solid state control circuits. Modules usually consist of RC (resistance capacitance) circuits and are designed to suppress the voltage transients to approximately 200% of peak coil supply voltage.

In certain cases a transient is generated when switching the integral control transformer that powers the coil control circuit. The transient suppression module, when used with devices wired for common control, is connected across the 120 volt transformer secondary—not across the coil.

Control Circuit Fuse Holder

Control circuit fuse holders can be attached to contactors or starters when either one or two control circuit fuses may be required.

ADDITIONAL ELECTRICAL CONTACTS

Four additional single circuit electrical interlocks can be added to any open or enclosed starter or contactor. All interlocks are front mounted with captive screws. Normally open or normally closed contacts, which can be easily converted in the field, are available.

POWER POLE ATTACHMENT- NEMA SIZES 1 and 2

One or two additional power poles can be added to a basic three pole Size 0, 1 or 2 contactor. The power pole adder is available with normally open or normally closed poles. Two captive front mounted screws secure the basic power pole to the contactor base. A four or five pole contactor can easily be built up using the power pole kits. An adaptor bracket allows these units to be mounted on contactor sizes 3-6.

PNEUMATIC TIMER ATTACHMENT

A pneumatic timer with a .2 sec. to 1 min. adjustable timing range can be mechanically operated from the contactor, saving the panel space normally required for mounting a separate timer. The timer is available in ON or OFF delay and can be easily converted. One single pole double throw contact is provided.

TRANSIENT SUPPRESSION MODULE

The transient suppression module is designed to be used where the transient voltage, generated when opening the coil circuit, interferes with the proper operation of nearby integrated or solid state control circuits. The module consists of an RC circuit and is designed to suppress the voltage transients to approximately 200% of peak line voltages. The transient suppression module is normally wired across the coil and is easy to add.

CONTROL CIRCUIT FUSE HOLDER

The fuse holder is designed to be used on Type S contactors when either one or two control circuit fuses, 600 volts maximum, are required. It mounts from the front with captive screws just like the electrical interlock.

The fuse holder will help satisfy the National Electrical Code requirements in Section 430-72. An additional benefit of this fuse holder is the easy removal of the cartridge to change fuses. With the cartridge removed, the fuse holder acts as an isolator for the control circuit.

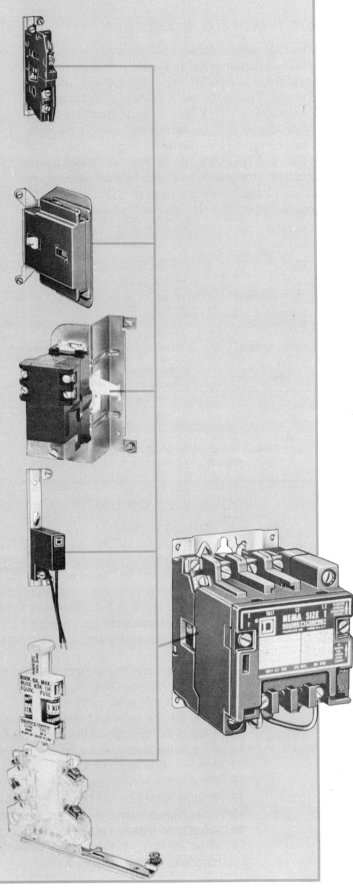

Figure 6-33 Several optional devices which may be added to a contactor or magnetic motor starter. (Square D Company)

INTERNATIONAL STANDARDS

The International Electro-Technical Commission (IEC), headquartered in Geneva, Switzerland, is primarily associated with equipment used in Europe. The National Electrical Manufacturers Association (NEMA), headquartered in Washington D.C., is primarily associated with equipment used in North America.

IEC and NEMA both rate contactors and motor starters. This causes confusion because ratings are different for the same horsepower. IEC devices are smaller in size for the equivalent rated contactor.

IEC devices are built with materials required for average applications. NEMA devices are built for a higher level of performance in a variety of applications. IEC devices are less expensive, but more application sensitive. NEMA devices are more costly but less application sensitive. IEC devices are commonly used in OEM (Original Equipment Manufacturer) machines where machine specifications are known and will not change. NEMA devices are commonly used where machine requirements and specifications could vary. Figure 6-34 illustrates a comparison of IEC and NEMA devices.

COMPARISON CHART OF IEC AND NEMA DEVICES		
CONSIDERATION	IEC	NEMA
1. SIZE	Smaller per horsepower rating than NEMA	Larger per horsepower rating than IEC
2. COST	Lower price per horsepower	Higher price per horsepower
3. PERFORMANCE	Generally an electrical life of 1 million operations is acceptable	Electrical life will typically be 2.5 to 4 times higher for same equivalent IEC device
4. APPLICATIONS	Application sensitive with greater knowledge and care necessary	Application easier with fewer parameters to consider
5. OVERLOADS	Fixed heaters that are adjustable to match different motors at same horsepower; heaters are not field changeable	Field changeable heaters allow adjustment to motors of different horsepowers
	Reset/stop dual function operation mechanism typical	Reset only mechanism typical
	Hand/auto reset typical	Hand reset only typical
	Typically designed for use with fast acting, current limiting European fuses	Designed for use with domestic time delay fuses and circuit breakers

Figure 6-34. The difference between IEC and NEMA devices is based on size, cost, performance, applications, and overloads.

QUESTIONS ON UNIT 6

1. What is a contactor?
2. What is meant by "two-wire" control?
3. What is meant by "low voltage release"?
4. What is meant by "three-wire" control?
5. What is meant by "low voltage protection"?
6. What are some of the differences between AC and DC contactors?
7. Why is "arc suppression" needed?
8. Why is it harder to extinguish an arc on contacts passing DC than on contacts passing AC?
9. What are some of the methods used to reduce an arc across a contact?
10. How are AC and DC contactors rated?
11. What is a magnetic motor starter?
12. What are the different types of overloads used with magnetic motor starter?
13. What is the function of the overloads?
14. What is the purpose of a "current transformer"?
15. What general factors must be considered when selecting the overloads?
16. What is the service factor rating of a motor?
17. What is meant by "ambient temperature"?
18. What is meant by ambient temperature compensation?
19. What is an inherent motor protector?
20. What are some of the optional devices that can be added to contactors and magnetic motor starters?

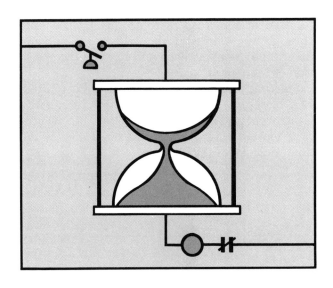

7 Time Delay and Logic Applied to More Complex Line Diagrams and Control Circuits

As line diagrams become more complex, cross referencing systems are required to trace the action in the circuit. A numbering system and specific symbols are necessary for consistency in developing a system. Complex diagrams, which include time delay and logic functions, require the accurate use of numbering systems and symbols. For example, time delay logic allows the circuit to pace itself and respond to a variety of load devices. In addition, timers can be used to provide protection circuits and sequence control.

NUMERICAL CROSS REFERENCING SYSTEMS

In Unit 3, "Introduction to Logic as Applied to Line Diagrams and Basic Control Circuits," we limited our line diagrams to two or three lines by showing only simple control circuits. It would, however, be misleading for you to believe that all diagrams are this simple.

Figure 7-1 represents more realistically the type of line diagrams you will be expected to read and troubleshoot. Since a circuit like that of Figure 7-1 may at first seem overwhelming, we are going to give you a few rules which will help you quickly simplify the operation of this complex circuit and others like it.

Numerical Cross Referencing System (NO Contacts)

The first thing to realize with a circuit like that of

Figure 7-1 is that relays, contactors, and magnetic motor starters usually have more than one set of auxiliary contacts and that these contacts may appear at several different locations in the line diagrams. Since it could take quite some time to locate each contact to determine how it is being used, we are going to show you a cross referencing system which quickly identifies the location and type of contacts controlled by a given device.

Starting with line 1 of Figure 7-1, we see that depressing master start PB2 will energize control relay coil CR1. Control relay CR1 in turn controls three sets of normally open contacts as denoted by the numerical codes (2, 3, 4) on the far right side of the line diagram. Each number (2, 3, 4) indicates the line in which the NO contacts will be located. Looking at line 2 we see that the NO contacts in this line form the holding circuit or MEMORY for maintaining the coil CR1 after master start PB2 is released. Looking at line 3, we note that the NO contacts in this line energize pilot light PL1, indicating that the circuit has been energized. Looking to line 4, we note that the NO contacts in this line allow the remainder of the circuit to be brought on line by connecting L1 to the remainder of the circuit. Thus, in a very short time, we have been able to determine through the use of this simple numerical cross referencing system, the

location of all contacts controlled by coil CR1 through line 4. In addition, we have been able to determine the effect each will have on the operation of the circuit.

Starting with line 5, we see that if float switch FL1 closes, it will energize control relay CR2. Control relay CR2 in turn will close the NO contacts located in lines 8 and 10 as indicated by the numerical codes (8, 10).

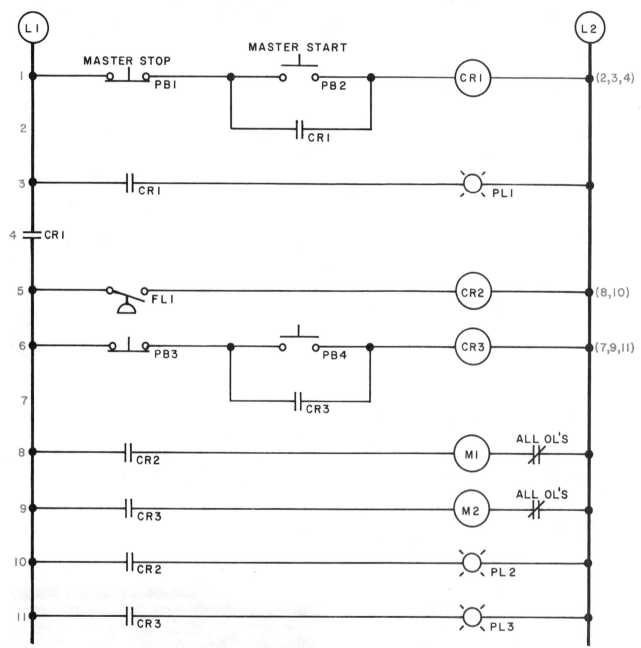

Figure 7-1 The location of NO contacts controlled by a device can be determined by the number found on the right side of the line diagram. Each number relates to the line in which the NO contacts will be found.

With the NO contacts of line 8 closed, the magnetic motor starter controlled by coil M1 will be energized. With the NO contacts of line 10 closed, the pilot light PL2 will go on indicating the motor has started.

Moving to line 6, we again see several NO contacts (7, 9, 11) being used to control other parts of the circuit through control relay CR3. Looking at line 7, we note that the NO contacts here form the MEMORY circuit for maintaining the circuit to control relay coil CR3 after the pushbutton PB4 is released.

With coil CR3 energized, the NO contacts of line 9 will close, energizing the magnetic motor starter controlled by coil M2. Simultaneously, NO contacts of line 11 close, causing pilot light PL3 to light as an indicator that the motor has started.

Using this cross reference system, complex line diagrams can quickly be simplified. All you need to do is see that each NO contact is clearly marked because each set of NO contacts are numbered according to the line that they appear in.

Numerical Cross Reference System (NC Contacts)

In addition to NO contacts, there will also be normally closed (NC) contacts in a circuit. To differentiate between NO and NC, we will indicate a NC contact as a standard number which will be underlined (6, 8, 10). Referring to Figure 7-2, you will note lines 9 and 11 contain devices which control NC contacts in lines 12 and 13 as indicated by their being underlined (12, 13).

Since the circuits of Figures 7-2 and 7-1 are identical up to line 4, we will not repeat that sequence of the operation of the circuit. If you are unsure of the sequence up to that point, review the operation of Figure 7-1 up to line 4.

Starting with line 5, we see that energizing control relay coil CR2 will cause NO contacts in lines 8, 9, and 12 to close. Closing these contacts energizes coils M1 and M2 and completes part of the circuit going to the pilot light in line 12.

In our previous circuit, closing these contacts to the pilot light caused the light to go on. In this case, however, the light will not glow because the NC contacts controlled by coil M2 line 12 are opened at the same time that the NO contacts are closed, leaving the circuit open. It may seem strange having one set of contacts opening and one set closing in the same line, but there is a logical reason for the sequence.

With the NO contacts of CR2 closed and the NC contacts of M2 open, the light will stay off unless

something happens to shut down line 9 with coil M2 in it. If coil M2 represents a safety cooling fan protecting the motor controlled by M1, then the light would indicate positively the loss of cooling to the operator. Study this sequence carefully so that you fully understand it before proceeding with the rest of the circuit.

Starting back at line 6, we will see somewhat the same sequence of events as took place when line 5 was energized. In this case, depressing pushbutton PB4 energizes control relay coil CR3, which in turn closes NO contacts in line 7, 10, 11, and 13. With 7, 10, 11 and 13 closed, a MEMORY circuit is formed in line 7; coils M3 and M4 are energized in lines 10 and 11, and part of the circuit to the pilot light in line 13 is completed. Since coil M4 is energized, the NC contacts it controls will open in line 13, forming the alarm circuit discussed previously for line 12.

If coil M4 should drop out for any reason, the NC contacts would return to their normally closed position and the pilot light alarm signal would be turned on. This circuit could be used where it is extremely important for the operator to know when something is not functioning.

It should be pointed out that the use of this circuit at this point was not meant for a complete discussion of its operating characteristics. Rather, it was intended to show how simply more complex circuits can be analyzed using the numerical cross referencing system. For your own information, however, it might be helpful to study this circuit in depth and try to determine what will or could happen to the circuit if various overloads tripped or devices failed.

(*Note*: Since this numerical cross reference system is so important to effective troubleshooting, we will use this sytem throughout the book on all circuits, no matter how simple they may seem.)

Cross Referencing Mechanically Connected Contacts

Control devices such as limit switches, flow switches, temperature switches, liquid level and pressure switches usually have more than one set of contacts operating when the device is activated. Usually these devices have at least one set of NO contacts and one set of NC contacts which operate simultaneously. For all practical purposes, the multiple contacts of these devices usually do not control other devices in the same lines of a control circuit.

There are two ways to illustrate how contacts found in different control lines belong to the same

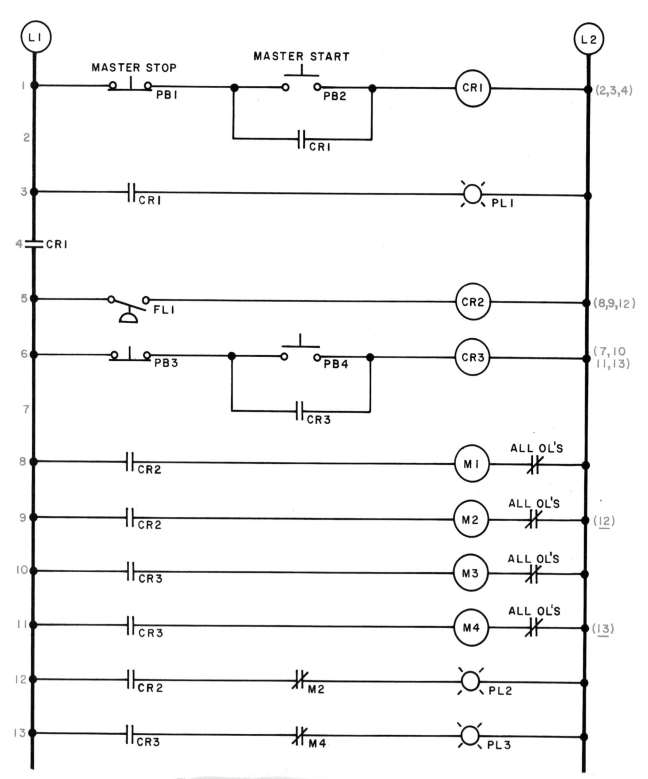

Figure 7-2 The location of NC contacts controlled by a device can be determined by the numbers found on the right side of the line diagram. Note that each NC is underlined to differentiate it from NO contacts. Each underlined number relates to the line in which the NC contacts will be found.

control switch. One method is to use a dashed line to illustrate that the contacts are mechanically but not electrically connected. The other method is a numerical cross referencing system.

Figure 7-3 illustrates how the dashed line method indicates the mechanical connection between the NO contacts and NC contacts of a typical limit switch. The dashed line between the NO and NC contacts means that both contacts will move from the normal position when the arm of the limit switch is moved. In this case, before the limit switch is actuated, the light is on and the coil to the motor starter is off. After the limit switch is actuated, the light is off and the coil to the motor starter is energized.

Using the dashed line method works very well when the control contacts are close by and the circuit is relatively simple. If, however, the dashed line had to cut across many individual lines, this would make the circuit hard to follow. To eliminate this confusion, a numerical cross referencing system like that of Figure 7-4 can be used.

In Figure 7-4, a pressure switch with a normally open contact, line 1, and a normally closed contact, line 5, is used to control a motor starter and a solenoid. Both the normally open and nor-

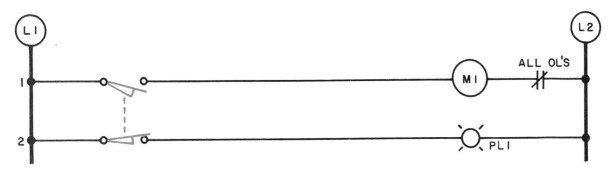

Figure 7-3 A dashed line between the sets of contacts indicates a mechanical connection between them. When the limit switch is actuated, both sets of contacts move.

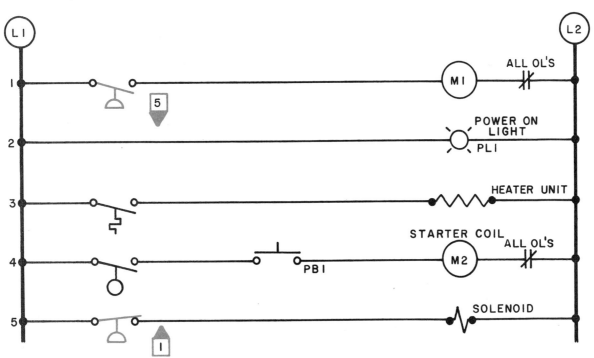

Figure 7-4 The numerical cross referencing system shown indicates which other contacts in which line are affected by the mechanical movement of one set of contacts. The solid arrow indicates the direction of movement exerted on the other contacts when one is moved.

mally closed contacts of the pressure switch are simultaneously actuated when a predetermined pressure is reached. To show this mechanical link between the two contacts, a solid arrow from the normally open contact will point down and be marked with a 5, for line 5. The normally closed contact line 5 will have a solid arrow pointing up and be marked with a 1, for line 1. This cross reference will eliminate the need for a dashed line cutting across lines 2, 3 and 4. This will make the circuit easier to follow and understand. This system may be used with any type of control switch found in a circuit.

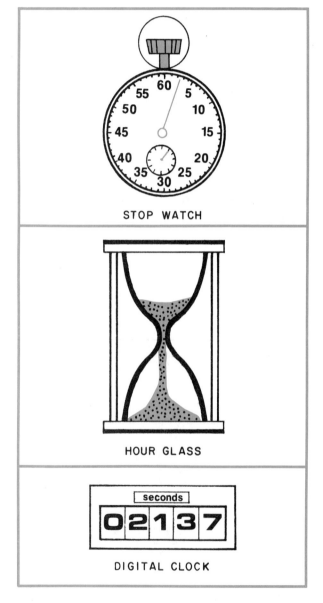

STOP WATCH

HOUR GLASS

seconds

DIGITAL CLOCK

Figure 7-5 In everyday life, we are accustomed to using timers that remind us when certain actions are to be initiated.

INTRODUCTION TO TIMERS

In everyday life we are accustomed to seeing and using timers. A timer called the alarm clock awakens us in the morning and a wall clock or time clock at work times the hours we spend on the job. In addition, we wear timers called wristwatches on our arms which help us keep appointments, remind us of coffee breaks, and tell us when it is time for lunch. Figure 7-5 illustrates the variety of timers used to signal us when certain actions are to be initiated.

Since personal timers are so essential in maintaining order and organization for individuals, it is not surprising that industry would find an even greater need for similar devices to help regulate the complex processes and multiple activities so common in production work.

Time Delay Logic

Like any controller, timers provide a logic function in a control circuit. In Unit 3 you learned that a control circuit is organized into three basic sections to accomplish the specific work desired. These three basic sections were the signal(s), the decision and the action. In Unit 3 you also learned that the decision part of the circuit determined in what order the work was to occur based on the logic of the circuit. In this unit you will learn how time delay logic not only determines what work is to be done and in what order, but also how long or in what time period it will take to do this work.

It is important to note that all of the other logic functions studied thus far (AND, OR, NOT, MEMORY, NOR and NAND) were instantaneous logic which responded immediately to the condition of the input signal. Time delay logic, however, is not instantaneous and in fact derives its major benefit from the fact that its decision-making capability can be varied from a few fractions of a second up to several hours.

Time delay logic is an essential function because it allows the circuit to pace itself and respond appropriately to a variety of load devices. Further in the unit we will show application circuits where timers can be used to provide protection circuits, sequence control and convenience.

Types of Timers

Basically there are three major categories of timers: dashpot timers, synchronous clock timers and electronic timers similar to those shown in Figure 7-6. Although each device accomplishes its task in a different way, all timers have in common the ability to introduce some degree of time delay into a control circuit.

Dashpot Timers. Dashpot timers provide time delays by controlling how rapidly air or a liquid is allowed to pass into or out of a container through an orifice (opening) that is either fixed in diameter or variable.

Figure 7-7 illustrates the concept of the dashpot timer quite simply. If you were to take a balloon with air by the neck and form a small opening, the balloon would take some time to deflate. If, how-

ever, the opening were completely released, the balloon would deflate quite rapidly. In other words, by restricting the size of the orifice (opening) we can determine how long it will take for the balloon to deflate. The use of this principle will be discussed more in depth as each timer is discussed in depth.

Synchronous Clock Timers. Figure 7-8 illustrates a very simple synchronous clock timer.

Figure 7-7 By restricting the size of the orifice (opening) of the balloon, we can determine how long it will take for the balloon to deflate. Dashpot timers operate on the principle of a restricted orifice.

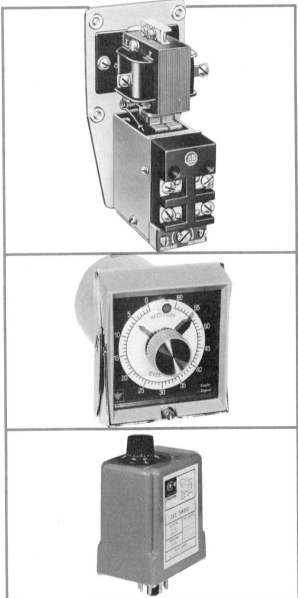

Figure 7-6 Basically there are three major categories of timers: dashpot timers, synchronous clock timers and electronic based timers. (Allen-Bradley Company, Eagle Signal Industrial Controls, Cutler Hammer)

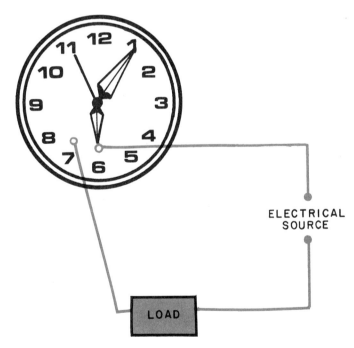

Figure 7-8 Synchronous clock timers have one or more contacts which open or close depending upon the position of the clock. Since most power companies maintain strict tolerances on the line frequency, this type of timer is quite accurate.

Basically, the timer will open and close the circuit depending upon the position of the hands of the clock. The timer may have one or more contacts where the circuit may be opened or closed. The time delay is provided by the speed at which the clock hands move around the perimeter of the face of the clock. In this case the contacts would be closed once every 12 hours. Synchronous clock motors are AC operated and maintain their speed based on the frequency of the AC power line which feeds them. Since power companies maintain strict tolerance on line frequency, the synchronous clock is quite an accurate timer.

Solid State Timers. Solid state timers derive their name from the fact that the time delay is provided by solid state electronic devices enclosed within the timing device. Since solid state timing devices like those in Figure 7-9 are almost always replaced and never repaired, we will not elaborate on how the device *electronically* provides time delay. Briefly, however, most solid state devices use an RC (resistance/capacitance) network along

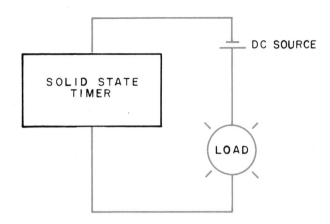

Figure 7-9 Solid state timers derive their timing ability from the solid state electronics built into the unit. Since these units are very inexpensive and often sealed in expoxy, they are most often replaced rather than repaired.

with transistor and SCRs (silicon controlled rectifier) to provide the time delay and switching characteristics.

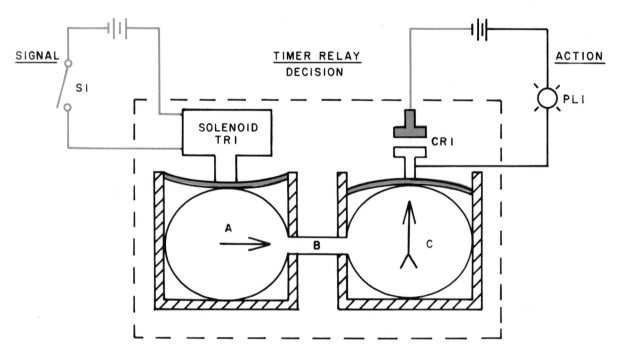

Figure 7-10 ON delay timed closed logic could be accomplished using two balloons. The length of the timing cycle is determined by the time it takes to fill balloon C.

These devices have been very popular and will replace any other timing devices because of their dependability and relatively low cost. When the units do go bad, it is cheaper to replace the device rather than to have it repaired. In many cases, these devices are encapsulated in epoxy resin for protection, making their repair impossible.

Time Delay as a Logic Function

Earlier in the unit we referred to a timer as a control device which was capable of decision making. Not only could the control device turn on and off another device, but it was also capable of determining how long the other device was turned on and off. In the next few paragraphs we will discuss the two major logic functions of a timer; namely, ON delay logic and OFF delay logic.

ON Delay Logic. ON delay derives its name from the fact that once the timer has received its signal to activate or turn ON, a predetermined time interval must take place before any action takes place in the load circuit. It is extremely important to realize that it is the turning on of the timer, *not* the load, which determines this logic function. Once the timer is ON, the timer is capa-

ble of turning on or turning off a load depending upon the way it is wired into a circuit.

ON Delay (Timed Closed). A simple example of ON delay timed closing is illustrated pictorially in Figure 7-10. When switch S1 is closed, the solenoid plunger will force air out of balloon A through orifice B into balloon C. When balloon C is filled, contacts CR1 will close, energizing the circuit to pilot light PL1, causing the lamp to light. If it takes five seconds for balloon C to be filled, then the ON delay function will be five seconds in duration.

Schematically, this circuit would appear like the line diagram of Figure 7-11. To help understand how the NO timing contacts work in ON delay, always remember that an arrow shown in the left hand side of Figure 7-11 is used to indicate the direction of time delay. On the line diagram, however, only half the arrow is ever shown. Note that even with only half the arrow showing, it is easy to determine that it is pointing in the direction of ON delay.

ON Delay (Timed Open). You will recall earlier that we said an ON delay timer could be designed to open or close a circuit after a predetermined

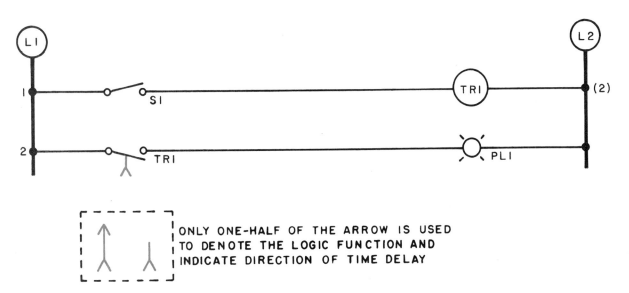

ONLY ONE-HALF OF THE ARROW IS USED TO DENOTE THE LOGIC FUNCTION AND INDICATE DIRECTION OF TIME DELAY

Figure 7-11 When switch contacts S1 close, the relay is turned on. After the timing cycle is completed, the contacts to the pilot light close, causing it to glow.

time delay. Figure 7-12 illustrates how a timed open ON delay function could be performed. With switch S1 closed, the solenoid plunger would again force air from balloon A to balloon C through orifice B. This time, however, after five seconds the contacts CR1 would open the circuit to the pilot light and the pilot light would go out.

Schematically this circuit would appear like the line diagram of Figure 7-13. Note again that only one half of the arrow is present, but it easily indicates that the NC contact will open after the ON delay function has taken place.

It should be pointed out that a synchronous clock timer or solid state timer could easily be substituted for the pneumatic timer used in the example. We chose the pneumatic timer at this point only because it is probably the easiest to understand in terms of mechanical operation.

OFF Delay. OFF delay, like ON delay, derives its name from the electrical condition of the *timer*. When a timer no longer receives an active signal or is in fact turned off, a predetermined time interval must take place before any action takes place in the load circuit. Again, it is extremely important to realize that it is the turning off of the timer, *not* the load, which determines the logic function. Once the timer is off, it is capable of

turning on a load or turning off a load, depending upon the way the timer is wired into the circuit.

OFF Delay (Timed Open). Figure 7-14 illustrates in line diagram format an OFF delay timed open contact circuit which could provide cooling for protection in a projector once the bulb has been turned off but has not had time to cool down. The circuit would operate in the following manner: Closing switch S1 will turn on the projector bulb and activate timer coil TR1. With coil TR1 energized, NO contacts TR1 will immediately close, energizing coil M1, which controls the cooling fan motor. Both the projector bulb and the cooling fan will remain on as long as switch S1 stays closed. When switch S1 is opened, however, the projector bulb will go off and power will be removed from coil TR1. The contacts TR1 will remain closed for a predetermined OFF delay and then open, causing the coil of the cooling fan M1 to drop out and shut down the cooling fan. This off delay timed open circuit is generally set to adequately cool the projector equipment before it is shut off.

OFF Delay (Timed Closed). Figure 7-15 illustrates in line diagram format an OFF delay timed closed contact circuit which provides a pumping system backspin protection and surge protection

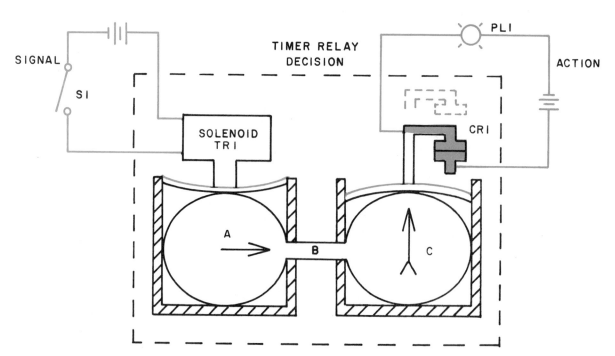

Figure 7-12 ON delay timed open logic could be accomplished using 2 balloons. In this case, after the timing cycle is complete, the balloon forces the contacts open.

on stopping. Surge protection is often necessary when the pump is turned off and the long column of water is stopped by a check valve. The force of the sudden stop may cause surges which operate the pressure switch contacts, thus subjecting the starter to "chattering." *Backspin* is the name given to the backward turning of a centrifugal pump when a head of water runs back through the pump just after it has been turned off. Starting the pump during backspin might damage the pump motor.

To eliminate the damage which could result from surges and backspin, the circuit is designed to respond as follows: Actuating pressure switch PS1 causes timer relay coil TR1 to be energized. With TR1 energized, the NC contacts of TR1

Figure 7-13 When switch contacts S1 close, the relay is turned on. After the timing cycle is completed, the contacts to the pilot light open, causing the pilot light to go out.

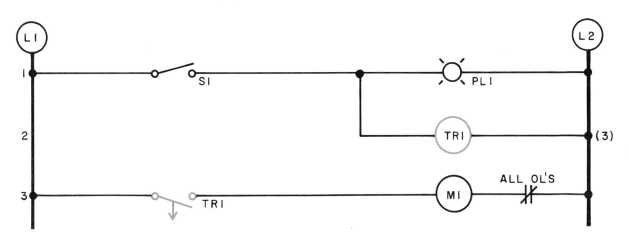

Figure 7-14 The line diagram illustrates an OFF delay timed open contact circuit which could provide cooling fan protection in a projector once the bulb has been turned off but has not had time to cool down.

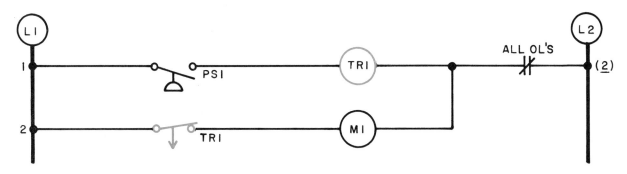

Figure 7-15 The line diagram illustrates an OFF delay timed closed circuit which could provide a pumping system backspin and surge protection on stopping.

immediately open and cause coil M1 to be de-energized, shutting off the pumping motor. Even if pressure switch PS1 opens, the coil and motor will remain off for a predetermined time (to allow surges and backspin to clear) and then restart once the OFF delay function has taken place. This OFF delay function provides that the system cannot be operated again until TR1 has timed out and its NC contacts return to their normally closed position.

Complex Timing Circuits

Figure 7-16 illustrates a more complex timing circuit using both NO timed ON delay contacts and NC timed ON delay contacts. The circuit would operate in the following manner. When pressure switch PS1 closes, coil M1 and coil TR1 will be energized or turned ON. We note immediately that TR1 also controls a set of NO contacts in line three and a set of NC contacts in line four. Since the timing function is ON delay, the action of the timer will take place at a time interval after it has been energized. Assuming the time delay function is 15 seconds, the following sequence will take place as long as the pressure switch PS1 remains closed: Coil M1 and coil TR1 will remain on; coil M2 will be energized and coil M3 will be de-energized. If pressure switch PS1 is opened, coil M1 and coil TR1 will be de-energized; at the same instant coil M2 will be de-energized and coil M3 will be energized, returning the circuit to its normal condition.

Adding a Code for ON Delay Circuits

From the circuit of Figure 7-16, you can see that it is sometimes very difficult to follow the conditions of the loads in a circuit when timers are used to control them. This difficulty exists because there are basically three stages the load is involved with in ON delay timing; namely, reset, timing and timed out. To simplify this procedure, we will add a coding system to the load which will help you more readily determine if the load is in reset, timing or timed out. We will also indicate whether this logic function is ON delay or OFF delay.

First, it will be helpful to define reset, timing and timed out. *Reset* can be defined as the beginning of the timing cycle when no signals are being introduced into the circuit and all contacts are in their normal position. *Timing* can be defined as the period of time from when the timer receives a signal until it makes a decision for action. *Timed out* can be defined as the period of time the action

is allowed to continue before reset occurs.

Next, it will be necessary to denote each point in the operating cycle (reset, timing and timed out) when the load is energized and when it is de-energized. We will use an X to denote when the load is energized and a O to denote when the load is de-energized. Reading from left to right, you will be able to tell by the X or O whether a load is energized or de-energized in reset, timing or timed out. For example, a sequence OOX indicates that the load is de-energized during reset, remains de-energized during timing, and becomes energized during timed out.

Applied to a circuit, the coding system would look like Figure 7-17. You will note that this is the same circuit we found in Figure 7-16. Now, however, it should be much easier to determine the condition of each load at each step in the timing cycle. Referring to line one and line two, we can see that coil M1 and coil TR1 will be energized as long as the pressure switch PS1 is energized (receives a signal). Referring to line three, we see that the coil M1 will be de-energized during reset (before the signal), will remain de-energized while timing, and will be energized after timed out. Referring to coil M3, we see that it will be energized during reset, remain energized while timing and will de-energize after timed out. From these codes, we can easily determine the condition of the load during reset, timing and timed out.

When the triangle around the O or X points in an upward direction, as in Figure 7-17, the timing function is ON delay logic; when the triangle points down, the function is OFF delay logic.

NOTE: Although adding codes to timing circuits is helpful, it is not standard practice. Therefore, not all circuits in this textbook have codes.

Adding a Code for OFF Delay Circuits

Most of the symbols used in ON delay logic can be applied to OFF delay logic; however, it is important to note that OFF delay logic is not the same as ON delay logic. The code used with OFF delay logic will have a different meaning from the code used for ON delay.

The most significant difference in OFF delay logic is that the action part of the circuit will take place after the timer has been shut off. With this in mind, let us look at the three phases an OFF delay timer goes through (reset, before timing and during timing) and define what each means. *Reset* can be defined as the beginning of the timing cycle when no signal is being introduced into the circuit and all contacts are in their normal position.

Before timing can be defined as the period of time that the control signal is applied to the timer and the contacts immediately change position. *During timing* can be defined as that period of time after which the control signal is removed until a decision for action has been initiated. At this point the circuit may automatically reset itself or must be manually reset.

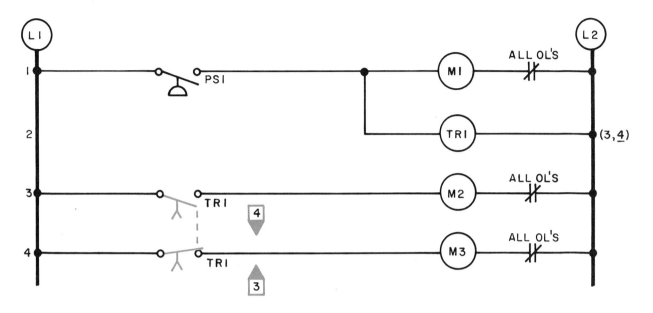

Figure 7-16 The line diagram illustrates an ON delay timing circuit using both normally open and normally closed contacts.

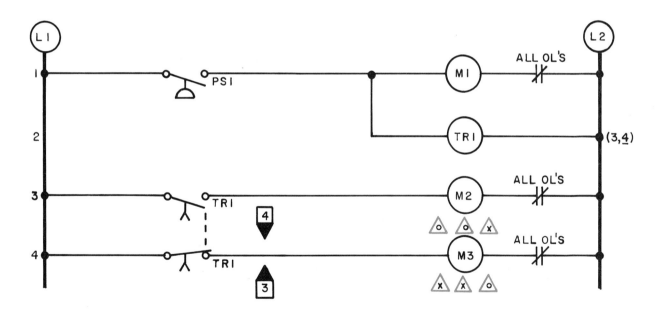

Figure 7-17 By using an X to indicate when a load is ON and an O to indicate when a load is OFF, it becomes very easy to determine the condition of the load in reset, timing and timed out in ON delay circuits. A triangle surrounding the X or O and pointing up denotes ON delay.

An example of an OFF delay timing circuit is illustrated in Figure 7-18. Referring to line one and line two, we can see that coil M1 and coil TR1 will be energized as long as pressure switch PS1 is actuated. Referring to our code in line three, we can see that coil M2 will be de-energized during reset (no signal), will energize before timing (signal), and will remain energized during timing (signal removed) until the predetermined time interval has passed. After the timing cycle has passed, coil M2 will de-energize as the NO contacts of TR1 return to their normally open position (reset). A square is now used during timing because the circuit automatically resets itself back to normal. If the square is not present, the unit would have to be manually reset.

Referring to line four, we can see that the load coil M3 will be energized during reset, will de-energize before timing and will remain de-energized during timing (signal removed). When the timing cycle is complete, however, the coil M3 will energize again as the NC contacts of TR1 return to their normally closed position. Again, a square is used during timing because the circuit automatically resets itself.

WIRING DIAGRAMS AND SPECIFICATIONS FOR TIMERS

Now that you understand the basic concepts of ON delay and OFF delay logic, it is important for you to learn how to properly connect various timers into practical circuits to perform both of these logic functions.

To aid the electrician or technician in selecting and installing timers properly, manufacturers provide catalog information and specification sheets to explain their product. In the remainder of this unit we will look at the three basic types of timers and determine what information needs to be found in manufacturers' specification sheets and wiring diagrams for proper installation.

Dashpot Timers Specifications

From an earlier discussion in this unit, we determined that a dashpot timer provided time delay by controlling how rapidly air or a liquid is allowed to pass into or out of a container through a controlled orifice (opening).

Figure 7-19 illustrates a specification sheet on a typical pneumatic timer used for many industrial applications. In the next few paragraphs we will examine these specifications to determine how they should be used in practical situations.

Special Features. When reading specifications, look for the special features of a device. In certain cases, the special features will make the device more useable and attractive. In other cases, the special features may drive up the price of the device and be unnecessary.

In the specification sheet of Figure 7-19, we

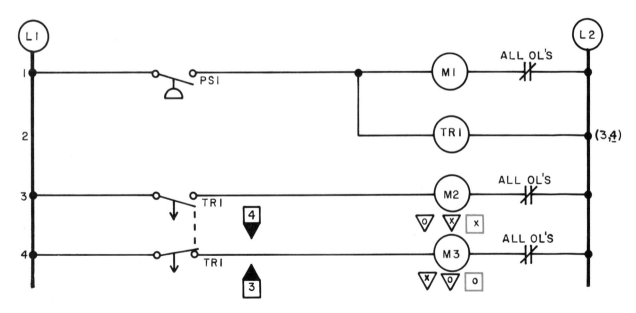

Figure 7-18 The line diagram illustrates an OFF delay timing circuit using both normally open and normally closed contacts.

STANDARD TIMER

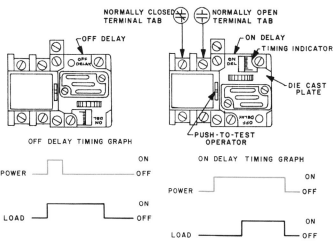

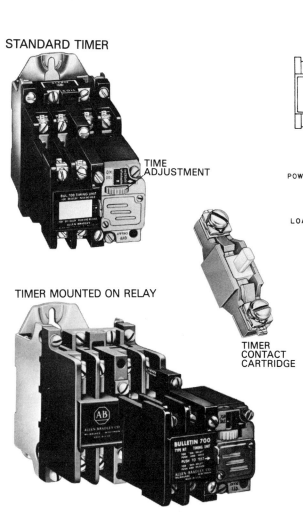

TIME ADJUSTMENT

TIMER MOUNTED ON RELAY

TIMER CONTACT CARTRIDGE

NORMALLY CLOSED TERMINAL TAB   NORMALLY OPEN TERMINAL TAB

OFF DELAY

ON DELAY

TIMING INDICATOR

DIE CAST PLATE

PUSH-TO-TEST OPERATOR

OFF DELAY TIMING GRAPH

POWER   ON / OFF

LOAD   ON / OFF

ON DELAY TIMING GRAPH

POWER   ON / OFF

LOAD   ON / OFF

Features

Each timer unit has two timed contact cartridges. These are individually convertible from normally open to normally closed simply by removing the cartridge, turning it over, and remounting it in the unit. The terminal tab on each end of the cartridge indicates the contact arrangement.

The timing unit is supplied with On-Delay action. It can easily be converted to Off-Delay by loosening two captive screws on the die cast plate, rotating the plate 180° and tightening the screws.

A push-to-test operator, located on top of the timing unit, is provided for manual operation of the timing contacts. This allows the timing range adjustment to be made without energization of the relay.

Adjustment

The timing range is adjusted by a front-mounted thumbwheel screw. As the timing range is increased or decreased, a timing indicator (located next to the adjustment screw) will move establishing a reference point in relation to a graduated scale.

Applications

The timing unit which is always used in conjunction with a Bulletin 700 Type N or NM relay can be utilized as a separate timer for any application within its operating characteristics or as part of a modular relay system on a control panel.

GENERAL SPECIFICATIONS

Minimum Time Delay —
.2 seconds
Maximum Time Delay —
60 seconds
Reset Time —
.075 seconds
Repeat of Accuracy —
± 15% of setting
Type of Timing Contacts —
Two convertible contact cartridges
Type of Instantaneous Contacts —
Up to 4 convertible contact cartridges for the Type N1 and up to 2 convertible contact cartridges for the Type NMT.
Operating Coils —
Coils can be supplied for voltages and frequencies up to 300 volts, 50-60 Hz. AC.
Operating Temperature Range —
0° F to 104° F
Note:
Minimum temperature is based on the absence of freezing moisture or water.
Contact Ratings — in voltage and current.

NEMA A 300-MAX AC CONTACT RATING PER POLE					
MAX AC VOLTAGE 50 OR 60 HZ	AMPERES		CONTINUOUS CARRYING CURRENT	VOLTAMPERES	
	MAKE	BREAK		MAKE	BREAK
120	60	6	10	7200	720
240	30	3	10	7200	720

NEMA P 300-MAX DC CONTACT RATING PER POLE			
125	1.1	5	138
250	0.55	5	138

Figure 7-19   Specification sheet for a typical pneumatic timer. (Allen-Bradley Company)

should note the following features: (1) Each timer unit has two timed contact cartridges that are individually convertible from normally open to normally closed simply by removing the cartridge, turning it over, and remounting the unit. The advantage of this conversion is the possibility of changing contact arrangements without changing the timer; (2) The unit is convertible from ON delay to OFF delay on the job site. This interchangeability is one major advantage of pneumatic timers and synchronous motor timers over solid state timers, since solid state timers generally will be either ON delay or OFF delay and are not usually convertible. It should be pointed out, however, that there are solid state timers which are convertible and adjustable; (3) A push-to-test operator is provided for manual operation of the timing contacts, allowing the unit to be time adjusted without energizing the relay in a line circuit.

Adjustment. Each timer must have some type of adjustment to establish its timing range. In this unit adjustment is made by a front-mounted thumbwheel screw. As the timing range is increased or decreased, a timing indicator next to the thumbwheel will move to establish a reference point in relation to a graduated scale. This scale indicates the percentage of time delay.

General Specifications. Manufacturers' timer specifications will generally state the time range for the device. For the timer of Figure 7-19, the range is from 0.2 seconds to 60 seconds with a reset time of 0.075 seconds and an accuracy of plus or minus 15% of the setting. What this means is that the timer can provide a time delay as short as 0.2 seconds or as long as 60 seconds but may vary as much as 15% from this value on a repeat basis. The device will take 0.075 seconds to reset and be ready for the next signal.

The specifications for this timer indicate that the input coil voltage can be up to 300 volts at either 50 or 60Hz. These voltages make the timer more adaptable to a variety of industrial circuits in the United States and abroad.

The operating temperature range of the timer is quite good, provided there is no freezing moisture or water.

Note that the current rating of the contacts for this timer in the contact rating chart is dependent upon the voltage which is applied to the contacts. You will note that for 120 volts the contacts make 60 amperes, break 6 amperes and have a continuous current of 10 amperes. The difference is that any time a contact breaks a circuit, the arc formed can have a destructive burning effect on the contacts. Since the arc is not as prevalent when the contacts make, they can handle much more current on making than on breaking. In some circuits with special loads, this must be taken into consideration; however, in most cases the continuous current rating is generally used. Note also that the DC rating on these contacts is much lower than the same voltage rating of AC. This is because DC is continuous and delivers more power to the contacts in a given moment than an

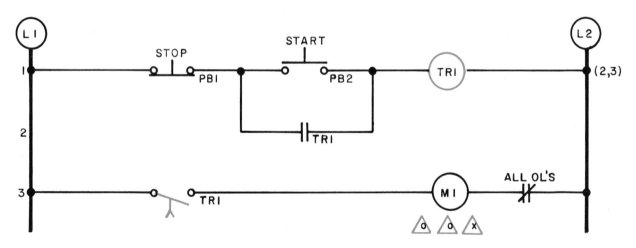

Figure 7-20 A standard start/stop station controlling an ON delay timer with a set of timed closed contacts.

equivalent AC. The effect of AC and DC on contacts can be reviewed in Unit 6.

Timing Graphs

Since timers may offer a variety of timing sequences, it has become quite necessary and popular to graph out how the timer affects the load during the timing sequence. You will note in Figure 7-19 that a graph is shown for ON delay and OFF delay timing, since the device is convertible. In cases where it is one or the other, only one graph will be shown. These timing graphs are important enough to justify a closer look as to what information they contain.

ON Delay Timing Graph. Using the circuit of Figure 7-20, we see a standard start/stop station controlling an ON delay timer with a set of timed

closed contacts. From our coding on the load we see that the load is de-energized during reset, de-energized during time and is energized after timed out.

From our timing graph (Figure 7-21) we can also visually follow the same sequence. From the control circuit graph, we can see that reset is shown by power off (no signal) and that during timing and timed out are shown by power on, since this is ON delay. Looking at the load graph, we see that the load is off in reset, off during timing and on after timed out. The load will remain ON until manually reset by pushing the stop button which will de-energize the coil. You will note from our coded triangles added to the load graph that this graph supplies the same timing information about what will happen to the load without showing a timer in an actual circuit.

ON-DELAY – TIMED CLOSED

CONTROL CIRCUIT TIMING GRAPH

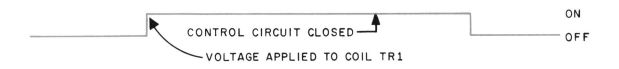

LOAD CIRCUIT TIMING GRAPH

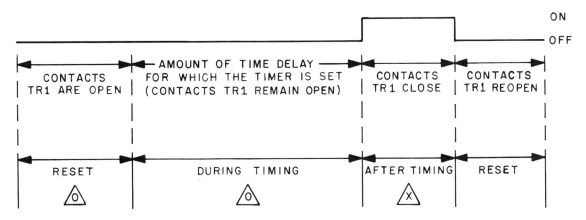

Figure 7-21 Timing graph illustrating how the control circuit and load circuit perform on a typical ON delay timer with a set of timed closed contacts.

OFF Delay Timing Graph. Using the circuit of Figure 7-22, we see an OFF delay timer with a set of timed open contacts controlling the coil M1. From our load coding system, we see that the load is de-energized at reset, energizes before timing and remains energized during timing. Once the unit has finished timing, the load is de-energized, or returned to reset. Since the contacts return to their normal condition, the circuit can be thought of as automatically resetting after performing its normal function. Automatic reset is denoted by the square with an X in it.

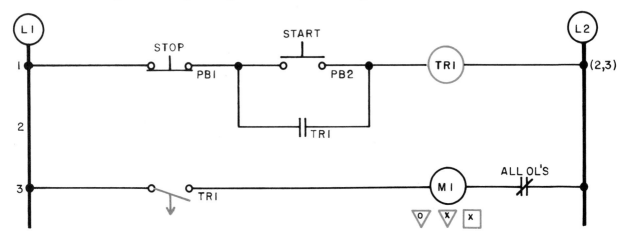

Figure 7-22 A standard start/stop station controlling an OFF delay timer with a set of timed open contacts.

OFF - DELAY - TIMED OPEN

CONTROL CIRCUIT TIMING GRAPH

LOAD CIRCUIT TIMING GRAPH

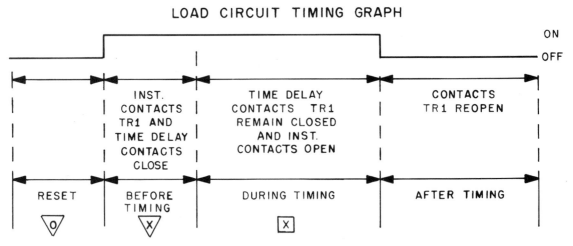

Figure 7-23 Timing graph illustrating how the control circuit and load circuit perform on a typical OFF delay timer with a set of timed open contacts.

Looking at the control circuit graph in Figure 7-23, we can see that pressing the start pushbutton applied a signal to the coil TR1, and pressing the stop pushbutton removed the signal from coil TR1. From the load graph we see the load de-energized during reset, energized before timing, and energized during timing. Once the during timing sequence finishes, the load de-energizes and returns to reset. Again the timing graph reveals the operating characteristics of the circuit during reset, before timing and during timing much the same as the load code but without the use of a circuit. This type of information is particularly helpful because the circuit may contain other devices of control than pushbutton, yet the logic of the circuit shown by the graph will not. These graphs will also be helpful to illustrate more complex timing sequences toward the end of this unit.

Motor Driven Timer

Figure 7-24 illustrates a typical motor driven timer used in many industrial applications such as heating, conveyor and machine tool control. This type of timer has the advantage of containing both instantaneous and time delay contacts. Many timers of this type are also convertible from ON delay to OFF delay. This timer, like most solid state timers, can be plugged into and taken out of the circuit without removing any wiring. This feature makes troubleshooting much easier. In this unit we will learn how to wire motor driven timers into simple control systems.

Wiring Diagrams. Wiring diagrams are very important on motor driven timers, since you must locate and wire to several different locations on the timer for it to perform properly.

Locating Contacts on Motor Driven Timer. The wiring diagram for a timer is generally located on the back of the timer. Figure 7-25 illustrates a common wiring diagram for a timer. By

Figure 7-24 Typical synchronous clock timer. (Eagle Signal Industrial Controls)

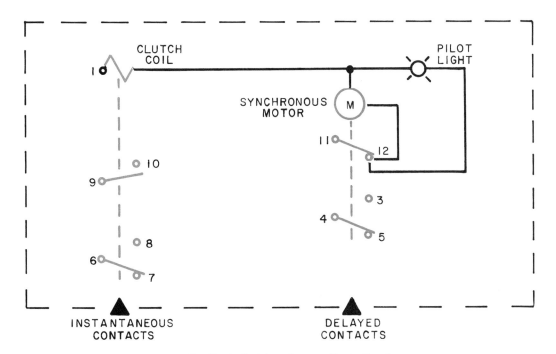

Figure 7-25 Typical wiring diagram illustrating the connection points on a synchronous clock timer.

quickly surveying the diagram, we can locate the clutch coil, the synchronous motor and the pilot light. The clutch coil in this timer operates much the same as that found in an automobile to engage and disengage the motor. In this case, however, the clutch coil engages and disengages contacts 9-10 and 6-7-8. When the clutch is engaged, contacts 9, 10 and 6, 8 instantaneously close. Simultaneously, contacts 6-7 open. In other words, the clutch controls two NO contacts and one NC contact instantaneously. The timing motor controls contacts 11, 12 and 3-4-5 through a time delay. When the motor times out, contacts 4, 5 and 11, 12 open and contacts 3, 4 close. For a very practical reason soon to be illustrated, you should also know that contacts 11, 12 open slightly later than contacts 4, 5 after the motor times out. The pilot light, obviously wired in parallel with the motor indicates when the motor is timing. When the

timer is reset, contacts 4, 5 and 11, 12 close and contacts 3, 4 open.

Wiring a Motor Driven Timer. Figure 7-26 illustrates how a motor-driven timer can be wired into L1 and L2 to provide power to the circuit. Tracing the circuit from L1 to terminal one we see that current would flow through the clutch coil on to L2. Also from terminal one current would flow through closed contacts 11, 12, feeding the parallel circuit provided by the motor and pilot light and then return to L2.

Operation of Motor Driven Timer on Loads. Figure 7-27 illustrates what effect a motor driven timer activated by a limit switch would have on various loads wired into the circuit.

As with any circuit, this timer can be controlled by a manual, mechanical or automatic input. Figure 7-27 illustrates the wiring of the timer to

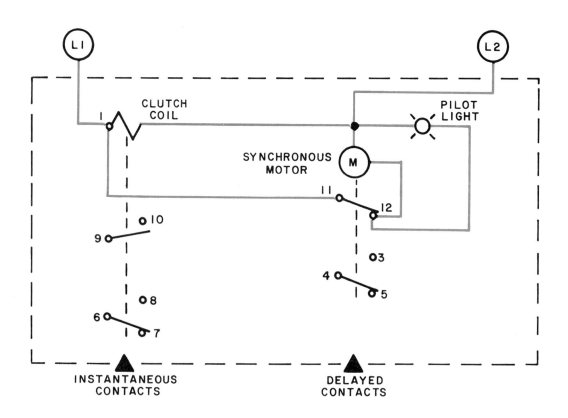

Figure 7-26 Wiring diagram of a synchronous clock timer as it would be connected into the power circuit.

achieve control using a "sustained" mechanical input. This input, shown as a limit switch, must remain closed to energize the clutch solenoid and power the timing motor. The timer illustrated is connected for ON delay, requiring input power to close the clutch starting the timer. The contacts—both instantaneous and time delay—are connected to four loads marked, A, B, C, and D. These loads can represent any load such as a solenoid, magnetic starter, light, etc. A code is added above each load to illustrate the sequence during reset, timing and timed out. Loads C and D are strictly relay type responses, using only the instantaneous contacts. Loads A and B utilize the combined action of the instantaneous and delay contacts to achieve the desired sequence. In this circuit the timing motor is wired through the delay contacts 11, 12, to insure motor cut off after the timer times out. This is required because the limit switch is still closed after time out. The limit switch otherwise would have to be opened to reset the timer. This also means that loss of plant power will reset the timer, as the clutch will open when power is lost.

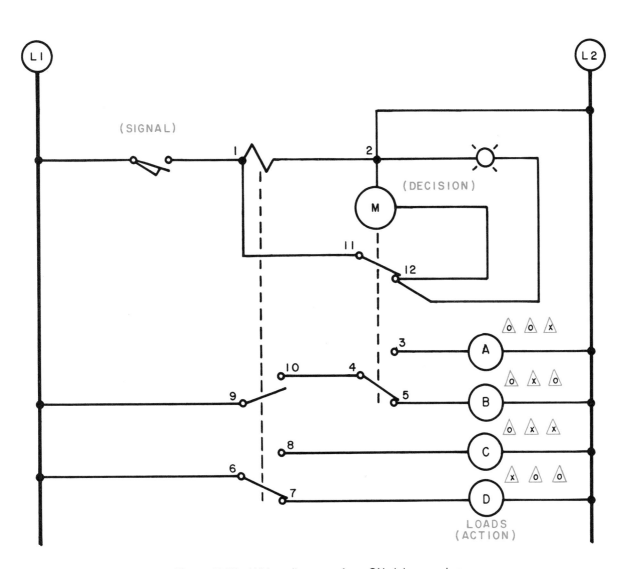

Figure 7-27 Wiring diagram of an ON delay synchronous motor timer controlling several loads when actuated by a limit switch.

Practical Application of Motor-Driven Timer. A practical application of using this timer with a "sustained" input to control a circuit is illustrated in Figure 7-28, top. In this application the boxes coming down the conveyor belt are to be filled with detergent. Each box must be filled with the same amount of detergent and the process must be automatic. To accomplish this, the timer circuit of Figure 7-28, bottom, is used to control the time it takes to fill one box. A limit switch "sustained input" is used to detect the box. A motor is used to drive the conveyor belt, and a solenoid is used to open or close the hopper full of detergent.

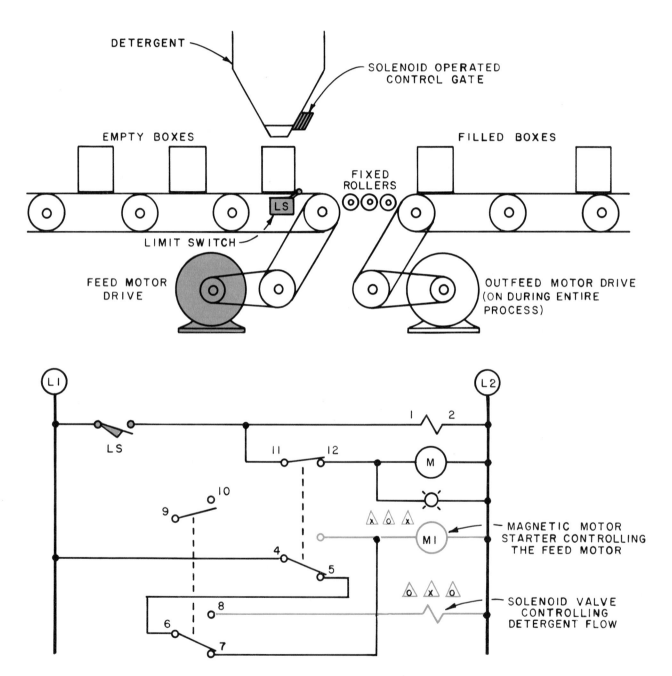

Figure 7-28 Top: Practical application of a synchronous motor timer used on an assembly line to fill boxes of detergent. Bottom: Wiring diagram of a synchronous motor timer used on an assembly line to fill boxes of detergent.

The sequence for this circuit would be as follows:

Sequence	Load	Action
1	Motor on (X)	Boxes coming down conveyor
	Solenoid Off (O)	No detergent fill
Box Contacts Limit Switch		
2	Motor Off (O)	Conveyor stops
	Solenoid on (X)	Box being filled
Box is Filled—Time Out		
3	Motor on (X)	Removing filled box, open limit switch resetting timer
	Solenoid Off (O)	No detergent fill

To accomplish this sequence, the timer, limit switch, motor starter and solenoid are connected as illustrated in Figure 7-28. The program for each load is added above that load. This program accomplishes the desired sequence for this application.

If a circuit is required to have a "momentary" input, such as a pushbutton to initiate timing, then MEMORY could be added to the circuit. Figure 7-29 illustrates a circuit using a pushbutton as the input signal. As with any MEMORY circuit, a NO instantaneous contact must parallel the pushbutton if MEMORY is to be added into the circuit. To accomplish this, a NO instantaneous contact 9-10 is paralleled with the pushbutton so that power is maintained when the control switch is released. Just as with all MEMORY circuits, a separate reset (stop) switch is used to reset the timer. After the timer has timed out, the timer

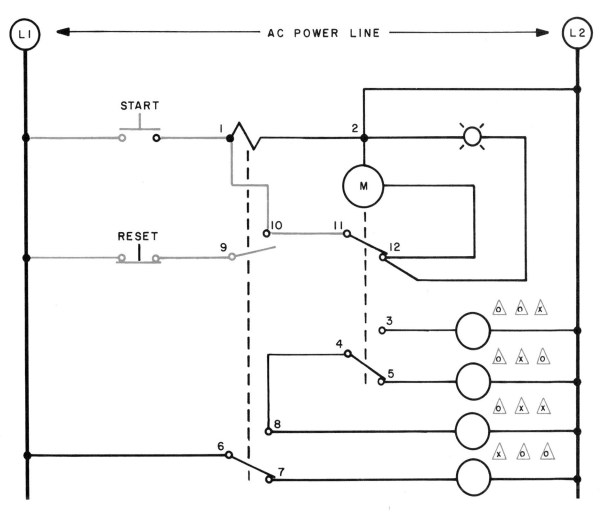

Figure 7-29 Wiring diagram of a synchronous motor timer circuit with MEMORY added similar to that found on standard start/stop control.

will reset only if the reset pushbutton is pressed. The sequence of each load is illustrated by the code given. These loads illustrate some of the many conmbinations possible using instantaneous and time delay contacts. As you may have noticed, these diagrams are not in pure line diagram form. Remember that line diagrams are standards used with any manufacturer, but wiring diagrams are the actual diagrams matching the logic of the line diagram to the manufacturers' product designed to perform that logic. Since this was the actual diagram found on the timer, this circuit diagram was used.

Solid State Timer

Solid state timers have become very popular because they are both dependable and economical. As mentioned earlier, solid state timing can be accomplished in many ways and how it is done electronically is not of prime importance. Since solid state timers are relatively inexpensive, they are seldom repaired. Rather, they are replaced as a unit. Therefore, discussion here will center on their selection and application.

Wiring Diagram. Figure 7-30 illustrates a typical solid-state timer with the manufacturer's specifications, wiring diagram (pin arrangement) and operation chart. This timer uses an eight pin plug-in relay socket. This enables the timer to be replaced without disconnecting any wires once the socket has been wired. To read the pin numbers, find the identification "notch" (see Figure 7-31) and count clockwise starting with one. You will find that not all manufacturers mark the numbers on the pins, so remembering this is important.

Locating Contacts. To find the NO and NC contacts in Figure 7-30, you must now use the pin numbers. You will find that pin one is common to both pins three and four. This is because pins three and four have an arrow pointing to pin one. So pins one and three are NO contacts and pins one and four are NC contacts. This type of switch is also called an SPST (Single-pole, single-throw) switch. Likewise, pins eight and six are NO and pins eight and five are NC. Pins eight, five, and six also make up an SPST switch. Together, these two switches (pins 1, 3, 4, and 8, 5, 6) would make a DPDT (Double-pole, double-throw) switch, as listed under the specifications for this timer.

General Operating Specifications. If you refer to the specifications in Figure 7-30, you will find the current rating for the contacts (10 amps at 115 VAC, and 6 amps at 230 VAC). This is the amount of current that each contact can safely handle. It determines the maximum size load the timer can turn on or off. It is important for you to *not* confuse the rating of the contacts with the coil voltage needed to run the timer. The coil voltage for this model can be anywhere from 12 to 230 volts and may be AC or DC (see ordering information), but the contact rating does not change. Thus it is possible that a coil voltage or 12 VDC (pins 2 and 7) can control a 120 VAC load (through pins 1 and 3).

Timing Graph. To help the electrician understand the logic of a timer, most manufacturers of solid-state timers provide an operating graph. The operating graph tells the electrician that this model timer is an ON delay timer. To understand this, refer to the operating graph in Figure 7-30. The graph is referring to the condition of the contacts (normal condition or switched) at reset, during timing and after timing. On the graph, the OFF level is on the bottom and represents the contacts in their "normal" condition. The ON level is on the top and represents the contacts in their "changed or switched" condition. For example, let us use pins one and three (NO) (Figure 7-30). When the coil voltage is applied, (pins two and seven) by closing the switch, contacts one and three remain open for whatever the "delay" is set for. After the delay, pins one and three close (turning on the load), and remain closed for as long as the voltage is applied to the coil. When the coil voltage is removed from the coil (switch open), pins one and three instantly re-open (load off). Thus the logic of this timer is ON delay and can be used in any ON delay circuit as long as it meets the specifications (contact rating, time duration, etc.)

Timing Adjustment. The timer in Figure 7-30, like most solid-state timers, is furnished with a knob mounted on top to adjust the time delay. This knob can be set between 0 and 10. The numbers (0 to 10) on the timer do not indicate the time delay directly. They represent the percentage of the total time the timer can be set for. Thus if the total range of the timer is 1 to 60 seconds, a setting of 5 on the timer dial would indicate a 30-second delay (5 represents 50%, and 50% of 60 seconds is 30 seconds). If the total range of the timer was 2 to 300 seconds, a setting of 5 on the timer dial would indicate a 150-second delay. Manufacturers use this system of making and setting the time range when repeatability can generally be about ±0.1%. If greater repeatability is required, a digital adjustable switch can be used. The repeatability of this type of timer is generally about ±0.005%.

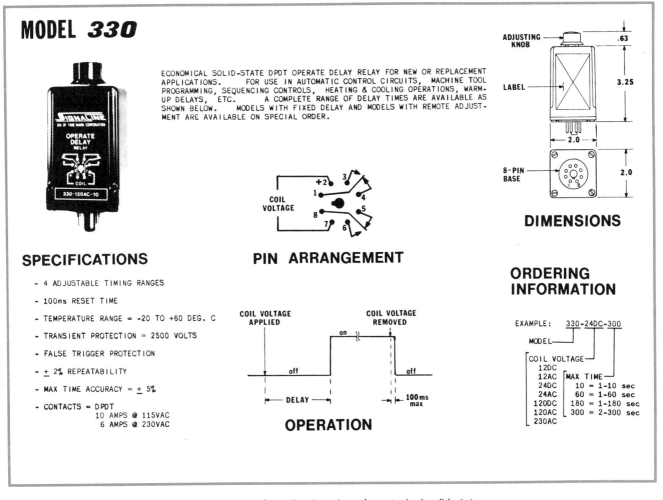

MODEL 330

ECONOMICAL SOLID-STATE DPDT OPERATE DELAY RELAY FOR NEW OR REPLACEMENT APPLICATIONS. FOR USE IN AUTOMATIC CONTROL CIRCUITS, MACHINE TOOL PROGRAMMING, SEQUENCING CONTROLS, HEATING & COOLING OPERATIONS, WARM-UP DELAYS, ETC. A COMPLETE RANGE OF DELAY TIMES ARE AVAILABLE AS SHOWN BELOW. MODELS WITH FIXED DELAY AND MODELS WITH REMOTE ADJUSTMENT ARE AVAILABLE ON SPECIAL ORDER.

PIN ARRANGEMENT

OPERATION

DIMENSIONS

SPECIFICATIONS

- 4 ADJUSTABLE TIMING RANGES
- 100ms RESET TIME
- TEMPERATURE RANGE = -20 TO +60 DEG. C
- TRANSIENT PROTECTION = 2500 VOLTS
- FALSE TRIGGER PROTECTION
- ± 2% REPEATABILITY
- MAX TIME ACCURACY = ± 5%
- CONTACTS = DPDT
  10 AMPS @ 115VAC
  6 AMPS @ 230VAC

ORDERING INFORMATION

EXAMPLE: 330-24DC-300

MODEL

COIL VOLTAGE
  12DC
  12AC
  24DC
  24AC
  120DC
  120AC
  230AC

MAX TIME
  10 = 1-10 sec
  60 = 1-60 sec
  180 = 1-180 sec
  300 = 2-300 sec

Figure 7-30 Specification sheet for a typical solid state timer. (Time Mark Corporation)

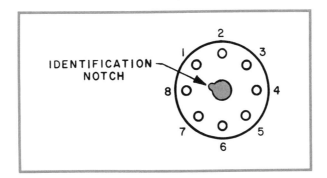

Figure 7-31 To read the pin numbers on a typical solid state relay, find the identification notch and count clockwise starting with one.

Timing symbols and codes used with both ON delay and OFF delay timers are reviewed in Figure 7-32.

NOTE: Although adding codes to timing circuits is helpful, it is not standard practice. Therefore, not all circuits in this textbook have codes.

TIMING SYMBOLS

NOTC:	**On-Delay Timed Closed Contact**—*Timer contact normally open — timed closed upon relay energization — opens immediately upon relay de-energization.*	
NCTO:	**On-Delay Timed Open Contact**—*Timer contact normally closed — timed open upon relay energization — closes immediately upon relay de-energization.*	
NOTO:	**Off-Delay Timed Open Contact**—*Timer contact normally open — closes immediately upon relay energization — timer contact times open upon relay de-energization.*	
NCTC:	**Off-Delay Timed Closed Contact**—*Timer contact normally closed — opens immediately upon relay energization and timer contact times closed upon relay de-energization.*	
NO:	**Instantaneous Contact**—*Normally open — Upon relay energization this contact closes immediately and opens immediately upon relay de-energization.*	
NC:	**Instantaneous Contact**—*Normally closed — Upon relay energization this contact opens immediately and closes immediately upon relay de-energization.*	

LOAD SEQUENCE FOR "ON DELAY"			
LOAD MARKING	RESET	DURING TIMING	AFTER TIMING
ENERGIZED	△X	△X	△X
DE-ENERGIZED	△O	△O	△O

LOAD SEQUENCE FOR "OFF DELAY"			
LOAD MARKING	RESET	BEFORE TIMING	DURING TIMING
ENERGIZED	▽X	▽X	▽X
DE-ENERGIZED	▽O	▽O	▽O

☐ SQUARE AROUND o OR x DENOTES AUTOMATIC RESET

Figure 7-32 Timing reference and review material.

Programmable timers that offer other timing functions in addition to ON delay and OFF delay are available. For example, Figure 7-33 shows a programmable timer that can be set for ON delay, interval (also called one-shot), or recycle timing by setting dip switches. The time range is programmed by a selection knob.

Universal Timer
Types S 117, S 217

electromatic

- Multi-function. Multi-voltage
- 4 selectable functions: Delay on operate, interval timer, symmetrical recycler (ON/OFF first)
- 4 selectable time ranges: 0.8 s to 60 min
- Automatic start
- Knob-adjustable time within range
- Oscillator-controlled time circuit
- Repeatability deviation: ≤ 1%
- Output: 10 A SPDT or 8 A DPDT relay
- Plug-in type module
- S-housing
- LED-indication for relay on
- AC and DC power supply

Wiring Diagrams

Power supply
S 117 156, S 117 166

Power supply
S 217 156, S 217 166

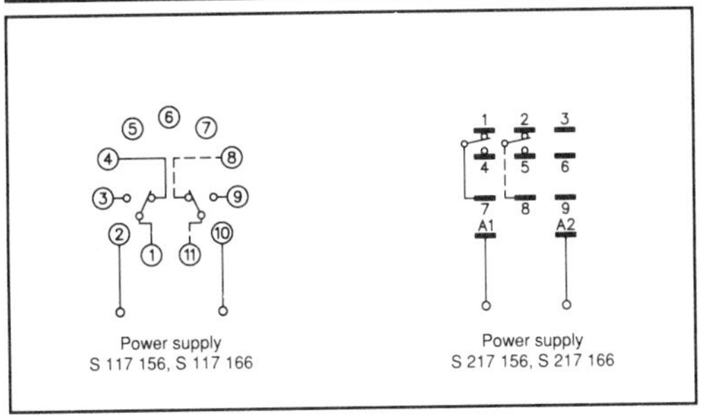

Function/Time Setting

Selection of function
DIP-switch selector (3 & 4).

1. Delay on operate

2. Interval timer

3. Recycler, OFF-time first

4. Recycler, ON-time first

Selection of time ranges
DIP-switch selector (1 & 2).

0.8 s - 15 s

3 s - 60 s

24 s - 480 s

3 m - 60 m

Time setting
Knob-adjustable on scale in per cent of max. time.

DIP-switches for selecting function and time range are placed behind a small removable frontplate on the time relay.

Accessories

Sockets◊	S 411, D 411 B
Hold down spring◊	HF
Mounting rack	SM 13
Socket covers	BB 4, BB 5B
Potentiometer lock	PL 3

For further information refer to "Accessories", p. 138 ff. For other AC/DC voltages refer to "General Information".

Operation Diagram

Power supply

1. Relay on
2. Relay on
3. Relay on
4. Relay on

Figure 7-33   Programmable timer. (Electromatic Controls Corp.)

1. How are the NO contacts of a relay, contactor or motor starter cross referenced in the line diagram for easy identification and location?
2. How are the NC contacts of a relay, contactor or motor starter cross referenced in the line diagram for easy identification and location?
3. What are the two different methods of illustrating how mechanically connected contacts found in different control lines belong to the same control switch?
4. What are the three basic types of timers used?
5. What is meant by ON delay logic?
6. What are the symbols for ON-delay NO and ON-delay NC contacts?
7. What is meant by OFF delay logic?
8. What are the symbols for OFF-delay NO and OFF-delay NC contacts?
9. When a load is coded (X-X-O, O-O-X, etc.) and an ON-delay timer is controlling the load, what does each part of the three-part code stand for?
10. When a load is coded (X-X-O, O-O-X, etc.) and an OFF-delay timer is controlling the load, what does each part of the three part code stand for?
11. What do timing graphs illustrate to the electrician?
12. The motor-driven timer illustrated in Figure 7-25 includes how many NO and NC instantaneous and timed delayed contacts?
13. In the wiring diagram illustrated in Figure 7-26, are the clutch, synchronous motor and pilot light connected in series, parallel, or a series/parallel combination?
14. If the pin numbers were not marked on a timer socket, how could the pin numbers be identified?
15. On a timer with an adjustable time setting of 1 to 300 seconds, what do the numbers (0 to 10) on the time stand for?

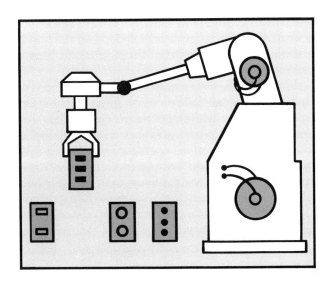

8 Applications and Installation of Control Devices

Control devices range from simple pushbutton switches to more complex solid state sensors. The type of control device selected depends upon the specific application. Specification sheets from the manufacturer detail required amperage, voltage, and sizing information for control devices. The technician must interpret this information when determining the proper application. Installation of control devices requires proper position and location for safety and function in the intended environment.

INDUSTRIAL PUSHBUTTONS

One of the simplest and most common forms of control is the manual pushbutton. Even in the most complicated control systems, you will find a variety of pushbuttons used to start, stop and change operations in a system.

Although the basic pushbutton is a simple device, industrial applications have required the pushbutton to become more complex in its electrical and mechanical options. No other control device is being manufactured in so many different combinations, offering almost unlimited selection.

Figure 8-1 illustrates the assembly of a typical pushbutton unit. The pushbutton may consist of one or more contact blocks, an operator of some type, a legend plate to denote function and mounting rings or washers to adjust and hold the device in an enclosure.

Contact Blocks

Figure 8-2, left, illustrates a closer view of the contact block which forms the switching mechanism of the pushbutton station. Inside these plastic molded blocks are sets of double-break contacts arranged in specific configurations. Figure 8-2, right, illustrates the configurations available, namely NC /NO, NO or NC. With these basic configurations used in combination, almost any type of switching arrangement is possible. Electrical connection to the contact block is made by compression type connectors extending from the sides of the contact block.

Legend Plates

Figure 8-3 illustrates typical legend plates which are quite common and can be ordered from stock. Other legend plates can, however, be special ordered or may be made up from blanks usually available from manufacturers or wholesalers. Basically, there are three sizes of legend plates: standard, large and jumbo. The large and jumbo allow more space for printing; but when you use them, you cannot mount the control elements as closely together as when using regular legend plates.

Operators

Figure 8-4 illustrates some of the different types of operators available. Shown are the flush button, half shrouded, extended button, jumbo mushroom, and illuminated pushbutton. Each of these

buttons is available in multiple colors. The selection of a particular style will depend on the application and requirements of the circuit and person operating the circuit.

The advantages of some of the different types of operators are as follows. The flush pushbutton operator has a full length guard ring which surrounds the button and prevents accidental operation. The half shrouded operator has a guard ring which extends to the bottom face on the top, but drops back on the bottom to expose the lower half of the button, making it more accessible for thumb operation. The extended button operator has a button extension beyond the guard ring, making

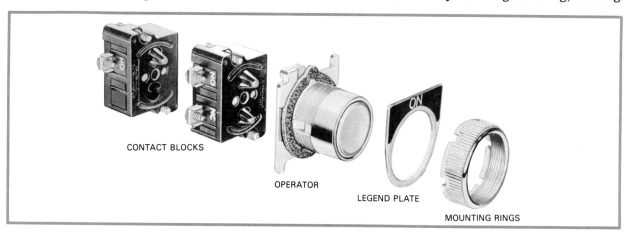

CONTACT BLOCKS

OPERATOR

LEGEND PLATE

MOUNTING RINGS

Figure 8-1 A complete pushbutton control unit consists of contact blocks, operator, legend plate and mounting rings. (Cutler Hammer)

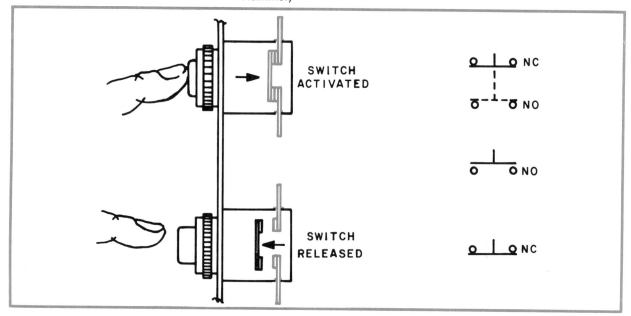

SWITCH ACTIVATED

SWITCH RELEASED

NC

NO

NO

NC

Figure 8-2 Left: Double break contacts form the switching mechanism of the contact block. Right: There are several switching configurations available.

the button easily accessible and button color visible from all angles. The mushroom head operator is particularly well suited for stop or emergency stop, or other operations that may require a fast operation of the pushbutton.

The illuminated pushbutton can be used for easily locating the unit or used to indicate when a device is on or off. The key operated unit can provide security access to a limited number of people on critical control applications.

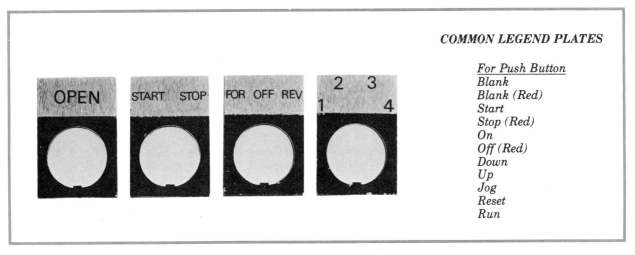

COMMON LEGEND PLATES

For Push Button
Blank
Blank (Red)
Start
Stop (Red)
On
Off (Red)
Down
Up
Jog
Reset
Run

Figure 8-3 Typical legend plates. (Cutler Hammer)

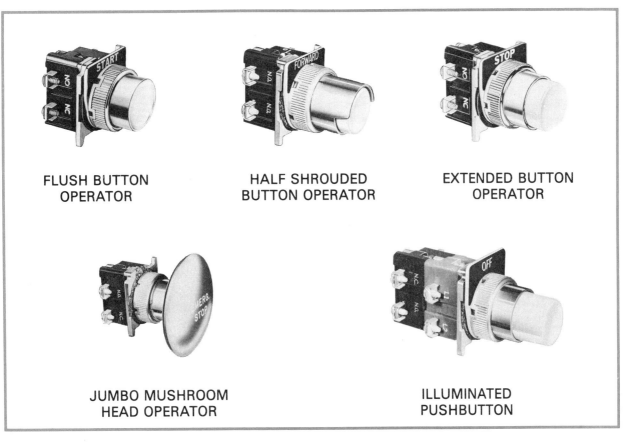

FLUSH BUTTON OPERATOR

HALF SHROUDED BUTTON OPERATOR

EXTENDED BUTTON OPERATOR

JUMBO MUSHROOM HEAD OPERATOR

ILLUMINATED PUSHBUTTON

Figure 8-4 Many different operators may be attached to contact blocks, depending upon the application. (Cutler Hammer)

After assembling the individual parts of the pushbutton, the electrician places the pushbutton or pushbuttons in enclosures called the pushbutton station. Figure 8-5 illustrates such a pushbutton station. These enclosures come punched in sizes from one to almost unlimited holes, for mounting the operators. Since pushbutton stations need to be mounted where it is convenient to operate them, every basic NEMA enclosure type is available. It is important to place the pushbutton in the proper enclosure type for continuous and safe operation. Pushbuttons often may be required to operate in environments where dust, dirt, oil, vibration, corrosive material, extreme variations of temperature and humidity along with other damaging factors are present. Always match the correct components and enclosure with the environment in which they will operate.

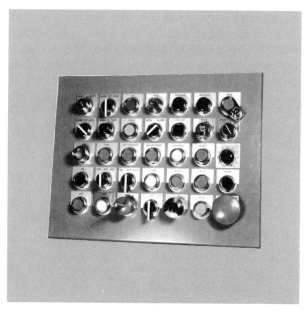

Figure 8-5 Once assembled, pushbutton stations are usually mounted in some type of enclosure. NEMA specifies different types of enclosures for different applications. (Furnas Electric Company)

SELECTOR SWITCHES

Figure 8-6 illustrates a selector switch operator and several common types of selector switches.

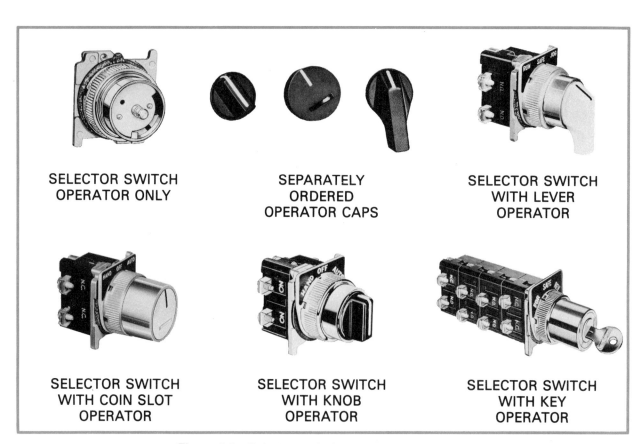

SELECTOR SWITCH
OPERATOR ONLY

SEPARATELY
ORDERED
OPERATOR CAPS

SELECTOR SWITCH
WITH LEVER
OPERATOR

SELECTOR SWITCH
WITH COIN SLOT
OPERATOR

SELECTOR SWITCH
WITH KNOB
OPERATOR

SELECTOR SWITCH
WITH KEY
OPERATOR

Figure 8-6 Selector switch operator and several common types of selector switches. (Cutler Hammer)

You will note from the construction that they are assembled in the same fashion as a pushbutton switch. The only basic difference between a pushbutton and a selector switch is the operator. With a selector switch, the operator is rotated to open and close contacts, while a pushbutton is a direct acting device which you push. Contact blocks used on pushbuttons are usually interchangeable with those on selector switches, providing they are made by the same manufacturer.

Selector switches are used to select or determine one of several different circuit possibilities. This selection process is needed because most machine operations are designed to operate in any one of several ways. Typical machine operations include manual or automatic, low or high, up or down, right or left and stop or run. It is the selector switch that controls which operation the machine will perform, and it is the operator that controls the position of the selector switch.

Selector switch controls are usually available in two, three, and four positions with a knob, lever, or key as an operator. Two-position selector switches are available with both positions maintained or with a spring return from one position or the other. Three-position units are available with

all three positions maintained or with spring return to center from either or both side positions. Four-position selection switches usually come only with maintained positions.

Circuit Operations: Two-Position Selector Switch

Figure 8-7 illustrates how a two-position selector switch would operate. In position one, Figure 8-7, the NC controls of the selector switch would remain closed and load number one would be activated. When the selector switch is moved to position two, indicated by the arrowhead, the NC contacts would be forced open, turning off load one, and the NO contacts would close and activate load two. In this circuit either load one or load two would be on at all times. If only load one were in the circuit, then the selector switch could be used as an on-off switch.

Circuit Operations: Three-Position Selector Switch

One disadvantage of the circuit in Figure 8-7 is the fact that both loads could not be shut off; either one or the other would be on. To eliminate

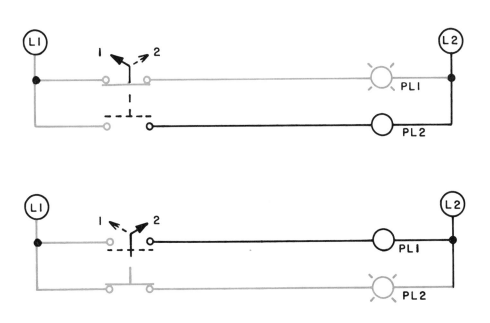

Figure 8-7 In the operation of two-position selector switch, as the selector switch is moved, load one is de-energized and load two is energized.

this problem, we would use a three-position selector switch like that of Figure 8-8. Figure 8-8 illustrates a typical hand/off/auto selector switch, its contact arrangements and a typical circuit application for controlling a pump manually or automatically.

In the hand (manual) position, the pump motor will start when the run pushbutton is depressed; the motor can be stopped by moving the selector switch to the off position. The liquid level switch will have no effect on the circuit in either the hand or off position.

When the selector switch is in auto, however, the circuit can be activated only by the liquid level switch as it closes. The advantage of this circuit is that an electrican has the ability to manually test the circuit if he or she feels the automatic liquid level is not operating or may have malfunctioned.

For additional safety, a circuit like this could have a selector switch which is key operated so that it could be locked in the off position so the machine could be safely worked on without being started.

Target Tables

There are two accepted methods of indicating contact position on a selector switch. In one method,

shown in Figures 8-7 and 8-8, solid lines, dashed lines and a series of small circles are used to denote contact position. In another method, a truth table (also called a target table) is used instead (see Figure 8-9). Each contact on the line diagram is marked A, B, etc.; and each position of the selector switch is marked 1, 2, etc. Then the truth table is made and positioned near the switch to illustrate each position and each contact. If a contact is closed in any position, an "X" is placed in the table. Consequently, the table is easily read as to what contacts are closed in what positions. When a selector switch has more than two contacts or more than three positions, the truth table illustrates the switch more clearly than the method of using solid and dashed lines and small circles.

JOY STICK

Figure 8-10, left, illustrates a typical joy stick and how it would be used in an industrial application. The joy stick operator can be used to provide a single operating control or multiple controls up to eight positions.

The advantage of the joy stick is that many operations can be performed by one person, without that person removing his or her hand from the

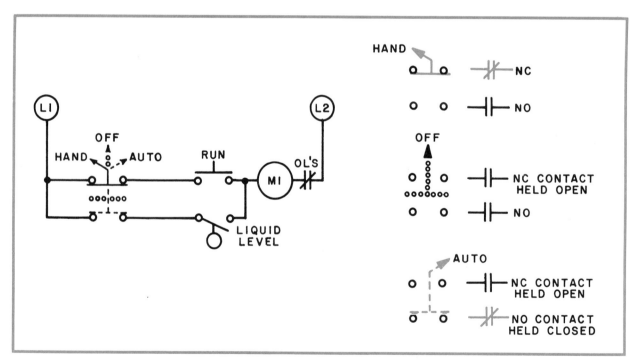

Figure 8-8 The operation of a maintained three-position selector switch used for hand/off/auto control of a pump motor.

joy stick controller. This is especially helpful where the operator should not be looking at individual controls but rather looking at the load that the controls are manipulating; for example, the overhead crane in Figure 8-10, right.

The joy stick can be compared to a stick shift in a car or truck, enabling the driver to change speeds without taking his eyes off the road.

The joy stick is available in two basic styles. One is the standard operator which is free to move

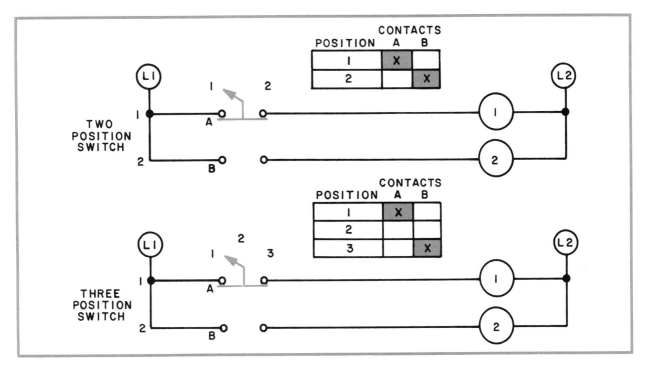

Figure 8-9 Use of a truth table (also called a target table) to illustrate the operation and position of selector switch contacts.

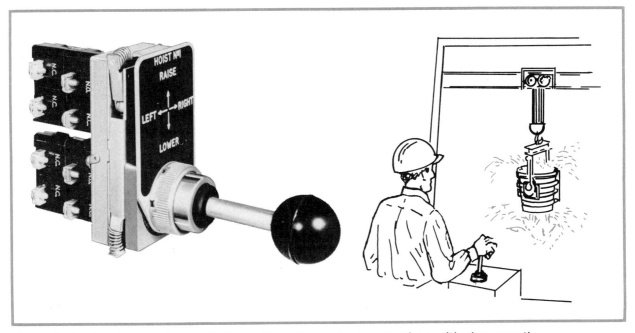

Figure 8-10 A typical joy stick controller and how it would be used by an operator in a foun-dry to control a positioning operation. (Cutler Hammer)

from position to position, while the other is a latched lever type. The latched lever type has a locking ring that must be pulled up to free the lever before the lever may move into another position. Each position on the joy stick is usually available with momentary or maintained operation or a combination of both.

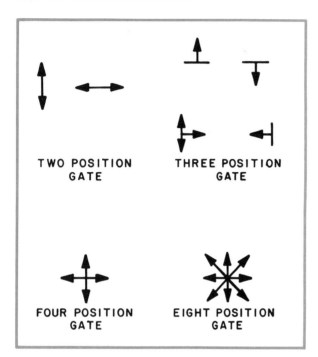

Figure 8-11 Many different combinations of contact positions are available when contact blocks are arranged in different configurations. Although the center position is not counted as a control position on the joy stick, it may still act as an off or stop position in the control circuit.

Joy Stick Construction

The joy stick uses the same standard contact blocks used with pushbuttons and selector switches. The difference is that the joy stick uses many contact blocks stacked together to provide anywhere from two to eight individual contacts, and sometimes more.

It is the responsibility of the electrician to make the adjustment as to how many positions the switch will have. This is accomplished by the electrician placing any one of four metal plates (usually called gates), which are furnished with the joy stick, between the contact blocks and the stick operator. Figure 8-11 illustrates the many different combinations of positions available from most joy sticks. On each of the illustrations, the center position is not counted as a position because, in the center position, no contacts are changed from their normal position. Although the center position is not counted as a control position on the joy stick, it may still act as an off or stop position in the control circuit.

Joy Stick Operation

As with the selector switch, there are two methods used to illustrate the different positions of a joy stick. Figure 8-12 illustrates one method used to show each position and the contact that it controls. A circle with a dot in it is drawn above the contact illustrating the position that closes that contact. To operate load one, the joy stick must be in the right hand position. Likewise, the joy stick must be in the left hand position to operate load two, the up position to operate load three, and the down position to operate load four. With the cir-

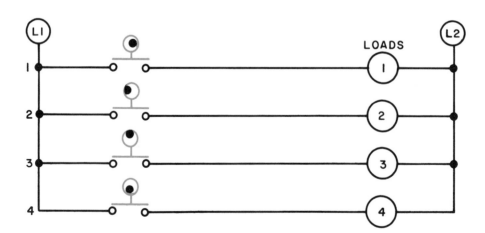

Figure 8-12 A black dot in a circle shows how each position of the joy stick is capable of controlling a set of contacts and a variety of loads.

cuit illustrated, four normally open contacts would be used. Any combination of contact blocks could be used or added to the circuit for different circuit combinations.

A second method used to illustrate the different positions of a joy stick is the truth table. Figure 8-13 illustrates the use of a truth table in determining what contacts close in which position.

Circuit Application

The circuit of Figure 8-14 illustrates schematically the operation of the joy stick controlling the

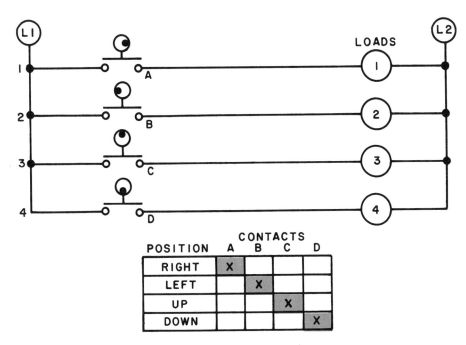

POSITION	CONTACTS			
	A	B	C	D
RIGHT	X			
LEFT		X		
UP			X	
DOWN				X

Figure 8-13 The truth table method of indicating the various contact positions of a joy stick.

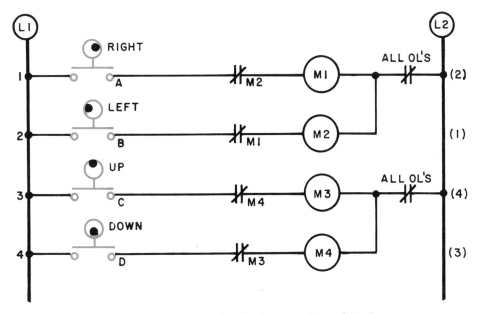

Figure 8-14 Schematically the operation of the joy stick controlling the foundry ladle illustrated in Figure 8-10.

foundry ladle illustrated previously in Figure 8-10 right. Here, an overhead hoist is controlled by two motors. Motor number one, which is controlled by starter M1 and M2, is used to move the hoist back and forth on the hoist beam. This back and forth movement is labeled left and right for the operator's convenience. The second motor, which is controlled by starter M3 and M4, is used to move the winch of the hoist back and forth. This back and forth movement is labeled up and down in reference to the load's movement, which the winch is controlling.

This same circuit could be used with any overhead hoisting operation where safety is important. If additional safety precautions are required to prevent unauthorized persons from using the hoist, a key lock stop pushbutton or selector switch could be added to the circuit to disconnect the power. Additional safety is built into this circuit by using the electrical interlocks illustrated in the line diagram. Although no two starters can be turned on simultaneously by the controller (because the joy stick cannot be in two positions at once), the interlocks are still included. They serve the purpose of a short delay before changing the direction of the motor. This helps to prevent a jerky action of the load that may spill or drop the load.

LIMIT SWITCHES

Figure 8-15 illustrates one of the most common types of limit switches used in industrial application. The limit switch is used to convert mechanical motion into an electrical signal. Limit switches accomplish this conversion by using some type of lever to force open or closed a set of contacts within the limit switch enclosure.

In Figure 8-15, the lever used is called a roller type and is used in about 85% of all applications. Some typical applications of limit switches include conveyors, elevator control, counting, positioning, detecting, sequencing, and monitoring. In these applications, the limit switch is usually tripped to start, stop, forward, reverse, recycle, slow down or speed up some operation.

Limit Switch Construction
The limit switch is made up of two main parts—the actuating mechanism and the electrical contacts enclosed in the switch's body. The actuator, or lever, is the part of the limit switch that transfers the signal from the moving part to be detected

to the electrical contacts. The electrical contacts (Figure 8-16) are usually sealed in the limit switch body and are accessible only through screw terminal connectors within the enclosure. The standard internal contacts used on most limit switches are of the quick make, quick break, snap action type. They are usually silver coated and have a high contact pressure to minimize rebound and arcing.

Contact arrangements come in several configurations. They may be NO, NC, or combinations of both NO and NC. The standard limit switch usually comes with one NO contact and one NC contact. The various schematic symbols for limit switch contacts are shown in Figure 8-17.

One point should be emphasized, especially when wiring limit switches in DC circuits. It is absolutely essential that polarity be observed when the contacts are connected. In all cases, the contacts should be connected to the proper polarity. The importance of polarity is illustrated in Figure 8-18. As long as the contacts are at the same polarity, then there will be no arcing between the contacts when the contacts energize and de-energize the load. If opposite polarity is used, however, then there may be arcing or even welding of the contacts from a possible short circuit.

The other consideration the electrician must take into account is the electrical rating of the contacts. Contacts must be selected according to

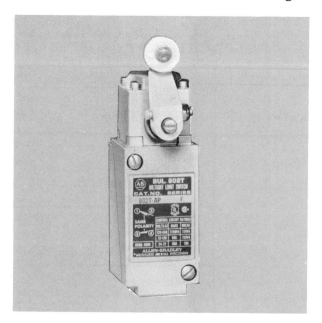

Figure 8-15 A common type of limit switch used to convert mechanical motion into an electrical signal. (Allen-Bradley Company)

proper voltage and current size according to the load and manufacturer's specifications. Figure 8-19 illustrates that if the load current exceeds the contact rating, then a relay, contactor or motor starter must be used in order to interface the switch with the load.

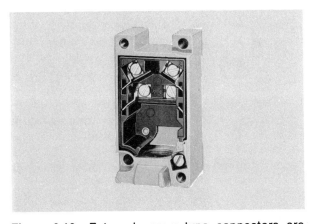

Figure 8-16 External screw type connectors are found in a limit switch. Internal electrical contacts for the control circuit are usually sealed in the limit switch. (Cutler Hammer)

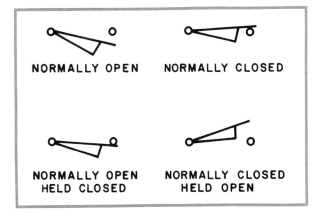

Figure 8-17 The schematic symbols used to represent limit switch contacts and operation.

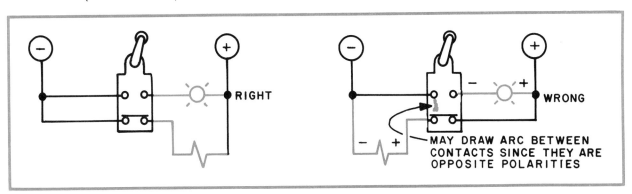

Figure 8-18 Polarity must be observed when wiring limit switches. Failure to observe polarity may result in arcing between contacts.

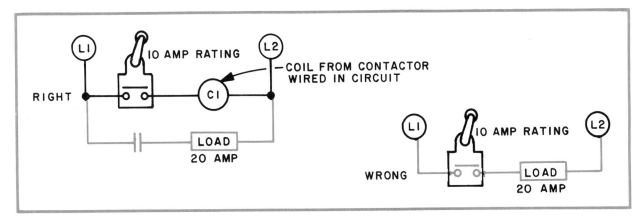

Figure 8-19 Use of a contactor in the control circuit when the rating of the limit switch contacts cannot directly handle the load.

Types of Actuators

There are many variations of the basic rotary actuator, designed for the many applications of the limit switch. Figure 8-20, A illustrates the adjustable roller lever designed for applications that may require the length of the roller travel to be adjusted; B illustrates an actuator with a pushroller designed for applications where a direct thrust with a limited travel is accomplished (here any side thrust or excess travel must be minimal to avoid damage); C illustrates the fork type roller lever. This type of actuator is usually used on limit switches with maintained contact switching. Maintained contact switches require a movement in the opposite direction to reset the contacts back to normal. Thus, any time a cam or part continues travel beyond the actuator, the contacts stay actuated. Applications of the fork lever and maintained contacts then are used, mostly with reciprocating movement. Finally in Figure 8-20, D illustrates the rod lever actuator. The advantage of the levers are that the length can be shortened or formed to angles for operating around corners or through small openings. Deflecting the rod through a specified angle in any direction away from the center will cause the electrical contacts to operate.

The installation of limit switches requires the electrician to pick the best actuator and mount the limit switch in correct position in relationship to the moving part. It is important that the limit switch not be operated beyond the manufacturer's recommended specifications of travel. The specifications of travel distance for the actuator can be found by referring to the manufacturer's bulletins for each switch. Figure 8-21 illustrates the travel terminology used by most manufacturers of limit switches in their bulletins.

Installation of Limit Switches

Figure 8-22 indicates the manufacturer's recommendations for the best way of installing a limit switch on a rotary type cam. In this arrangement, the roller lever should not be allowed to snap back freely. The cam should be tapered to allow a slow release of the lever. This will help eliminate roller bounce, switch wear, and allow for better repeat accuracy.

Figure 8-23 illustrates the correct way of installing a limit switch where relatively fast motions are involved. Here the cam arrangement should be such that the limit switch's lever does not receive a severe impact. The cam should be

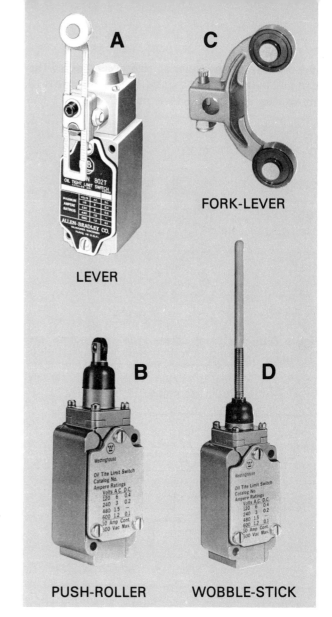

Figure 8-20 Several common types of actuators can be attached to a standard limit switch to perform a variety of functions.

tapered to extend the time it takes to engage the electrical contacts. This not only prevents wear on the switch, but allows the contacts longer closing time. This will assure that the circuit is completed.

For limit switches using the push roller lever, it is important that the switch not be operated beyond its travel in emergency conditions. Figure 8-24 illustrates the use of a rotary lever instead of a push lever in applications where an override may happen. Here the push lever has a leeway in an overtravel and the limit switch and its mountings would not be damaged. Figure 8-24 illustrates that a limit switch should never be used as a stop. This is especially true with the push lever limit switch. A stop plate should always be added to protect the limit switch and its mountings from any damage due to overtravel.

In considering the installation of limit switches, you must keep in mind some other general considerations. One of these is that the limit switch is designed to be used as an automatic type controller that is mechanically activated, and care should be taken to avoid any human error. This is illustrated in Figure 8-25. Here the limit switch is mounted so that the operator cannot accidently actuate the lever.

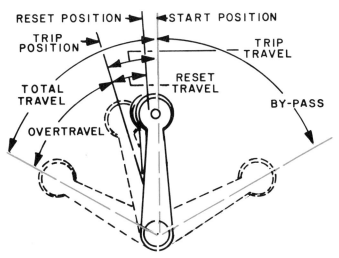

Figure 8-21 The travel terminology used by most manufacturers of limit switches. Limit switches must not be operated beyond the manufacturer's recommended travel. For specifications of a specific switch, refer to manufacturer's bulletins.

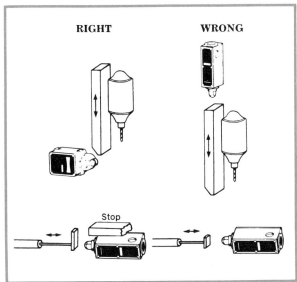

Figure 8-24 There is a right way and a wrong way to install limit switches when they are used as push levers in an over-travel position.

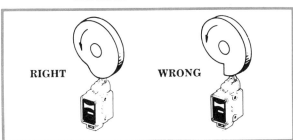

Figure 8-22 Be sure to use the correct type of rotary cam when installing one with a limit switch.

Figure 8-23 There is a right way and a wrong way to install limit switches where relatively fast motions are involved.

Figure 8-25 Limit switches should be installed in such a way that an operator can not accidently actuate the device.

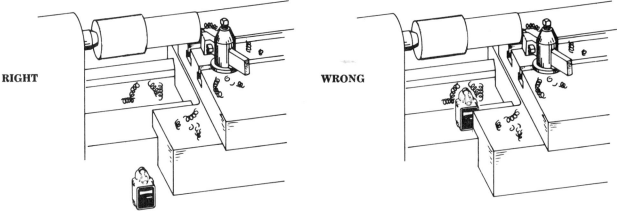

RIGHT

WRONG

Figure 8-26 There is a right way and a wrong way to install a limit switch when the atmos-

phere surrounding the limit switch may be a problem.

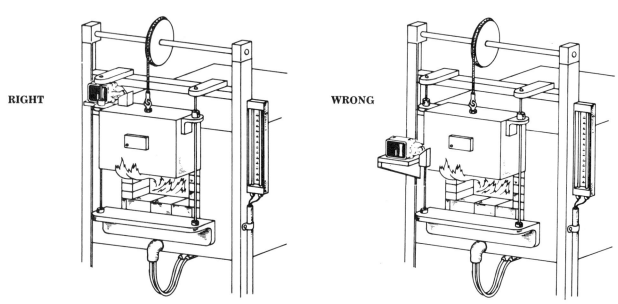

RIGHT

WRONG

Figure 8-27 There is a right way and a wrong way to install a limit switch when heat may be a factor.

A second consideration includes the atmosphere and surroundings of the limit switch. Figure 8-26 illustrates that a limit switch must be placed in a location where machining chips or other materials do not accumulate. These could interfere with the operation of the lever and cause circuit failure. This application also applies to splashing or submerging the limit switch with oils, coolants, or other liquids. Since heat beyond the specified limits of the switch could also damage it, this too must be avoided. Figure 8-27 illustrates a correct way of positioning a limit switch to avoid any excessive heat.

PRESSURE CONTROLS

Pressure switches are control devices that respond to pressure changes in a medium such as

air, water or oil. This medium of air, water or oil is called a *fluid*. The pressure switch opens or closes electrical contacts in response to these pressure changes. The electrical contacts may be used to start or stop motors or fans, open or close dampers or louvers, and signal a warning light or alarm. For small loads or some fractional horsepower motors, the pressure switch may handle the current directly. For large loads, the pressure switch is used to energize relays, contactors, or motor starters, which then energize the load.

Operation of Pressure and Vacuum Controls

Figure 8-28 illustrates the basic operating mechanism behind pressure and vacuum controls. The electrical contacts are usually operated by the movement of a diaphragm (or sometimes a piston

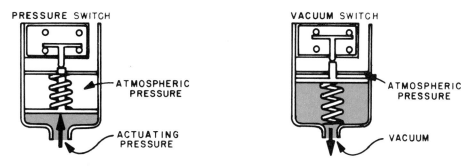

Figure 8-28 Basic operating mechanisms behind pressure and vacuum controls.

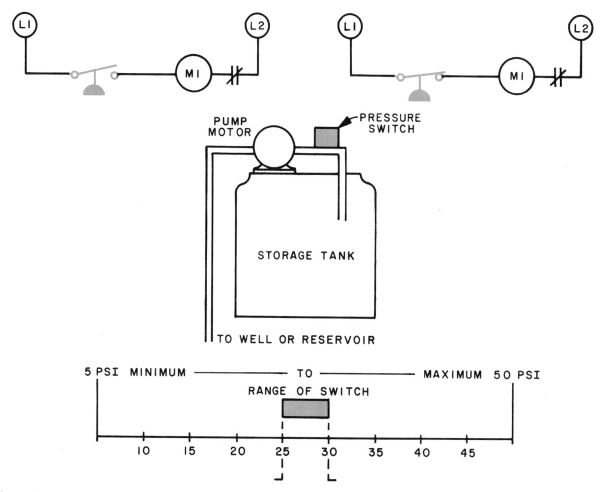

Figure 8-29 A control circuit can utilize a pressure switch to control a fractional horse-power pump motor which must maintain a given air pressure range in a tank.

or bellows) against the force of a pre-calibrated spring. When a difference in pressure exists between the fluid and the spring, the spring is forced to move. This movement can be started by adding to the pressure or reducing the pressure in the case of a vacuum switch. It is the spring setting that determines the pressure or vacuum at which the contacts open or close. The most common application of a pressure switch is in main-

taining a pre-determined pressure in a storage tank. Storage tanks that require a maintained pressure include the water well storage tank found in your home or community, the air compressor storage tank found at your local service station and the storage tanks used for commercial and industrial applications.

Figure 8-29 illustrates the proper way of connecting a pressure switch into a circuit to energize a fractional horsepower pump motor that is used to maintain pressure in the tank. As illustrated, the NC contacts must be used to maintain the pressure in the storage tank. Anytime the pressure falls below the lower limit (set at 25 psi), the pump motor starts and runs until the upper limit (set at 30 psi) is reached. The minimum and maximum limits of the pressure switch illustrated are given; these limits can be found directly on most pressure switches or will be listed on the specification sheet. The difference between these limits are adjustable by the differential pressure setting.

Pressure Differential

One of the difficult concepts for the electrician to grasp is the pressure differential setting on a pressure or vacuum switch. When a change in the fluid pressure occurs, causing the diaphragm to move and actuate the contacts, some of that pressure must be removed before the contacts will reset back to normal. This pressure difference is called the *differential*. The differential, then, is the pressure that must be removed before the contacts reset back to normal. The differential setting (sometimes called dead band) is the pressure range between the rising pressure and the falling pressure required to actuate the contacts. For example, if a pressure switch is set to actuate when the rising pressure reaches 30, then as the pressure rises to 30, the contacts are switched. As the pressure begins to drop, the contacts remain actuated until a lower pressure limit is reached. If this lower limit is, say 25, then the contacts will remain actuated to 25 and then open. The difference then is called the differential (5) and is adjustable on most pressure switches. The electrician will set the pressure switch for both the rising and falling pressure settings. This is accomplished by turning a screw or screws usually marked on the pressure switch as *differential adjustment, cut-in pressure, cut-out pressure, rising pressure, falling pressure,* or *pressure range.* Many problems related to pressure switch circuits are the result of the electrician not properly setting the differential pressure setting. A too small differential will cause the contacts to "chatter," while a too large differential will allow for a wider pressure range than is necessary. Chatter results from the contacts opening and closing rapidly. Care must be exercised in maintaining the proper balance.

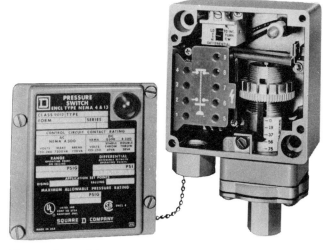

Figure 8-30 The internal parts and adjustments of a typical pressure switch. (Square D Company)

Pressure Switch Specifications

Figure 8-30 illustrates a typical industrial pressure switch. This pressure switch comes with two sets of contacts, one NO and one NC, as is standard on many pressure switches. The pressure setting range is adjustable with an adjustable differential pressure range. This pressure switch also offers an optional pilot light that can indicate a rising or lowering pressure.

In selecting a pressure switch for a particular application, the electrician would have to take into consideration:

The Adjustable Range. This is the span of pressures within which the pressure sensing element (diaphragm) of the switch can be set to actuate the contact on the switch. For example, a pressure switch may have a range of 20 to 100 psi.

The Adjustable Differential. This is the span of pressure between the higher pressure limit, which changes the electrical contacts, and the lower pressure limit, which returns the electrical contacts to their normal condition. For example, a pressure switch may have an adjustable range of 20 to 100 psi with an adjustable differential of 5 to 15 psi.

The Rating of the Electrical Switch. This includes the type of switch used, such as single-pole, single-throw; single-pole, double-throw; etc., and the amount of voltage and current the contacts can switch. It is this rating that determines whether the pressure switch can actuate the load directly or whether some interfacing device such as a relay or motor starter is required.

Repetitive Accuracy. This is the ability of the switch to operate repetitively at its set point under consistent conditions. This is typically ±1% of the maximum operating pressure for industrial standards and up to ±5% for economical type switches.

Fluid Used. The pressure switch must be capable of working with the type of fluid used. Most pressure switches are designed to work with air, water, or oil. Consideration must be given to any fluid that may be corrosive. One of the most corrosive fluids that is often misunderstood is salt water. If corrosive fluids are used, a special type of

pressure switch must be used. Other considerations would be required in problem or hazardous atmospheres. Here the type of enclosure used with the pressure switch is very important, since the atmosphere could be explosive as well as corrosive.

Pressure Switch Applications

Although the most common application of pressure switches is to maintain a predetermined pressure in a tank or reservoir, as illustrated previously, there are many more applications of pressure switches found in industry. For example, Figure 8-31, top, illustrates how a pressure switch could be used to sequence the return of pneumatic or hydraulic cylinder. When the operator presses the start pushbutton, the two-

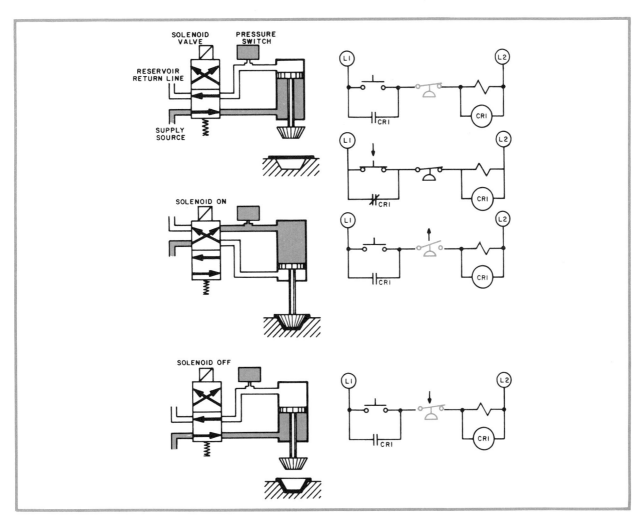

Figure 8-31 A pressure switch could be used to sequence the return of a hydraulic or pneumatic cylinder. A process similar to this could be used in the formation of stamped metal pie plates.

position, four-way, directional control valve solenoid is energized. This changes the directional control valve from the spring position to the solenoid actuated position as illustrated in Figure 8-31, middle. Since the flow of pressure is changed in the cylinder, as illustrated by the arrows, the cylinder advances. Since a control relay is paralleled with the solenoid in the electrical circuit, it also is energized and its NO contact closes, adding MEMORY to the circuit. The operator can release the pushbutton and the cylinder will continue to advance. The cylinder will advance until the preset pressure is reached on the pressure switch, which signals the return of the cylinder by de-energizing the solenoid and relay. The de-energizing of the solenoid will return the directional control valve back to the spring position. The return back to the spring position of the directional control valve will reverse flow in the cylinder, thus returning it. Figure 8-31, bottom, illustrates the opening of the pressure switch and the return of the cylinder. The return of the cylinder is dependent on the setting of the pressure switch. The setting of the pressure switch is dependent upon the application of the cylinder. This setting may be very low for packing fragile materials or very high for forming metals, as illustrated.

The advantage of using a pressure switch over a pushbutton for returning the cylinder is that the load will always receive the same amount of pressure before the cylinder returns. An emergency stop could be added to the control circuit for manual return of the cylinder.

Figure 8-32 illustrates the use of a low range pressure switch used with a metal tubing arrangement in a fluidic eye. In this application a constant low pressure stream of air is directed at a sheet of material through the metal tubing. As long as the material in process is present, the air stream is deflected. If the material should break, the stream of air will be sensed in the receiver tube; the pressure switch would signal corrective action through the control relay. Although this application can be accomplished completely electrically by using a photoelectric eye instead of a fluidic eye, the fluidic eye has certain advantages. For example, the air stream flowing over the material in normal operation may perform a second function such as cooling, cleaning or drying the material. The fluidic eye is also very inexpensive compared to the photoelectric eye since the only cost is the metal tube and the pressure switch.

TEMPERATURE CONTROLS

Temperature switches are control devices that respond to temperature changes. The temperature control may be used to maintain a specified temperature within a process or to protect against overtemperature conditions. They are available for a wide range of temperature control.

Bi-metallic Controls

There are several standard methods of actuating electrical contacts in response to temperature changes. Figure 8-33 illustrates a bi-metallic temperature switch used in some thermostats. The bi-metallic strip, as mentioned earlier, is made of two different metals securely bonded together. These two metals expand at different rates when the temperature changes. This causes

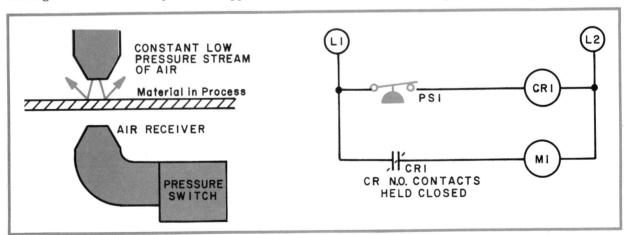

Figure 8-32 A low range pressure switch can be used as a fluid eye. In the event the material were to break, the pressure switch would signal the operation to halt.

the metal strip to bend into an arc, as illustrated in Figure 8-33. With a contact added to the bi-metallic strip, the movable contact will make and break contact with a stationary contact, opening and closing the circuit according to temperature changes.

When a spring is added to this basic device, the temperature range of the switch can be controlled. The temperature range and sensitivity can also be increased by making the bi-metallic strip into a spiral, as illustrated in Figure 8-34.

Both of these designs have limited use in industrial applications but are common in most homes. The non-spiral bi-metallic strip is used in many kitchen appliances such as coffee makers, toasters, electric irons and other common heating devices. The spiral bi-metallic strip is the most common type used on thermostats controlling central heating and cooling systems.

Bi-metallic thermostats are slow in action and will arc and burn if used to make and break larger currents than the contacts can safely handle at a slow speed. For this reason, many applications require a set of contacts that will snap open or closed. Figure 8-35 illustrates a bi-metallic temperature switch with snap-action contacts that will open quickly. Here the bi-metallic strip is formed into a concave disc. When the temperature changes, the disc changes. Since expansion of the metal is greater on one side than on the other, the disc moves upward. This movement must overcome the stress applied by the concave disc. When this force is overcome, the disc will invert with a snap. This snap action quickly opens the contacts. As the temperature changes in the opposite direction, this action is reversed.

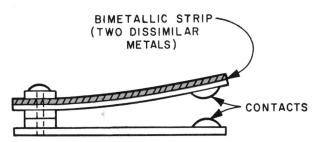

Figure 8-33 A bi-metallic strip made of two dissimilar metals will warp when heated and cause sets of contacts to open.

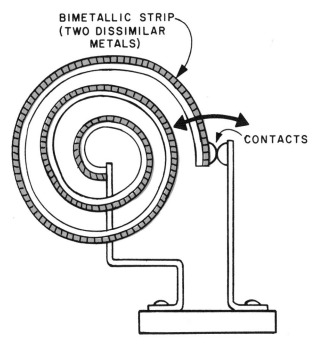

Figure 8-34 The temperature range and sensitivity range of a bi-metallic temperature controller can be increased by forming the strip into a spiral.

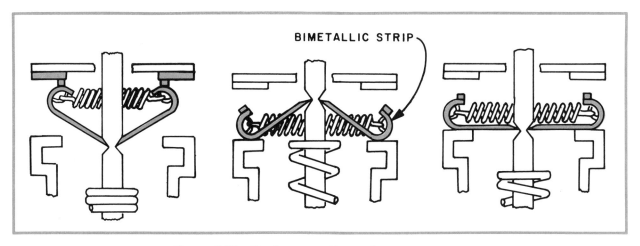

Figure 8-35 Basic operation of a bi-metallic temperature switch with snap action.

Capillary Tube Control

A second type of temperature switch used for industrial applications is illustrated in Figure 8-36. Here the temperature switch is equipped with a bulb that is used as the sensing element and a length of capillary tube for removing the electrical contacts some distance from the sensing element and heat.

These units use a temperature sensitive liquid as the active element in the bulb. The pressure in the system changes in proportion to the temperature of the bulb. As the temperature of the bulb rises, the vapor pressure within the bulb increases. As the temperature of the bulb decreases, the vapor pressure within the bulb decreases. The pressure changes are transmitted to the bellows or diaphragm through the capillary tube. This pressure is then used to actuate the electrical contacts in the switch. The switches are designed for precision snap action such that when the temperature creates the right pressure the contacts close. In controlling refrigeration units where the temperatures are usually below normal room temperature, a gas or vapor instead of a liquid is used to sense the temperature in the bulb.

Thermocouple Control

If a faster response time is required from the tem-

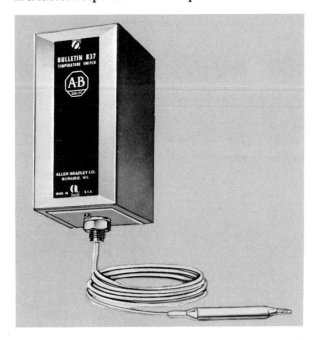

Figure 8-36 Type of temperature controller which uses a bulb and capillary tube to activate a set of contacts. (Allen-Bradley Company)

perature switch, then a thermocouple controller like that of Figure 8-37 is usually used. A thermocouple is a temperature sensing device whose electrical characteristics vary with a change in temperature. The thermocouple is made of two wires of different metals joined together to form a closed loop. The type of wire used depends upon the temperature range to be measured. This may be copper-constantan for temperatures of about 40 to 150°C (100° to 300°F), iron-constantan for temperatures of about −20° to 640°C (0° to 1200°F), and chromelalumel for temperatures of about −20° to 1300°C (0° to 2400°F). An electrical current proportional to the temperature flows from the thermocouple when it is heated. This conversion of heat into electricity is measured, sometimes amplified and applied to control circuits within the temperature controller to open and close contacts.

Application of Solid State Temperature Control Using Thermocouple

Figure 8-37 illustrates a solid state temperature controller that uses a thermocouple as the temperature sensing element. Figure 8-38 illustrates the wiring diagram found on the back of this type of solid-state timer, along with a typical circuit connection. It is this wiring diagram that the electrician uses in connecting the temperature controller into a circuit. Figure 8-38 shows a typical industrial circuit for maintaining temperature in a tank or process room. In this application, a thermocouple is used to measure and

Figure 8-37 Solid state temperature controller that uses a thermocouple as the temperature sensing element. (Eagle Signal Industrial Controls)

transmit the temperature of the enclosure while a heating contactor is used to turn on and off the heater's coils to provide uniform temperature.

In this circuit, the power lines are connected to terminals one and two of the temperature controller. These power lines must be at the same voltage rating of the timer and must maintain a constant voltage supply to the timer at all times. Terminals eleven and twelve are used to connect the thermocouple to the temperature controller. Terminal eleven is marked positive and terminal twelve is marked negative so that you will be able to match the polarity marking of the thermocouple. Terminal three is used for grounding and must be properly connected to a suitable ground. Terminals six, seven, and eight are the switching contacts that will be used to control the load. Terminals eight and six are the NC contacts and terminals eight and seven are the NO contacts. For the

application illustrated, the NO contacts (terminal eight and seven) are used to control the magnetic contactor.

Terminals nine and ten can be used for an on-off control or can be set up for time proportioning control. Time proportioning control is used when it is necessary to reduce temperature overshoot that occurs with heating elements that continue to produce heat for a period of time after they have been shut off. This excess heat will be added to the environment, sending the actual temperatures beyond the setting of the control even with the power disconnected.

To compensate for this, a time proportioning feedback circuit is used so that when the temperature of the controlled load is within a proportional bandwidth, the temperature controller will cycle at a frequency and duty cycle which are dependent upon the actual temperature of the

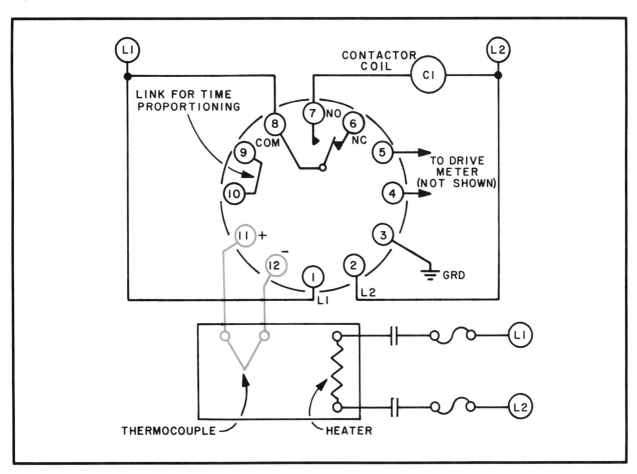

Figure 8-38 This wiring diagram would be found on the back of the solid state temperature controller illustrated in Figure 8-37. Also illustrated is the wiring diagram the electrician would use in installing this device into a circuit.

thermocouple and the reading of the set point on the controller's indicator dial.

This, in effect, allows the heating element to decrease in temperature as the set point is almost reached. The average power supplied to the load will then be dependent upon the difference between the actual and desired temperatures, thus maintaining a closer control than otherwise achieved.

Terminals four and five are to be connected to a meter drive if used. A meter drive is a single meter that is capable of calling up and reading the temperature of many individual temperature controllers. Figure 8-39 illustrates an example of a typical meter drive that is used to monitor six temperature controllers. Through the use of a selector switch or thumbwheel switch, this single meter drive can read the temperature of any one

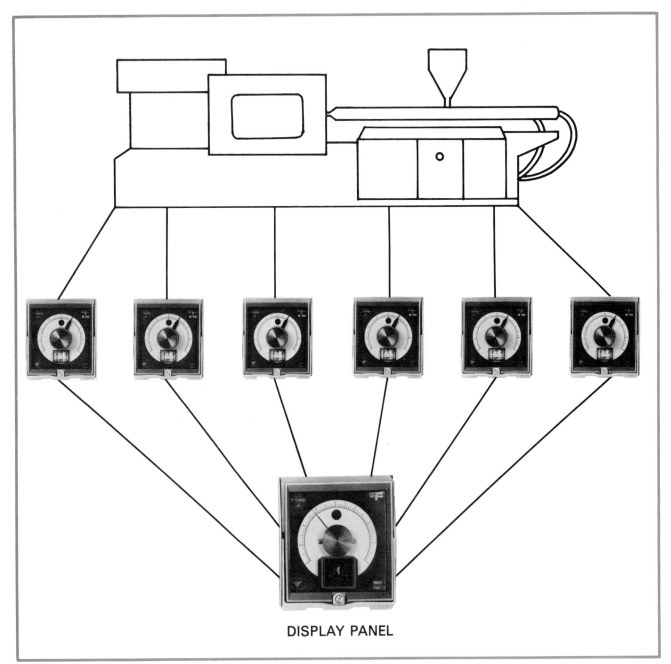

DISPLAY PANEL

Figure 8-39 A meter drive temperature controller can be used to monitor many temperature controllers. In this way any one of the individual controllers can be monitored or called up at any time. (Eagle Signal Industrial Controls)

of the individual controllers. Although the meter drive cannot change any of the individual temperature settings, it has great value in being able to call up and read individual temperatures. The meter drive can be connected at a central control station that can routinely call up and check the individual temperatures. This is useful in applications where the individual temperature controllers are not easily accessible or are some distance from the control station.

Application of Temperature Control for Liquids, Air or Surface Temperatures

As illustrated in Figure 8-40 temperature sensors are available to monitor the temperature of a liquid, the air or surface temperature. Depending upon the application any one or all types may be used.

A temperature sensor is used to monitor and control liquids in many applications. This is one of the most common process controls used in industry.

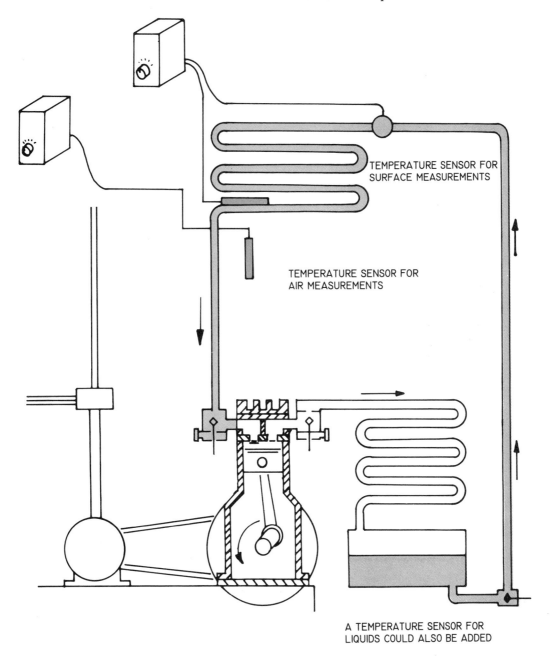

Figure 8-40 Temperature sensors are commonly used in industry to monitor liquid, air and surface temperatures.

The temperature of many liquids must be at a set point before a mixing or fill process can start. In other applications, a process may have to be stopped if the liquid is too cool or too hot. This type of temperature sensor is available in a wide range of temperatures with extreme ranges from −400 °F to 3200 °F (−240 °C to 1760 °C).

Air temperature sensors are available in a wide range of temperatures, with the most common in the −30 °F to 150 °F ranges. These sensors are used mostly in heating and air conditioning control to maintain a desired temperature of a room or inside a storage unit.

Surface type temperature sensors are designed to attach typically to a metallic surface. They can be used to detect an ice build-up in an air conditioning system or heat build-up in many other processes. The advantage of a surface type sensor is usually in the fact that they are easy to install on pipes without having to open the system.

FLOW CONTROL

A flow switch is a control switch that is inserted into a pipe or duct to sense the movement of a fluid. This fluid may be air, water, oil or some other gas or liquid. The sensing element is a valve or paddle that extends into the pipe or duct. This paddle will move and actuate electrical contacts whenever the fluid flow is sufficient to overcome the spring tension on the paddle as illustrated in Figure 8-41. The spring tension is adjustable on many flow switches, allowing the electrician to adjust the switch to actuate at a given flow rate. The electrical contact may be used to start or stop motors or pumps, open or close valves and light or sound a warning alarm.

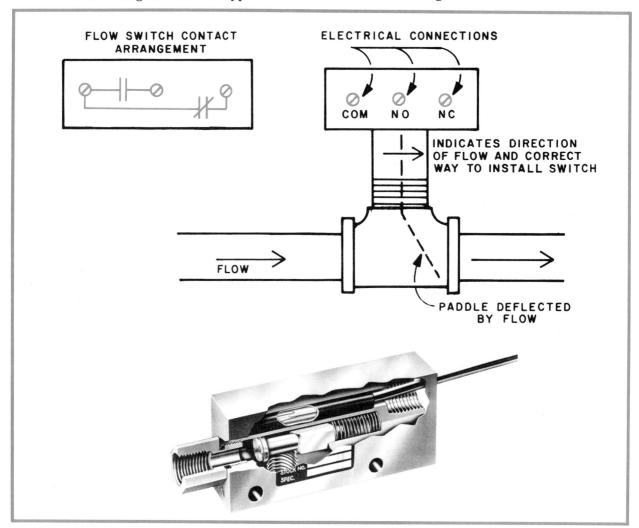

Figure 8-41 Schematic symbol and operating mechanism for a typical flow switch.

Flow Control Applications

One type of application of a flow switch is illustrated in Figure 8-42. Here the flow switch is used to signal a fire alarm when the head of a sprinkling system has sensed a fire and opened. The fire's heat automatically opens the sprinkling head, and the flow switch is used to automatically signal that there is flow in the pipe and indicate that corrective action is needed.

Although the fire could be sensed automatically through the use of heat and smoke detectors, the flow switch provides an additional advantage. If the sprinkling head were to open accidently for some other reason than fire, the water could flow undetected, causing much damage before being detected.

Another application of flow switches is illustrated in Figure 8-43. Here the flow switch is used to detect air flow across the heating elements of an electric heater. If sufficient air flow is not present, the heating coils would burn out. The flow switch is used as an economical way to turn the heater off any time there is not enough air flow. This circuit application can also be applied to an air conditioning or refrigeration system. In this type of circuit the flow switch is used to detect insufficient air flow over the refrigeration coils. The restricted air flow is usually caused by the icing of the coils, which blocks the air flow. In this case, the flow switch would automatically start the defrost cycle of the refrigeration unit.

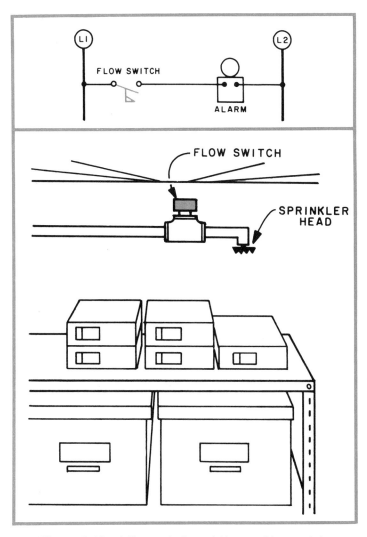

Figure 8-42 A flow switch could be used in a sprinkler system to sound an alarm.

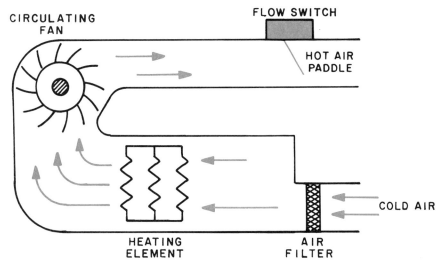

Figure 8-43 A flow switch could be used to determine sufficient air flow across the heating elements of an electric heater.

Another application of flow switches is illustrated in Figure 8-44. Here the flow switch is used to detect the proper flow of air that is required in an exhaust system. Often the exhaust system may be directing dangerous gases away from the operator; if there is insufficient air flow, poisonous gases could overcome the operator or damage could occur to the process involved.

Flow switches are often used to protect the large motion picture projector used in theaters where poor air flow would cause heat build-up and reduce the life expectancy of the very expensive bulbs used in projectors. Air flow may be restricted from a large draft caused by high winds outside of the building or by clogged air filters in the intake system.

To detect a clogged air filter, a flow switch could be used as illustrated in Figure 8-45. Here the flow switch is used not only to detect a clogged or dirty air filter, but also to take automatic corrective measures. The flow switch is used to start a gear reduced motor that would slowly advance the roll of filter material until sufficient air flow is present.

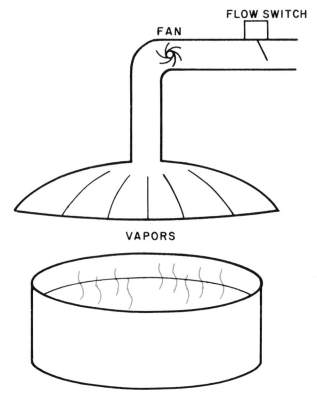

Figure 8-44 A flow switch could be used to maintain a critical ventilation process. In the event the flow was shut off, the alarm would be sounded.

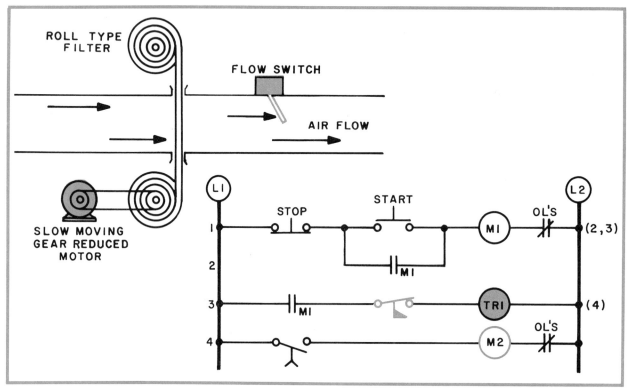

Figure 8-45 A circuit for automatically advancing a clogged filter based on restricted air flow.

LIQUID LEVEL CONTROL

Liquid level switches are control devices that measure and respond to the level of material in a container. Here liquid level refers to materials such as water, oil, paint, granulates, etc., and not to gases, since gases have to be measured by other techniques.

Liquid level controls can be used to start and stop motors and other devices either directly or by interfacing with contactors and starters.

Because there are so many different types of liquids found in industrial processes, many different methods of detection have become available. It is important for the electrician to understand these different liquids since it is the type of liquid that determines the best method of control.

Float Switch

The most popular method of sensing the level of a liquid is illustrated by the float switch in Figure 8-46. You will recall from Unit 2, "Industrial Electrical Symbols and Line Diagrams," that a float switch can be used for pump or sump operation, depending upon whether the NO or NC contacts of the float switch are used. Figure 8-47 illustrates the electrical control circuit for pump and sump operation using a float switch.

Electrode Sensing

Although the float switch is the most common method used to measure the level of a liquid, it has limited use for some industrial application. For example, if the liquid in the tank was paint, then the paint would add a thin layer of paint to the float every time the tank was filled or emptied. After many layers had built up on the float, the float would become heavier, changing its level setting. Or the liquid may be of an acid type material, in which case a liquid level switch that came in physical contact with the liquid could not be used.

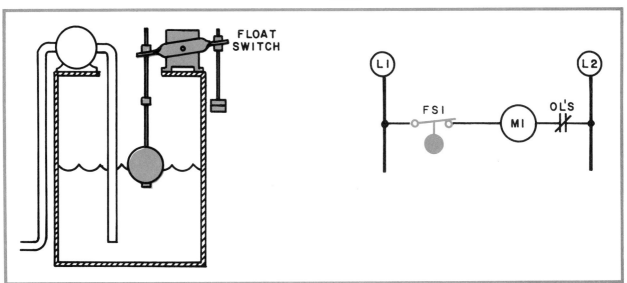

Figure 8-46 A typical float switch application along with the operating circuit.

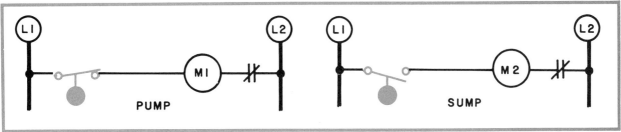

Figure 8-47 Line diagram control circuit for pump and sump operation using a float switch.

Figure 8-48 illustrates another method for sensing the liquid level in a tank. Here the liquid is used as a conductor to complete the path for current to flow, energizing the load. To accomplish this, the tank, which must be metal, is used as one current carrying conductor, and an electrode insulated from the tank is the other.

The liquid is used to complete the electrical connection between the tank conductor and the electrode conductor. Figure 8-49 illustrates the advantage of this method of measuring the level.

Here several electrodes and a selector switch are used to allow for many different levels to be maintained. In both these applications, two considerations must be given when using the tank and liquid as current carrying conductors. The first consideration is that the liquid must be a good conductor and not of any flammable material. The second consideration is that the current and voltage used to pass through the tank must be low enough and grounded properly to avoid any danger.

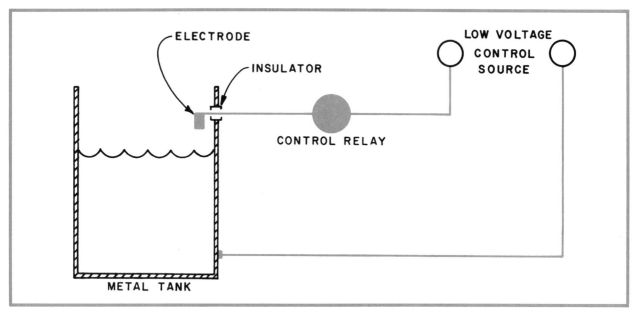

Figure 8-48 A liquid level can be utilized as a conductor between two electrodes to complete a circuit.

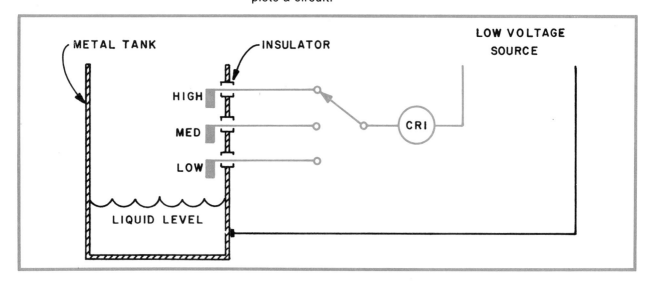

Figure 8-49 Electrodes can be arranged in a tank to provide multilevel control.

Transmission Detection

Another method of level sensing is illustrated in Figure 8-50. Here an x-ray, gamma-ray or radiation transmitter and receiver are used to detect the liquid level in the tank. The advantage here is that any liquid or solid can be sensed with this application. The liquid is used to cut the transmitted beam and actuate a set of electrical contacts. The sensitivity is so adjusted that this cutting of the beam will actuate the contacts even though the beam will still pass through the liquid. These contacts are then connected into the circuit to energize or de-energize the load.

Other methods of sensing the liquid level are:

1. Using a pressure switch at the bottom of the tank. Here the pressure at the bottom of the tank and on the diagram of the pressure switch changes at different levels because of the weight of the liquid in the tank. The pressure switch can be set to sense a level if the density of the liquid is known.

2. By using photoelectric or proximity detectors to sense the liquid level in a tank. Here the liquid is used to cut a beam of light or change the capacitance of the sensor. Photoelectric and proximity are both covered in the next unit.

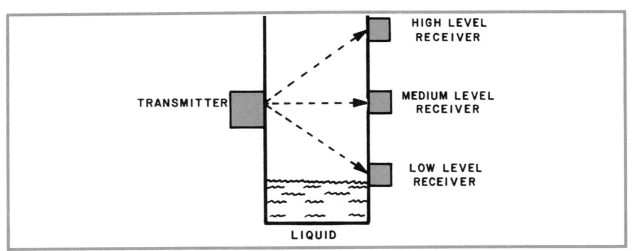

Figure 8-50 A transmitter and receiver are used to detect the liquid level in a tank.

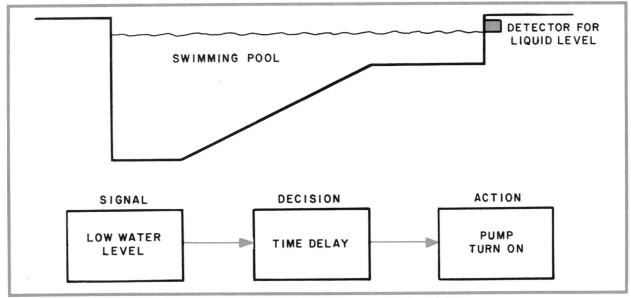

Figure 8-51 A liquid level detector with time delay can be used to monitor the depth of a swimming pool.

Logic Module Sensing

We learned very early in this book that all control circuits can be broken down into signals, decisions and actions. This concept has been recognized by major manufacturers of electronic control devices and has led to the development of hundreds of specific plug-in modules which can create signals, make decisions and implement action in a variety of control applications. These modules make it very easy to extend the capability of simple control circuits merely by inserting the proper module.

For example, it may be necessary to add a timing logic circuit along with a liquid level control to eliminate the interference of false turn-ons by disturbances in the liquid (such as waves). The swimming pool of Figure 8-51 is just such an application. In this circuit, when the sensor is in touch with the liquid, the relay is de-energized. If the liquid level falls, however, the relay energizes after a delay of about one minute, calling for the pump to fill the pool. This same relay and level sensor could also be used for similar industrial problems.

Figure 8-52 illustrates a typical plug-in module along with its wiring diagram and specifications. Once an electrician determines what it is he wishes to accomplish in a circuit, he can merely look through the manufacturer's data sheets until he locates the one which fits his needs. This type of system lends itself well to simple design, maintenance and troubleshooting.

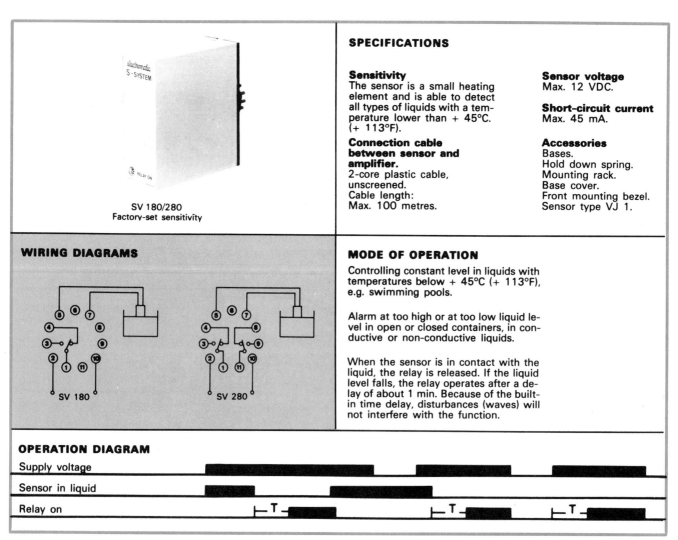

Figure 8-52 The specifications and wiring diagrams for an electronic liquid level time delay logic module used on swimming pools. (Electromatic Controls Corp.)

Control of Granulates

Not all levels to be measured or maintained need be of the liquid level. Figure 8-53 illustrates a level control relay module and sensor used with granulates (sand, pebbles, sugar or chemicals) or non-conductive liquids. This illustration shows the level control connected for maximum and minimum control for registration of two levels. The relay operates when the contents reach the maximum probe (pins 5-6) provided the mini-

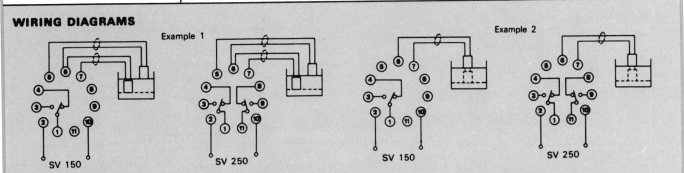

SPECIFICATIONS

Sensitivity
The sensitivity depends on the type of sensor and the material to be detected.

Short-circuit current
Max. 45 mA.

Sensor voltage
Max. 24 VDC.

Sensor current
Max. 20 mA.

Connection cable between sensor and amplifier.
2-core plastic cable, normally unscreened.
Cable length:
Max. 100 metres.

In certain cases it is recommended to use screened cable between sensor and amplifier, e. g. where the cable is placed parallel to load cables.
The screen is connected to pin 6.

Accessories
Bases.
Hold down spring.
Mounting rack.
Base cover.
Front mounting bezel.
Sensors type VR.
Nut type VM 1,5.

SV 150/250
Factory-set sensitivity

WIRING DIAGRAMS

Example 1

Example 2

SV 150

SV 250

SV 150

SV 250

MODE OF OPERATION

Max. and/or min. control of solid, fluid or granulated substances, e.g. sand, gravel, sugar or chemicals.

Example 1
The diagram shows the level control connected as max. and min. control, i.e. registration of 2 levels.
The relay operates when the contents reach the max. sensor (pins 5 and 6), provided that the min. sensor (pins 6 and 7) is in contact with the contents. The relay releases when the min. sensor is no longer in contact with the contents.

Example 2
The diagram shows the level control connected as max. or min. control, i.e. registration of 1 level.
The relay operates when the sensor (pins 6 and 7) is in contact with the contents.

OPERATION DIAGRAM Ex. 1.

Supply voltage						
Max. sensor, pins 5-6, in contents						
Min. sensor, pins 6-7, in contents						
Relay on						

OPERATION DIAGRAM Ex. 2.

Supply voltage			
Min. sensor, pins 6-7, in contents			
Relay on			

Figure 8-53 The specifications and wiring diagrams for an electronic level control that must monitor granulates (sand, pebbles, sugar or chemicals or non-conductive liquids. (Electromatic Controls Corp.)

mum probe (pins 6-7) is in touch with the contents. The relay releases when the minimum probe is no longer in contact with the contents. In certain cases, it is recommended to use screened cable between the sensor and level relay where the sensor cable may be placed parallel to load cables.

Figure 8-54 illustrates a typical application of using a non-liquid level controller. In this application the granulate sensor illustrated in Figure 8-53 can be used to automatically keep the hop-imum/minimum level control for two level registration. Since the relay is turned on when the max-

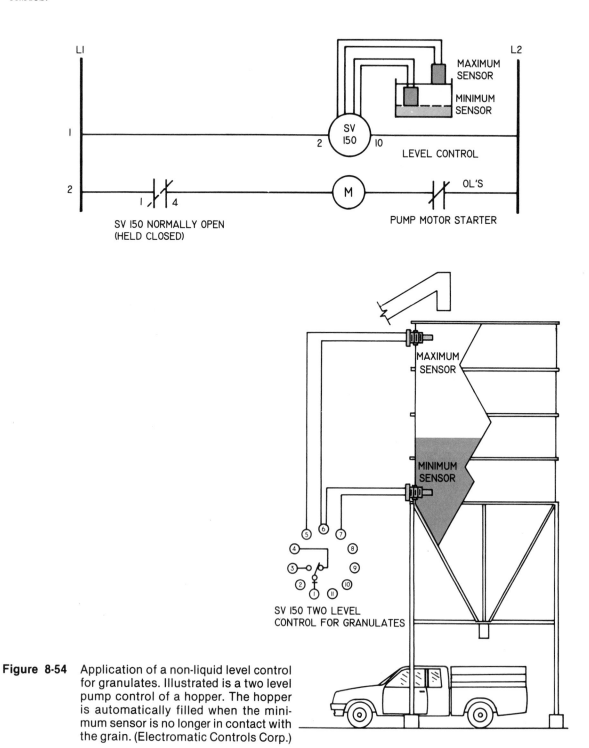

Figure 8-54 Application of a non-liquid level control for granulates. Illustrated is a two level pump control of a hopper. The hopper is automatically filled when the minimum sensor is no longer in contact with the grain. (Electromatic Controls Corp.)

imum sensor is in contact, a normally closed (pins 1 and 4) held open contact is used for this pump application. If a sump application is required, the normally open contacts would be used.

Combination of Temperature and Level Control

Figure 8-55 illustrates the combination of a temperature and level control. When using a heating element a contactor is used in the power circuit with a temperature module used in the control circuit. When using a pump motor a magnetic motor starter is used since overload protection is required. If a solenoid is used to open and close a valve that fills the tank, a contactor can be used. Circuit interlocking may be required to prevent the heating element from turning on, unless the level in the tank is at the maximum point.

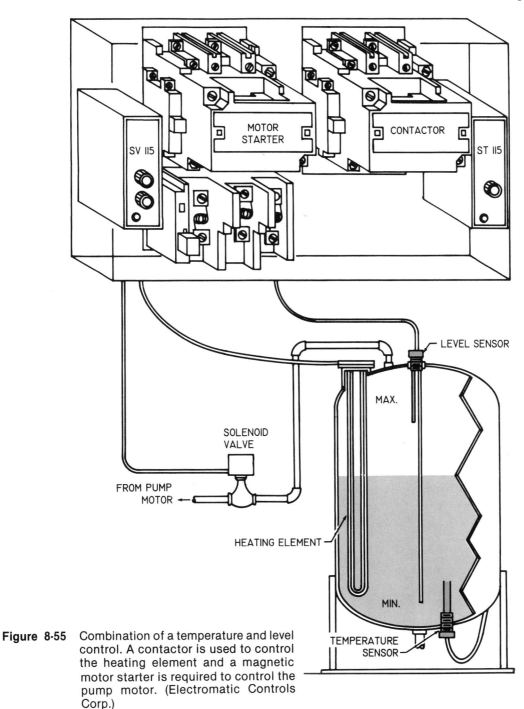

Figure 8-55 Combination of a temperature and level control. A contactor is used to control the heating element and a magnetic motor starter is required to control the pump motor. (Electromatic Controls Corp.)

Solar Heating Control

Interest in and use of solar heating is becoming one of the fastest growing areas of study in the world. This is simply because solar energy is plentiful enough to provide substantial amounts of energy that can be used to replace other forms of more expensive or less available fuels. Although solar energy has been used in the past for many applications, such as supplying power in space, its main use is in space heating and hot water heating. In each of these applications, some type of temperature control is needed.

In controlling a solar heating system, the temperature controllers must be able to measure the temperature inside the solar collector at the water storage area, and circulate the heat accordingly. This circulation of heat must be accomplished based on the temperature setting and differential setting of the controller. To meet these requirements, the temperature controller must be able to take two separate signal inputs (collector and heated area) and make a decision that will properly give the correct heat circulation.

Organizing the Control Circuit. Figure 8-56 illustrates the requirements of a solar heating control system. As illustrated, the control system needs two signal inputs (temperature sensor 1 and 2), a decision circuit (to compare inputs and control the action) and an action part to start or stop the pump motor.

Although the parts of this system, when combined, may appear to be a complicated circuit to connect, it is rather simple. This is because the decision part of the circuit can be a logic module that need have only the input and output connected to the module for the circuit to work. The electrician's job is made much easier if the decision part of the circuit can be identified and a control module used to accomplish the requirement. This leaves only the inputs and outputs to be connected.

Typical Temperature Control Circuit for Solar Heating Systems. Figure 8-57 and 8-58 shows a typical control circuit used with a solar heating system. Figure 8-58 shows the wiring diagram for this heating system. In this application, a water storage tank is used to hold hot water. A solar heat collector is used to absorb the sun's heat and a pump motor is used to circulate the water. This system does not include an auxiliary heating system such as gas or electric to supplement the solar system, but one may be added.

The temperature relay illustrated is operated in conjunction with two semi-conductor temperature sensors. These temperature sensors are available for use in liquids, gases, or may be constructed to measure the temperature of a metallic surface. These sensors may come in either a semi-conductor (usually thermistor) or bulb type construction. The relays react to certain temperature

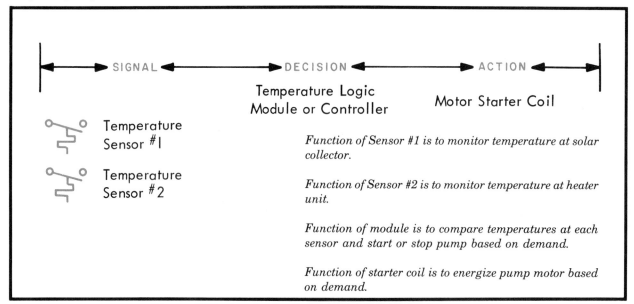

Figure 8-56 A solar heating control system can be organized into signals, decisions and action.

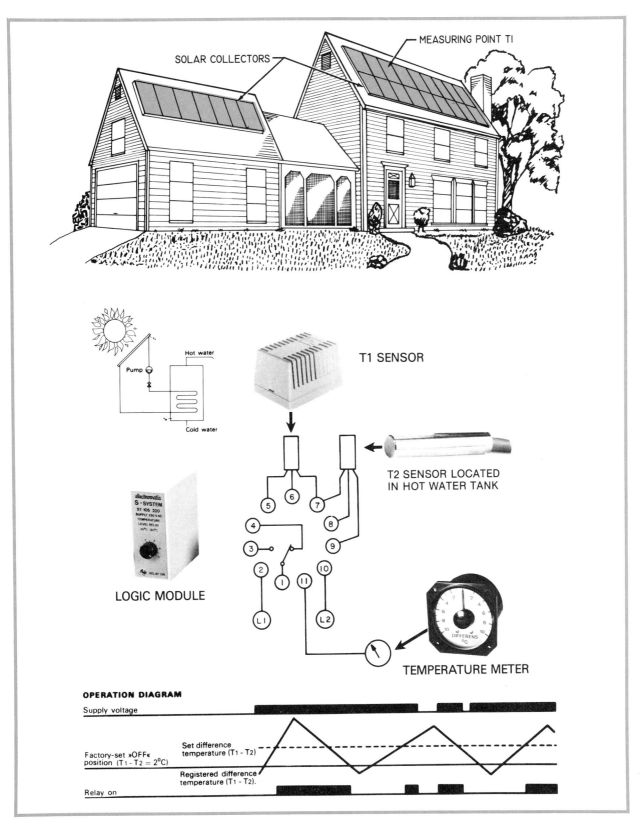

SOLAR COLLECTORS

MEASURING POINT TI

T1 SENSOR

T2 SENSOR LOCATED
IN HOT WATER TANK

LOGIC MODULE

TEMPERATURE METER

OPERATION DIAGRAM

Supply voltage

Set difference
temperature (T1 - T2)

Factory-set »OFF«
position (T1 - T2 = 2°C)

Registered difference
temperature (T1 - T2).

Relay on

Figure 8-57 In a solar heating system, the circulation pump is activated by the tempera-
ture differential between T1 and T2.

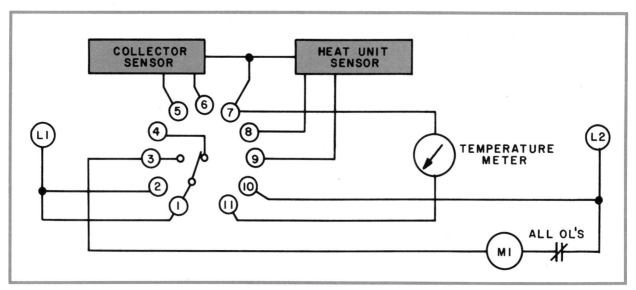

Figure 8-58 This is the wiring diagram that an electrician would follow in order to properly install the control circuit for a solar heating system.

differences as set on the adjustable knob. When the temperature in the solar panel (measuring point T1) exceeds the temperature in the water tank used to accumulate the heat (measuring point T2) by a predetermined valve (set on the relay), the relay operates.

The smallest differential temperature (T1 − T2) at which the relay shall operate can be set to any desired value between 3° to 10°C. The relay will not release until the temperature difference is reduced to 2°C (T1 − T2 = 2°C) as set by the factory. For example, if the relay is set for a 6°C temperature difference between the two measuring points (solar collector and water tank), then the pump will turn on at a 6° temperature change between the two measuring points. The pump would then circulate the hot water from the solar collector to the storage tank. This circulation would continue until there is a 2°C temperature difference between the two measuring points, in which case the relay would turn off the pump.

A moving coil instrument for indication of the temperature may be added to this circuit as illustrated. In cases where it is required to display also a negative temperature difference (T1 − T2), an instrument with its neutral position in the middle of the scale must be used.

Air Pollution Control

With major emphasis on air-pollution control today, the need for an additional automatic control device has emerged. That is a control device that can detect and react to gases and smoke in the air. Such a system is similar to the smoke detectors found in your home. Industrial applications, however, often require the detection of other pollutants besides smoke. These applications may require the detection of gases or carbon in the air. Such detection may be used for safety or in measuring such pollutants for meeting set minimum levels.

In monitoring air pollution, the controller must be able to measure the pollutants in the air and compare them to a set level. If the level is exceeded, the controller must also take appropriate action. This action may be the sounding of an alarm or some corrective measure such as turning on an exhaust fan. In meeting these requirements, the system must include a pollution detector (signal input), a logic module or control circuit that can react to the detector (decision) and an output (action). Figure 8-59 illustrates just such a typical circuit using an air pollution detector and matching relay to detect flammable gases, smoke and carbon in the air.

The relay is used with an air pollution detector

that detects very small concentrations of all reducing or flammable gases such as hydrogen, carbon dioxide, methane, propane, butane, acetylene, and sulphur dioxide. Even very small concentrations, such as .02% [200 parts per million (ppm)], are detected with this model. The detector also reacts to a smoke-filled carbonaceous atmosphere.

When the connecting supply voltage is applied, the relay immediately operates. The relay releases again when the registered pollution level exceeds the set level and the relay does not operate again until the registered level is about 10% lower than the set level. The activation of the relay can be used to signal an alarm or record the time and duration of high levels of pollution.

When placing a detector in the room to be supervised, the electrician should consider whether the gases present are heavier than air.

Several methods are used to measure the amount of smoke or gas in the air. The method used makes little or no difference to the electrician, provided the method measures what the application calls for. The only concern is that of connecting the signal inputs to the decision part of the circuit and connecting the output device to be controlled.

WIND METERING

As the need for electricity increases, alternative methods of producing it, such as windmill genera-

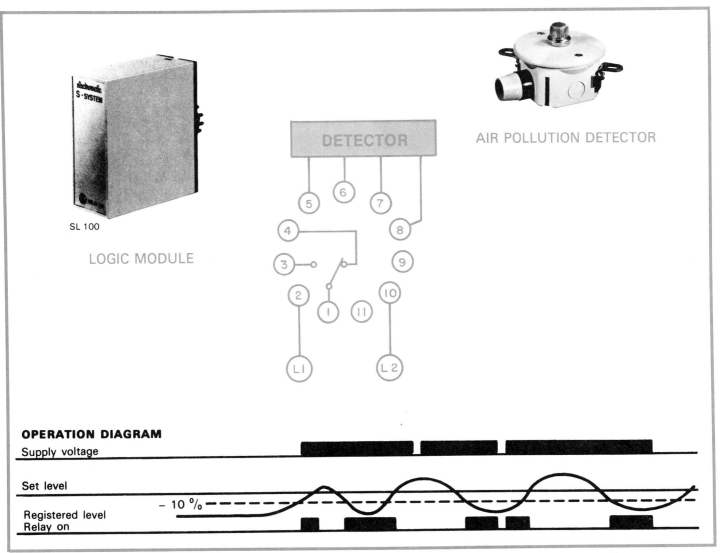

Figure 8-59 The operating and wiring diagram for an electronic air pollution detector circuit using logic modules and sensors. (Electromatic Controls Corp.)

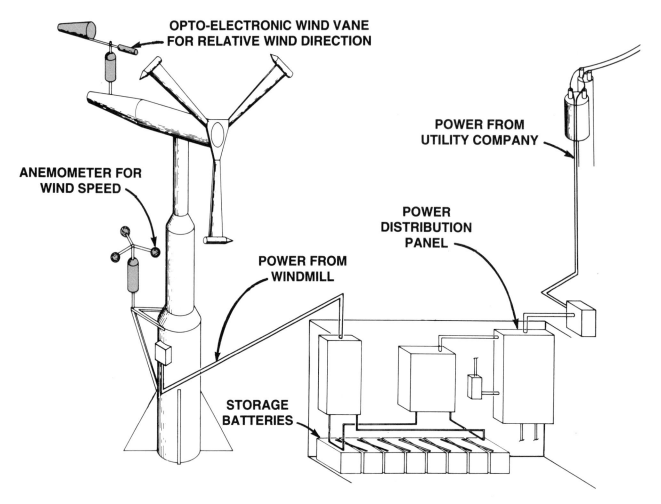

OPTO-ELECTRONIC WIND VANE FOR RELATIVE WIND DIRECTION

POWER FROM UTILITY COMPANY

ANEMOMETER FOR WIND SPEED

POWER DISTRIBUTION PANEL

POWER FROM WINDMILL

STORAGE BATTERIES

Figure 8-60 A windmill generating electricity must be controlled for maximum efficiency in all wind conditions.

tion, become more popular. A typical windmill consists of one or two generators depending upon the size. A windmill with two generators commonly has a small generator with synchronous speed of 1,000 RPM, and a large generator with a synchronous speed of 1,500 RPM. The small generator is first connected to the system. When the maximum power of this generator is reached it is connected and the larger one is connected after a short time delay. Figure 8-60 illustrates a typical windmill configuration.

On a windmill, a control device is required to measure and react to wind velocity. This device is called an anemometer. On some windmill applications a control device is also required to determine relative wind direction. This device is called a wind vane.

The anemometer is used to stop the windmill at too low and too high of wind velocity. Stopping the

windmill at low wind velocities prevents the constant connection and disconnection of the small generator. Stopping the windmill at high wind velocities helps to protect the windmill against damage and wear from the high wind speeds. To stop the windmill the anemometer and wind velocity relay control a mechanical brake inside the windmill. The specifications for the anemometer and wind velocity relay are illustrated in Figure 8-61.

The wind vane is used to control the yawing function of the windmill. In yawing, the windmill is deviated from the set course. This is required to turn the windmill into the direction of the wind for windmills that are not freely allowed to turn. To do this the wind direction relay incorporated a relay with a neutral center position. In this position the top of the windmill is kept still, while the two working positions of the contacts cause a turning either to the right or to the left of the windmill. The specification for the wind vane and wind direction relay are illustrated in Figure 8-62.

In addition to the controls for wind speed and wind direction other controls may be required depending on the application of the generated power. In a simple windmill application, the generated power can be directly used without synchronizing with other power. An example of this is an application that connects the generated power to a set of heating elements.

In an application that connects the generated power to the utility lines, synchronization of frequency and voltages must be built in. In connecting the generated power to a set of batteries a rectifier circuit will be required.

PV 01

- Opto-electronic anemometer.
- Measuring range:
  2–30 metres/second (5–67 miles/hour).
- Pulse generator for wind velocity relay, type SP 115.
- Rotor consists of hemispherical stainless steel cups.
- Supply voltage 11–27 VDC.
  Normally supplied from the S-system.

SP 115
Two adjustable
knobs

SP 115

- Wind velocity relay with outputs for instruments.
- Measuring range: 2–30 m/s in conjunction with opto-electronic pulse generator type PV 01.
- The relay operates, when the velocity of the wind has been within the set limits for 10 minutes.
- The relay releases instantly when one of the limits is exceeded.
- 10 A SPDT output relay.
- LED-indication of relay position.
- AC- or DC supply voltage.

WIRING DIAGRAM

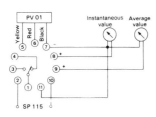

MODE OF OPERATION

In conjunction with an anemometer, type PV 01 being an opto-electronic pulse generator, the wind velocity relay, type SP 115 can be used for remote display of velocities up to 30 m/s (equalling 67 miles/h).
Instantaneous value and average value of the wind velocity can be displayed on two moving-coil instruments with a full-scale deflection of 1 mA.
The SP 115 has a minimum scale (2 to 10 m/s) and a maximum scale (10 to 30 m/s). Average value is measured in the minumum range, and instantaneous value in the maximum range.

The relay operates, when the velocity of the wind has been within the set minimum and maximum limits for 10 minutes.
The relay releases instantly when one of the limits is exceeded. The relay will not operate again, until the wind speed has been within the set limits for 10 minutes.

This mode of operation is applicable for the avoidance of both commuting and over-loading of wind-driven generators (windmills).
Other S-systems which are applicable for the control of wind-driven generators (windmills) are the SF 140 and SF 160.

OPERATION DIAGRAM

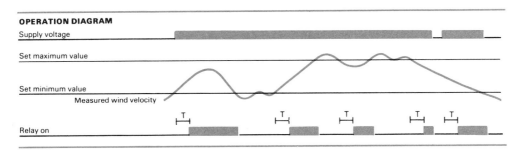

Figure 8-61 On a windmill, the ammeter measures the wind speed to control the function of the windmill. (Electromatic Controls Corp.)

OD 02

* Opto-electronic wind vane for relative wind direction.
* Signal for change of wind >7° and direction of change (left/right).
* Signal generator for wind direction relay, type SO 115.
* Supply voltage 11-27 VDC. Normally supplied from the S-system.

SO 115

* Windmill relay for relative wind direction.
* Operates in conjunction with opto-electronic wind vane, type OD 02.
* Delay on yawing and time for forced yawing are separately adjusted (0.8-18 secs).
* Input for forced yawing.
* 10 A SPDT output relay with neutral centre position.
* LED-indication of both working positions.
* AC- or DC supply voltage.

Two adjustable knobs

SPECIFICATIONS

Common technical data and ordering key
Pages 10-12.

Control range
±7° from the present starting point.
While measuring deviations from the present starting point, the S-system registers whether the wind veers or backs.

Hysteresis
3.5°.

Delay on yawing
The delay period is set by a built-in potentiometer (0.8 to 18 secs), and the timing starts

when a change of wind has caused a deflection of more than 7° compared to the present position of the top of the windmill. If the wind returns to its starting point ±7° before the set time has elapsed, the timing is reset.

Forced yawing
On connection between pins 7 and 9 the relay operates (pins 1-3), and the motor that turns the top of the windmill is activated. The motor works as long as pin 7 is connected with pin 9, through maximum during the set period. Adjustment on built-

in potentiometer (0.8-18 secs) dependent on the time required for the motor to yaw the top of the windmill e.g. 90°.

Adjustment
2 potentiometers with absolute scales:
0.8 to 18 seconds.

Upper knob: Adjustment of delay on yawing.
Lower knob: Adjustment of forced yawing.

Signal generator
Opto-electronic wind vane type OD 02. Supply voltage from the SO 115.

Inputs for OD 02
Pin 5: Centre signal.
Pin 6: + 24 VDC.
Pin 7: Negative.
Pin 8: Signal for direction of deflection (left/right).

Accessories
Bases.
Hold down spring.
Mounting rack.
Base covers.
Front mounting bezel.

Opto-electronic wind vane, type OD 02.
See catalogue on accessories.

WIRING DIAGRAMS

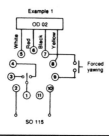

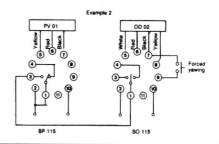

MODE OF OPERATION

S-system, type SO 115 is used in conjunction with wind vane, type OD 02 for automatic control of the yawing of windmills.
SO 115 incorporates a relay with neutral centre position. In this position the top of the windmill is kept still, while the two working positions cause a turning either to the right or to the left of the windmill.
In order to avoid incessant turning of the top of the windmill at changing directions of the wind, the S-system incorporates adjustable delay on yawing (upper knob). The SO 115 also has a built-in potentiometer (lower knob) for adjustment of yawing time (equal-

ling a turning of the top of the windmill of e.g. 90°) in case of error in conditions.

If the position of a windmill deviates 180° from the wind direction on starting, it may be necessary to force the start of the yawing by connecting pin 7 with pin 9 until the windmill stands app. 45° towards the wind direction.

Example 1
At a change of the wind in excess of 7°, the set delay period starts. When the wind vane has had a constant deflection of more than 7° for the set time, the relay operates, thus causing the top of the windmill to turn towards the new wind direction.

Example 2
As example 1 but here wind velocity relay, type SP 115 and anemometer, type PV 01 are connected too. A deflection of the wind vane requires a certain minimum velocity of wind (app. 2 metres/sec), and this example shows how incessant yawing in calm can be avoided. The required minimum wind speed is adjusted on S-system, type SP 115. See descriptions on SP 115/PV 01.

Other S-systems which are applicable for the control of wind-driven generators (windmills) are the SF 140 and SF 160.

OPERATION DIAGRAM

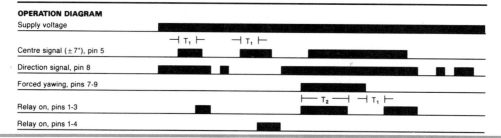

Figure 8-62 The wind vane is used to control the position of the windmill in different wind directions. (Electromatic Controls Corp.)

AUTOMATED SYSTEM

An automated system includes manual, mechanical and automatic control devices. These control devices are interconnected to provide the required inputs to make the system function as designed. Figure 8-63 illustrates a typical automated system. This operation is very common and could be used for paint, food products, beverages, cleaners or other products put in a container.

In this operation, manual inputs such as pushbuttons and selector switches are required at the individual and main control stations. Automatic inputs such as pressure, temperature, flow, and level controls are required to control each step of the process from start to finish. Mechanical inputs are used to detect position as required.

All of these basic control devices provide a method for controlling the product or operation. Each control device must be selected and installed as if the entire operation depends upon that one control input. In an automated system, the failure of any one control device could shut down the entire process.

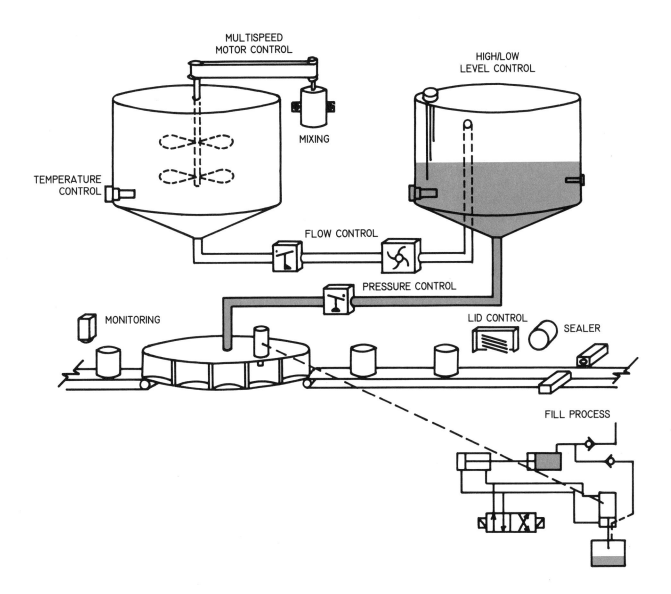

Figure 8-63 An automated system uses several interconnected control devices to make the system function.

1. What are some of the different contact configurations that are available on pushbuttons?
2. What are some of the different types of operators that are available on pushbuttons?
3. What type of operator is best suited for an emergency stop pushbutton?
4. What type of NEMA enclosures are available for pushbutton stations?
5. What is a selector switch?
6. What is a target table used for?
7. What is a "joy stick"?
8. What are the two basic styles of joy sticks?
9. What are the two methods used to illustrate the different positions of a joy stick?
10. What are limit switches used for?
11. What are the two main parts of a limit switch?
12. What types of actuators are available with limit switches?
13. What is a pressure switch?
14. How is the pressure setting of a pressure switch determined?
15. What is the "pressure differential" setting of a pressure switch?
16. What may happen if the pressure differential setting is too small?
17. What are some typical applications of pressure switches?
18. What is a temperature switch?
19. What are some of the different methods used to switch electrical contacts in response to a temperature change?
20. What is a flow switch?
21. What are two functions of the temperature sensor?
22. What are some typical applications of flow switches?
23. What is the difference between a sump and a pump operation?
24. What are some of the different methods of detecting a change in a liquid level?
25. What is a logic module?
26. What is the advantage of a logic module?
27. What is a anemometer used for?
28. What is an automated system?

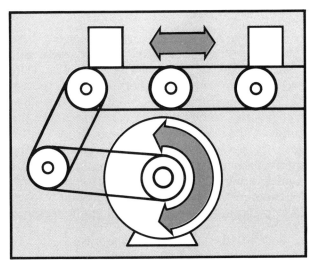

9 Reversing Circuits Applied to Single-Phase, Three-Phase and DC Motor Types

Motors are constructed as single phase, three phase, and DC types. Although motors generally run in one direction only, there are applications when a motor may need to be operated in the reverse direction. Reversing circuits provide the means for safely and conveniently reversing the direction of a motor.

Reversing the electrical connections to a motor can be accomplished with a manual reversing starters, magnetic reversing starters, or drum switches. Interlocking for safety can be done mechanically, electrically, or both mechanically and electrically.

MOTORS

Almost 85% of the industrial machines being used today are driven by electric motors. It is no wonder, then, that the industrial electrician must give electric motors some consideration when servicing these machines on a regular basis.

There are many factors to consider in selecting, wiring, and maintaining electric motors. Emphasis for the electrician, however, is generally on the electrical and mechanical considerations for a particular job. In other words, the electrician is not expected to design a drive system, but rather to see that it is properly installed mechanically and electrically.

Satisfying the electrical needs of a motor means that the unit will be installed according to manufacturer's wiring diagrams and that adequate protection for the motor has been provided according to the *National Electrical Code®*.

Satisfying the mechanical needs of a motor means that the unit selected meets the intended specifications of enclosurers, bearings, frame size and insulation. These specifications are usually established by the original manufacturer but must be understood by the electrician in selecting a proper replacement. Basically this means that the electrician must be able to read the manufacturer's motor nameplate information to the point where he knows that the motor he is choosing for replacement is equal to the one he has removed.

In this unit we will deal primarily with meeting the electrical needs of a motor when operating in forward and reverse. Overload protection for motors was covered in Unit 6, "AC/DC Contactors and Magnetic Motor Starters." Further consideration will be given to motor maintenance throughout this unit and in more detail in Unit 15, "Preventive Maintenance and Troubleshooting Techniques and Applications."

Three-Phase Motor Construction

All three-phase motors are constructed internally with a number of individually wound electrical coils. Regardless of how many individual coils there are in a three-phase motor, the individual coils will always be wired together (series or parallel) to produce three distinct windings, which are called phases. Each phase will always contain one-third of the total number of individual coils. These composite windings or phases are usually referred to as phase A, phase B and phase C.

Three-phase motors vary from fractional horsepower size to several thousand horsepower. These motors have a fairly constant speed characteristic but a wide variety of torque characteristics. They are made for practically every standard voltage and frequency and are very often dual-voltage motors. The three-phase motor is probably the simplest and most rugged of all electric motors. To get a proper perspective on how important the three-phase motor is, remember that this motor is used in nine out of ten industrial applications. Fortunately, it is the easiest type of motor to learn how to properly wire for forward and reversing.

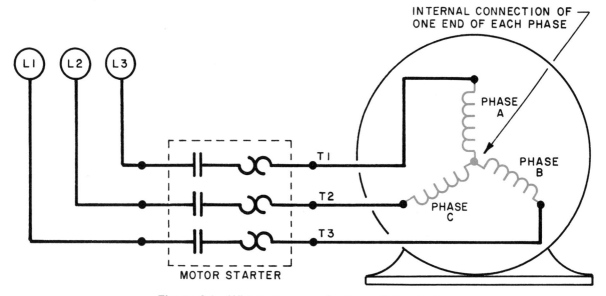

Figure 9-1 Wiring diagram of a three-phase motor wired in a Wye (Y) configuration.

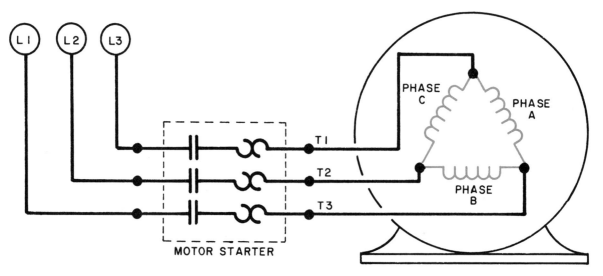

Figure 9-2 Wiring Diagram of a three-phase motor wired in a Delta (Δ) configuration.

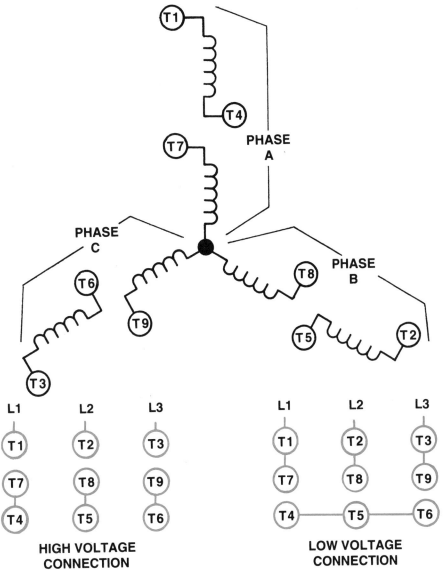

HIGH VOLTAGE CONNECTION

LOW VOLTAGE CONNECTION

Figure 9-3 Wiring diagram and terminal connections for a dual-voltage Wye connected three-phase nine-lead motor.

Wye Connected Three-Phase Motors. All three-phase motors are wired so that the phases are connected in either a Wye (Y) or Delta (Δ) configuration. Figure 9-1 illustrates a motor wired in the Wye (Y) configuration. In a Wye (Y) configuration, one end of each of the three phases is connected to the other phases internally. The remaining end of each phase is then brought out externally and connected to the power line. The leads which are brought out externally are labeled terminals one, two and three (T1, T2 and T3). When connected, terminals (T1, T2, T3) are matched to the three-phase power lines labeled lines one, two, and three (L1, L2 and L3). For the motor to operate properly, the three-phase line supplying power to the motor must have the same voltage and frequency ratings as the motor.

Delta Connected Three-Phase Motors. Figure 9-2 illustrates a motor wired in the Delta (Δ) configuration. In a Delta configuration each winding is wired end to end to form a completely closed loop circuit. At each point where the phases are connected, however, leads are brought out externally to form terminals one, two and three (T1, T2 and T3). These terminals, like those of a Wye connected motor, are correspondingly attached to the power lines—line one, line two and line 3 (L1, L2

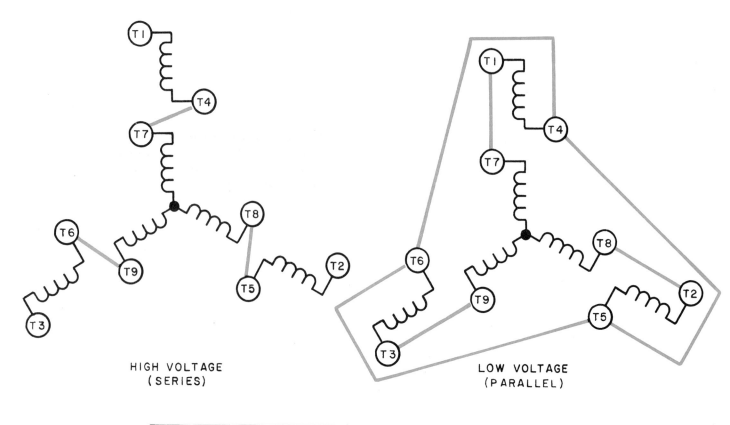

HIGH VOLTAGE
(SERIES)

LOW VOLTAGE
(PARALLEL)

VOLTAGE	LI	L2	L3	TIE TOGETHER		
LOW	TIT7	T2T8	T3T9		T4T5T6	
HIGH	TI	T2	T3	T4T7	T5T8	T6T9

Figure 9-4 Wiring diagram and terminal connections for a dual-voltage Wye connected three-phase nine-lead motor. (*Note:* In the low voltage configuration, the coils are wired in parallel; in the high voltage configuration, the coils are wired in series.)

Dual Voltage Motors

Many three-phase motors are made so that they can be connected for either of two voltages. The purpose in making motors for two voltages is to enable the same motor to be used with two different power line voltages. Usually the dual voltage rating of industrial motors is 230/460V; however, the nameplate must always be checked for proper voltage ratings.

When the electrician has the choice of deciding which voltage to use, the higher voltage is preferred. The motor will use the same amount of power, giving the same HP output for either high

and L3). Again, the three-phase lines supplying power to the Delta motor must have the same voltage and frequency rating as the motor.

or low voltage, but as the voltage is doubled (230 to 460), the current will be cut in half. With half the current, wire size can be reduced and a savings is created on installation.

Dual Voltage Wye Connected Motor. Figure 9-3 illustrates the wiring diagram and terminal connections for a dual voltage Wye connected three-phase motor. Note from the wiring diagram that the windings are arranged to bring nine leads out of the motor. These leads are marked T1 through T9 and may be connected externally for either of the two voltages. The terminal connections for high and low voltage are shown below the wiring diagram. The information provided on this terminal connection chart is usually provided on the nameplate of the motor.

Figure 9-4 illustrates electrically what takes

place in the motor when the terminals are connected for high and low voltage. To connect a Wye configuration for high voltage, connect line 1 to terminal 1, line 2 to terminal 2, line 3 to terminal 3, and tie together terminals 4 and 7, 5 and 8, and 6 and 9. Electrically, we have merely series-connected the individual coils in phases A, B and C so that each coil receives 50% of the line-to-neutral voltage.

To connect the Wye motor for low voltage, we connect line 1 to terminals 1 and 7, line 2 to terminals 2 and 8, line 3 to terminals 3 and 9, and then tie together terminals 4, 5 and 6. Electrically, we have merely parallel-connected the individual coils in phases A, B and C so that each coil receives 100% of the line-to-neutral voltage.

Dual Voltage Delta Connected Motor. Figure 9-5 illustrates the wiring diagram and terminal connection for a dual voltage Delta connected three-phase motor. Like the Wye configuration, the leads are marked T1 through T9 and a terminal connection chart is provided for wiring high and low voltage operation.

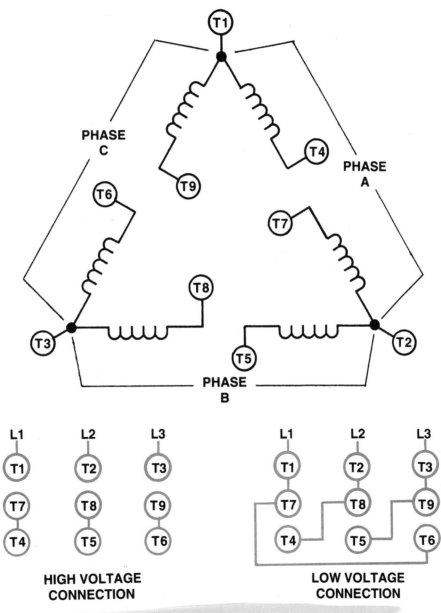

Figure 9-5 Wiring diagram and terminal connection for a dual-voltage Delta connected three-phase nine-lead motor.

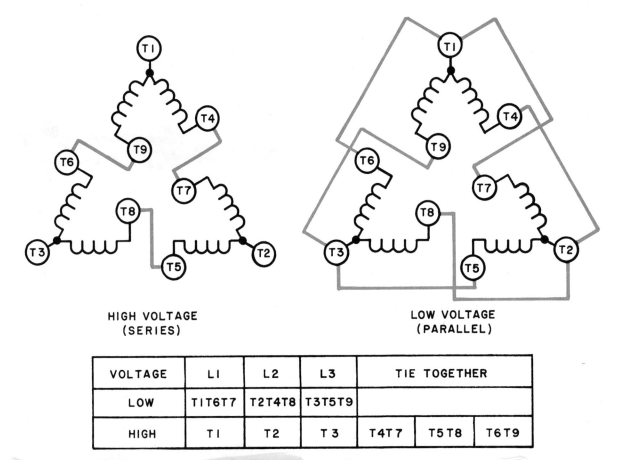

VOLTAGE	L1	L2	L3	TIE TOGETHER		
LOW	T1T6T7	T2T4T8	T3T5T9			
HIGH	T1	T2	T3	T4T7	T5T8	T6T9

Figure 9-6 Wiring diagram and terminal connections for a dual-voltage Delta connected three-phase nine-lead motor. (*Note:* In the low voltage configuration, the coils are wired in parallel; in the high voltage configuration, the coils are wired in series.)

Figure 9-6 illustrates electrically what takes place in a Delta motor when the terminals are connected for high and low voltage. In the high voltage configuration the coils are wired in series, and for low voltage the coils are wired in parallel to distribute the voltage according to the individual coil rating.

Reversing Three-Phase Wye and Delta Motors

Reversing three phase Wye and Delta motors is very simple. The direction of rotation of either of these motors can be reversed merely by interchanging any two of the three main power lines to the motor. Although any two lines can be interchanged, the industrial standard is to interchange lines one (L1) and three (L3). This standard holds true for all three-phase motors including three, six, and nine lead, Wye and Delta motors. Regardless of the type of three-phase motor, for forward rotation connect L1 to T1, L2 to T2 and L3 to T3; for reverse rotation connect L1 to T3, L2 to T2 and L3 to T1. If a three-phase motor has more than

three leads (T1, T2, T3) coming out, these leads will have to be connected according to their wiring diagrams.

SINGLE-PHASE MOTORS

Single-phase motors are most often found where a fractional horsepower motor drive is called for or where no three-phase power is available. Single-phase large horsepower motors generally are not used because they are inefficient compared to three-phase motors and thus cost more to operate. In addition, single-phase motors are not self-starting on their running windings, as are three-phase motors. With the exception of the universal type motor, all single-phase motors must have an auxiliary means for developing starting torque. The basic types of single-phase motors include the split-phase, shaded pole, capacitor, universal, repulsion and synchronous.

Since there are so many different types and combinations of single-phase motors, it is an absolute must that the electrician refer to the wiring

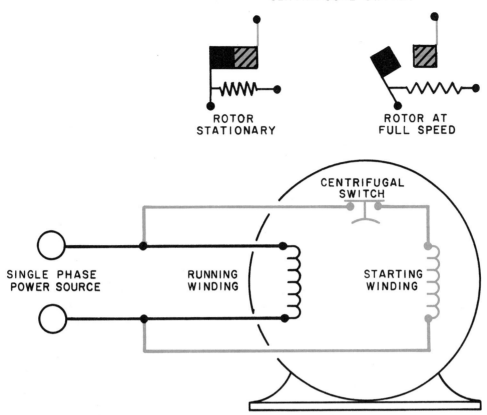

CENTRIFUGAL SWITCH

ROTOR
STATIONARY

ROTOR AT
FULL SPEED

CENTRIFUGAL
SWITCH

SINGLE PHASE
POWER SOURCE

RUNNING
WINDING

STARTING
WINDING

Figure 9-7 Wiring diagram of a split-phase motor with running and starting windings. The centrifugal switch will disconnect the starting winding at about 75% of motor full speed.

diagram provided by the manufacturer. Although there are many similarities in single-phase motor types, minor differences between manufacturers have led to hundreds of different wiring diagrams for single-phase motors. We will take a look at some of the more common types of single-phase motors and their circuits.

Split-Phase Motors

The split-phase motor is an AC motor of fractional horsepower size, usually 1/20 to 1/3 horsepower. It is commonly used to operate washing machines, oil burners, and small pumps and blowers. The motor has three main parts: the rotating part, called the *rotor;* the stationary part called the *stator* (the stator is made up of the running and starting winding); and a *centrifugal switch* that is located inside the motor to disconnect the starting winding at about 75% of full-load speed.

Figure 9-7 illustrates an example of a split-phase motor. When starting, both the running windings and the starting windings are in parallel across the line. The running winding is usually

made up of a heavy insulated copper wire and the starting winding is made of fine insulated copper wire. When the motor reaches approximately 75% of full speed, the centrifugal switch opens, disconnecting the starting winding from the circuit, allowing the motor to operate on the running winding only. The running winding is also often called the main winding and the starting winding is often called the auxiliary winding.

Reversing a Split-Phase Motor. Reversing the rotation of a split-phase motor is accomplished by interchanging the leads of the starting winding. It should be noted, however, that whenever possible the manufacturer's wiring diagram should be referred to for the exact wires to interchange.

If manufacturer's information is not available or is lost, the electrician can measure the resistance of the starting winding and running windings to determine which leads are connected to which windings. The running windings are made of a heavier gauge wire than the starting winding, so the running windings will show a much lower resistance than the starting windings.

Dual Voltage Split-Phase Motors. Figure 9-8 illustrates the wiring diagram and terminal connections for a dual-voltage split-phase motor. Electrically all the windings receive the same voltage when wired for low voltage operation. Wiring for high voltage puts the running windings in series; the starting winding is parallel with only one of the running windings to provide proper voltage distribution. Since each motor may be different, always refer to nameplate or manufacturer's wiring diagrams.

Capacitor Motors

The capacitor motor is an AC motor made in sizes ranging from 1/8 to 10 horsepower. It is widely used to operate such machines as refrigerators, compressors, washing machines, and air conditioners. The construction of the capacitor motor is similar to that of the split-phase motor, but an additional component called a capacitor is connected in series with the starting winding. This addition of a capacitor in the starting winding gives the capacitor start motor more torque than the split-phase motor. There are three types of capacitor motors. They are the capacitor start motor, permanent-split capacitor motor and the two-value capacitor motor.

Capacitor Start Motor. Figure 9-9 illustrates a typical example of a capacitor start motor. The capacitor start motor operates much the same as a split-phase motor in that it uses a centrifugal switch that operates at approximately 75% full speed. In this case the starting winding and the capacitor are removed when the circuit opens. The capacitor used in the starting winding gives the capacitor start motor a high starting torque.

Permanent-Split Capacitor Motor. Figure 9-10 illustrates a typical example of a permanent-split capacitor motor. The term "permanent-split" comes from the fact that the starting winding with the capacitor connected in series will remain

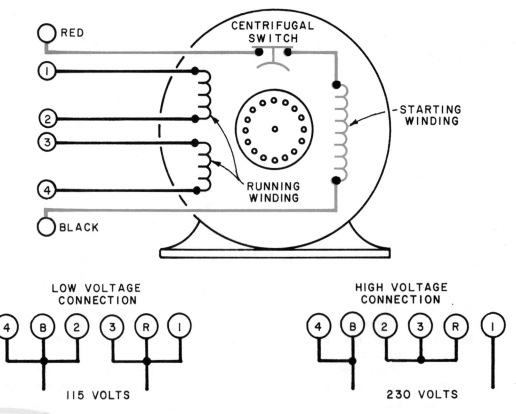

Figure 9-8 Wiring diagram of a dual voltage split-phase motor with running and starting windings. The centrifugal switch will disconnect the starting winding at about 75% of motor full speed.

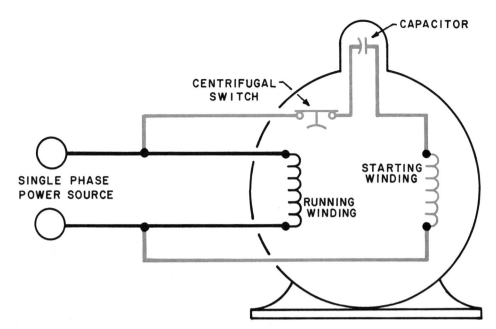

Figure 9-9 Wiring of a typical capacitor start motor. The capacitor used in the starting windings gives the capacitor start motor a high starting torque. The centrifugal switch will disconnect the starting winding and capacitor at about 75% of motor full speed.

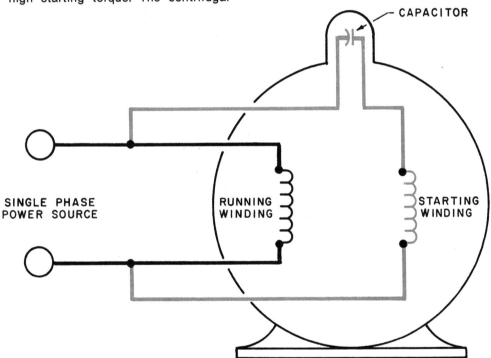

Figure 9-10 Wiring diagram of a typical permanent-split capacitor motor. The capacitor will remain permanently in the circuit, both during starting and during running. Permanent-split capacitor motors have medium starting torque and relatively high running torque compared to a capacitor start motor.

continuously or permanently connected into the circuit both during starting and during running. A lower value capacitor is used in the permanent-split capacitor motor than in the capacitor start motors because it will remain in the circuit at full speed. This gives the permanent-split capacitor motor medium starting torque and somewhat higher running torque than a capacitor start motor.

Two Value Capacitor Motor. Figure 9-11 illustrates a typical example of a two-value capacitor motor. The two-value capacitor motor starts with a high and low value capacitor connected in parallel with each other but in series with the starting winding to provide a very high starting torque. The centrifugal switch then disconnects the high value capacitor at about 75% full load speed, leaving the low value capacitor in the circuit. The net result is that the two-value capacitor combines the advantage of the capacitor start motor (high starting torque) with that of the split capacitor motor (high running torque). In each case, the loads and the amount of torque (both starting and running) determine which type of capacitor start motor to use.

Reversing Capacitor Start Motor. The direction of rotation of a capacitor motor can be changed by reversing connection to the starting winding. Whenever possible, the manufacturer's wiring diagram should be referred to for the exact wires to interchange.

DC MOTORS

Direct current motors are usually found where the load requires an adjustable speed and simple control of torque. Typical application for DC motors are printing presses, cranes, elevators, shuttle cars and automobile starters.

The DC motor electrically consists of two circuits—called the *field circuit* and the *armature circuit*. The *field circuit* consists of stationary windings and the *armature circuit* is found in the rotating winding.

There are basically three types of DC motors: the series motor, the shunt motor and the compound motor. Internally and externally, all three types of DC motors are practically the same. The main difference between the motors is in the way

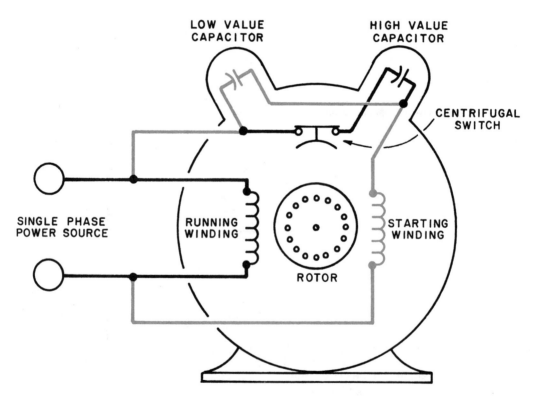

Figure 9-11 Wiring diagram of a typical two-value capacitor motor. The two-value capacitor motor combines the advantage of high starting torque and high running torque due to the way in which the windings are wired into the circuit.

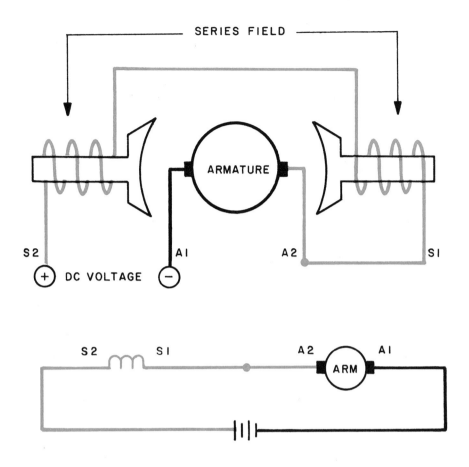

Figure 9-12 Wiring diagram of a typical series DC motor.

Series DC Motor

Figure 9-12 illustrates the electrical circuit involved in a typical series DC motor. The series field coils are composed of a few turns of heavy gauge wire connected in series with the armature. The wires extending from the series coil are marked S1 and S2. The wires extending from the armature are marked A1 and A2.

The DC series motor has a high starting torque and a variable speed characteristic. What this means is that the motor can start very heavy loads, but the speed will increase as the load is decreased. The series DC motor can develop this high starting torque because the same current that passes through the armature also passes through the field. Thus, if the armature calls for more current (developing more torque), then this current also passes through the field, increasing the field strength.

This inherent ability to draw more current is an advantage as long as a load is applied to the motor, since it will tend to hold the speed under control. If, however, the load is removed, the variable speed characteristic will allow the motor to keep increasing speed after the load is moved; eventually, if left unchecked, the motor will actually throw itself apart at uncontrolled speeds. For this reason it is necessary to positively connect DC series motors to loads through couplings or gears that do not allow slip. Belt drives should not be used on DC series motors.

In small DC series motors brush friction, bearing friction and winding loss may provide sufficient load to hold the speed down to a safe level.

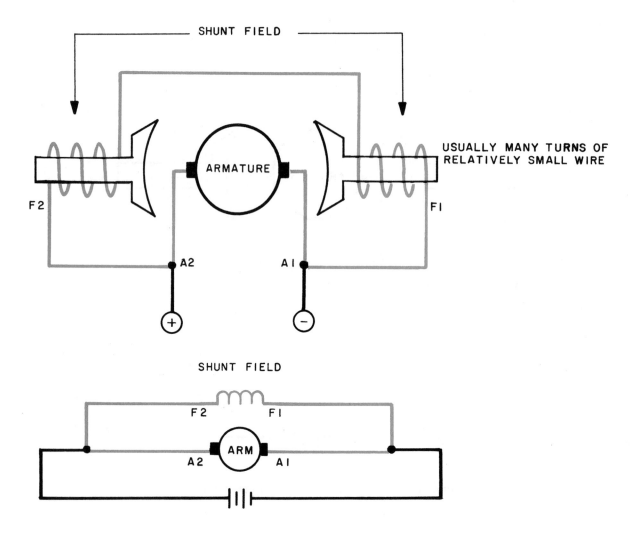

SHUNT FIELD

USUALLY MANY TURNS OF
RELATIVELY SMALL WIRE

F2

F1

ARMATURE

A2

A1

+

−

SHUNT FIELD

F2

F1

ARM

A2

A1

Figure 9-13 Wiring diagram of a typical shunt DC
motor.

Shunt DC Motor

Figure 9-13 illustrates the electrical connection
involved in a typical shunt motor. The shunt
motors have their armatures and field circuits
wired in parallel, giving essentially constant field
strength and motor speed. The windings extending from the shunt field are marked F1 and F2.

The shunt field can be either connected to the
same power supply as the armature (called self-
excited) or may be connected to another power
supply (separately excited).

When DC shunt motors are separately excited,
the motor can be speed controlled by varying the
field current. Field speed control can be accomplished by inserting external resistances in series
with the shunt field circuits. As resistance is
increased in the field circuit, field current is
reduced and the speed is increased. Conversely, as

resistance is removed from the shunt field, field
circuit goes up and the speed goes down. By selecting the proper controller, you can set the motor for
a specific speed control range. Standard DC shunt
field voltages are 100, 150, 200, 240 and 300 volts
DC.

DC Compound-Wound Motor

Figure 9-14 illustrates the electrical connection
involved in a typical DC compound-wound motor.
Note that the field coil is a combination of the
series and shunt fields. The series field is connected in series with the armature and the shunt
field is connected in parallel with the series field
and armature. This arrangement combines the
characteristics of both the series and shunt motor.
Compound DC motors have high starting torque
and fairly good speed torque characteristics at

rated load. Because of the complicated circuits needed to control compound motors, only large bidirectional motors of this type are usually built. Some smaller DC motors, however, may be slightly compounded to improve starting characteristics. The DC compound motor is used for drives needing fairly high starting torque and reasonably constant running speed.

Reversing DC Motors

The direction of rotation of DC motor, series, shunt or compound may be reversed by reversing the direction of the current through the field without changing the direction of the current through the armature, or by reversing the direction of the

current through the armature, but not both. It is the industrial standard to reverse the current through the armature. It is important in the compound-wound motor that the series and shunt field relationship to the armature be left unchanged. The shunt must be connected in parallel and the series field in series with the armature. Reversal is accomplished by reversing the armature connections only. If the motor has commutating pole windings, these windings are considered a part of the armature circuit and the current through them must be reversed when the current through the armature is reversed. Commutating windings (also called interpoles) are used to prevent sparking at the brushes in some DC motors.

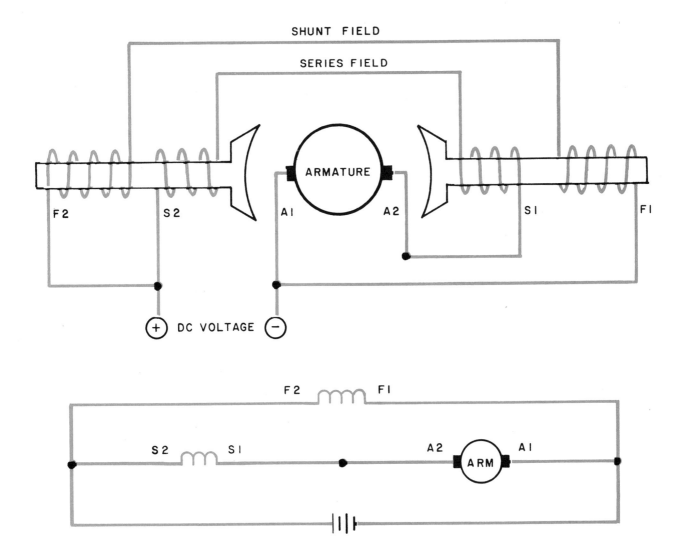

Figure 9-14 Wiring diagram of a typical compound DC motor.

REVERSING THREE-PHASE, SINGLE-PHASE AND DC MOTORS USING A MANUAL STARTER

Manual reversing starters used to change the direction of rotation in three-phase, single-phase and DC motors are made by connecting two manual starters together (Figure 9-15). Electrically, the manual starter would look like Figure 9-16 internally. This type of manual starter is used to run lower horsepower motors (such as those found on fans, small machines, pumps and blowers) in forward and reverse.

You will note from Figure 9-16 that the individual manual starters are marked start/stop instead

Figure 9-15 A typical manual pushbutton reversing motor starter. (Square D Company)

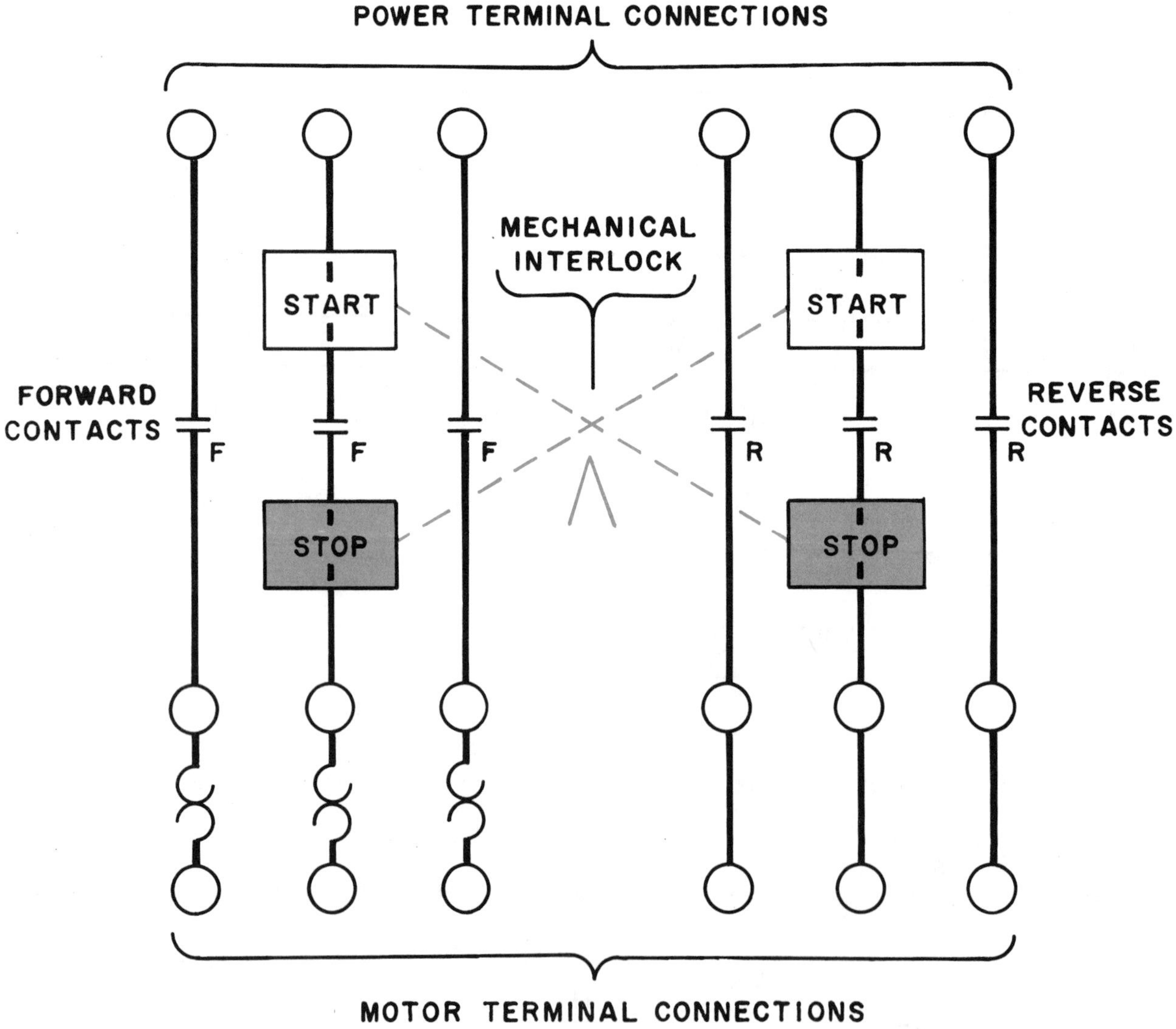

POWER TERMINAL CONNECTIONS

MECHANICAL INTERLOCK

START   START

FORWARD CONTACTS

F   F   F   R   R   R

STOP   STOP

REVERSE CONTACTS

MOTOR TERMINAL CONNECTIONS

Figure 9-16 Electrically a manual pushbutton reversing motor starter consists of two sets of electrical contacts which are mechanically interlocked.

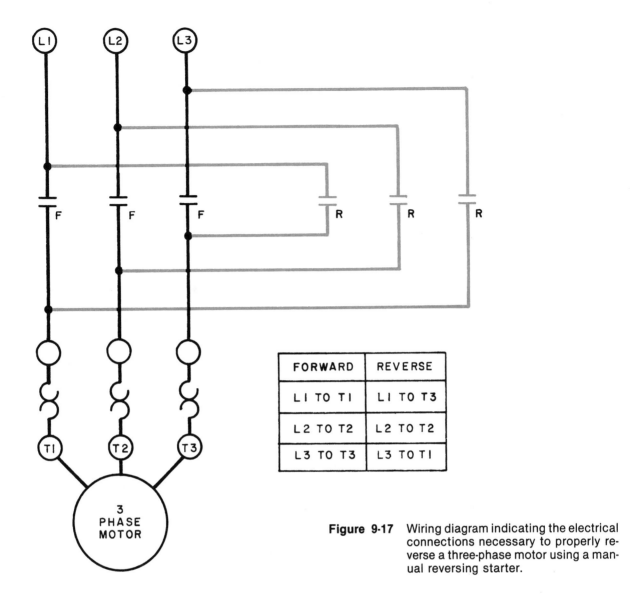

FORWARD	REVERSE
L1 TO T1	L1 TO T3
L2 TO T2	L2 TO T2
L3 TO T3	L3 TO T1

Figure 9-17 Wiring diagram indicating the electrical connections necessary to properly reverse a three-phase motor using a manual reversing starter.

of forward/stop or reverse/stop. This is quite common when two manual starters are placed in the same enclosure to make up a manual reversing starter. It is up to the electrician to properly label this unit once it is wired properly.

In the wiring diagram of Figure 9-16 you should also have noticed the use of crossing dashed lines between the manual starter, indicating a mechanical interlock. Since the motor cannot run in both directions at the same time, some means must be included to prevent both starters from energizing at the same time. The manual reversing starter uses the same type of mechanical interlock used in Unit 4 ("AC Manual Contactors and Motor

Starters") for separating the contacts on a two-speed manual starter. These mechanical devices are inserted between the two starters as they are installed to assure that both switching mechanisms cannot be energized at the same time. If the unit is not pre-assembled, it is up to the electrician to see that this interlock is provided.

Reversing a Three-Phase Motor Using a Manual Starter

Figure 9-17 illustrates a wiring diagram indicating the electrical connections necessary to properly reverse a three-phase motor using a manual reversing starter. Note from the way the unit is

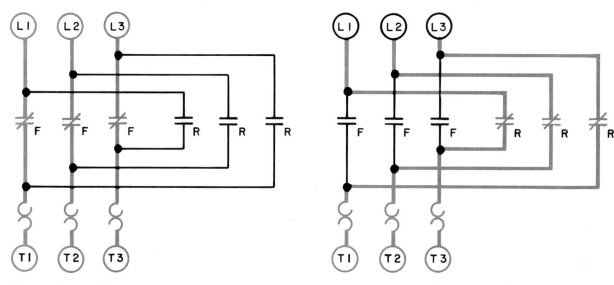

Figure 9-18 Left: Wiring diagram indicating how current would flow through a three-phase motor when the forward contacts were closed. Right: Wiring diagram indicating how current would flow through a three-phase motor when the reverse contacts were closed.

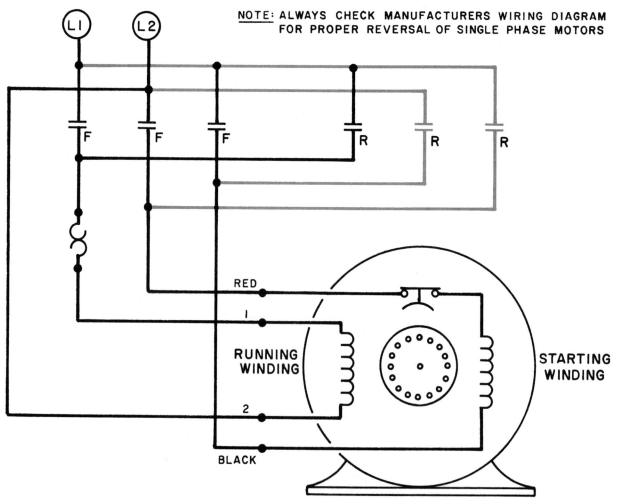

NOTE: ALWAYS CHECK MANUFACTURERS WIRING DIAGRAM FOR PROPER REVERSAL OF SINGLE PHASE MOTORS

Figure 9-19 Wiring diagram indicating the electrical connections necessary to properly reverse a single-phase motor using a manual reversing starter.

wired that only one set of overloads need be installed. Referring to Figure 9-18, left, and 9-18, right, the circuit would operate as follows: When forward contacts F close, L1, L2 and L3 are connected directly in sequence to T1, T2 and T3. When the reverse contacts R close and the forward contacts F open, L1 is connected to T3, L2 is connected to T2, and L3 is connected to T1. Since it is necessary to interchange only two leads on a three-phase motor to reverse rotation, the motor will change directions each time forward or reverse is depressed.

Reversing a Single-Phase Motor Using a Manual Starter

Figure 9-19 illustrates one type of wiring diagram indicating the wiring necessary to properly reverse one type of single-phase motor using a manual reversing starter.

In Figure 9-20 the circuit would operate as follows: When forward contacts F close, line L1 is connected to the black lead of the starting winding and to side one of the running winding; L2 is connected to the red lead of the starting winding and to side two of the running windings. When the reverse contacts R close and the forward contacts F open, line L1 is connected to the red lead of the starting winding and side one of the running winding; L2 is connected to the black lead of the starting winding and side two of the running winding. In other words, the starting windings are interchanged while the running windings remain the same. Since it is necessary to interchange only the starting windings on a single-phase motor to reverse rotation, the motor will change direction each time forward or reverse is depressed. (*Note:* It is extremely important when reversing single-phase motors to check manufacturer's wiring diagrams to determine which leads are connected to the starting winding. Most often the red and black wires are the ones used for reversal.)

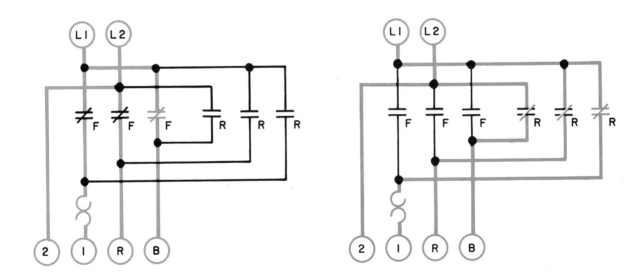

NOTE: ALWAYS CHECK MANUFACTURERS WIRING DIAGRAM FOR PROPER REVERSAL OF SINGLE PHASE MOTORS.

Figure 9-20 Left: Wiring diagram indicating how current would flow through a single-phase motor when the forward contacts were closed. Right: Wiring diagram indicating how current would flow through a single-phase motor when the reverse contacts were closed.

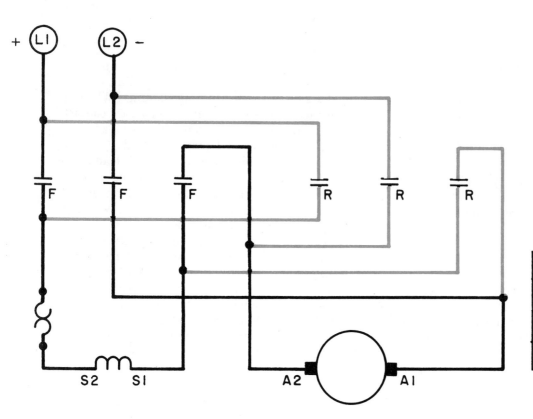

FORWARD	REVERSE
+ TO S2	+ TO S2
S1 TO A2	S1 TO A1
A1 TO −	A2 TO −

Figure 9-21 Wiring diagram indicating the electrical connections necessary to properly reverse a series DC motor using a manual reversing starter.

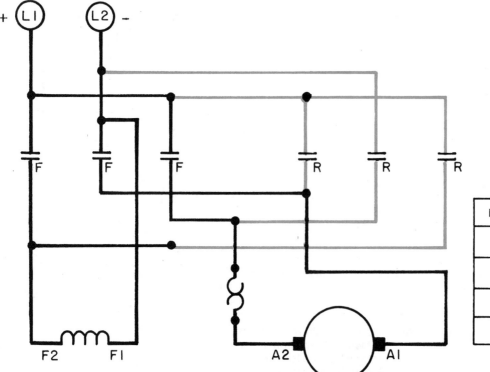

FORWARD	REVERSE
+ TO A2	+ TO A1
− TO A1	− TO A2
− TO F1	− TO F1
+ TO F2	+ TO F2

Figure 9-22 Wiring diagram indicating the electrical connections necessary to properly reverse a shunt DC motor using a manual reversing starter.

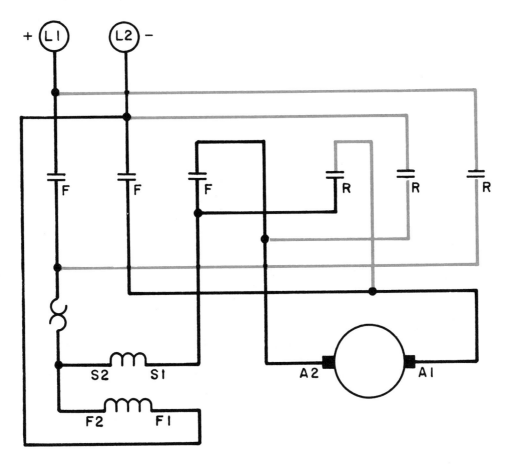

FORWARD	REVERSE
+ TO S2	+ TO S2
+ TO F2	+ TO F2
− TO A1	− TO A2
− TO F1	− TO F1
S1 TO A2	S1 TO A1

Figure 9-23 Wiring diagram indicating the electrical connections necessary to properly reverse a compound DC motor using a manual reversing starter.

Reversing DC Motors Using a Manual Starter

Figures 9-21, 9-22, and 9-23 illustrate the wiring diagrams for properly reversing a DC series motor, a DC shunt motor, and a DC compound motor respectively. By following each wiring diagram carefully, you can determine in each case how the voltage polarity and current flow is reversed to the armature circuit while the field voltage and current remain constant. Since it is necessary to reverse polarity only to the armature in a direct current motor in order to reverse rotation, the motor will change direction each time forward or reverse is depressed.

REVERSING THREE-PHASE, SINGLE-PHASE AND DC MOTORS

A magnetic reversing starter like the one illustrated in Figure 9-24 performs the same function as a manual reversing starter. Electrically the only difference between manual and magnetic starters is the addition of forward and reversing coils and the use of auxiliary contacts. The forward and reversing coils replace the pushbuttons of a manual starter and the auxiliary contacts provide additional electrical protection and circuit flexibility. Since the reversing circuit is the same for both manual and magnetic starters, we will concentrate on the electrical protection and circuit flexibility that magnetic reversing starters provide.

Mechanical Interlocking

Figure 9-25 illustrates a line diagram of a magnetic reversing starter controlled by a forward and reverse pushbutton. Recall that a line diagram does not show the power contacts and that the power contacts are found in the wiring diagram.

From Figure 9-25 we see broken lines running from the forward coil to the reverse coil, indicating that the coils are mechanically interlocked

much like those of a manual reversing starter. This mechanical interlock is usually factory installed by most manufacturers who supply forward and reversing starters. The circuit of Figure 9-25 would operate in the following manner: Depressing forward pushbutton PB2 completes the forward coil circuit from L1 to L2, energizing coil F. Coil F in turn energizes auxiliary contacts F1, providing MEMORY. Mechanical interlocking keeps the reversing circuit from closing. Depressing stop button PB1 opens the forward coil circuit, causing coil F to de-energize and contacts F1 to return to their normally open position. Depressing reverse pushbutton PB3 completes the reverse coil circuit from L1 to L2, energizing coil R. Coil R in turn energizes auxiliary contacts R1, providing MEMORY. Mechanical interlocking keeps the forward circuit from closing. Depressing stop button PB1 opens the reverse coil circuit, causing coil R to de-energize and contacts R1 to return to their normally open position. Overload protection is provided both in forward and reverse by the same set of overloads.

Auxiliary Contact Interlocking

Although most magnetic reversing starters provide mechanical interlock protection, some circuits are provided with a secondary back-up or

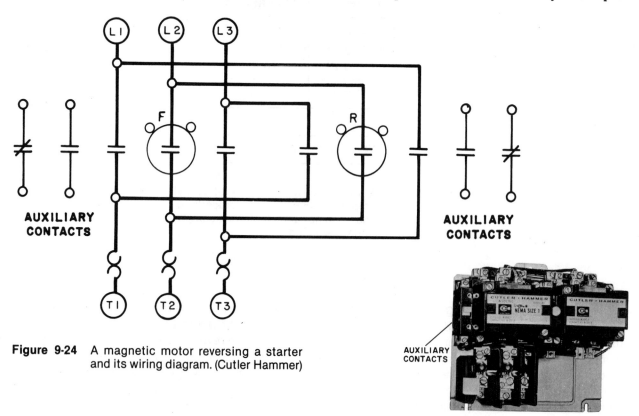

Figure 9-24 A magnetic motor reversing a starter and its wiring diagram. (Cutler Hammer)

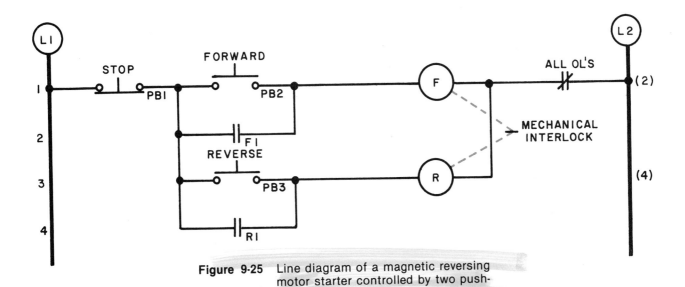

Figure 9-25 Line diagram of a magnetic reversing motor starter controlled by two push-buttons. Mechanical interlocking is provided in forward and reverse.

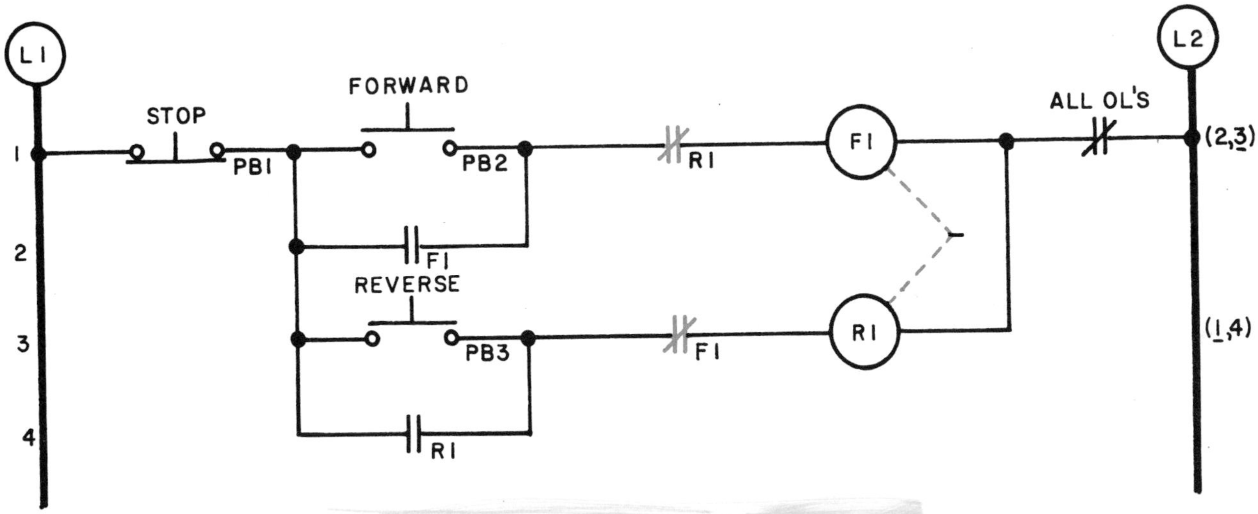

Figure 9-26 Line diagram of a magnetic reversing motor starter controlled by two push-buttons. Mechanical and auxiliary contact electrical interlocking is provided in forward and reverse.

safety back-up system to insure electrical interlocking. Figure 9-26 illustrates a circuit using auxiliary contacts to provide electrical interlocking. When the forward coil circuit is energized in this example, two sets of F1 contacts will be activated: one NO and one NC. The NO will close, providing MEMORY and the NC will open, providing electrical isolation in the reverse coil circuit. In other words, when the forward coil circuit is energized, the reverse coil circuit is automatically opened or isolated from the control voltage. Even if the reverse button is closed, no electrical path would be available in the reverse circuit. For the reverse circuit to operate, the stop button would need to be pressed so that the forward circuit would de-energize and return the contacts to their normal position. With the forward contacts in their normal position, depressing the reverse button would provide the same electrical interlock for the reverse circuit.

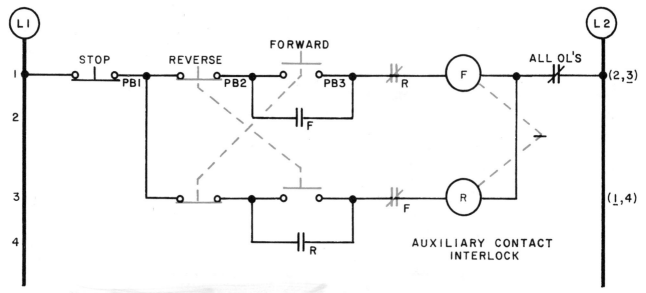

Figure 9-27 Line diagram of a magnetic reversing motor starter controlled by two push-buttons. Mechanical, auxiliary contact and pushbutton electrical interlocking are provided in forward and reverse.

Pushbutton Interlocking

A third method of electrical interlocking is illustrated in Figure 9-27. Pushbutton interlocking may be used in conjunction with either or both mechanical and auxiliary interlocking. Pushbutton interlocking utilizes both normally open and normally closed contacts mechanically connected on each pushbutton.

When NO contacts on the forward pushbutton closes to energize the F coil circuit in Figure 9-27, the NC contacts wired into the R coil circuit open, providing electrical isolation. Conversely, when the NO contacts on the reverse pushbutton close to energize the R coil circuit, the NC contacts wired into the F coil circuit open, providing electrical isolation. Mechanical and auxiliary contact interlocking are also provided in the circuit.

A word of caution should be expressed at this point regarding the rapid reversal of a motor under full load. In many cases motors or the equipment they are powering cannot withstand a rapid reversal of direction. Care must be exercised to determine what can be safely reversed under load. One must also consider what type of braking must be provided to slow the machine to a safe speed before reversal. Braking will be covered in detail in a future unit.

PRACTICAL APPLICATION USING MAGNETIC REVERSING STARTERS

Since magnetic reversing starters are controlled electrically, many different practical applications can be built around the basic starter. In the next few paragraphs several practical applications using different circuits will be shown and analyzed.

Starting and Stopping in Forward and Reverse with Indicator Lights

When control circuits operate equipment with reversing motors, it is often helpful for the operator to know which direction the rotation is at a given moment. This is extremely important, for example, if the motor is controlling a crane which will raise and lower a load. Figure 9-28 illustrates a line diagram capable of indicating, through lights, in which direction the motor is operating. By the electrician adding nameplates, these lights could indicate up and down directions of the hoist. Electrically the circuit would operate as follows:

Pushing the momentary contact forward pushbutton causes the NO and NC contacts to move simultaneously. While this button is depressed, the NO contacts close, energizing coil F. Coil F in turn causes the memory contacts F to close and the NC electrical interlock to open, isolating the reversing circuit. With holding contacts F closed, the forward pilot light will go on.

It should also be noted that for the period of time the pushbutton is depressed, the NC contacts of the forward button open and effectively isolate the reversing coil R.

Pushing the momentary contact reverse pushbutton causes the NO and NC contacts to move simultaneously. The opening of the NC contacts

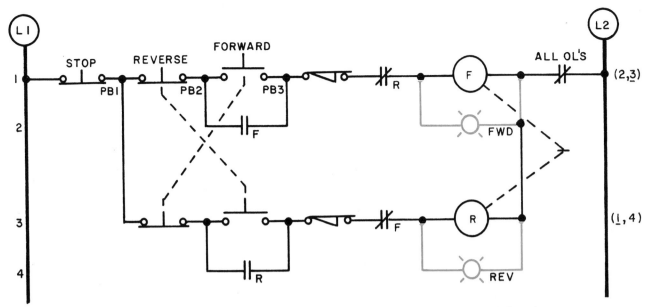

Figure 9-28 Line diagram illustrating a circuit for starting and stopping a motor in forward and reverse. Indicator lights have been added to show intended direction of rotation of motor.

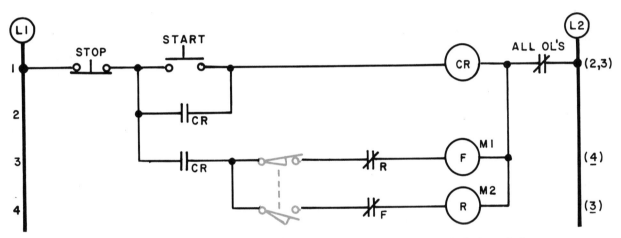

Figure 9-29 Line diagram illustrating a circuit for starting and stopping a motor in forward and reverse with limit switches controlling reversing.

de-energizes coil F. With coil F de-energized, the memory contacts F1 open and the electrical interlock F closes. The closing of the NO contacts energizes coil R. Coil R in turn causes the holding contacts R1 to close and the NC electrical interlock to open, isolating the forward circuit. With the memory contacts R closed, the reverse pilot light will go on. Pushing the stop button with the motor running in either direction stops the motor and returns the circuit to its normal position.

Overload protection for the circuit is provided by the heater coils. Operation of the overload contacts breaks the circuit, thus opening contacts OL. The motor cannot be restarted until the overloads are reset and the forward or reverse button is again depressed.

This circuit provides protection against low voltage or a power failure; a loss of voltage de-energizes the circuit and hold-in contacts F or R open. This design prevents the motor from starting automatically after the power returns.

Starting and Stopping in Forward and Reverse with Limit Switches Controlling Reversing

Limit switch control is very popular in providing several types of automatic control. The circuit of Figure 9-29 uses limit switches and a control relay to automatically reverse the direction of a machine at predetermined points. This circuit

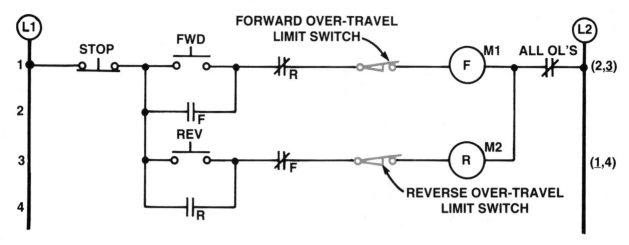

Figure 9-30 Line diagram illustrating a circuit for starting and stopping a motor in forward and reverse with limit switches providing over-travel protection.

could be found controlling the table of an automatic grinding machine where the operation must be periodically reversed. Electrically, the circuit would operate as follows.

Pushing the start button causes control relay CR to become energized. When control relay CR energizes, the auxiliary CR contacts close. One set of contacts form the holding contacts and the other contacts connect the limit switch circuit. With the limit switch circuit activated, the motor will run. If the forward limit is closed, the motor will run in the forward direction. If the reverse limit is closed, the motor will run in the opposite direction.

Overload protection for the circuit is provided by the heater coils. Operation of the overload contacts breaks the circuit, thus opening contacts OL. The motor cannot be restarted until the overloads are reset and the start button is again depressed.

This circuit provides protection against low voltage or a power failure; a loss of voltage deenergizes the circuit and hold-in contacts CR open. This design prevents the motor from starting automatically after the power returns.

Starting and Stopping in Forward and Reverse with Limit Switch Acting as Safety Stop at Certain Points in Either Direction

For safety reasons, it may be necessary to insure that a load controlled by a reversing motor does not go beyond certain operating points in the system. We would not, for example, want a box to go too far down a conveyor system or a hydraulic lift

to raise too high. To provide this, we would incorporate limit switches to shut the operation down if it went far enough to be unsafe. The circuit of Figure 9-30 provides over-travel protection through the use of limit switches. Electrically the circuit would operate as follows.

Pushing the forward button activates coil F. Coil F in turn pulls in the holding contacts F and opens the electrical interlock F, isolating the reversing circuit. The motor will run in the forward mode until either the stop button is pushed or the limit switch is activated. If either control is activated, the circuit will be broken and the holding contacts and electrical interlock will return to their normal state.

Pressing the reverse button activates coil R. Coil R in turn pulls in the holding contacts R and opens the electrical interlock R, isolating the forward circuit. The motor will run in the reverse mode until either the stop button is pushed or the limit switch is activated. If either control is activated, the circuit will be broken and the holding contacts and electrical interlock will return to their normal state. It should be noted that if either limit switch is opened, the circuit may still be reversed to clear the jam or undesirable situation.

Overload protection for the circuit is provided by the heater coils. Operation of the overload contacts breaks the circuit, thus opening contacts OL. The motor cannot be restarted until the overloads are reset and the start button is again depressed.

This circuit provides protection against low voltage or a power failure in that a loss of voltage

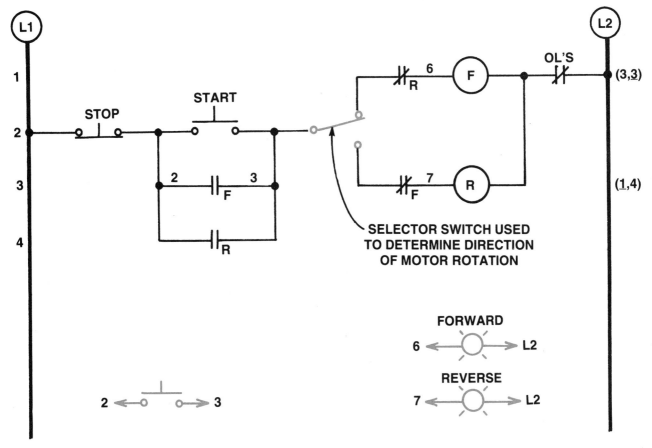

Figure 9-31 Line diagram illustrating a circuit for reversing the direction of a motor using a selector switch.

de-energizes the circuit and hold-in contacts F or R open. This design prevents the motor from starting automatically after the power returns.

Selector Switch Used to Determine Direction of Motor Travel

It is not always necessary to use several pushbuttons when setting up a reversing circuit. Figure 9-31 illustrates a circuit in which a selector switch and a basic start/stop station can be used to provide reversing. The motor can be run in either direction, but the desired direction must be set by a selector switch before starting. Electrically the circuit would operate as follows.

Pushing the start button with the selector switch in the forward position energizes coil F. Coil F in turn closes the holding contacts F and opens the electrical interlock F, isolating the reversing circuit. Pushing the stop button de-energizes coil F, which in turn releases the holding contacts and the electrical interlock. Pushing the start button with the selector switch in the

reverse position energizes coil R. Coil R in turn closes the holding contacts R and opens the electrical interlock R, isolating the forward circuit.

Overload protection for the circuit is provided by the heater coils. Operation of the overload contacts break the circuit, thus opening contacts OL. The motor cannot be restarted until the overloads are reset and the start button is again depressed.

This circuit provides protection against low voltage or a power failure in that a loss of voltage de-energizes the circuit and hold-in contacts F or R open. This design prevents the motor from starting automatically after the power returns.

This circuit also illustrates the proper connections for adding a forward and reverse indicator lights. The forward indicator light would connect to wire number six and line two (L2). The reverse indicator light would connect to wire number seven and line two (L2). If additional start pushbuttons were needed, they would connect to wire number two and three as shown. It is standard industrial practice to mark the NO MEMORY

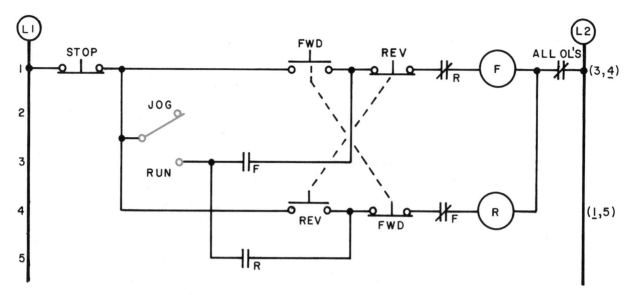

Figure 9-32 Line diagram illustrating a circuit which provides for starting, stopping and jog- ging in forward and reverse with jogging controlled through a selector switch.

contacts 2 and 3. It is also standard to mark the wire coming from the forward coil and leading to the NC reverse contact (used for interlocking) as wire 6. Likewise, the wire coming from the reverse coil and leading to the NC forward contact is marked 7. These numbers are usually printed on the magnetic starters to help in wiring the circuit.

Starting, Stopping, and Jogging in Forward and Reverse with Jogging Controlled Through a Selector Switch

In certain industrial operations it may be necessary to reposition equipment a little at a time for small adjustments. A jogging circuit allows the operator to start the motor for short times without MEMORY. In the circuit of Figure 9-32 we have a circuit which allows us to make small adjustments in forward and reverse motor rotation or to run continuously, depending upon the position of the selector switch. Electrically the circuit would operate as follows: Pressing the forward push button with the selector switch in the run position activates coil F. Coil F in turn pulls in the NO holding contacts F and opens the NC electrical interlock F, isolating the reverse circuit. The motor starts and continues to run. Pressing the reverse pushbutton with the selector switch in the run position activates Coil R. Coil R in turn pulls in the NO holding contacts R and opens the NC electrical interlock R, isolating the forward circuit. The motor starts in the reverse direction and continues to run. Pressing the stop button in either direction breaks the circuit and returns the circuit contacts to their normal positions. Press-

Figure 9-33 A typical drum switch used to select forward and reverse directions by three-phase, single-phase and DC motors. (Furnas Electric Company)

ing the forward pushbutton with the selector switch in the jog position activates Coil F and the motor only for the period of time that the forward pushbutton is depressed. In addition, the NC electrical interlock F opens and isolates the reverse circuit. Pressing the reverse pushbutton with the selector switch in the jog position activates Coil R and the motor only for the period of time that the reverse pushbutton is depressed. In addition, the NC electrical interlock R opens and isolates the forward circuit.

Overload protection for the circuit is provided by the heater coils. Operation of the overload contacts breaks the circuit, thus opening contacts OL. The motor cannot be restarted until the overloads are reset and the start button is again depressed.

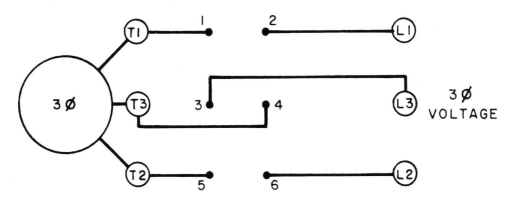

3 Ø VOLTAGE

HANDLE END

REVERSE	OFF	FORWARD
1 — 2	1 2	1 2
3 — 4	3 4	3 4
5 — 6	5 6	5 — 6

MOTOR CONNECTIONS FOR

REVERSE	FORWARD
L1 TO T1	L1 TO T3
L2 TO T2	L2 TO T2
L3 TO T3	L3 TO T1

INTERNAL SWITCHING FOR REVERSING DRUM SWITCH

Figure 9-34 Wiring diagram illustrating a three-phase motor connected to the contacts of a drum switch for forward and reversing rotation of the motor. The charts provide internal switching of the drum controller and the resultant wiring seen by the motor connections for forward and reverse.

This circuit provides protection against low voltage or a power failure in that a loss of voltage de-energizes the circuit and hold-in contacts F or R open. This design prevents the motor from starting automatically after the power returns.

REVERSING THREE-PHASE, SINGLE-PHASE AND DC MOTORS USING A DRUM SWITCH

A drum switch like that in Figure 9-33 is a manual switch made up of moving contacts mounted on an insulated rotating shaft. The moving contacts make and break contact with stationary contacts within the controller as the shaft is rotated. Because the moving contacts are rotated much like a drum, the term "drum-switch" seemed appropriate.

Drum switches are totally enclosed, and an insulated handle provides the means for moving the contacts from point to point. Drum switches are available in several sizes and can have different numbers of poles and positions. Drum switches are usually used where the operator must keep his eyes on a particular operation such as a crane raising and lowering a load.

The drum switch may be purchased with maintained contacts or spring return contacts. In either case, when the motor is not running in forward or reverse, the handle is in the center (off) position. To reverse a running motor, the handle must first be moved to the center position until the motor stops and then moved to the reverse position.

Drum switches are not motor starters because they do not contain protective overloads. Usually separate overload protection is provided by placing a non-reversing manual or magnetic starter in line before the drum switch. This starter provides the required overload protection and acts as a second disconnecting means. The drum switch then is used only as a means for controlling the direction of a motor by switching the leads of a motor.

Reversing a Three-phase Motor Using a Drum Switch

Figure 9-34 illustrates a three-phase motor connected to the contacts of a drum switch for forward

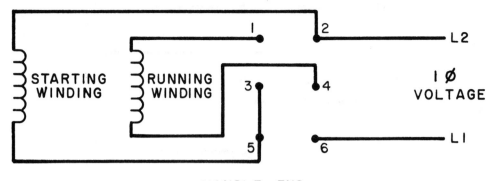

HANDLE END

INTERNAL SWITCHING FOR REVERSING DRUM SWITCH

Figure 9-35 Wiring diagram illustrating a split-phase single-phase motor connected to the contacts of a drum switch for forward and reverse rotation of the motor.

Always check manufacturer's wiring diagram when reversing single-phase motors.

and reversing rotation of the motor. In addition, two charts are provided to indicate the internal switching of the drum controller and the resultant wiring seen by the motor connections for forward and reverse. Careful observation of this switching arrangement will reveal that L1 and L3 are interchanged as the drum controller is moved from the forward to the reverse position. Since it is necessary only to interchange two leads on a three-phase motor to reverse rotation, the motor will change directions each time the drum switch is moved to forward or reverse.

Reversing a Single-phase Motor Using a Drum Switch

Figure 9-35 illustrates a split-phase, single-phase motor connected to the contacts of a drum switch for forward and reversing. The chart indicating the internal switching of the drum controller is typical of that found on many controllers of this type. Remember, this is only a typical example of a single-phase motor, and manufacturers' wiring diagrams must be consulted to assure proper wiring. In this case, however, since it is necessary only to interchange terminals of the starting

winding on a single-phase motor to reverse rotation, the motor will change direction each time the drum controller is moved to forward or reverse.

Reversing a DC Motor Using a Drum Switch

You have already learned that the direction of rotation of any DC motor series, shunt, or compound may be reversed by reversing the direction of the current through the field or fields without changing the direction of the current through the armature, or by reversing the direction of the current through the armature without changing the direction of the currents through the fields. Figure 9-36 illustrates a drum switch connected to forward and reverse, the DC series, shunt, and compound motors. In each circuit, the current through the armature is changed. Remember that some DC motors have commutating windings (also called interpoles) that are used to prevent sparking at the brushes in the motor. For this reason it is important on all DC motors with commutating windings to reverse the armature circuit (armature and commutating windings) to reverse the motor.

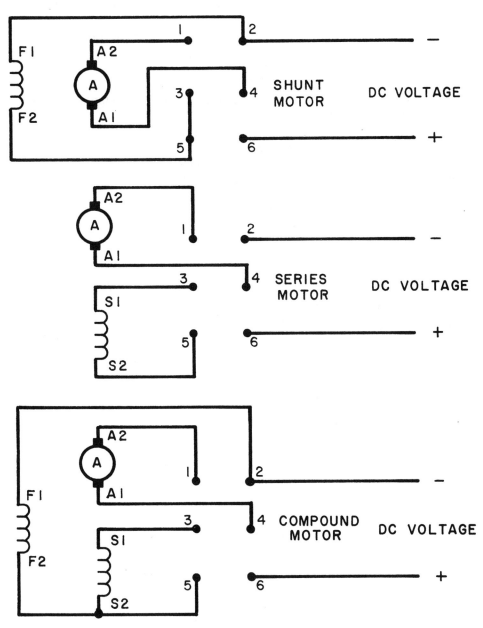

HANDLE END

REVERSE	OFF	FORWARD

INTERNAL SWITCHING FOR REVERSING DRUM SWITCH

SHUNT MOTOR — DC VOLTAGE

SERIES MOTOR — DC VOLTAGE

COMPOUND MOTOR — DC VOLTAGE

Figure 9-36 Wiring diagram illustrating the proper connection for forward and reverse wiring series, shunt and compound DC motors.

1. How are the motor windings connected to form a Wye motor?
2. How are the motor windings connected to form a Delta motor?
3. Why is the higher voltage of a dual-voltage motor preferred over the lower voltage?
4. How are the nine leads of a dual voltage Wye motor connected for low voltage?
5. How are the nine leads of a dual voltage Wye motor connected for high voltage?
6. How are the nine leads of a dual-voltage Delta motor connected for low voltage?
7. How are the nine leads of a dual voltage Delta motor connected for high voltage?
8. How is a three-phase motor reversed?
9. What is a split-phase motor?
10. What function does the centrifugal switch perform in the motor circuit?
11. How is the split-phase motor reversed?
12. What function does the capacitor perform in the capacitor motor?
13. Why are two capacitors used with some motors?
14. How is a capacitor start motor reversed?
15. What are the three basic types of DC motors?
16. How are DC motors reversed?
17. What is the reason for having a mechanical interlock on forward and reversing starter combinations?
18. What is auxiliary contact interlocking?
19. What is pushbutton interlocking?
20. Why are indicator lights often used with forward and reversing circuits?
21. When a limit switch is used to stop a motor that is running in one direction, are the NO or NC contacts used?
22. What is meant by "jogging" a motor?
23. Why is a drum switch not considered to be a motor starter even though it can start and stop a motor in either direction?

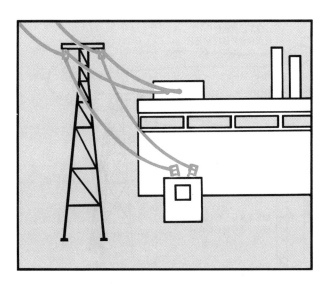

10 Power Distribution Systems, Transformers, Switchboards, Panelboards, Motor Control Centers and Busways

The electrical power that is created in a generating station must go through many stages of distribution before use by electrical loads.

Electrical power from a generating station starts out as very high voltage (in excess of 200,000 volts) and is transmitted along power lines to substations. The substation contains step-down transformers that reduce the voltage. Some substations reduce the voltage to the final end user voltage, while others reduce it only enough for further distribution. Within the plant unit, substations distribute power to motor control centers, switchboards, and power panels through bus ducts.

POWER DISTRIBUTION SYSTEMS

Today's residential, commercial and industrial requirements for electricity are growing at a faster rate than ever. Our dependency on electricity is increasing more and more, ranging from our everyday needs and comforts, jobs, entertainment and medical requirements to discoveries yet to come. As the need for electricity grows, so does the need for a distribution system that can safely and efficiently deliver the power required.

It is important that the electrical power produced at the source and delivered to the load be adequate and uninterrupted. This means that the distribution system from the source to the plant and within the plant itself must be in good working order and properly maintained. People working in the electrical field must have a working knowledge of power distribution systems.

Power distribution is the delivering of electrical power to where it is needed. This, however, is not always as simple as it may sound. Much control, protection, transformation and regulation must take place before any power can be delivered. Figure 10-1 illustrates the many stages and steps that the distribution system must go through in delivering the power required. For industrial users, Figure 10-2 illustrates the various stages necessary for providing power to the mining industry. To help you understand this system, we will start with the source and follow the electrical power through the distribution system.

SOURCES OF ELECTRICAL ENERGY

Although there are several minor sources of electrical energy (static, chemical, thermal, solar,

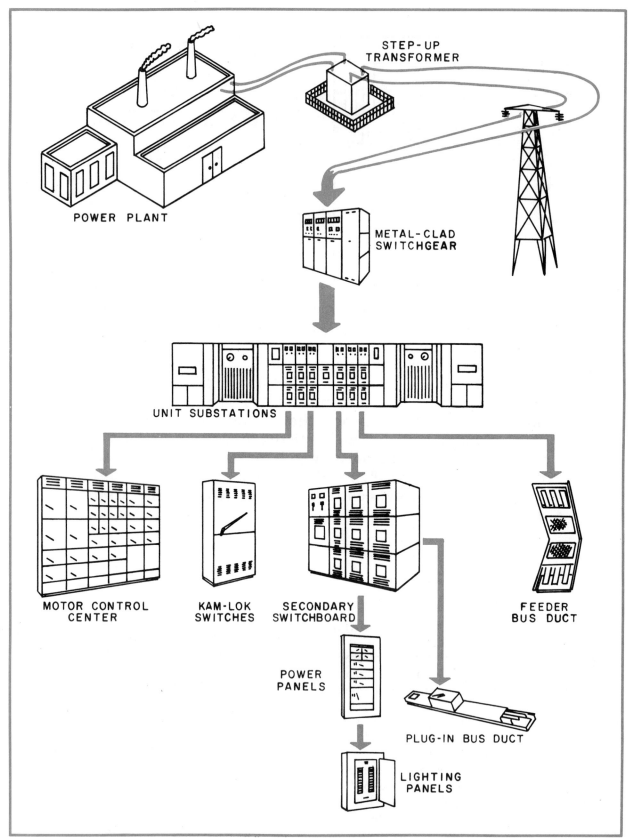

STEP-UP
TRANSFORMER

POWER PLANT

METAL-CLAD
SWITCHGEAR

UNIT SUBSTATIONS

MOTOR CONTROL
CENTER

KAM-LOK
SWITCHES

SECONDARY
SWITCHBOARD

FEEDER
BUS DUCT

POWER
PANELS

PLUG-IN BUS DUCT

LIGHTING
PANELS

Figure 10-1 A distribution system goes through many stages and steps in delivering the power required for an industrial user.

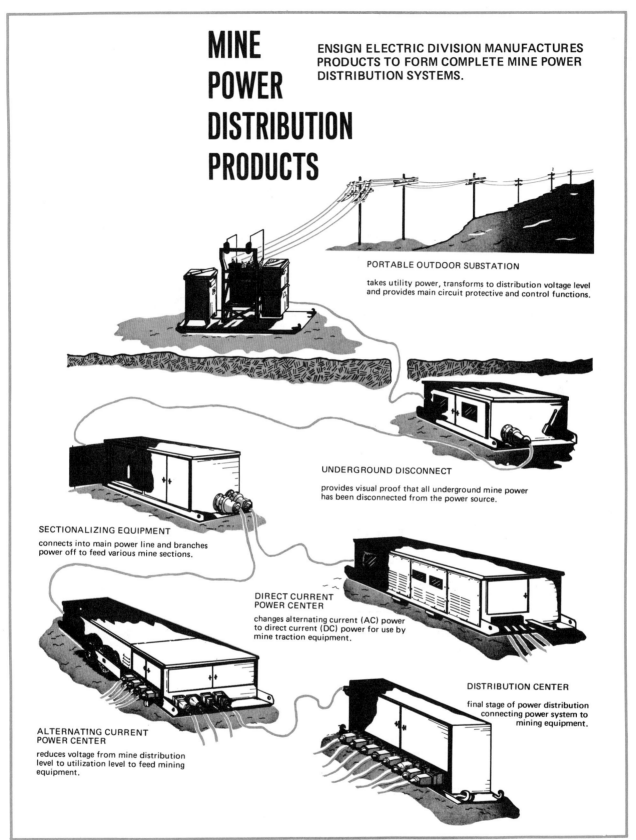

MINE POWER DISTRIBUTION PRODUCTS

ENSIGN ELECTRIC DIVISION MANUFACTURES PRODUCTS TO FORM COMPLETE MINE POWER DISTRIBUTION SYSTEMS.

PORTABLE OUTDOOR SUBSTATION

takes utility power, transforms to distribution voltage level and provides main circuit protective and control functions.

UNDERGROUND DISCONNECT

provides visual proof that all underground mine power has been disconnected from the power source.

SECTIONALIZING EQUIPMENT

connects into main power line and branches power off to feed various mine sections.

DIRECT CURRENT POWER CENTER

changes alternating current (AC) power to direct current (DC) power for use by mine traction equipment.

DISTRIBUTION CENTER

final stage of power distribution connecting power system to mining equipment.

ALTERNATING CURRENT POWER CENTER

reduces voltage from mine distribution level to utilization level to feed mining equipment.

Figure 10-2 Various steps are necessary for providing power to the mining industry. (Ensign Electric Division, HARVEY HUBBELL INCORPORATED)

etc.), the major source of electrical power is the alternator. The alternator, with its connecting parts, provides an efficient means for the conversion of energy. Energy is converted from fuel to heat, heat to mechanical energy and mechanical energy to electrical energy. In fact, all the other equipment located in a generating power plant is there simply to control, protect, transport and monitor this conversion process. Figure 10-3 illustrates the process by which basic fuels are converted into electrical energy.

Today's generating plants produce alternating current (AC). This is because AC permits efficient transmission of electrical power between the power station and desired location. Direct current (DC) is limited by the distance over which it can be transmitted economically. If DC is required at the desired location or equipment, the AC power supply is easily rectified (changed) from AC to DC. It will not be our intent to cover the complete design of generators in this unit, but to explain basically how the alternating currents and voltages are created and distributed.

AC ALTERNATOR OPERATION

Alternators operate on the theory of electromagnetic induction, which specifies that when conductors are moved through magnetic fields, voltages are induced into the conductors.

In Unit 6, we learned how the right-hand rule for motors can be applied as an arc suppression technique by disrupting an electric current through a magnetic field. In this unit, we will see how the left-hand rule, as applied to generators,

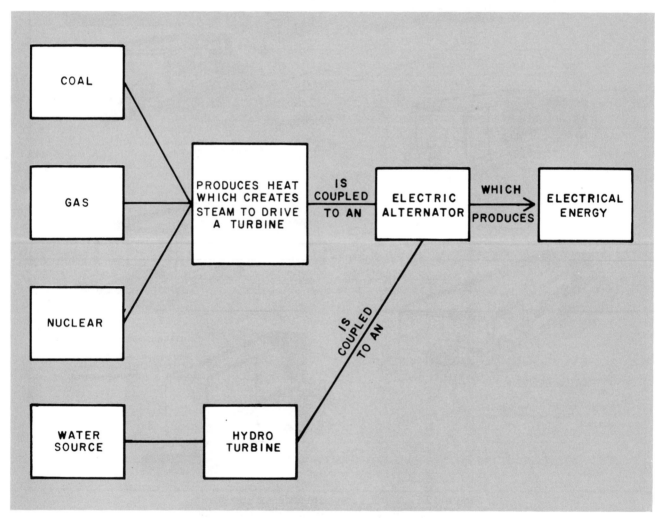

Figure 10-3 Basic fuels can be converted into electrical energy.

can be used to describe generator action (Figure 10-4). The left-hand rule simply states that if a conductor is moved at a right angle through a magnetic field, then a voltage will be generated in that conductor. To determine the direction of current flow, the thumb should indicate the direction of conductor motion and the index finger should indicate the direction of the magnetic field. The middle finger will then indicate the direction of current flow. Reversing the direction of motion will reverse the direction of current flow providing the magnetic field remains constant.

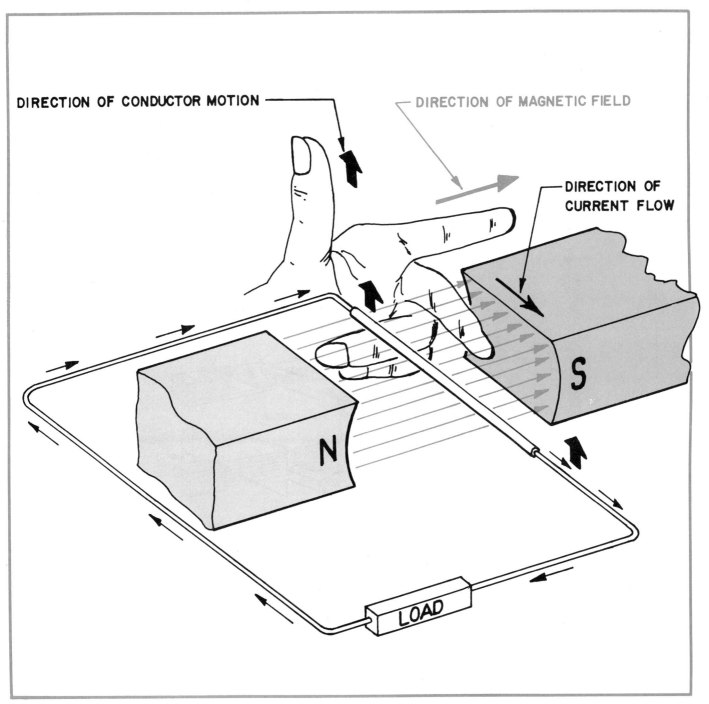

Figure 10-4 The left-hand rule as applied to generators can be used to create an electrical current. (Energy Concepts, Inc.)

Instead of a single conductor, imagine a coil (or loop) of wire rotating between the poles of a permanent magnet like that in Figure 10-5. What we have created is a simple yet functional AC alternator.

Figure 10-5 allows us to trace the action of the AC alternator through one complete revolution or cycle. Position one shows the rotor just before it begins to rotate in a clockwise direction. There is no current flow at this point because the rotor is not cutting any lines of magnetic flux. As the rotor rotates from position one to position two, the rotor begins to cut across lines of magnetic flux. The voltage in segments AB and CD increases as the rotor turns. Notice that the maximum of lines of force is cut when the rotor is in position two, so the induced voltage is greatest in this position.

From position two to position three, the voltage decreases to zero since the rotor is cutting less and less lines of flux and finally, none at the end of position two.

As the rotor continues to rotate from position three to position four, the voltage increases, but in the opposite direction. The voltage reaches maximum negative at position four, then returns to zero at position five. If the rotor is allowed to rotate, a continuous sine wave of alternating voltage shown in the last sequence will continue to be produced.

You will notice from our simple example the presence of slip rings. Slip rings are used on an AC alternator so that the external load may be easily attached to the rotor without interfering with its rotation. In an alternator with two mag-

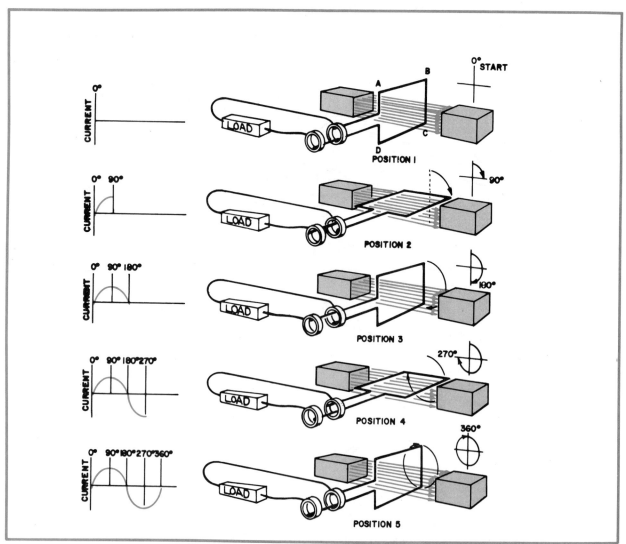

Figure 10-5 An AC alternator produces alternating current. (Energy Concepts, Inc.)

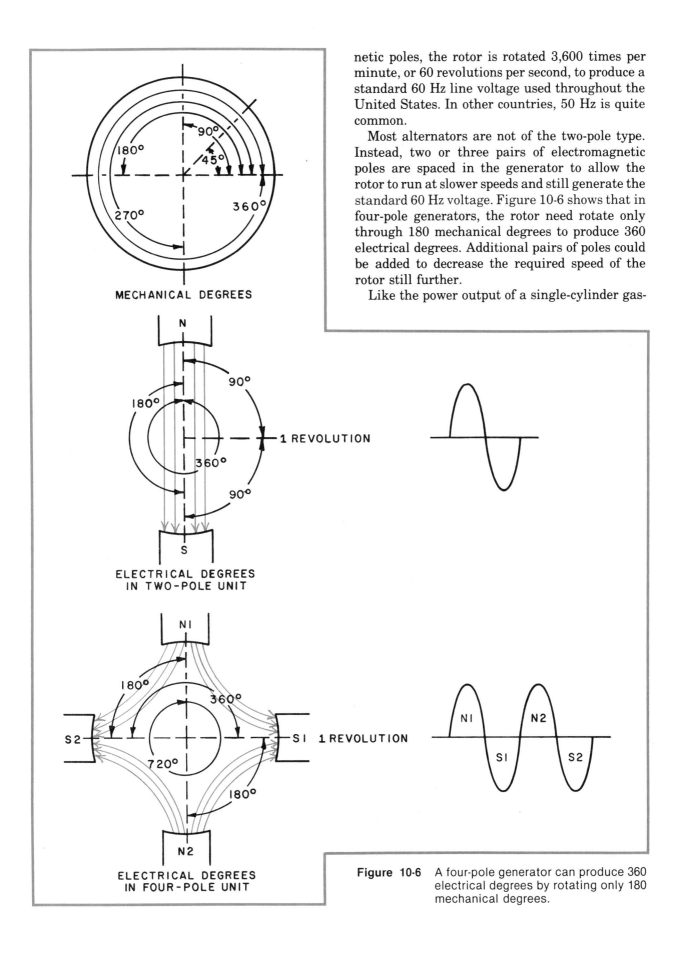

netic poles, the rotor is rotated 3,600 times per minute, or 60 revolutions per second, to produce a standard 60 Hz line voltage used throughout the United States. In other countries, 50 Hz is quite common.

Most alternators are not of the two-pole type. Instead, two or three pairs of electromagnetic poles are spaced in the generator to allow the rotor to run at slower speeds and still generate the standard 60 Hz voltage. Figure 10-6 shows that in four-pole generators, the rotor need rotate only through 180 mechanical degrees to produce 360 electrical degrees. Additional pairs of poles could be added to decrease the required speed of the rotor still further.

Like the power output of a single-cylinder gas-

MECHANICAL DEGREES

ELECTRICAL DEGREES
IN TWO-POLE UNIT

ELECTRICAL DEGREES
IN FOUR-POLE UNIT

Figure 10-6 A four-pole generator can produce 360 electrical degrees by rotating only 180 mechanical degrees.

oline engine, the power output of the single-phase generator previously illustrated occurs in pulses. For small power demands, this is satisfactory; but for larger power outputs, the physical size needed and the pulsating power become a problem. The single-phase generator, like the single-cylinder engine, is not practical for producing large amounts of power. For this reason, three single-phase generators are coupled to form one three-phase generator. This three-phase generator will produce smoother power and provide for more economical use of the magnetic field and space. It is for this reason that practically all large amounts of power are generated and transmitted by three-phase.

Manufacturers of gasoline engines for cars solved their problem by adding more individual cylinders. They arranged the cylinders in such a manner that the power is delivered by the individual cylinders at evenly spaced intervals per revo-

lution of the drive shaft. This allowed for more and smoother power in a smaller engine. Figure 10-7 illustrates how three-phase power is generated. Three separate coils are spaced 120 electrical degrees apart. By having three separate coils equally spaced around the armature by 120 electrical degrees, each of the coils' generated voltages will also be spaced 120 electrical degrees apart. These individual generated phases are called Phase 1, Phase 2 and Phase 3, or Phase A, Phase B and Phase C. The symbol ø is sometimes used to indicate phase.

CONNECTING THE PHASES OF AN ALTERNATOR

We have seen that in the three-phase alternator three individual separate phases are present. Since each phase coil has a beginning and end, six wires extend out of the generator. These three

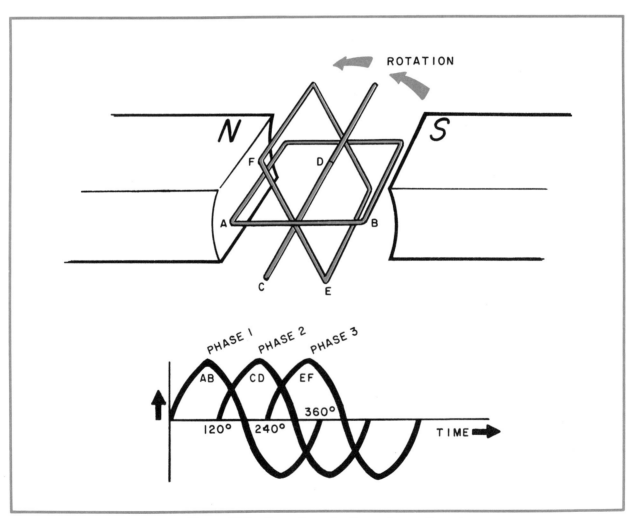

Figure 10-7 Operation of a three-phase alternator.

coils with six wires can be connected internally or externally in two different ways. Special names are given to the connections; namely, Wye (star) or Delta connection.

THE WYE (Y) CONNECTION

Figure 10-8 illustrates three lights connected to each separate phase of a Wye alternator. Each light will light from the generated single-phase power delivered from each phase. Notice that the A2, B2 and C2 wires all return to the generator together. This circuit can be simplified by using only one wire, connecting it to the A2, B2 and C2 phase ends. This is illustrated in Figure 10-9. This common wire is called the *neutral* wire.

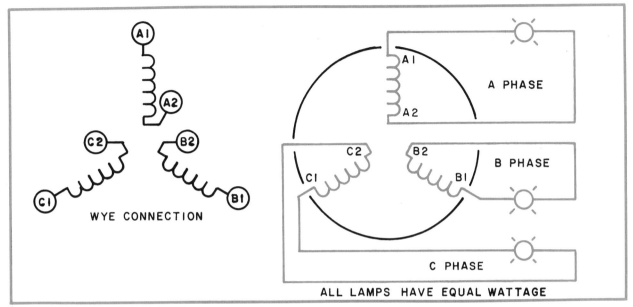

Figure 10-8 Three lights may be connected to separate phases of a Wye alternator.

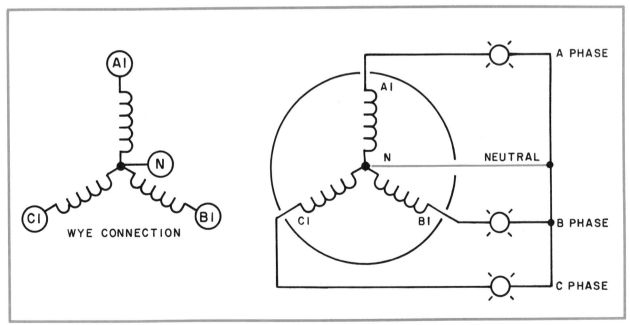

Figure 10-9 A common neutral wire can safely connect the internal leads of a Wye alternator to form a common return for the lighting loads.

We can safely connect the three ends together at the neutral point because no voltage difference exists between them. This is illustrated in Figure 10-10.

From Figure 10-10, left, we can see that as Phase A is maximum, Phases B and C will be opposite to or opposing A. If we add the equal opposing values of B and C together vectorially as in Figure 10-10, right, we see that the opposing force of B and C combined is exactly equal to A. If you can imagine three people pulling on ropes as indicated by Figure 10-10, right, you will see that the resulting

forces cancel each other out and the resultant force is zero in the center or neutral point. The net effect is a large voltage (pressure difference) between the A1, B1 and C1 coil ends, but no pressure difference at all between A2, B2 and C2 coil ends.

In the circuit illustrated (Figure 10-9), the three-phase circuit is said to be balanced because the loads are all equal in power consumption. In such a balanced circuit, there will be no flow of current in the neutral wire because the sum of all the currents will be zero.

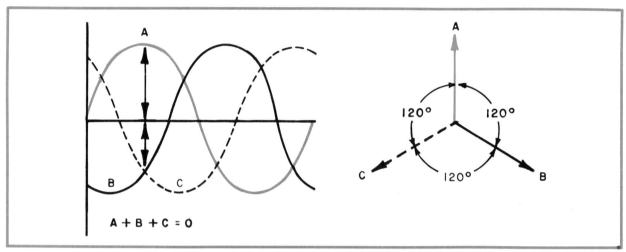

Figure 10-10 The three phase voltages of a Wye-connected alternator effectively cancel each at the neutral point, allowing these three leads of the alternator to be connected together.

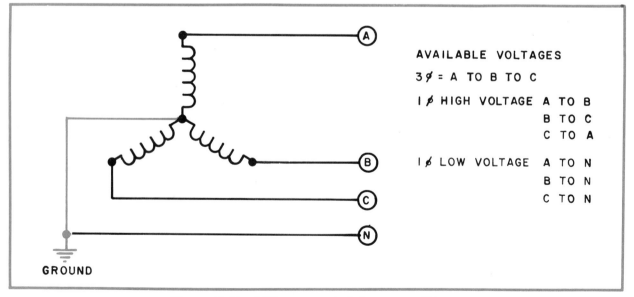

Figure 10-11 A Wye-connected alternator will have the neutral wire connected to ground.

All large power distribution systems are designed as three-phase systems with the loads balanced across the phases as closely as possible. The only current that flows in the neutral wire is the unbalanced current. This is usually kept to a minimum because most systems can be kept fairly balanced. It is customary to connect the neutral to a *ground* such as the earth itself, as illustrated in Figure 10-11. The voltage available from such a system is phase-to-phase, phase-to-neutral or phase-to-phase-to-phase. Figure 10-12 illustrates a Wye-connected system with ammeters and voltmeters connected throughout the system. We will use this illustration to determine the voltage and current characteristics of a Wye system.

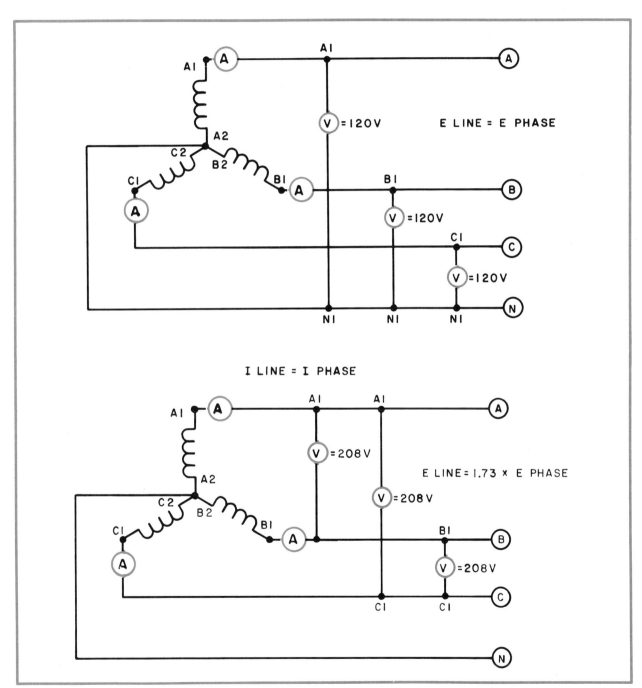

Figure 10-12 Voltages available from a Wye-connected alternator.

Phase-to-Neutral Voltage. In a three-phase Wye system, the phase-to-neutral voltage is equal to the voltage generated in each coil (Figure 10-12, top). For example, if in this system the generator produced 120 volts from A1 to A2, the equivalent 120 volts would be present from B1 to B2 and C1 to C2. Thus, in a three-phase Wye system the output voltage of each coil will appear between each phase and the neutral. In the circuit illustrated, the phase-to-neutral voltages would be:

$$E_{phase\ (A)}\ to\ N = 120\ volts$$
$$E_{phase\ (B)}\ to\ N = 120\ volts$$
$$E_{phase\ (C)}\ to\ N = 120\ volts$$

Phase-to-Phase Voltage. In the three-phase Wye system, you might believe that a connection with a voltmeter from one phase to another would give a voltage reading that would be the sum of the two coil voltages. This is not the case, however. Since the coils are set 120 electrical degrees apart, it is necessary to add them vectorially. Since vectoring is beyond the scope of this study, we need only know that, in such an arrangement, the phase-to-phase voltage can be obtained by multiplying the phase-to-neutral voltage by 1.73. In the circuit illustrated, the phase-to-phase voltage would be:

$$E_{(\emptyset A)}\ to\ E_{(\emptyset B)} = 120\ volts \times 1.73 = 208V$$
$$E_{(\emptyset B)}\ to\ E_{(\emptyset C)} = 120\ volts \times 1.73 = 208V$$
$$E_{(\emptyset C)}\ to\ E_{(\emptyset A)} = 120\ volts \times 1.73 = 208V$$

Similarly, on larger Wye distribution systems, a phase-to-neutral voltage of 2400 volts creates 4160 line-to-line, and 7200 phase-to-neutral becomes 12,470 volts line-to-line.

One of the benefits Wye distribution brings to the utility company is that even though its generators are rated at 2400 or 7200 volts per coil, they can transmit at a higher phase-to-phase voltage with a reduction in losses and better voltage regulation. This is because the higher the transmitted voltage, the less voltage losses. This is especially important in longer rural lines.

In the Wye system, the neutral connection point is grounded and a fourth wire is carried along the system and grounded at every distribution trans-

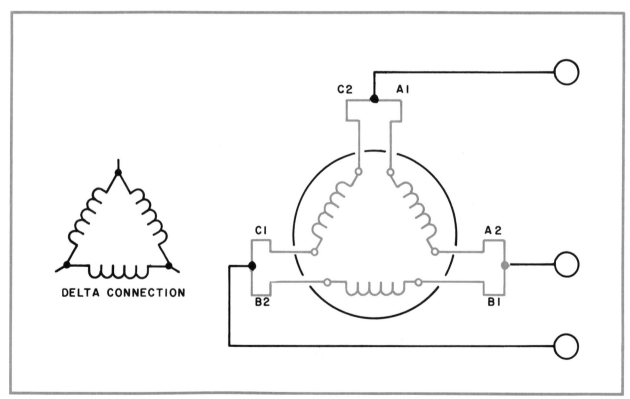

Figure 10-13 The alternator coil windings of a three-phase system can be Delta connected.

former location. This solidly grounded system is regarded as the safest of all distribution systems.

Current in Three-Phase Balanced Wye System. In the three-phase Wye system, the current in the line is the same as the current in the coil (phase) winding. This is because the current in any series circuit is the same throughout all parts of the circuit. For the circuit illustrated, the current relationship would be:

$$I_{line} = I_{coil \, (phase)}$$

The Delta (△) Connection

The alternator coil windings of a three-phase system can also be connected as illustrated in Figure 10-13. As in the Wye system, the coil windings are spaced 120 electrical degrees apart.

Voltage in the Delta System. In a Delta system, the voltage measured across any two lines is equal to the voltage generated in the coil winding (Figure 10-14). This is because the voltage is being measured directly across the coil winding. If,

for example, the generated coil voltage was equal to 240 volts, then the voltage between any two lines would be equal to 240 volts. The voltage for a balanced Delta system would then be:

$$E_{line \, to \, line} = E_{coil \, (phase)}$$

Current in the Delta System. If you follow any line in a Delta system back to the connection point, you can see that the current supplied to that line is supplied by two coils. (Phase A can be traced back to connection point A1, C2.) However, as in the Wye system, the coils are 120 electrical degrees apart; therefore, the line current will be the vector sum of the two coil currents. In a balanced circuit, the phase currents would be equal. In a balanced system, the line current is then equal to 1.73 times the current in one of the coils. In a balanced three-phase Delta system, the line current would be:

$$I_{line} = 1.73 \times I_{coil}$$

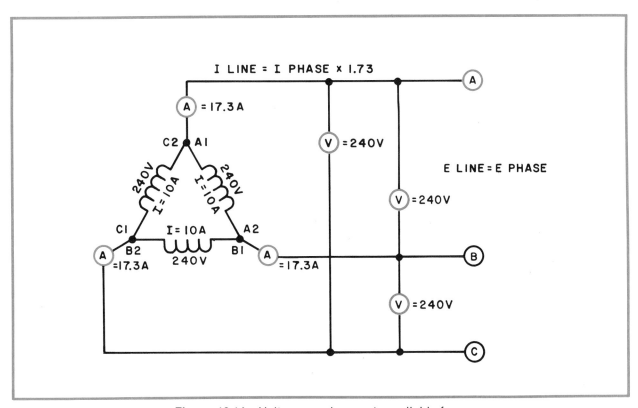

Figure 10-14 Voltages and currents available from a Delta-connected system.

For example, if each coil current is equal to 10 amps, the line current is equal to 1.73×10 or 17.3 amperes.

In the illustration of the Delta system, only three wires appeared on the system. None of the three lines are usually connected to a ground. However, when a Delta system *is not* grounded, it is possible for one phase to accidentally become grounded without anyone being aware of this. Not until another phase also grounds will the problem be apparent. For this reason, some plants deliberately ground one corner of their Delta system so that inadvertent faults on the other two phases will cause a fuse or breaker to trip. Although it is not common, some plants also may make a ground in a Delta system by grounding the midpoint of one of the phases. This is illustrated in Figure 10-15.

The Delta system delivers some different voltage possibilities. Three-phase power (240 as illustrated) is available as normal between A, B and C. Single-phase 120 volt power is available from B to N and C to N. Single-phase 240 volt power is available from A to B, B to C and C to A. Also available is 195 volt (approximate) single-phase power from A to N. This voltage is called the

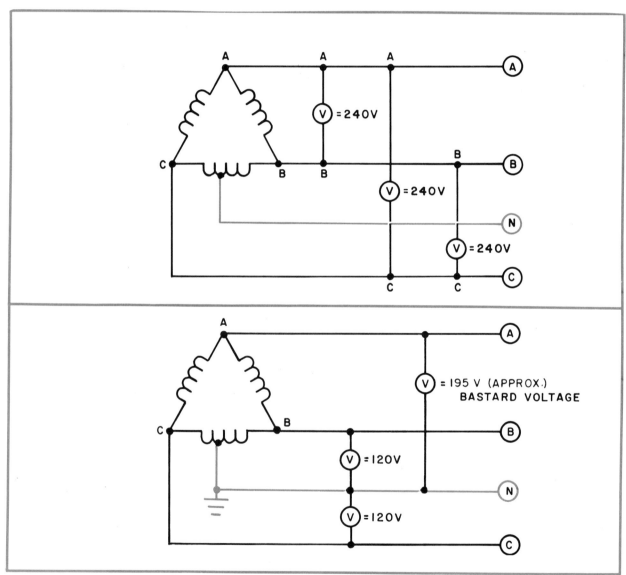

Figure 10-15 Voltage available from a Delta-connected system when one phase is grounded in the center of the phase. The 195V (approximate) high, or bastard, phase should be avoided.

"high" or "bastard" voltage, and should be avoided, because it is an unreliable source of voltage that could damage the equipment.

A Delta system connection known as "open Delta" makes three-phase power available anywhere along the distribution line with only two transformers rather than the usual three. Although this system delivers only 57.7% of the nominal full load capability of a full bank of three transformers, it has the advantage of lower initial cost if the extra power is not needed now. Another advantage is that in an emergency, this system would allow both reduced single- and three-phase power at a location where one transformer burned out.

USE OF TRANSFORMERS IN POWER DISTRIBUTION SYSTEMS

Transformers are used in the distribution system to increase or decrease the voltage and current safely and efficiently. Transformers are first used to increase the generated voltage to a high level for transmission across country and then to decrease it to a low level for use by electrical loads. This allows the power companies to distribute large amounts of power at a reasonable cost.

In order to understand the advantage and use of a transformer, let us first look at some basic transformer principles. Figure 10-16 illustrates the basic parts of a transformer.

The transformer has two windings—called the primary winding and the secondary winding—wound around an iron core. The primary winding is the coil that draws power from the source. The secondary winding is the coil that delivers the energy at a transformed or changed voltage to the load.

The transformer core is used to provide a controlled path for the magnetic flux generated in the transformer by the current flowing through the windings. The core is not a solid bar of steel, but is

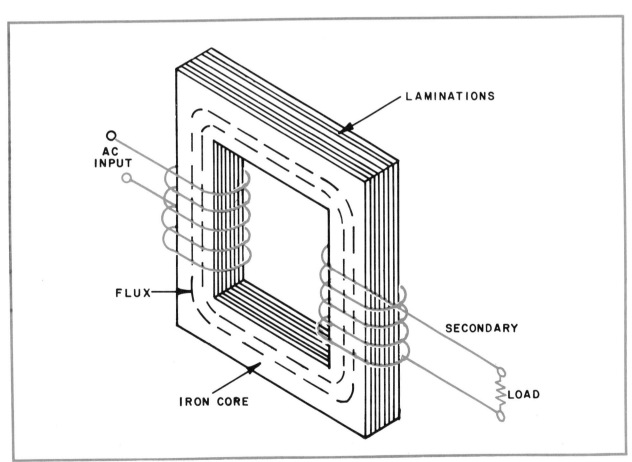

Figure 10-16 Basic construction of a transformer.

constructed of many layers (laminations) of thin sheet steel. The core is laminated to help reduce heating, which creates power losses. Since the two circuits are not electrically connected, the core serves the very important part of transferring electrical power into the secondary winding through magnetic action.

Principle of Transformer Operation

Figure 10-17 illustrates how a transformer transfers AC energy from one circuit to another. This energy transfer is made magnetically through the iron core. Figure 10-17, top, illustrates that a magnetic field builds up around a wire when current is passed through the wire. This magnetic field will

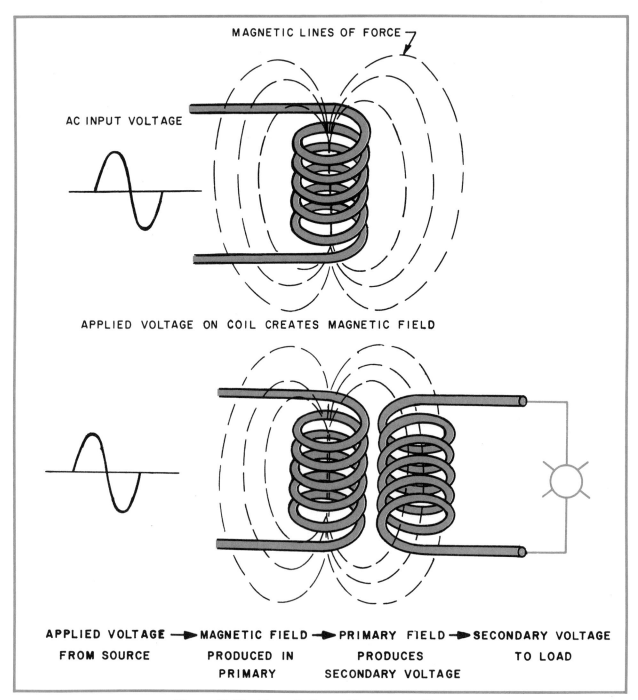

Figure 10-17 Magnetic lines of force created by one coil may induce a voltage into a second coil.

build up and collapse each half cycle because the wire is carrying AC.

Figure 10-17, bottom, illustrates that this primary magnetic field will induce a secondary voltage in any wire that is within this magnetic field. This induced voltage will also be alternating since the magnetic flux field reverses direction every half cycle in the primary. In other words, the alternating voltage connected to the primary coil produces an alternating magnetic field in the iron core. This magnetic field cuts through the secondary coil and induces an alternating voltage on the transformer's secondary.

The primary coil of the transformer supplies the magnetic field for the iron core. The secondary winding supplies the load with an induced voltage which is proportional to the number of conductors cut by the magnetic flux of the core. Depending on the number of lines cut, the transformer will either be a "step-up" or "step-down" transformer.

Figure 10-18 illustrates the relationship between the induced voltage and the number of turns. Figure 10-18, top, illustrates that if only half as many turns are on the secondary, only half the voltage is induced on the secondary. The ratio of

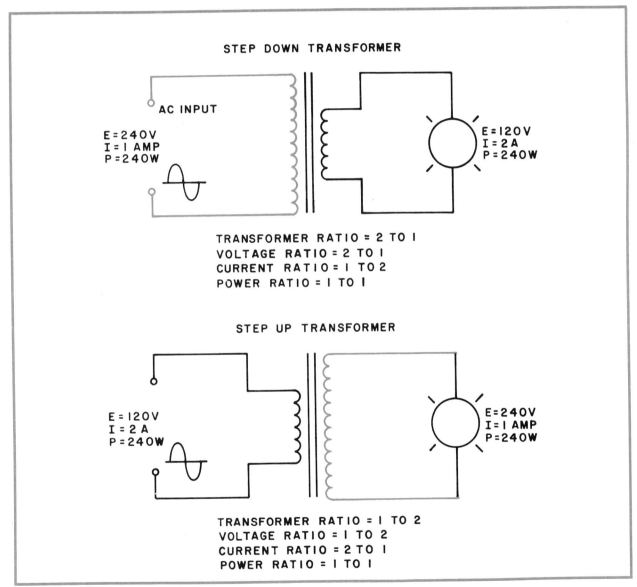

Figure 10-18 Voltage and current change from primary to secondary in "step-up" and "step-down" transformers.

primary to secondary would be 2 to 1, making this transformer a step-down transformer. Figure 10-18, bottom, illustrates that if twice as many turns are on the secondary, then twice the voltage is induced on the secondary. The ratio of primary to secondary would be 1 to 2, making this transformer a step-up transformer.

In the example of the step-up transformer, we saw that with a 1 to 2 ratio, the voltage would double. This, at first, might seem that we are gaining or multiplying voltage without sacrificing anything. This, of course, cannot be true. Ignoring small losses, the amount of power transferred in the transformer is equal on both the primary and secondary.

Since power is equal to voltage times current (P = E × I) and power is always equal on both sides of a transformer, we cannot change the voltage without changing the current. To illustrate the relationship between voltage and current in step-up and step-down transformers, again refer to Figure 10-18. In these examples, we can see that when voltage is stepped down from 240V to 120V in a 2 to 1 ratio, the current will be increased from 1 to 2 amps, keeping the power equal on each side of the transformer. By contrast, when the voltage is stepped up from 120V to 240V in a 1 to 2 ratio, the current is reduced from 2 to 1 amp to maintain power balance. In other words, voltage and current may be changed for particular reasons; but power is merely transferred from one point to another.

One big advantage of increasing voltage and reducing current is that power may be transmitted through smaller gauge wire, thus reducing the cost of power lines. For this reason, the generated voltages are stepped up very high for distribution across large distances, and then stepped back down to meet the consumer needs.

It should be noted that even though both the voltage and current can be stepped up or down, the words "step-up" or "step-down" when used with transformers always apply to the voltage.

Transformers Connected for Wye and Delta Distribution Systems

As we had seen earlier, large amounts of power are generated using a three-phase system. This generated voltage will be stepped up and down many times before it reaches the loads in your home or plant. This transformation can be accomplished by using Wye or Delta connection trans-

formers or a combination of both Wye and Delta transformers along with differing voltage ratio transformers.

Transformers Connected for a Wye-to-Wye Installation. Figure 10-19 illustrates how three single-phase transformers are connected for a Wye-to-Wye, step-down transformer bank. Remember that the line voltage is equal to 1.73 × the coil voltage and that the line current and coil current are equal in a balanced Wye system.

As illustrated, the Wye-connected secondary provides three different types of service: (1) a three-phase 208 volt service for the three-phase motor load; (2) single-phase 120 volt service for the lighting loads; and (3) single-phase 208 volts for the 208 single-phase motors. This type of system is commonly used in schools, commercial stores and offices.

On the primary side of the transformer bank, the grounded neutral wire is connected to the common points of all three high-voltage primary coil windings. These coil windings are marked H1 and H2 on each transformer to indicate the high (H) side (higher voltage side) of the transformer. The lower voltage side is marked with X1 and X2. The voltage from the neutral to any phase of the three power lines is 2,400 volts. The voltage across the three power lines is 4,152 volts (1.73 × 2,400).

As on the primary side, the grounded neutral is connected to the common points of all three low voltage secondary coil winding. This allows for a 120 volt output on each of the secondary coils. The voltage across the three secondary power lines is 208 volts (1.73 × 120).

To help maintain a balanced transformer bank, the electrician should always try to connect loads so as to divide them evenly among the three transformers. This will naturally be done when connecting a three-phase motor, since a three-phase motor will draw the same amount of current from each line. Care should then be taken to balance the single-phase loads. Three single-phase transformers of the same power ratings are used in most Wye to Wye systems. The capacity of transformers is rated in kilovolt-amperes (KVA). The total KVA capacity of a transformer bank is found by adding the individual KVA ratings of each transformer in the bank. If, for example, each transformer in Figure 10-19 were rated at 50 KVA, the total capacity of the bank would be 150 KVA (50 + 50 + 50).

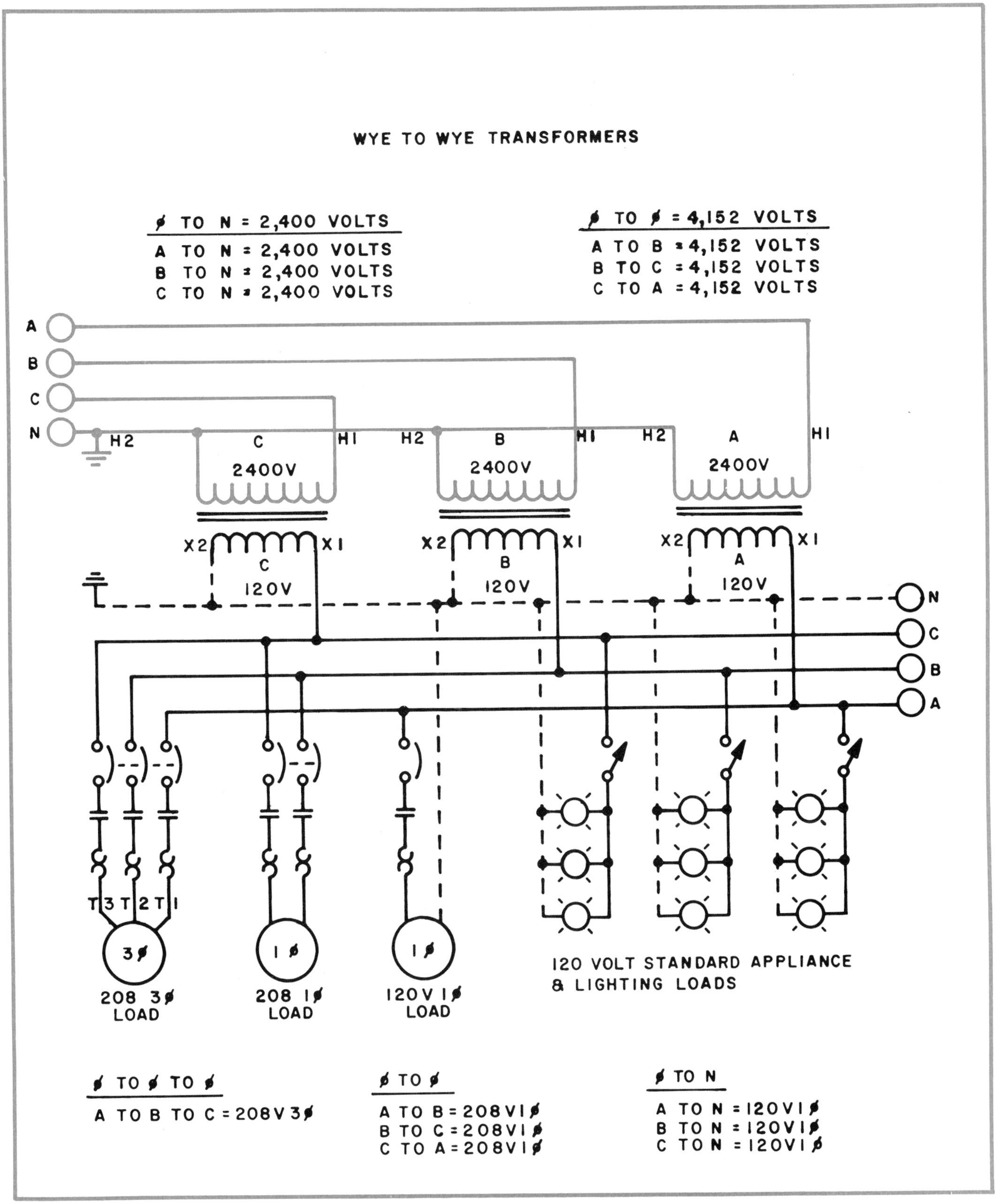

WYE TO WYE TRANSFORMERS

ϕ TO N = 2,400 VOLTS

A TO N = 2,400 VOLTS
B TO N = 2,400 VOLTS
C TO N = 2,400 VOLTS

ϕ TO ϕ = 4,152 VOLTS

A TO B = 4,152 VOLTS
B TO C = 4,152 VOLTS
C TO A = 4,152 VOLTS

A
B
C
N

H2 C 2400V H1   H2 B 2400V H1   H2 A 2400V HI

X2 C X1   X2 B X1   X2 A X1
120V   120V   120V

N
C
B
A

T3 T2 TI

3ϕ

208 3ϕ
LOAD

1ϕ

208 1ϕ
LOAD

1ϕ

120V 1ϕ
LOAD

120 VOLT STANDARD APPLIANCE
& LIGHTING LOADS

ϕ TO ϕ TO ϕ

A TO B TO C = 208V 3ϕ

ϕ TO ϕ

A TO B = 208V1ϕ
B TO C = 208V1ϕ
C TO A = 208V1ϕ

ϕ TO N

A TO N = 120V1ϕ
B TO N = 120V1ϕ
C TO N = 120V1ϕ

Figure 10-19 Three single-phase transformers are connected for a Wye-to-Wye step-down transformer bank.

Transformers Connected for a Delta-to-Delta Installation. Figure 10-20 illustrates how three single-phase transformers are connected for a Delta-to-Delta step-down transformer bank. Remember from our previous discussion that the line voltage is equal to the coil voltage and that the line current is equal to 1.73 times the coil current in a balanced Delta system.

As shown in Figure 10-20, the Delta connected secondary, with one coil centered tapped at the mid-point, provides for three different types of service: (1) a three-phase 240 volt service for the three-phase motor load; (2) single-phase 120 volt service for the lighting loads; and (3) single-phase 240 volts for the 240 single-phase motor. Not

illustrated is the high-phase of line C to N. This is because the high-phase is not to be used.

As with the Wye transformer bank, a closed Delta bank would deliver a total KVA output equal to the sum of the individual transformer ratings. However, we have mentioned that one advantage of a Delta system is that if one transformer is damaged or removed from service, the other two can be connected in an open-delta connection (also called the V-connection). This type of connection enables power to be maintained at a reduced level. This level is reduced to 57.7 percent of a full Delta-connected transformer bank not to 66.6%, as might be thought in a reduction of 3 to 2 transformers.

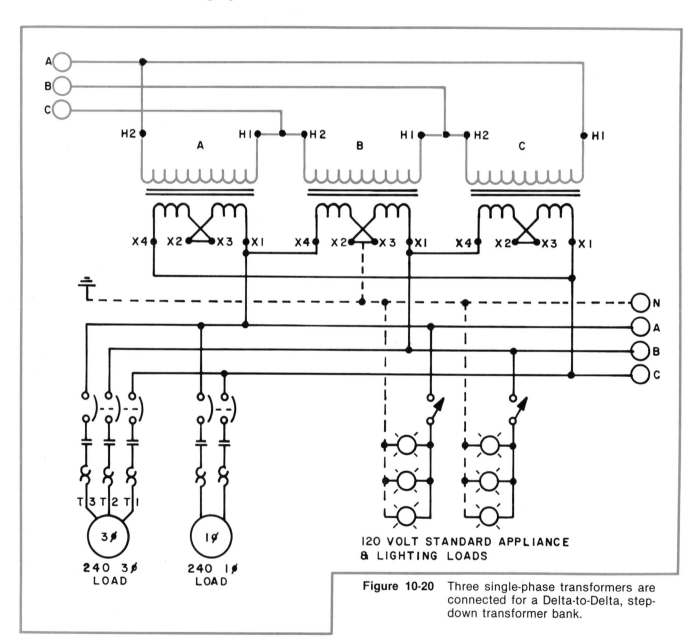

Figure 10-20 Three single-phase transformers are connected for a Delta-to-Delta, step-down transformer bank.

SUBSTATIONS

Substations serve as a source of voltage transformation and control along the distribution system. Their function is to:

1. Receive the voltage generated and increase it to a level appropriate for transmission.

2. Receive the transmitted voltage and reduce it to a level appropriate for customer use.

3. Provide a safe point in the distribution system for disconnecting the power in the event of trouble.

4. Provide for a place to adjust and regulate the outgoing voltage.

5. Provide a convenient place to take measurements and check the operation of the distribution system.

6. Provide for a switching point where different connections may be made between various transmission lines.

Substations have three main parts, as illustrated in Figure 10-21. These three parts are the

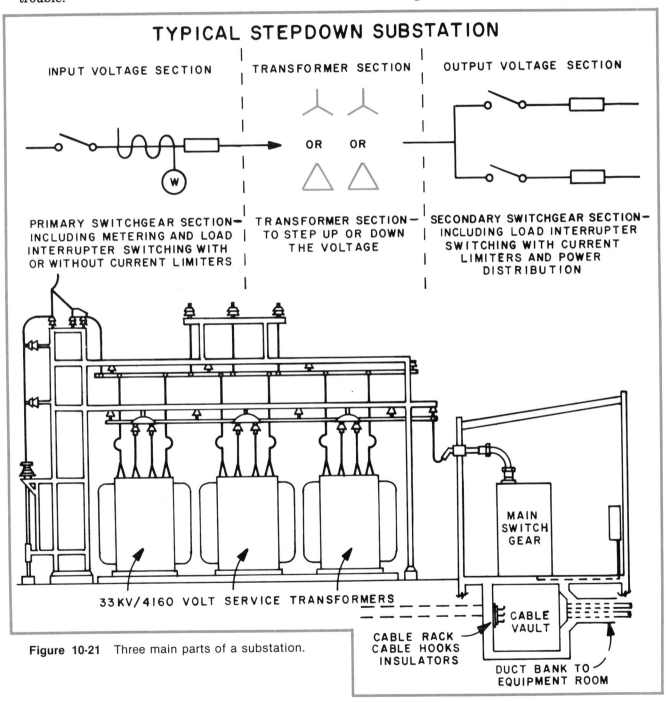

Figure 10-21   Three main parts of a substation.

input voltage section, transformer, and output voltage section. Depending on the function of the substation (stepping up or down the voltage), the input and output section will be called the high-voltage or low-voltage section. In step-up substations, the input section will be called the low-voltage section and the output called the high-voltage section. In step-down substations, the input section will be called the high-voltage section and the output called the low-voltage section. The substation sections usually include breakers, junction boxes and interrupter switches.

Substations may be entirely enclosed in a building or totally in the open, as in the case of outside substations located along the distribution system. The location for a substation is generally selected so that the station will be as near as possible to the area to be served.

Substations can be built to order on site or purchased from factory-built, metal-enclosed units. These purchased units are called *unit substations*. The unit substation offers standardization and flexibility for future changes when quick replacements are needed.

The transformer's function in the substation is the same as that of any transformer to step up or step down the voltage. These transformers are broadly classified as wet or dry. In wet types, oil or some other liquid serves as a heat-transfer medium and insulation. Dry types use air or inert gas in place of liquid. Sometimes fans are used for forced-air cooling on transformers in order to: (1) provide additional power for peaks, with the normal loads handled by natural circulation; (2) Lower oil and copper temperatures where the surrounding air is hot and the transformers could not

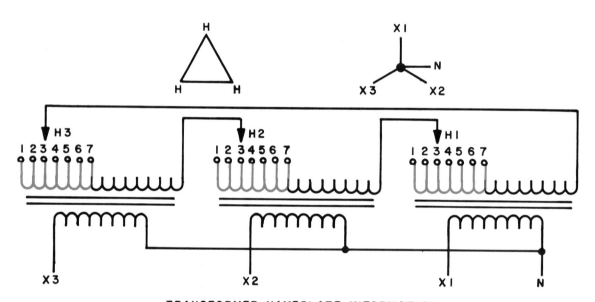

TRANSFORMER NAMEPLATE INFORMATION

HIGH VOLTAGE 480 LINE TO LINE		JUMPER CONNECTIONS EACH PHASE		
	VOLTS	TAP	VOLTS	TAP
HIGH VOLTAGE 480 LINE TO LINE	503	1	456	5
LOW VOLTAGE 208 LINE TO LINE				
LOW VOLTAGE 120 LINE TO NEUTRAL	493	2	443	6
	480	3	433	7
OTHER INFORMATION FOUND ON NAMEPLATE				
KVA = _ _ _ _ _ _ TYPE _ _ _ _ _ CLASS _ _ _ _				
TEMPERATURE = _ _ _ _ _ °C	466	4		
MODEL & SERIAL NO.				

Figure 10-22 Taps are built into a transformer to compensate for voltage differences.

handle their full rated load without exceeding their recommended temperature.

Most transformers used in substations include a voltage regulator. The regulator on a transformer has taps that allow for a variable output. These taps are needed because transmission lines rarely deliver the transformer input voltage for which they are rated because of transmission line losses, line loading and other factors.

Figure 10-22 illustrates how taps are built into the transformer to compensate for the voltage difference. The taps are usually provided at 2-1/2% increments for adjusting above and below the rated voltage. This adjustment could be manual or automatic. In the automatic regulator, a control circuit automatically changes the tap setting on the transformer's windings. This allows the outgoing voltage to be kept approximately constant even though the incoming primary voltage or load demands may vary.

SWITCHBOARDS

We have seen that electrical power is delivered to industrial, commercial and residential buildings through an orderly distribution and transmission system. Once this power is delivered to a building, it is up to the building electrician to further distribute the power to where it is required within the building or plant. It is the switchboard that is the link between the power delivered to the building and the start of the local distribution system throughout the building or plant. It can be said that, for all practical purposes, the switchboard is the last point on the power distribution system as far as the power company is concerned and the beginning of the distribution system as far as the building electrician is concerned.

The switchboard can simply be defined as the electrical piece of equipment in which a large block of electric power is delivered from the substation and broken down into smaller blocks for distribution throughout the building. Figure 10-23 illustrates the function of the switchboard in dividing the incoming power into smaller branch circuits. In this function, the switchboard serves the same function as a heart that takes large amounts of blood and feeds smaller arteries for proper distribution throughout the body's transmission system.

In addition to dividing the incoming power, the switchboard may contain all the equipment needed for controlling, monitoring, protecting and recording the functions of the substation.

Switchboards are designed for use in three categories: (1) service entrance, (2) distribution, and (3) combination of service entrance and distribution.

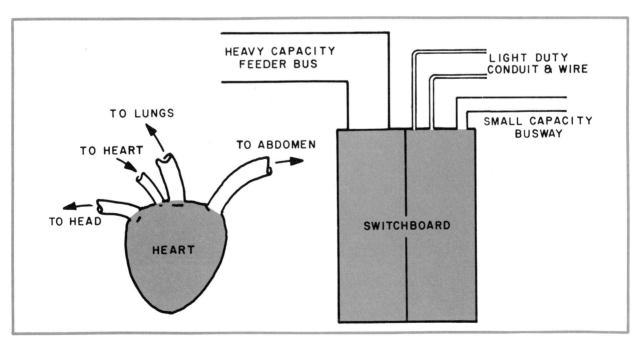

Figure 10-23 The switchboard is used to divide incoming power into smaller branch circuits.

The service entrance switchboard has space and mounting provisions required by the local power company for metering equipment as well as both overcurrent protection and disconnect means for the service conductors. Provision for grounding the service neutral conductor when a ground is needed is also provided. The distribution switchboard contains the protective devices and feeder circuits required to distribute the power throughout the building. The distribution switchboard may contain either circuit breakers or fused switches.

Figure 10-24 illustrates a combination service entrance switchboard with a distribution section added to it as illustrated. The switchboard has the space and mounting provisions required by the local power company for metering their equipment and incoming power. To meter the incoming energy, the switchboard must have a Watthour meter to measure energy usage. Metering is always located on the incoming line side of the disconnect. The compartment cover is sealed in order to prevent tapping power ahead of the power company's metering equipment.

Many other meters and indicating lights, such as ammeters and voltmeters, may also be built into the meter compartment. These are not requirements in most cases but options, depending on the application and the plant's requirements. The voltmeter is used to indicate to the operation or maintenance person the various incoming and outgoing voltages. The ammeter is used to indicate the various currents throughout the system. The Wattmeter is used to indicate the power (voltage × current × power factor) being used throughout the system. Each of these instruments can be of the indicating type, recording type, or both types. The recording instrument is used to keep track of the various values over a period of time.

In addition to measuring the voltage, current and power of the system, the switchboard also controls the power. This control is achieved through the use of switches and over-current and over-voltage relays that are used to disconnect the power. These devices protect the distribution system in the event of a fault.

Switchboards that have more than six switches or circuit breakers must include a *main switch* to protect or disconnect all circuits. Switchboards

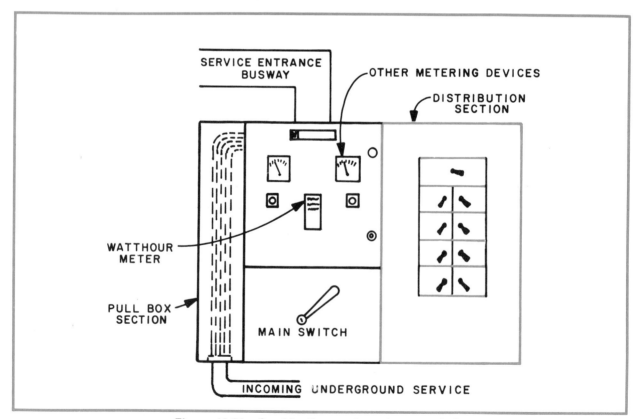

Figure 10-24 Combination service entrance switchboard with a distribution section added to it.

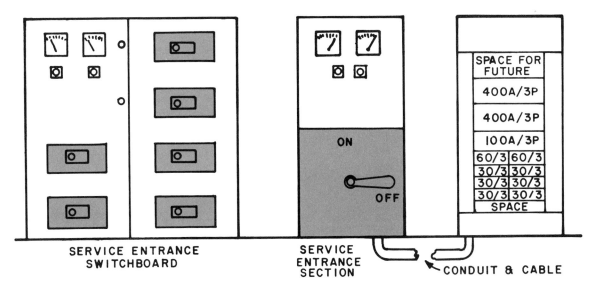

Figure 10-25 Service entrance switchboards may be set up in different ways.

with more than one, but not more than six switches or circuit breakers, do not require a main switch. Figure 10-25 illustrates each of these examples.

Figure 10-25, left, illustrates a switchboard with the allowable six switches or circuit breakers without a main disconnect. In this illustration, each of the switches or circuit breakers actually could be called sub mains.

In Figure 10-25, right, a switchboard with more than six switches or breakers is illustrated. In this illustration, the service entrance section of the switchboard may have any number of feeder circuits added, up to the rated capacity of the main. A switchboard with a main section can easily contain more than one single distribution section. This all depends on how many feeder circuits are required in addition to the blank spaces needed for future expansion.

In addition to distributing the power throughout the building, the distribution section of a switchboard may contain the provisions for motor starters and other control devices, as illustrated by the photo in Figure 10-26. The addition of starters and controls to the switchboard allows for motors to be connected to the switchboard. This combination can be used when the motors to be controlled are located near the switchboard. This combination allows for the high-current consumption loads motors, for example, to be connected to the source of power without further power distribution.

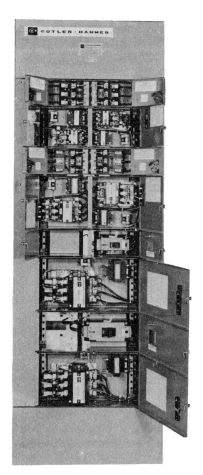

Figure 10-26 The distribution section of a switchboard may contain the provisions for motor starters and other control devices. (Cutler Hammer)

PANELBOARDS AND BRANCH CIRCUITS

A panelboard is a wall-mounted distribution cabinet containing a group of overcurrent and short circuit protective devices for lighting and appliance or power distribution branch circuits. The wall-mounted feature distinguishes the panelboard from a switchboard, which usually stands or rests on the floor as illustrated in Figure 10-27. The panelboard is usually supplied from the switchboard and further divides the power distribution system into smaller parts. Panelboards are the part of the distribution system that provides the last centrally located protection for the final power run to the load and its control circuitry. Panelboards are classified according to their use in the distribution system. The three types of panelboards are lighting and appliance, power, and distribution.

Figure 10-28 illustrates some typical examples of panelboard interiors. The panelboard provides the required circuit control and overcurrent protection needed, with all circuits and power consuming loads connected to the distribution system. These panelboards are located throughout the plant or building, providing the necessary protection for the branch circuits feeding the loads.

The branch circuit is that portion of the distribution system between the final overcurrent device protecting the circuit and the outlet or load connected to them. The basic requirements for panelboards and overcurrent protection are given in the *National Electrical Code®* and must be used as the guidelines for individual applications. In addition, it is good to check local power company, city and county regulations, which may be even stricter.

The overcurrent devices used for protecting the branch circuit may be fuses or circuit breakers. The overcurrent device must provide for proper overload and short circuit protection. The size (in amperes) of the overcurrent device is based on the rating of the panelboard and load. The overcurrent device must protect the load and be within the rating of the panelboard. If the overcurrent

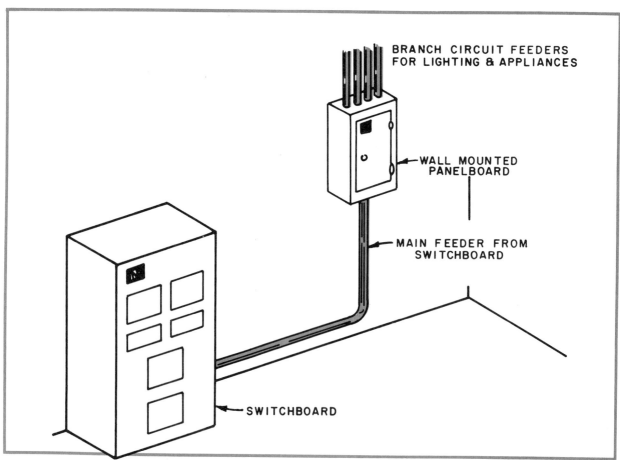

Figure 10-27 Difference between a panelboard and a switchboard.

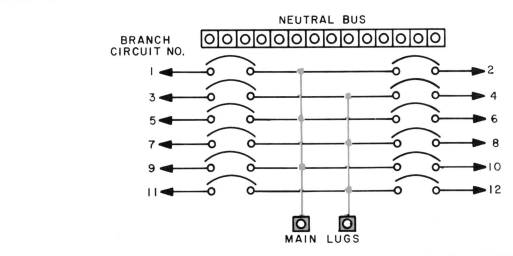

TYPICAL SINGLE-PHASE, 3-WIRE PANELBOARD WITH CIRCUIT BREAKERS

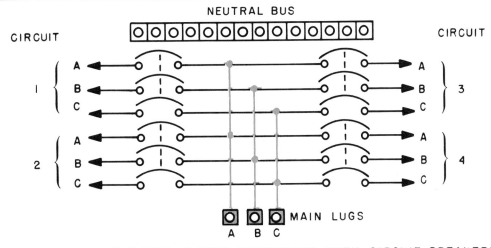

TYPICAL THREE-PHASE, 4-WIRE PANELBOARD WITH CIRCUIT BREAKERS

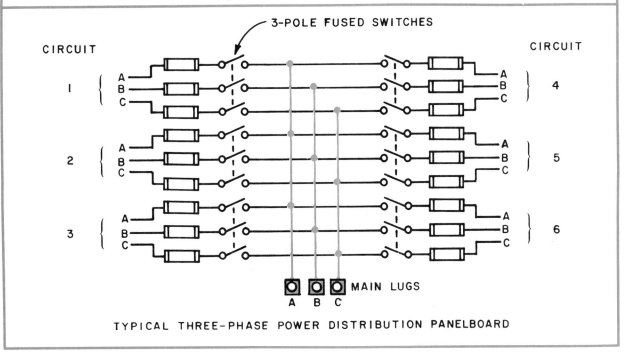

TYPICAL THREE-PHASE POWER DISTRIBUTION PANELBOARD

Figure 10-28 Typical examples of panelboard interiors.

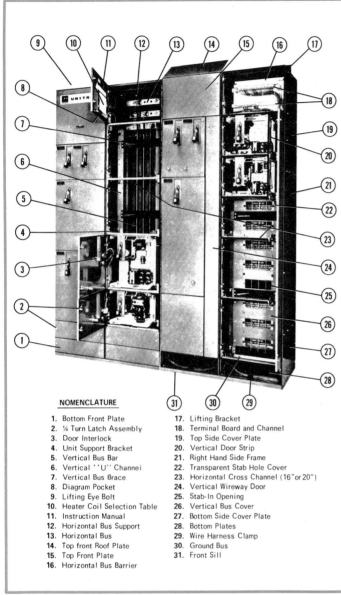

NOMENCLATURE

1. Bottom Front Plate	17. Lifting Bracket
2. ¼ Turn Latch Assembly	18. Terminal Board and Channel
3. Door Interlock	19. Top Side Cover Plate
4. Unit Support Bracket	20. Vertical Door Strip
5. Vertical Bus Bar	21. Right Hand Side Frame
6. Vertical ''U'' Channel	22. Transparent Stab Hole Cover
7. Vertical Bus Brace	23. Horizontal Cross Channel (16"or 20")
8. Diagram Pocket	24. Vertical Wireway Door
9. Lifting Eye Bolt	25. Stab-In Opening
10. Heater Coil Selection Table	26. Vertical Bus Cover
11. Instruction Manual	27. Bottom Side Cover Plate
12. Horizontal Bus Support	28. Bottom Plates
13. Horizontal Bus	29. Wire Harness Clamp
14. Top front Roof Plate	30. Ground Bus
15. Top Front Plate	31. Front Sill
16. Horizontal Bus Barrier	

Figure 10-29 Typical motor control center used to take incoming power, control circuitry, required overload and overcurrent protection and combine them in one convenient location. (Cutler Hammer)

device exceeds the ampacity of the bus bars in the panelboard, the panelboard is undersized for the load or loads that will be connected.

The panelboard can be compared to the load center found in your home. It performs about the same function. The load center in your home contains the fuses or breakers which control the individual branch circuits throughout the house.

Although the panelboard and load center perform about the same function, they are separated by certain distinct features when used in commer-

cial and industrial applications. The panelboard has some features not shared by load centers. These are:

1. A box made of Underwriters Laboratory (UL) code gauge corrosion-resisting galvanized steel.

2. A minimum of a 4″ wiring gutter on all sides.

3. Combination catch and lock in addition to the hinges.

4. Bus bars listed to 1200 amps (load center's main bus bars are generally 200 amps maximum).

5. Greater enclosure depth accommodates conduit over 2-1/2 inches.

6. Main and branch terminal lugs.

MOTOR CONTROL CENTERS

In the power distribution system there are many different kinds of loads that are to be connected to the system. These loads vary considerably from application to application, as does their degree of control. For example, a light may be connected to the system, requiring only a switch for control (along with proper protection). However, other loads, such as motors, may require complicated and lengthy control and protection circuits. The more complicated a control circuit becomes, the more difficult it becomes to wire into the system.

By far the most common load that requires control from simple to complex is the electric motor. Since the electric motor is the backbone of almost all production and industrial applications, the need to simplify and consolidate motor control circuits is only logical. To do this, a control center is used to take the incoming power, control circuitry, required overload and overcurrent protection and any transformation of power, and combine them into one convenient center. This center is called the *motor control center*. The motor control center combines individual control units into standard modular structures joined together on formed sills for a convenient control center in one location. The motor control center is usually supplied from a panelboard or switchboard. The motor control center is different from a switchboard containing motor panels in that the motor control center is a modular structure designed specifically for plug-in type control units and motor control.

Figure 10-29 illustrates a typical motor control center. The function of the control center is to take

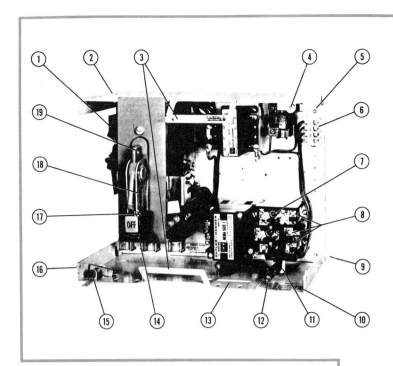

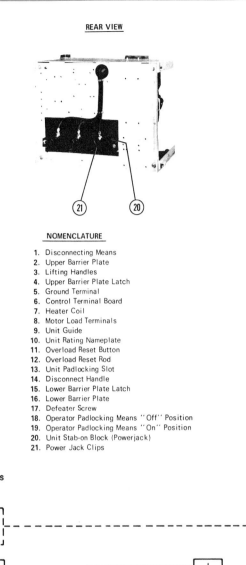

NOMENCLATURE

1. Disconnecting Means
2. Upper Barrier Plate
3. Lifting Handles
4. Upper Barrier Plate Latch
5. Ground Terminal
6. Control Terminal Board
7. Heater Coil
8. Motor Load Terminals
9. Unit Guide
10. Unit Rating Nameplate
11. Overload Reset Button
12. Overload Reset Rod
13. Unit Padlocking Slot
14. Disconnect Handle
15. Lower Barrier Plate Latch
16. Lower Barrier Plate
17. Defeater Screw
18. Operator Padlocking Means "Off" Position
19. Operator Padlocking Means "On" Position
20. Unit Stab-on Block (Powerjack)
21. Power Jack Clips

the incoming power and deliver it to the control circuit and motor loads that are being controlled. The center provides space for the control and load wiring in addition to providing required control components. The control inputs into the center are the control devices such as pushbuttons, liquid level, limit and other devices that provide a signal. The output of the control center is the wire connecting the motors. All other control devices are located in the centers. This includes relays, control transformers, motor starters, overload and overcurrent protections, timers, counters and any other control devices required.

The advantage of a control center is that it provides one convenient place for installing and troubleshooting the control circuits. This is especially useful in applications that require individual control circuits to be related to other control circuits. Examples of this are on any assembly line in which one machine feeds the next.

A second advantage is that individual units can easily be removed, replaced, added to and interlocked at one central location. Manufacturers of motor control centers make available factory preassembled units to meet all the standard motor functions such as start/stop, reversing, reduced voltage starting and speed control. This leaves only the connecting of the control devices (limit switches, etc.) and the motors to the center.

Figure 10-30 illustrates a typical pre-assembled motor control unit panel that is available from the factory, along with a typical schematic diagram.

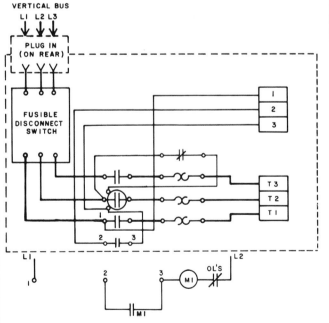

Figure 10-30 Typical pre-assembled motor control unit panel that is available from the factory, along with a typical schematic diagram. (Cutler Hammer)

Notice that the only required wiring by the electrician is the connection to control inputs, terminal blocks and the motor. The motor is connected to T1, T2 and T3 as illustrated in Figure 10-31. The control inputs are connected to the terminal blocks marked 1, 2 and 3. If, for example, the standard start/stop pushbutton station were required, it would be connected as illustrated. If a two-wire control like a liquid level switch were to be connected to the circuit, this would be connected to terminals 1 and 3 only. Also provided on each unit are predrilled holes to allow for easy additions to the circuit. These holes match the manufacturer's standard devices, and most manufacturers provide templates for easy layout and circuit designs.

FEEDER AND BUSWAYS

The industrial electrical distribution system in a plant must transport the electrical power from the source of supply to the electrical loads. In today's industry, that may mean distribution over large areas with many different electrical requirements as illustrated in Figure 10-32. In many cases, the distribution system must be changed from time to time where shifting of production machinery is common. A busway is a metal enclosed distribution system of bus bars available in prefabricated sections for simple installation. Prefab fittings, tees, elbows and crosses simplify the connecting and reconnecting of the distribution system.

From the illustration of Figure 10-33, we can see that busways are somewhat like the railroad

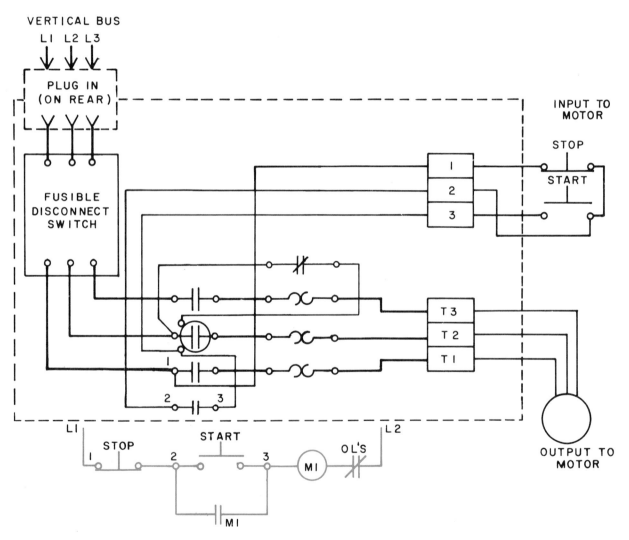

Figure 10-31 The only control circuitry necessary to make the pre-assembled motor control unit functional.

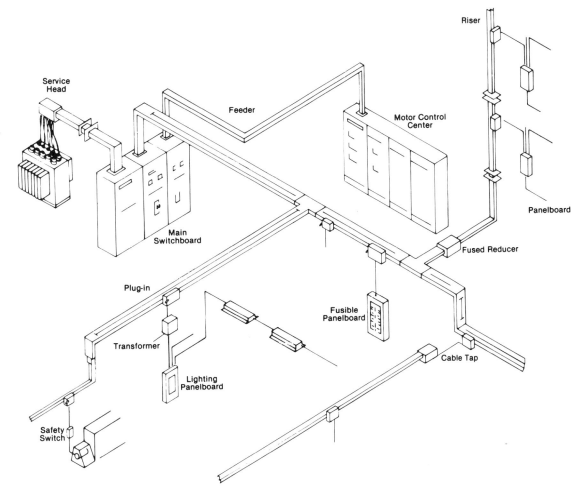

Figure 10-32 Typical distribution system over a large area servicing many different electrical requirements.

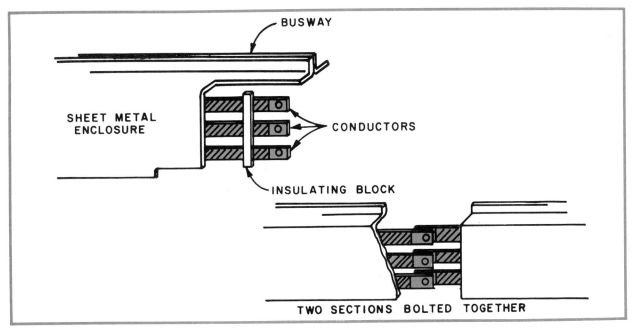

Figure 10-33 Busway sections can be bolted together to make electrical power available at many locations along the system.

tracks of a toy electric train (which also carry electricity). By plugging sections together, you can make the electrical power available at many locations and throughout the system. If the train track needed to be changed, it could easily be disconnected and reconnected.

The major difference between the train tracks and a busway is that the busway does not have exposed conductors. This is because the power in a plant distribution system is at a much higher level than that of a small electrical train. To offer protection from the higher voltage, the conductors of a busway are supported with insulators and covered with an enclosure in order to prevent accidental contact.

Figure 10-34 illustrates a typical busway distribution system. This system provides for fast con-

nection and disconnection of machinery. Plants can be retooled or revamped without major changes in the distribution system.

The most common length of busways is ten feet. Shorter lengths are used as needed. Prefab elbows, tees and crosses make it possible to go up, down, around corners and tap off from the distribution system. This allows the distribution system to have maximum flexibility with simple installation.

There are two basic types of busways: the feeder type and the plug-in type Figure 10-34 illustrates the use of both types. Feeder type busways deliver the power from the source to a load-consuming device. Plug-in type busways serve the same function but also allow load-consuming devices to be conveniently added along the bus structure. Fig-

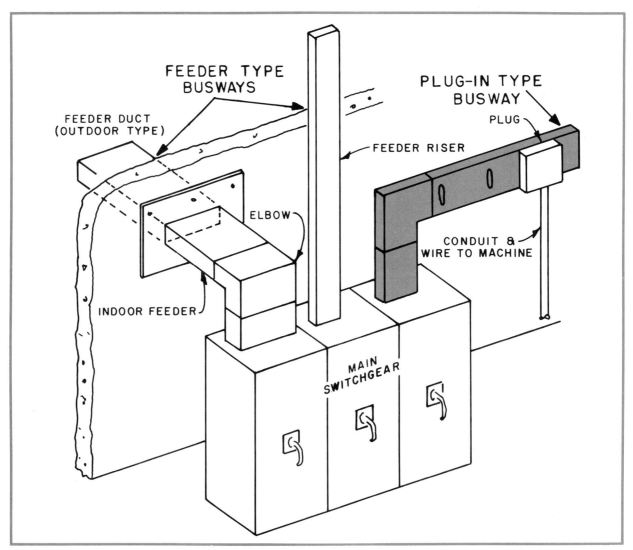

Figure 10-34 Typical busway distribution system with straight feeders and plug-in duct.

ure 10-35 illustrates how a plug-in power module is used on a plug-in type busway system.

There are three general types of plug-in power panels used with busways. These panels are fusible switches, circuit breakers and specialty plugs (duplex receptacles with circuit breakers, twistlock receptacles, etc.). It is from these fusible switches and circuit breaker plug-in panels that the conduit and wire is run to the machine or load. Generally, cords may be used only for portable equipment.

Oftentimes, the loads connected to the power distribution system are either portable or unknown at the time of installation. For this reason, the power distribution system often must termi-

nate in such a manner as to provide for a quick connection of a load at some future time. To do this, the electrician will install receptacles throughout the building or plant in the same manner as they are found in your home. With these receptacles, different loads can be connected easily.

Since the distribution systems wiring and protection devices determine the size of load that can be connected to it, there must be some method for distinguishing the rating in voltage and current of each termination. This is especially true in industrial applications which require a variety of different currents, voltages and phases.

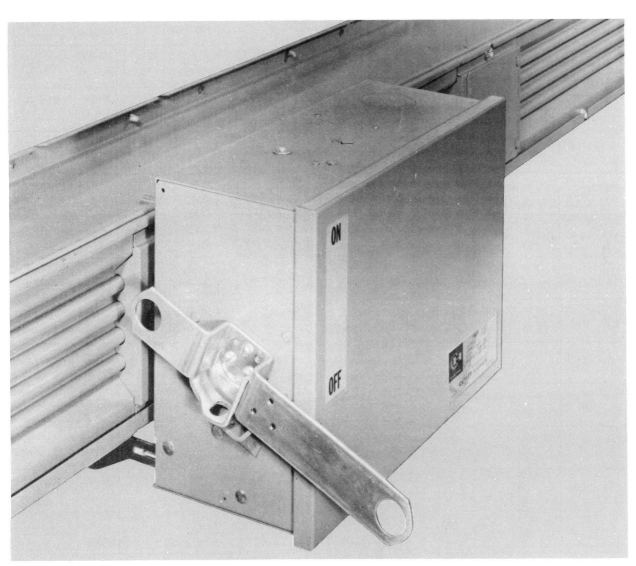

Figure 10-35 Typical plug-in power panel module used in a busway system.

The National Electrical Manufacturers Association (NEMA) has established a set of standard plug and receptacle configurations that clearly tell the type of termination. These standards are illustrated in Figure 10-36. The electrician who becomes familiar with these standard configura-

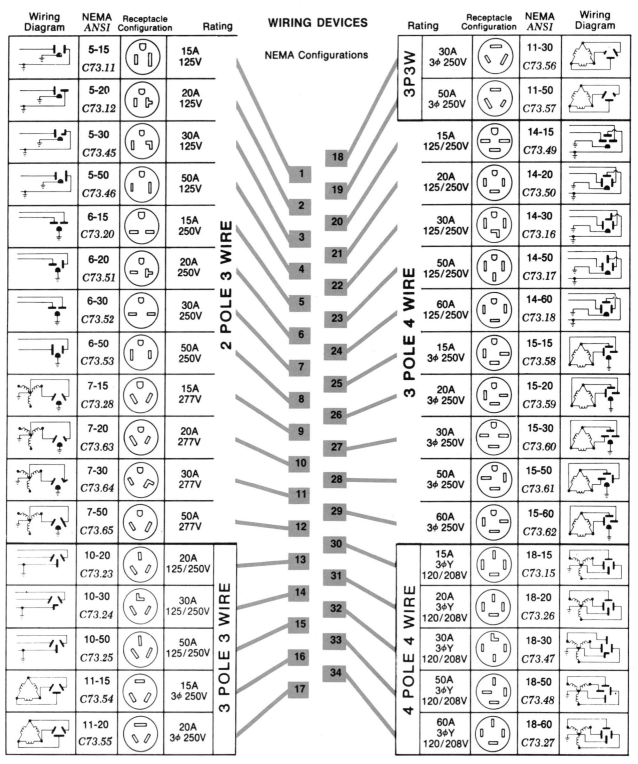

Figure 10-36 A set of standard receptacle configurations which will help the electrician in determining the proper type of termination. (Bryant Wiring Device Division, Westinghouse Electric Corporation, 500 Sylvan Avenue, Bridgeport CT 06606)

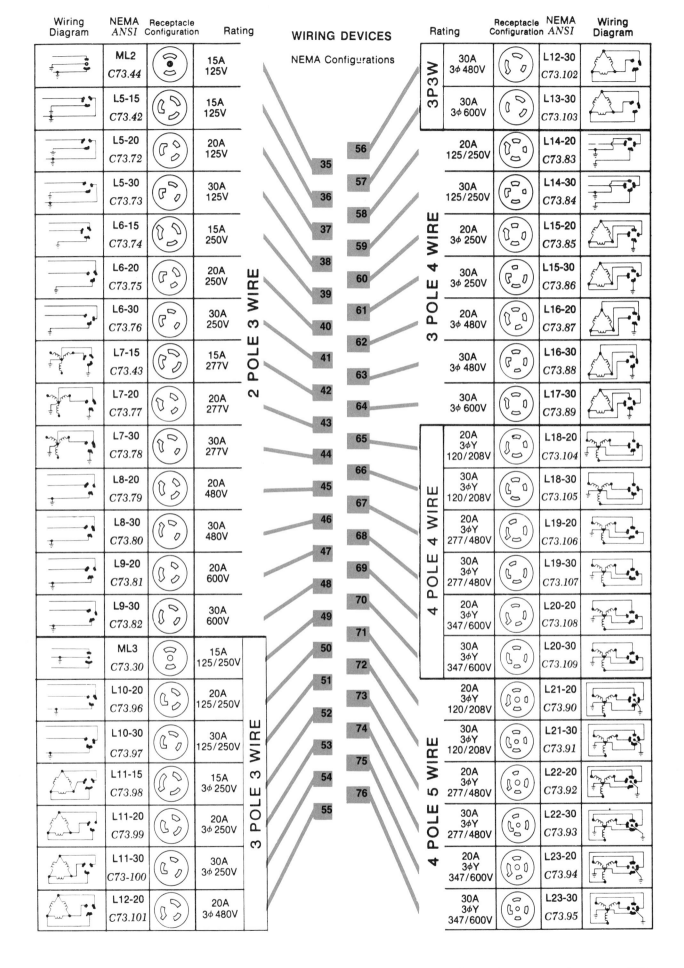

WIRING DEVICES

NEMA Configurations

Wiring Diagram	NEMA ANSI	Receptacle Configuration	Rating
	ML2 C73.44		15A 125V
	L5-15 C73.42		15A 125V
	L5-20 C73.72		20A 125V
	L5-30 C73.73		30A 125V
	L6-15 C73.74		15A 250V
	L6-20 C73.75		20A 250V
	L6-30 C73.76		30A 250V
	L7-15 C73.43		15A 277V
	L7-20 C73.77		20A 277V
	L7-30 C73.78		30A 277V
	L8-20 C73.79		20A 480V
	L8-30 C73.80		30A 480V
	L9-20 C73.81		20A 600V
	L9-30 C73.82		30A 600V

2 POLE 3 WIRE

Wiring Diagram	NEMA ANSI	Receptacle Configuration	Rating
	ML3 C73.30		15A 125/250V
	L10-20 C73.96		20A 125/250V
	L10-30 C73.97		30A 125/250V
	L11-15 C73.98		15A 3φ 250V
	L11-20 C73.99		20A 3φ 250V
	L11-30 C73.100		30A 3φ 250V
	L12-20 C73.101		20A 3φ 480V

3 POLE 3 WIRE

NEMA configuration numbers: 35, 36, 37, 38, 39, 40, 41, 42, 43, 44, 45, 46, 47, 48, 49, 50, 51, 52, 53, 54, 55, 56, 57, 58, 59, 60, 61, 62, 63, 64, 65, 66, 67, 68, 69, 70, 71, 72, 73, 74, 75, 76

Rating	Receptacle Configuration	NEMA ANSI	Wiring Diagram
30A 3φ 480V		L12-30 C73.102	
30A 3φ 600V		L13-30 C73.103	

3P3W

Rating	Receptacle Configuration	NEMA ANSI	Wiring Diagram
20A 125/250V		L14-20 C73.83	
30A 125/250V		L14-30 C73.84	
20A 3φ 250V		L15-20 C73.85	
30A 3φ 250V		L15-30 C73.86	
20A 3φ 480V		L16-20 C73.87	
30A 3φ 480V		L16-30 C73.88	
30A 3φ 600V		L17-30 C73.89	

3 POLE 4 WIRE

Rating	Receptacle Configuration	NEMA ANSI	Wiring Diagram
20A 3φY 120/208V		L18-20 C73.104	
30A 3φY 120/208V		L18-30 C73.105	
20A 3φY 277/480V		L19-20 C73.106	
30A 3φY 277/480V		L19-30 C73.107	
20A 3φY 347/600V		L20-20 C73.108	
30A 3φY 347/600V		L20-30 C73.109	

4 POLE 4 WIRE

Rating	Receptacle Configuration	NEMA ANSI	Wiring Diagram
20A 3φY 120/208V		L21-20 C73.90	
30A 3φY 120/208V		L21-30 C73.91	
20A 3φY 277/480V		L22-20 C73.92	
30A 3φY 277/480V		L22-30 C73.93	
20A 3φY 347/600V		L23-20 C73.94	
30A 3φY 347/600V		L23-30 C73.95	

4 POLE 5 WIRE

tions will have the ability to tell the voltage and current rating of any receptacle or plug simply by looking at the configuration.

GROUNDING

Equipment grounding is important throughout the entire distribution system. This means connecting to ground all noncurrent carrying metal parts including conduit, raceways, transformer cases and switch gear enclosures. The objective of this grounding is to limit the voltage between all metal parts and the earth to a safe level.

Grounding is accomplished by connecting the noncurrent-carrying metal to a ground bus with an approved grounding conductor and fitting. The grounding bus is a network that ties solidly to grounding electrodes. The grounding electrode is a conductor embedded in the earth to provide a good ground. The ground bus should surround the transmission station or building. This bus must be connected to the grounding electrodes in several spots. The size of the ground bus will be based on how much current may flow through the grounding system and for how long.

In addition to grounding all noncurrent-carrying metal, lightning arresters may be needed. A lightning arrester is a device which protects the transformers and other electrical equipment from voltage surges caused by lightning. The lightning arrester provides a path over which the surge can pass to the ground before it has a chance to damage the electrical equipment.

QUESTIONS ON UNIT 10

1. What is the major source of electrical power?
2. How is electricity generated in an alternator?
3. How is the external circuit connected to an alternator without interfering with the alternator's rotating rotor?
4. What is the effect of adding additional pairs of poles to an alternator?
5. Why is three phase power generated instead of single-phase power whenever possible?
6. What are the voltages available from a Wye-connected system with a common neutral?
7. If the phase-to-neutral voltage is 208 volts in a Wye system, what would the phase-to-phase voltage be?
8. What are the voltages available from a Delta-connected system without a neutral wire?
9. What are the voltages available from a Delta-connected system with a neutral connected at the midpoint of one of the phases?
10. What are the two windings of a transformer called?
11. What is the function of the transformer core?
12. Is a transformer with a ratio of 4 to 1 a step-up or step-down transformer?
13. If the voltage is doubled on the output of a transformer, what happens to the current on the output of the transformer?
14. How is the "high" side of a transformer identified?
15. How is the "low" side of a transformer identified?
16. How is a transformer rated for power output?
17. Why is it important to balance a transformer bank?
18. What are the three main parts of a substation?
19. What are the functions of the "taps" on a transformer?
20. As far as the power company is concerned, what is the last point on the power distribution system?
21. What is the difference between a service entrance switchboard and a distribution switchboard?
22. What is the function of a panelboard in the power distribution system?
23. What are the three types of panelboards?
24. What is the function of the motor control center in the power distribution system?

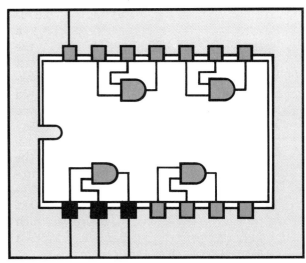

11

Solid State Electronic Control Devices

Control of heat, flow, pressure, and other industrial processes is critical for maximum productivity and safety. Solid state electronic control devices have contributed to the increased control of industrial processes. For example, in lighting circuitry, lights once were either on or off. The invention of the diode and SCR now allows a range from full intensity to very low intensity. Integrated circuits can turn lights on and off for energy efficiency. Microprocessor programmed control units allow timing on and off plus variation in intensity through the use of solid state sensors. Solid state devices have not changed the basics of the industrial process, but have provided more precise control over this process.

SEMICONDUCTOR DEVICES

Semiconductor devices often are found mounted on a PC board. A PC board is made of an insulating material such as fiberglass or phenolic with conducting paths secured to one or both sides of the board. PC boards are manufactured through a variety of processes. The purpose of all PC boards is to provide electrical paths of sufficient size to ensure a reliable electronic circuit. Each part of the PC board has a name. See Figure 11-1. *Pads* are the small round conductors to which component leads are soldered. *Traces,* or *foils,* are used to interconnect two or more pads. A large trace or foil extending around the edge to provide conduction from several sources is called a *bus.*

An *edge card* is a PC board with multiple terminations (terminal contacts) on one end. Most edge cards have terminations made from the same material as the foils—namely copper. In some instances, the terminations are gold plated, allowing for the lowest possible contact resistance. An *edge card connector* allows the edge card to be connected to the system's circuitry with the least amount of hardware.

Components are usually mounted on one side of the PC board. See Figure 11-2. However, in some cases where space is a premium, components may be mounted on both sides of the PC board. Component leads extend through the insulated board and are connected to the pads, traces, and bus with solder. To expedite troubleshooting procedures,

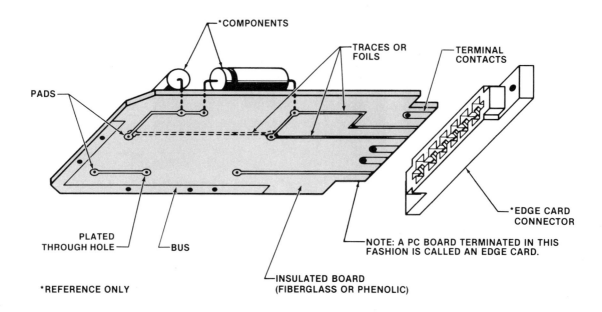

Figure 11-1 A printed circuit board (PC board) is constructed of an insulating material such as fiberglass or phenolic. Conducting paths are laminated on one or both sides of the board.

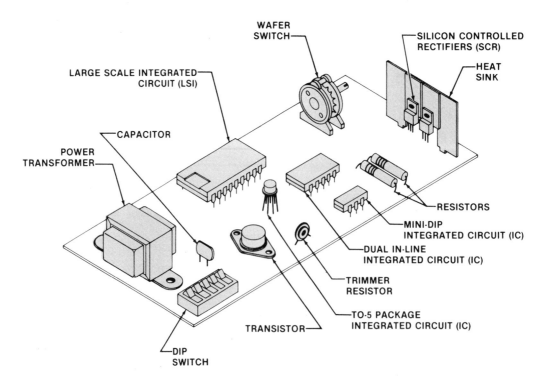

Figure 11-2 Semiconductor control devices can be mounted on a PC board. Component leads extend through the board and are connected to the pads, traces and buses with solder.

many PC boards have an appropriate marking next to the component. The marking identifies the component in relation to the schematic.

SEMICONDUCTOR THEORY

The world of matter is made of atoms. Everything around us is made up of an organized collection of atoms. Atoms contain three fundamental particles called electrons, protons, and neutrons. The easiest way to picture the atom is to think of it as having a center core, called the nucleus, with particles, called electrons, whirling about this nucleus in orbits or shells.

Nucleus

The nucleus consists of several different kinds of particles. The two most important ones are protons and neutrons. Protons carry a positive charge of one unit; neutrons carry no electrical charge.

Electrons

Electrons are negatively charged particles whirling around the nucleus at great speeds. They are arranged in orbits or shells. Each of these shells can hold a specific number of electrons. The innermost shell can hold two electrons, the second shell can hold eight, the third shell can hold eighteen, and so on. The shells are filled starting with the inner shell and working out so that when the inner shells are filled with as many electrons as they can hold, the next shell is started. The electron and the proton have equal amounts of opposite charge; the electron is a negatively charged particle and the proton is a positively charged particle. There are as many electrons as there are protons in a given atom, which leaves the entire atom electrically neutral.

Valence Electrons

Most elements do not have a completed outer shell—that is, they do not have the maximum allowable electrons in their outer shells. These outer electrons are called valence electrons. Valence electrons determine the conductive or insulative value of a given material. Conductors generally have only one or two valence electrons in their outer shell. See Figure 11-3. Insulators will generally have several electrons in their outer shell and will be filled completely or almost filled with valence electrons. Semiconductor materials fall somewhere between the low resistance offered by a conductor and the high resistance offered by an

insulator. Semiconductors are made from materials that have four valence electrons in their outer orbits.

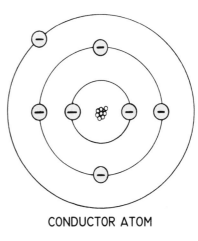

CONDUCTOR ATOM

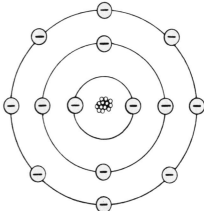

INSULATOR ATOM

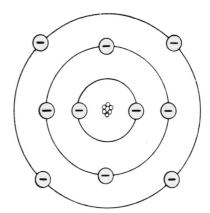

SEMICONDUCTOR ATOM

Figure 11-3 Valence electrons determine the amount of conductivity or insulating characteristics of a given material.

Doping

The basic material used in most semiconductor devices is either germanium or silicon. In their natural state, germanium and silicon are pure crystals. These pure crystals do not have enough free electrons to support a significant current flow. To prepare these crystals for use as a semiconductor device, their structure must be altered to permit significant current flow.

Doping is the process by which the crystal structure is altered. In doping, some of the atoms in the crystal are replaced with atoms from other elements. The addition of new atoms in the crystal creates *N-type material* and *P-type material*.

N-Type Material. N-type material is created by doping a region of a crystal with atoms from an element that has more electrons in its outer shell than the crystal. Adding these atoms to the crystal results in the possibility of more free electrons. Since electrons have a negative charge, the doped region is called N-type material.

As in any conductor, free electrons support current flow. When voltage is applied to N-type material, current flows from negative to positive through the crystal. See Figure 11-4. The free electrons help move current and are called *carriers*.

Some elements commonly used for creating N-type material are arsenic, bismuth, and antimony. The quantity of doping material used generally ranges from a few parts per billion to a few parts per million. By controlling even these small quantities of impurities in a crystal, the manufacturer can control the operating characteristics of the semiconductor.

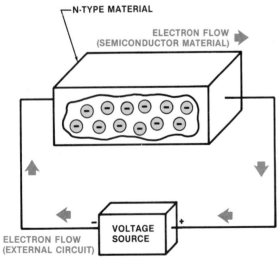

Figure 11-4 When a voltage is applied to N-type material, current flows, assisted by free electrons called carriers, from negative to positive.

P-Type Material. To create P-type material, a crystal is doped with atoms from an element that has fewer electrons in its outer shell than the natural crystal. This combination creates empty spaces in the crystalline structure. The missing electrons in the crystal structure are called *holes*. The holes are represented as positive charges. See Figure 11-5.

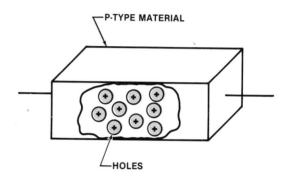

Figure 11-5 P-type material is created by doping the basic crystal with a material that has a deficiency of electrons. The missing electrons, called holes, are represented by a positive charge.

Typical elements used for doping a crystal to create P-type material are gallium, boron, and indium. In P-type material, the holes act as carriers. When voltage is applied, the holes are filled with free electrons as the free electrons move from negative potential to positive potential through the crystal. See Figure 11-6. Movement of the electrons from one hole to the next makes the holes appear to move in the opposite direction. Hole flow is equal to and opposite of electron flow.

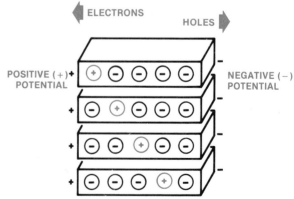

Figure 11-6 Movement of the electrons from one hole to the next makes the holes appear to move in opposite direction of the electrons.

Diodes are components that have the unique ability of allowing current to pass through them in only one direction. See Figure 11-50. This is made possible by the doping process, which creates N-type material (region) and P-type material. At the junction of the two materials, the P-type and N-type materials exchange carriers, creating a thin zone called the *depletion region*. See Figure 11-7. Because the depletion region is very thin, it responds rapidly to voltage changes.

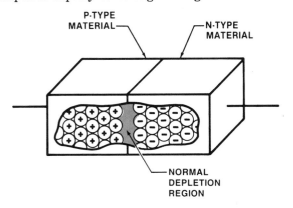

Figure 11-7 At the junction of the P-type and N-type materials, the materials exchange carriers, and a thin neutral area called a depletion region is formed.

The operating characteristics of a specific diode can be determined through the use of its operating characteristic curve. See Figure 11-8. When voltage is applied to the diode, the action occurring in the depletion region either blocks current flow or passes current.

Application of the proper polarity to a semiconductor diode is called a forward biased voltage, or *forward bias*. Forward bias results in *forward current*. When the opposite polarity is applied, it is called a reversed biased voltage, or *reverse bias*. Reverse bias results in a *reverse current* which should be close to zero or very small—usually a millionth of an ampere (μA).

Diodes are also rated for the maximum reverse bias voltage they can withstand. This rating is called *peak inverse voltage (PIV)*. The PIV ratings for most diodes used in industry range from a few volts to several thousand volts. If the reverse bias applied to the diode exceeds its PIV rating, the diode will break down and pass current freely. Current passed in this manner is called *avalanche current*, which can destroy diodes. To avoid avalanche, diodes with the correct rating must be used.

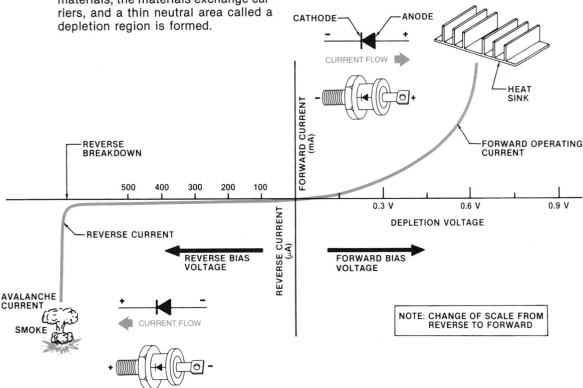

Figure 11-8 The operating characteristic curve indicates the response of a diode when subjected to a variety of forward and reverse bias voltage.

Rectification of Alternating Current

It is more efficient and economical to generate and transmit AC power than DC power. However, machinery and other loads often need DC to operate. In fact, the three-phase generator itself requires a DC voltage to create the magnetic field in the stationary field of the generator. For this reason, it is sometimes necessary to change the AC into DC as part of the distribution system. This changing of AC into DC is called rectification.

Single-Phase Rectifier. Figure 11-9 illustrates how a diode is used to convert AC into DC. The diode used in this circuit permits only the positive half-cycles of the AC sine wave to pass. For this reason, this circuit is called the half-wave rectifier. Half-wave rectification is accomplished because current is allowed to flow only when the anode terminal is positive with respect to the cathode. Current is not allowed to flow through the rectifier when the cathode is positive with respect to the anode.

The output voltage of the half-wave rectifier is considered pulsating DC, with half of the AC sine wave cut off. The rectifier will pass either the positive or negative half-cycle of the input AC sine wave, depending on the way the diode is connected into the circuit. Half-wave rectification is ineffi-

cient for most applications since one-half of the input since wave is not used.

To make use of both halves of the input AC sine wave, a full-wave rectifier circuit is used. Full-wave rectification can be obtained from a single-phase AC source by using two diodes with a center tap transformer or through a bridge rectifier circuit. Both of these circuits are illustrated in Figures 11-10 and 11-11.

The circuit, using two diodes, makes use of a transformer (Figure 11-10), with a tap at mid-point of the secondary. When voltage is induced in the secondary from point A to B (with A positive with respect to B) current will flow through half of the secondary from N to A through the load, then through rectifier number one. When the voltage across the secondary reverses during the next half-cycle of the AC sine wave, current will flow from N to B, through the load, then through rectifier number two. This procedure is repeated every single cycle of the AC input, allowing for a full-wave DC output as illustrated.

The bridge circuit illustrated in Figure 11-11 produces the same full-wave DC output. The bridge circuit requires four diodes but does not require a center-tapped transformer. This eliminates the need for a transformer. The bridge circuit is also more efficient than the center-tapped circuit because the bridge diodes need block only half as much reverse voltage as the center-tapped diodes for the same output voltage.

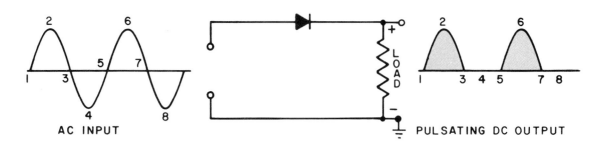

Figure 11-9 A half-wave rectifier can be used to convert alternating current to pulsating DC current.

In this circuit (Figure 11-12), when the AC supply voltage is positive (+) at point A and negative (−) at point B, current flows from point B to C and C through the load to D and from D to A. This means the current is passing through two diodes (D2 and D1). When the AC supply voltage is positive (+) at point B, and negative (−) at point A, current flows from point A to C and C through

the load to D and from D to B. This also allows the current to pass through two diodes (D4 and D3).

The output of a full-wave rectifier is pulsating DC and must be filtered or smoothed out before it can be used in most electronic equipment. This filtering is done by a filter circuit connected to the output of the rectifier circuit. This filter circuit usually consists of one or more capacitors, induc-

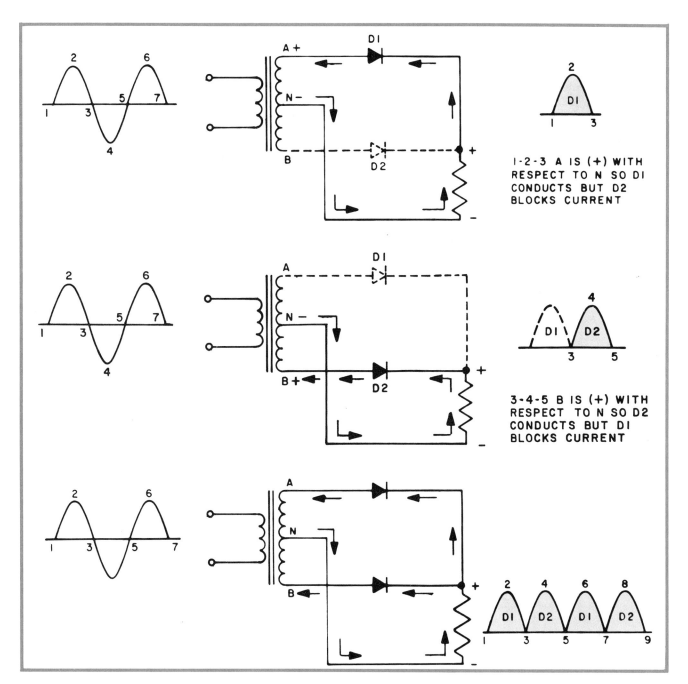

Figure 11-10 Operation of a transformer's center-tapped full-wave rectifier.

tors, or resistors connected in different combinations. The choice of a filter circuit is determined by the type of load (how much ripple it can take), cost and available space.

Figure 11-12 illustrates a filtered DC output (as compared to a pulsating output). The purpose of the filter is to eliminate the pulsations and thereby produce smooth direct current of constant intensity. This is accomplished because the pulsating voltage no longer drops to zero at the end of each pulsation. This results in the average voltage delivered by the rectifier circuit being higher in a filtered circuit. Thus, the purpose of a filter is to smooth out and increase the DC voltage output of the circuit.

Three-Phase Rectifier. A DC output can also be supplied from a three-phase power source. The three-phase power source has an advantage over the single-phase power source when used to rectify AC into DC. The advantage of using three-phase power is that it is possible to obtain a smooth DC output without the use of a filter circuit. This is possible in a three-phase circuit because when any one phase goes negative, at least one of the other phases is going positive. The result is a relatively smooth output without any filtering.

Figure 11-13 illustrates a three-phase rectifier circuit with the resulting DC output waveform. This circuit uses three rectifier diodes connected to a wye circuit with a neutral tap. Each diode con-

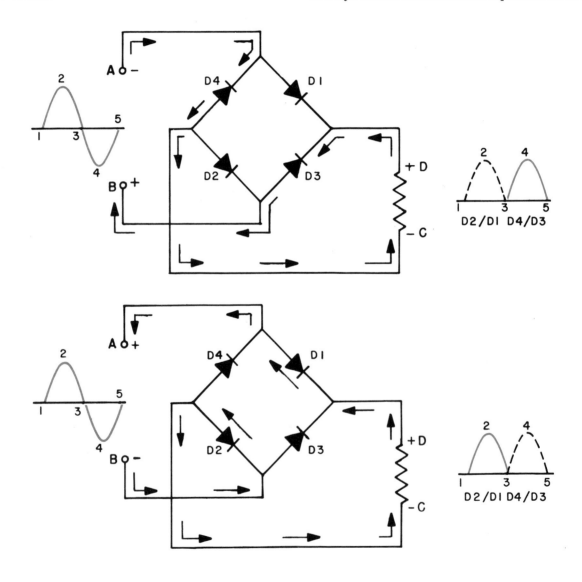

Figure 11-11 Operation of a full-wave bridge rectifier.

ducts in succession while the remaining two are blocking. As shown in Figure 11-13, the output voltage never goes below a certain voltage level. This circuit delivers the same smooth DC output as a filtered single-phase bridge circuit.

ZENER DIODE

The zener diode looks similar to other silicon diodes (See Figure 11-50). Its purpose is to act as a *voltage regulator* either by itself or in conjunction with other semiconductor devices. In a schematic, the zener diode symbol differs from a standard diode symbol in that the normally vertical cathode line is bent slightly at each end.

The zener diode is unique because it is most often used to conduct current under reverse bias conditions. Standard diodes usually conduct in forward bias and can be destroyed if the reverse voltage or bias is exceeded. Because the zener diode usually operates in reverse breakdown, it is often called an *avalanche diode.*

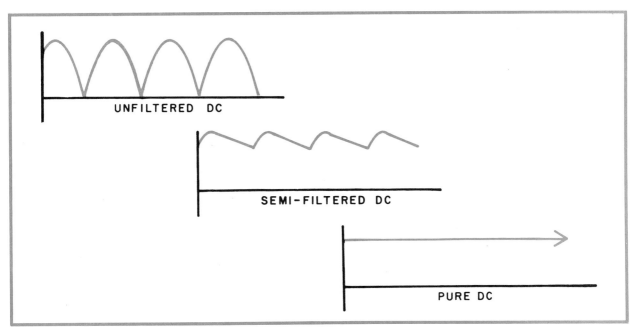

Figure 11-12 Filtered direct current eliminates pulsations and provides direct current at a constant intensity.

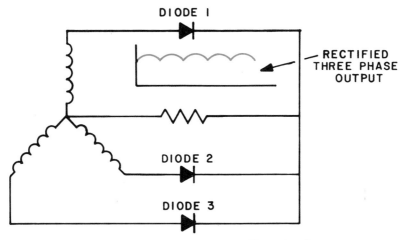

Figure 11-13 A three-phase rectifier circuit uses three rectifier diodes connected to a wye circuit with a neutral tap.

Zener Diode Operation

The operation of a zener diode is best understood through the use of an operating characteristic curve. See Figure 11-14. The overall forward and reverse characteristics of the zener diode are similar to the standard diode. When a source voltage is applied to the zener diode in the forward direction, there is a breakover voltage and forward current. When a source voltage is applied to the zener diode in the reverse direction, the current remains very low until the reverse voltage reaches reverse breakdown (zener breakdown). The zener diode then conducts heavily or avalanches. Reverse current flow through a zener diode must by limited by a resistor or other device to prevent diode destruction. The maximum current that may flow through a zener diode is determined by diode size. Like the forward voltage drop of a rectifier diode, the reverse voltage drop or zener voltage of a zener diode remains essentially constant, despite large current fluctuations.

The zener diode is capable of being a constant voltage source because of the resistance changes that take place within the PN junction.

When a source of voltage is applied to the zener diode in the reverse direction, the resistance of the PN junction remains high and should produce only leakage currents in the microampere range. However, as the reverse voltage is increased, the PN junction reaches a critical voltage and the zener diode avalanches. As the avalanche voltage is reached, the normally high resistance of the PN junction drops to a low value and the current increases rapidly. The current is limited generally by a circuit resistor or load resistance R_L.

THERMISTOR

A *thermistor* is a thermally sensitive resistor. The resistance of a thermistor changes with a change in temperature. See Figure 11-15. As a match is placed under the thermistor, its resistance decreases and current flow increases. When the match is removed, the resistance increases to its original state (resistance value).

The operation of a thermistor is based on the electron-hole theory. As the temperature of the semiconductor increases, the generation of electron-hole pairs increases due to thermal agitation. Increased electron-hole pairs causes a drop in resistance.

Solid State Control Devices

Thermistors are popular because of their small size. They can be mounted in places that are inaccessible to other temperature-sensing devices. Thermistors may be directly heated or indirectly heated. Figure 11-50 shows the schematic symbol for both.

Photoconductive Cell

The schematic symbol for a photoconductive cell is similar to that of a thermistor. See Figure 11-50. The difference is the absence of the letter *t* and the addition of two arrows. The photoconductive cell is formed with a thin layer of semiconductor material such as cadmium sulfide or cadmium selenide deposited on a suitable insulator. Leads are attached to the semiconductor material and the entire assembly is hermetically sealed with glass. The transparency of the glass allows light to reach the semiconductor material.

For maximum current-carrying capacity, the photoconductive cell is manufactured with a short conduction path having a large cross-sectional area.

Figure 11-14 The overall forward and reverse characteristics of the zener diode are similar to the standard silicon diode. However, the zener diode normally operates in reverse breakdown and is often called an avalanche diode.

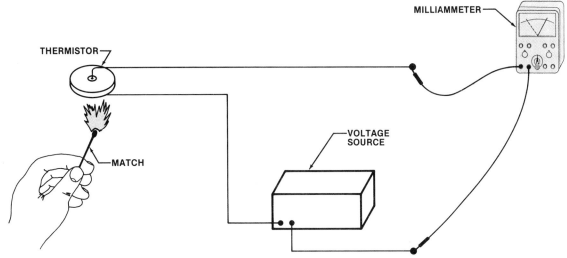

Figure 11-15 A thermistor is a transducer that acts as a thermally sensitive resistor. Its resistance changes with a change in temperature. An increase in temperature causes a decrease in resistance and an increase in current.

Photovoltaic Cell

The function of a photovoltaic cell (solar cell) is to convert solar energy to electrical energy. A photovoltaic cell is sensitive to light and it produces a voltage without an external source. There are several different types of photovoltaic cells. The schematic symbol for the photovoltaic cell is shown in Figure 11-50. The symbol indicates that the device is equivalent to a single-cell voltage source like those found in batteries.

The use of photovoltaic cells is increasing. As they are perfected and decrease in cost, their use as a remote power source will become more popular. Many manufacturers are designing into their products the use of photovoltaic cells on individual and multi-cell applications.

Photovoltaic Cell Operation. The photovoltaic cell generates energy by using a PN junction to convert light energy into electrical energy. See Figure 11-16. It produces a potential difference between a pair of terminals only when exposed to light.

At the junction of N-type material and P-type material, some recombination of the electrons and holes occurs, but the junction itself acts as a barrier between the two charges. The electrical field at the junction maintains the negative charges on the N-type material side and the holes or positive charges on the P-type material side.

If the load is connected across the PN junction, current flows with the light acting as a generator. When current flows through the load, electron-hole pairs formed by light energy recombine and return to the normal condition prior to the application of light. Consequently, there is no loss or addition of electrons to the silicon during the process of converting light energy to electrical energy. The photovoltaic cell should have no limit to its life span, provided it is not damaged.

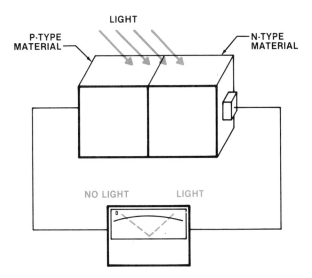

Figure 11-16 The photovoltaic cell uses a PN junction to convert light energy into electrical energy.

Photoconductive Diode

A *photoconductive diode (photodiode)* is internally much the same as a regular semiconductor diode. The primary difference is the addition of a lens in the housing for focusing light on the PN junction area. See Figure 11-50. The schematic symbol for a photoconductive diode is also shown along with one of the typical housings for photodiodes.

Photoconductive Diode Operation. With the photodiode, the conductive properties change when light strikes the surface of the PN junction. Without light, the resistance of the photodiode is high. When it is exposed to light, the resistance reduces proportionately.

Hall Effect Sensors

The Hall effect principle is shown in Figure 11-17. A constant control current passes through a thin strip of semiconductor material (*Hall generator*). When a permanent magnet is brought near, a small voltage, called Hall voltage, appears at the contacts that are placed across the narrow dimension of the strip. As the magnet is removed, the Hall voltage reduces to zero. Thus, the Hall voltage is dependent on the presence of a magnetic field and on the current flowing through the Hall generator. If either the current or the magnetic field is removed, the output of the Hall generator would be zero. In most Hall effect sensors, the control current is held constant the flux density is changed by movement of a permanent magnet.

NOTE: The Hall generator must be combined with an association of electronic circuits to form a Hall effect sensor. Since all of this circuitry is usually on an integrated circuit (IC), the Hall effect sensor can be considered a single device with a voltage output.

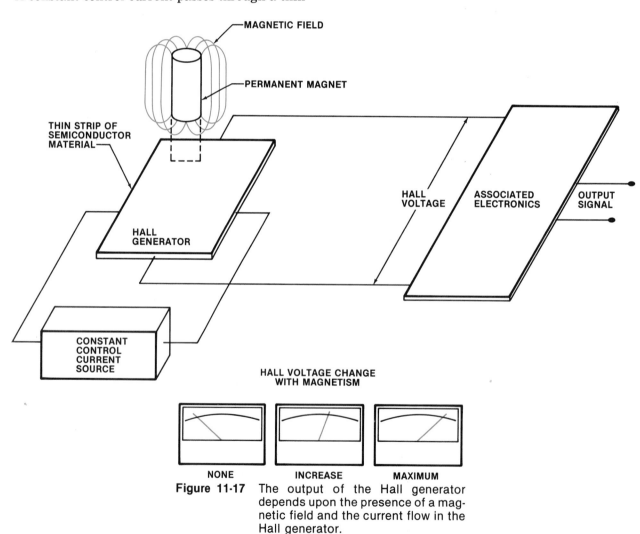

Figure 11-17 The output of the Hall generator depends upon the presence of a magnetic field and the current flow in the Hall generator.

SOLID STATE PRESSURE SENSOR

A solid state *pressure sensor* is a transducer that changes resistance with a corresponding change in pressure. See Figure 11-50. The pressure sensor is designed to activate or deactivate when its resistance reaches a predetermined value. The pressure sensor is used for high or low pressure control, depending on the switching circuit design. It is suited for a wide variety of pressure measurements on compressors, pumps, and other similar equipment.

A pressure sensor can detect low pressure, high pressure, or it can trigger a relief valve. Because a pressure sensor is extremely rugged, it is also used to measure compression in various types of engines.

LIGHT EMITTING DIODES

The *light emitting diode (LED)* produces light through the use of semiconductor materials. See Figure 11-50. A diode junction can emit light when an electrical current is present. An electrical current produces light energy because electrons and holes are forced to recombine.

The energy level of an electron as it passes through the junction of a semiconductor diode can be shown graphically. See Figure 11-18. To get

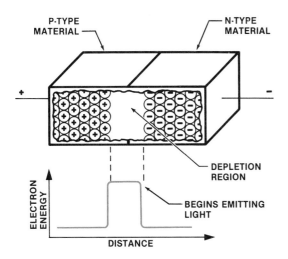

Figure 11-18 As the electron moves across the depletion region, the electron gives up its extra kinetic energy. The extra energy is converted to light.

through the depletion region, the electron must acquire some additional energy. This additional energy comes from the positive field of the anode. If the field is not strong enough, the electron will not get through the depletion region and no light will be emitted. For a standard silicon diode, a minimum of 0.6 volts must be present before the diode will conduct. For a germanium diode, 0.3 volts must be present before the diode will conduct. Most LED manufacturers make a larger depletion region that requires 1.5 volts for the electron to get across the depletion region. As the electron moves across the depletion region, it gives up its extra kinetic energy. The extra energy is converted to light.

LED Construction

Manufacturers of LEDs generally use a combination of gallium and arsenic with silicon or germanium to construct semiconductors. By adding and adjusting other impurities to the base semiconductor, different wavelengths of light can be produced. LEDs are capable of producing a light that is not visible to the human eye, called *infrared light,* or it may emit a visible red or green light. If other colors are desirable, the plastic lens may be of a different color.

Like standard semiconductor diodes, there must be a method for determining which end of an LED is the anode and which end is the cathode. The cathode lead is identified by the flat side of the device, or it may have a notch cut into the ridge.

A colored plastic lens focuses the light produced at the junction of the LED. Without the lens, the small amount of light produced at the junction would be diffused, and it would be virtually unusable as a light source. The size and shape of the LED package determines how it must be positioned for proper viewing.

The schematic symbol for an LED is exactly like that of a photodiode, but the arrows point away from the diode. The LED must be forward biased, and a current-limiting resistor is usually present to protect the LED from excessive current.

TRANSISTORS

The two types of transistors are PNP and NPN. Figure 11-50 shows their basic construction and their schematic symbol. The PNP transistor is formed by sandwiching a thin layer of N-type material between two layers of P-type material. The

NPN transistor is formed by sandwiching a thin layer of P-type material between two layers of N-type material. See Figure 11-19. Transistors are

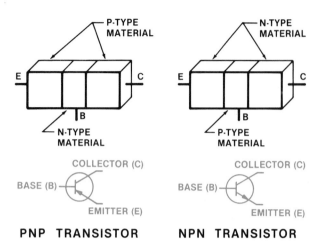

PNP TRANSISTOR   **NPN TRANSISTOR**

Figure 11-19 The two types of transistors are PNP and NPN.

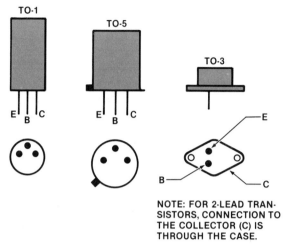

NOTE: FOR 2-LEAD TRANSISTORS, CONNECTION TO THE COLLECTOR (C) IS THROUGH THE CASE.

Figure 11-20 When a specific shaped transistor must be used, its TO number is used as a reference.

three-terminal devices. Their terminals are the *emitter (E)*, *base (B)*, and *collector (C)*. The emitter, base, and collector are located in the same place for both symbols. The only difference is the direction in which the emitter arrow points. However, in both cases, the arrow points from the P-type material toward the N-type material. (The term *bipolar* is often used when describing a transistor. It means that both holes and electrons are used as internal carriers for maintaining current flow in the transistor.)

Transistor Terminal Arrangements

Transistors are manufactured with either two or three leads extending from their case. These packages are accepted industry-wide, no matter which company manufactures them. When a specific shaped transistor must be used, its *transistor outline (TO) number* is used as a reference. See Figure 11-20. TO numbers are determined by individual manufacturers.

NOTE: The bottom view of transistor TO-3 shows only two leads (terminals). Frequently, transistors use the metal case as the collector-pin lead.

Spacing can also be used to identify leads. Usually, the emitter and base leads are close together and the collector lead is farther away. The base lead is usually in the middle. A transistor with an *index pin* must be viewed from the bottom. The

leads are identified in a clockwise direction from the index pin. The emitter is closest to the index pin.

NOTE: For detailed information on transistor construction and identification, refer to a transistor manual or to the manufacturer's specification sheets.

Biasing Transistor Junctions

In any transistor circuit, the *base-emitter junction* must always be forward biased and the *base-collector junction* must always be reverse biased. Figure 11-21 shows the base-emitter junction of an NPN transistor. The external voltage (bias voltage) is connected so that the positive terminal connects to the P-type material (the base), and the negative terminal connects to the N-type material (the emit-

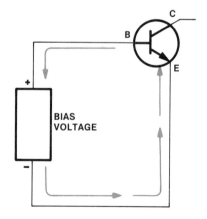

Figure 11-21 In any transistor circuit, the base-emitter junction must always be forward biased.

ter). This arrangement forward biases the base-emitter junction. Current flows in the external circuit as indicated by the arrows. The action that takes place is the same as the action that occurs for the forward biased semiconductor diode.

Figure 11-22 illustrates the base-collector junction of an NPN transistor. The external voltage is connected so that the negative terminal connects to the P-type material (the base) and the positive terminal connects to the N-type material (the collector). This arrangement reverse biases the base-collector junction. Only a very small current (leakage current) will flow in the external circuit as indicated by the dashed arrows. The action that takes place is the same as the action that occurs for a semiconductor diode with reverse bias applied.

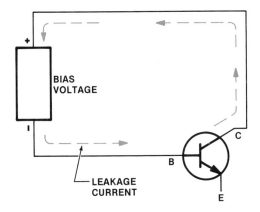

Figure 11-22 In any transistor circuit the base-collector junction must always be reverse biased.

Transistor Current Flow
Individual PN junctions can be used in combination with two bias arrangements. See Figure 11-23. The base-emitter junction is forward biased while the base-collector junction is reverse biased. This circuit arrangement results in an entirely different current path than the path that occurs with the individual circuits only. The heavy arrows indicate that the main current path is directly through the transistor. The main current path has a small amount of current through the base as indicated by the thinner arrows.

Figure 11-23 also shows that the forward bias of the emitter-base circuit causes the emitter to inject electrons into the depletion region between the emitter and the base. Because the base B is thin

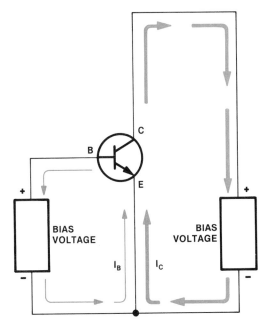

Figure 11-23 With both junctions biased, an entirely different current path is created than with the junctions biased separately. Because the base is so thin, the more positive potential of the collector pulls the electrons through the thin base material. This results in 95% of the current in the collector and 5% going through the base.

(less than .001″ for most transistors), the much more positive potential of the collector C pulls the electrons through the thin base. As a result, the greater percentage (95%) of the available free electrons (I_C) from the emitter passes directly through the base into the N-type material, which is the collector of the transistor.

Control of Base Current
The base current I_B is a critical factor in determining the amount of current flow in a transistor. It is critical because the forward biased junction has a very low resistance and could be destroyed by heavy current flow. Therefore, the base current must be limited and controlled.

Transistor as a DC Switch
The primary reason for the rapid development of the transistor was to replace mechanical switching. The transistor has no moving parts and can switch ON and OFF quickly.

Mechanical switches have two conditions: open and closed, or ON and OFF. The switch has a very high resistance when open and a very low resistance when closed. A transistor can be made to operate much like a switch. For example, it can be used to turn a pilot light PL1 ON or OFF. See Figure 11-24. In this circuit, the resistance between the collector C and the emitter E is determined by the current flow between the base B and E. When no current flows between B and E, the collector-to-emitter resistance is high, like that of an open switch. See Figure 11-24, left. The pilot light does not glow because there is no current flow.

If a small current does flow between B and E, the collector-to-emitter resistance is reduced to a very low value, like that of a closed switch. See Figure 11-26, right. Therefore, the pilot light will be switched ON and begin to glow.

A transistor switched ON is usually operating in the *saturation region*. The saturation region is the maximum current that can flow in the transistor circuit. At saturation, the collector resistance is considered zero and the current is limited only by the resistance of the load.

When the circuit reaches saturation, the resistance of pilot light PL1 is the only current-limiting factor in the circuit. When the transistor is switched OFF, it is operating in the *cutoff region*. The cutoff region is the point at which the transistor is turned OFF and no current flows. At cutoff, all the voltage is across the open switch (transistor) and the collector-to-emitter voltage V_{CE} is equal to the supply voltage V_{CC}.

Transistor as an AC Amplifier

Transistors can be used as AC amplification devices as well as DC switching devices. Amplification is the process of taking a small signal and making it larger. In control systems, transistor AC amplifiers are used to increase small signal currents and voltages so they can do useful work. Amplification is accomplished by using a small signal to control the energy output from a larger source, such as a power supply.

AMPLIFIER GAIN

The primary objective of the amplifier is to produce *gain*. Gain is a ratio of the amplitude of the output signal to the amplitude of the input signal. In determining gain, the amplifier can be thought of as a "black box." A signal applied to the input of the black box gives the output of the box. Mathematically, gain can be found by dividing output by input:

$$Gain = \frac{Output}{Input}$$

Sometimes, a single amplifier does not provide enough gain to increase the amplitude for the output signal needed. In such a case, two or more amplifiers can be used to obtain the gain required. Amplifiers connected in this manner are called *cascaded amplifiers*. For many amplifiers, gain is in the hundreds, and even thousands.

Gain is a ratio of output to input and has no unit of measure, such as volts or amps, attached to it. Therefore, the term gain is used to describe current gain, voltage gain, and power gain. In each case, the output is merely being compared to the input.

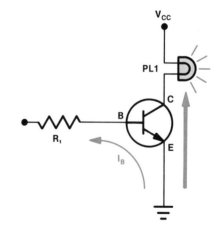

WITHOUT I_B
NO CURRENT FLOW

WITH I_B
CURRENT FLOW

Figure 11-24 With no emitter-base current I_B, the light remains OFF (left). With I_B flow-ing, the transistor is switched ON delivering current to the lamp (right).

TYPES OF TRANSISTOR AMPLIFIERS

The three basic types of transistor amplifiers are the *common-emitter, common-base,* and *common-collector.* See Figure 11-25. The amplifier is named after the transistor connection that is common to both the input and the load. For example, the input of a common-emitter circuit is across the base and emitter, while the load is across the collector and emitter. Thus, the emitter is common to both input and load.

Classes of Operation

The four main classes of operation for an amplifier are designated by the letters A, B, AB, and C. In each case, the letter is a reference to the level of an amplifier operation in relation to the cutoff condition. The cutoff condition is the point at which all collector current is stopped by the absence of base current.

SILICON CONTROLLED RECTIFIER (SCR)

The *silicon controlled rectifier (SCR)* is a four-layer (PNPN) semiconductor device. It uses three electrodes for normal operation. See Figure 11-26. The three electrodes are the anode, cathode, and gate. The anode and cathode of the SCR are similar to the anode and cathode of an ordinary semiconductor diode. The gate serves as the control point for the SCR.

The SCR differs from the ordinary semiconductor diode in that it will not pass significant current, even when forward biased, unless the anode voltage equals or exceeds the *forward breakover voltage.* However, when forward breakover voltage is reached, the SCR switches ON and becomes highly conductive. The SCR is unique because the gate current is used to reduce the level of breakover voltage necessary for the SCR to conduct or fire.

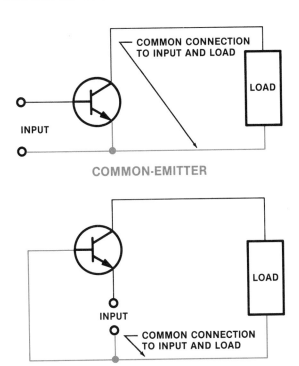

COMMON-EMITTER

COMMON-BASE

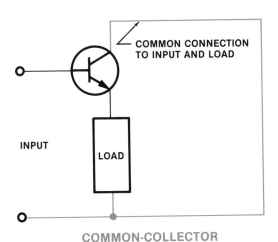

COMMON-COLLECTOR

Figure 11-25 The three basic types of transistors and amplifiers are the common-emitter, common-base, and common-collector. The transistor is named after the connection that is common to both the input and the load.

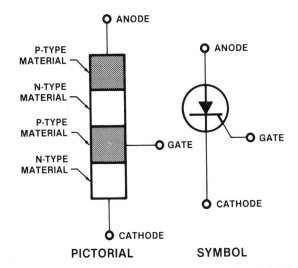

PICTORIAL   SYMBOL

Figure 11-26 The silicon controlled rectifier (SCR) is a four-layer semiconductor device. Its three electrodes are the anode, cathode, and gate.

There are many case styles for the SCR. See Figure 11-50. Low-current SCRs can operate with an anode current of less than one amp. High-current SCRs can handle load currents in the hundreds of amperes. The size of an SCR increases with an increase in its current rating.

SCR CHARACTERISTIC CURVE

Figure 11-27 shows the voltage-current characteristic curve of an SCR when the gate is not connected. In reverse bias, the SCR operates like a regular semiconductor diode. With reverse bias, there is a small current until avalanche is reached. After avalanche is reached, the current increases dramatically. This current can cause damage if thermal runaway begins.

When the SCR is forward biased, there is also a small forward leakage current called the *forward blocking currrent*. This current stays relatively constant until the forward breakover voltage is reached. At that point, the current increases rapidly and is often called the forward avalanche region. In the forward avalanche region, the resistance of the SCR is very small. The SCR acts much like a closed switch and the current is limited only by the external load resistance. A short in the load circuit of an SCR can destroy the SCR if overload protection is not adequate.

Operating States of an SCR

The SCR operates much like a mechanical switch. It is either ON or OFF. When the applied voltage is above the forward breakover voltage (V_{BRF}), the SCR fires, or is ON. The SCR will remain ON as long as the current stays above a certain value called the *holding current*. When voltage across the SCR drops to a value too low to maintain the holding current, it will return to its OFF state.

Gate Control of Forward Breakover Voltage

When the gate is forward biased and current begins to flow in the gate-cathode junction, the value of forward breakover voltage can be reduced. Increasing values of forward bias can be used to reduce the amount of forward breakover voltage (V_{BRF}) necessary to get the SCR to conduct.

Once the SCR has been turned ON by the gate current, the gate current loses control of the SCR forward current. Even if the gate current is completely removed, the SCR will remain ON until the anode voltage has been removed. The SCR will also remain ON until the anode voltage has been significantly reduced to a level where the current is not large enough to maintain the proper level of holding current.

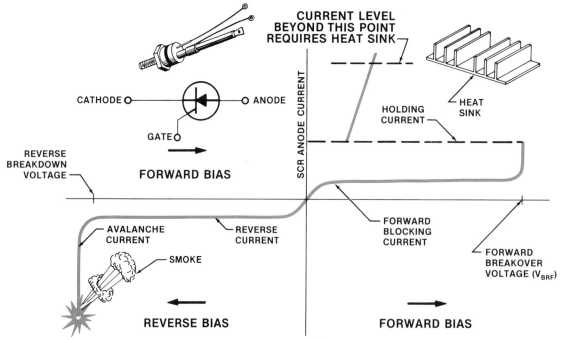

Figure 11-27 The voltage-current characteristic curve of an SCR shows that the SCR operates much like a regular diode in reverse bias. In forward bias, a certain value of forward breakover voltage must be reached before the SCR will conduct.

Process Control Using SCR

The SCR can be used in circuits to provide heat control. For example, it can bring a chemical mixture stored in a vat up to a specific temperature and maintain that temperature. See Figure 11-28. With the proper circuitry, the temperature of the mixture can be precisely controlled. Using a bridge circuit, the temperature can be maintained with 1 °F over a temperature range of 20 °F to 150 °F.

In the circuit of Figure 11-28, transformer T1 has two secondary windings, W1 and W2. W1 furnishes voltage through the SCR to relay coil K1. W2 furnishes AC voltage to the gate circuit of the SCR. Primary control over this circuit is accomplished through the use of the bridge circuit. The bridge circuit is formed by thermistor R1, fixed resistors R2 and R3, and potentiometer R4. Resistor R5 is a current-limiting resistor used to protect the bridge circuit. The fuse is used to protect the primary of the transformer.

When the resistance of R1 equals the resistance setting on R4, the bridge is balanced. None of the AC voltage introduced into the bridge by winding W2 is applied to the gate of the SCR. Hence, the relay coil K1 remains de-energized and its normally closed contacts apply power to the heating elements.

If the temperature increases above a preset level, the resistance of thermistor R1 decreases. The bridge becomes unbalanced such that a current flows to the gate of the SCR while the anode of the SCR is still positive. This turns ON the SCR and energizes the relay coil K1, thereby switching power from the load through the relay contact. If the temperature falls below the preset temperature setting, R1 will unbalance the bridge in the opposite direction. Therefore, a negative signal is applied to the gate of the SCR when the anode of the SCR is positive. The negative signal stops the SCR from conducting and allows current to continue to flow to the heating elements.

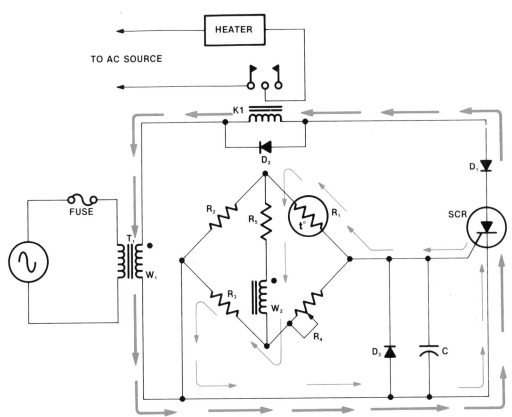

Figure 11-28 When the temperature increases, the resistance of thermistor R1 decreases, unbalancing the bridge so that a gate current flows to the SCR while its anode is still positive. This turns ON the SCR and energizes the relay coil K1, thereby disconnecting power from the load.

TRIAC

A *triac* is a three-electrode AC semiconductor switch. It is triggered into conduction in both directions by a gate signal in a manner similar to the action of an SCR. The triac was developed to provide a means for producing improved controls for AC power. Triacs are available in a variety of packaging arrangements. See Figure 11-50. They can handle a wide range of amperages and voltages. Triacs generally have relatively low current capabilities compared to SCRs. Triacs are usually limited to less than 50 amps and cannot replace SCRs in high-current applications.

Triac Construction

Although triacs and SCRs look alike, their schematic symbols are different. See Figure 11-50. The terminals of the triac are the gate, Terminal 1 (*T1*), and Terminal 2 (*T2*). There is no designation of anode and cathode.

Current may flow in either direction through the main switch terminals, T2 and T1. Terminal T2 is the case or metal-mounting tab to which the heat sink can be attached. The structure of the triac is complex. However, for all practical purposes, the triac can be considered two NPN switches sandwiched together on a single N-type material wafer.

Triac Operation

The triac blocks current in either direction between Terminal 1 (T1) and Terminal 2 (T2). A triac can be triggered into conduction in either direction by a momentary pulse in either direction supplied to the gate. The triac operates much like a pair of SCRs connected in a reverse parallel arrangement. If the appropriate signal is applied to the triac gate, it will conduct.

The characteristic curve of Figure 11-29 shows how a triac may be triggered into conduction. The triac remains OFF until the gate is triggered at point A. At point A, the trigger circuit pulses the gate and turns ON the triac, allowing current to flow. At point B, the forward current is reduced to

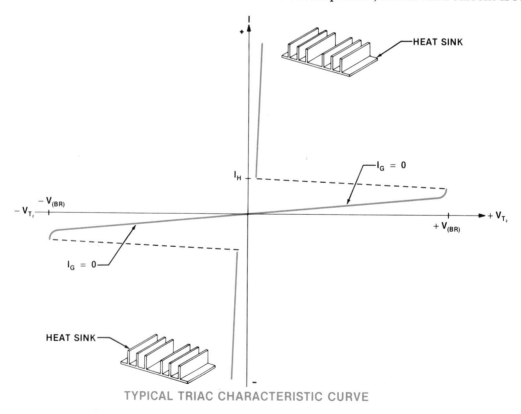

TYPICAL TRIAC CHARACTERISTIC CURVE

Figure 11-29 The characteristics of the triac are based on terminal 1 as the voltage reference point. The triac can be triggered into conduction in either direction by a gate current (I_G) in either direction.

zero, so the triac turns OFF. The trigger circuit can be designed to produce a pulse that varies at any point in the positive or negative half cycle. Therefore, the average current supplied to the load can vary.

One advantage of the triac is that virtually no power is wasted by being converted to heat. Heat is generated when current is impeded, not when current is switched OFF. The triac is either fully ON or fully OFF. It never partially limits current. Another important feature of the triac is the absence of a reverse breakdown condition of high voltages and high current, such as those found in diodes and SCRs. If the voltage across the triac goes too high, the triac will merely turn ON. When turned ON, the triac can conduct a reasonably high current.

UNIJUNCTION TRANSISTOR (UJT)

The *unijunction transistor (UJT)* consists of a bar of N-type material with a region of P-type material doped within the N-type material. The N-type material functions as the base and has two leads, Base 1 (B1) and Base 2 (B2). The lead extending from the P-type material is the emitter (E). The schematic symbol for a UJT is shown in Figure 11-50.

In the schematic symbol, the arrowhead represents the emitter (E). Although the leads are usually not labeled, they can be easily identified because the arrowhead always points to B1.

The UJT is used primarily as a triggering device because it serves as a step-up device between low-level signals and SCRs and triacs. Outputs from photocells, thermistors, and other transducers can be used to trigger UJTs, which in turn, fire SCRs and triacs. UJTs are also used in oscillators, timers, and voltage-current sensing applications.

UJT Biasing

In normal operation, B1 is negative, and a positive voltage is applied to B2. See Figure 11-30. The internal resistance between B1 and B2 will divide at the emitter (E), with approximately 60% of the resistance between E and B1. The remaining 40% of resistance is between E and B2. The net result is an internal voltage split. This split provides a positive voltage at the N-type material of the emitter junction, creating an emitter junction that is

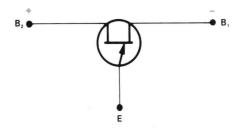

Figure 11-30 In normal operation, B1 is negative and a positive voltage is applied to B2.

reverse biased. As long as the emitter voltage remains less than the internal voltage, the emitter junction will remain reverse biased even at a very high resistance.

However, if the emitter voltage rises above this internal value, a dramatic change takes place. When the emitter voltage is greater than the internal value, the junction becomes forward biased. Also, the resistance between E and B1 drops rapidly to a very low value. The UJT characteristic curve shows the dramatic change in voltage due to this resistance change. See Figure 11-31.

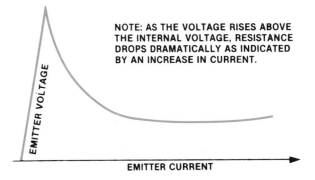

NOTE: AS THE VOLTAGE RISES ABOVE THE INTERNAL VOLTAGE, RESISTANCE DROPS DRAMATICALLY AS INDICATED BY AN INCREASE IN CURRENT.

EMITTER VOLTAGE

EMITTER CURRENT

Figure 11-31 When the emitter voltage rises above the internal voltage, the junction becomes forward biased and the resistance between the emitter and B1 drops rapidly to a very low value.

DIAC

Figure 11-50 shows a typical packaging of a *diac* and its schematic symbol. This diac is a three-layer, *bidirectional* device. Unlike the transistor, the two junctions are heavily and equally doped. Each junction is almost identical to the other.

Electrically, the diac acts much like two zener diodes that are series connected in opposite directions. The diac is used primarily as a triggering device. It accomplishes this through the use of its negative resistance characteristic. (Negative resistance characteristic means the current decreases with an increase of applied voltage.) The diac has negative resistance because it does not conduct current until the voltage across it reaches breakover voltage. See Figure 11-32. When a positive or negative voltage reaches the breakover voltage, the diac rapidly switches from a high-resistance state to a low-resistance state. Since the diac is a bidirectional device, it is ideal for controlling triacs, which are also bidirectional.

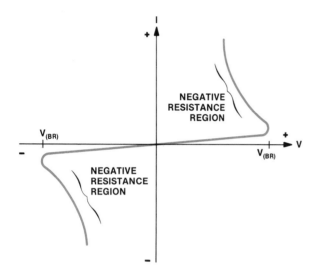

Figure 11-32 The diac is bidirectional and is used primarily as a triggering device. When a positive or negative voltage reaches the breakover voltage, the diac rapidly switches from a high-resistance state to a low-resistance state.

INTEGRATED CIRCUITS

Integrated circuits (ICs) are popular because they provide a complete circuit function in one small semiconductor package. (ICs are often called chips, which are actually a component part of the IC.) See Figure 11-33. Although many processes have been developed to create these devices, the end result has always been a totally enclosed system with specific inputs and specific outputs.

Because of the nature of ICs, the technician must approach ICs in an entirely different manner from individual solid state components. An IC is a system within a system. The entire system of an

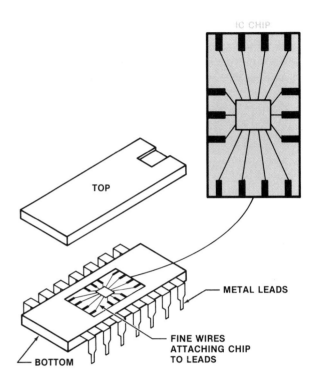

Figure 11-33 An integrated circuit (IC) provides a complete circuit function in one semiconductor package.

IC and what it does must first be understood. Data books and manufacturer's specification sheets can usually provide this information. When these are not available, the inputs and outputs of the system must be studied by using meters and an oscilloscope. Troubleshooting ICs also requires knowledge of how the system functions, and what the input and output should be. Since ICs cannot be repaired, they must be replaced if they are defective.

IC PACKAGES

Figure 11-50 shows various IC packaging. Their shapes and sizes range from standard transistor shapes, such as the TO-5 packages, to the latest in *large-scale integration (LSI)*. (Metal-oxide substrate, MOS, is a type of LSI.) When space is a premium, there are even ICs designed with *flat-pack* construction.

The *dual-in-line package (DIP)* with 14, 16, or 24 pins is the most widely used configuration. The *mini-DIP* is a smaller dual-in-line package with 8 pins. A modified TO-5 is available with 8, 10, or 12 pins.

The housings for ICs may be metal, plastic, or ceramic. Ceramic is used in applications where high temperatures may be a factor.

The appearance of these devices varies and may give an initial impression of being one particular type of device. Therefore, it is essential that part numbers and manufacturer's literature be reviewed before working on the device.

PIN NUMBERING SYSTEM

All manufacturers use a standardized pin numbering system for their devices. When unsure about pin numbering patterns, consult the manufacturer's data sheets.

Dual-in-line packages and flat-packs have index marks and notches at the top for reference. Before removing an IC, note where the index mark is in relation to the board or socket to aid in installation of the unit. The numbering of the pins is always the same. The notch is at the top of the chip. To the left of the notch is a dot that is in line with pin 1. The pins are numbered counterclockwise around the chip.

OPERATIONAL AMPLIFIER (OP AMP)

The *operational amplifier (OP amp)* is one of the most widely used ICs. An OP amp is a very high-gain, directly coupled amplifier that uses external feedback to control response characteristics. An example of this feedback control is gain. The gain of OP amp can be controlled externally by connecting feedback resistors between the output and input. A number of different amplifier applications can be achieved by selecting different feedback components and combinations. With the right component combinations, gains of 500,000 to 1,000,000 are common.

OP Amp Schematic Symbols

The schematic symbol for an OP amp may be shown in two ways. See Figure 11-34. In each case, the two inputs of the OP amp are the inverting (−), and the non-inverting (+). The two inputs are usually drawn as shown, with the inverting input at the top. The exception to the inverting input being at the top is when it would complicate the schematic. In either case, the two inputs should be clearly identified by polarity symbols on the schematic symbol.

Internal OP Amp Operation

Internally, an OP amp has three major parts. See Figure 11-35. It consists of a high-impedance differential amplifier, a high-gain stage, and a low-output impedance power-output stage. The differential amplifier provides the wide bandwidth and the high impedance. The high-gain stage boosts the signal. The power-output stage isolates the gain stage from the load and provides for the power output.

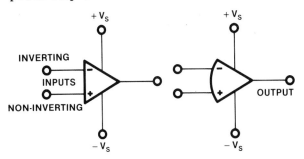

Figure 11-34 The schematic for an OP amp can be shown in two ways.

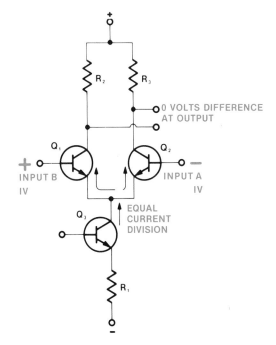

Figure 11-35 In this circuit, currents to the emitter-coupled transistors (Q_1 and Q_2) are supplied by the source Q3. As long as the two input voltages, A and B, are either zero or equal in amplitude and polarity, the amplifier is balanced because the collector currents are equal. When balanced, a zero voltage difference exists between the two collectors.

The operation of the differential amplifier is unique. Currents to the emitter-coupled transistors Q1 and Q2 are supplied by the source Q3. When manufactured, the characteristics of Q1 and Q2, along with their biasing resistors (R1, R2, and R3), are closely matched to make them as equal as possible.

As long as the two input voltages, A and B, are either zero or equal in amplitude and polarity, the amplifier is balanced because the collector currents are equal. When balanced, zero voltage difference exists between the two collectors.

The sum of the emitter currents is always equal to the current supplied by Q3. Thus, if the input to one transistor causes it to draw more current, the current in the other decreases and the voltage difference between the two collectors changes in a differential manner. The differential swing, or output signal, with be greater than the simple variation than can be obtained from only one transistor. Each transistor amplifies in the opposite direction so that the total output signal is twice that of one transistor. This swing will then be amplified through the high-gain stage and matched to the load through the power-output stage. By changing the OP-amp to different configurations, they can be made into oscillators, pulse generators and level detectors.

555 OP Amp Timer

The 555 timer consists of a voltage-divider network (R1, R2, and R3), two comparators (Comp 1 and Comp 2), a flip-flop, two control transistors (Q1 and Q2), and a power output amplifier. See Figure 11-36.

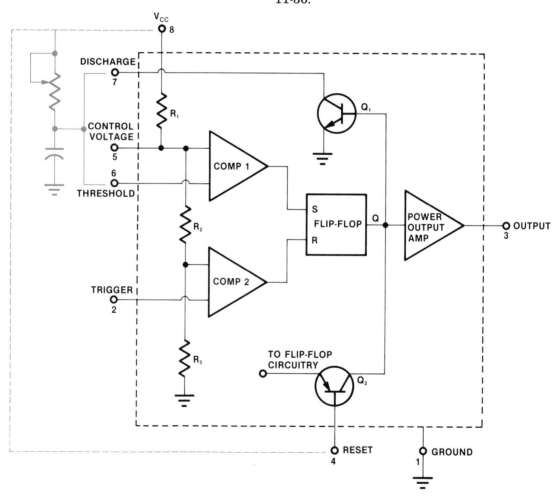

Figure 11-36 The 555 timer consists of a voltage divider network (R_1, and R_2, and R_3), two comparators (Comp 1 and Comp 2), a flip-flop, two control transistors (Q_1 and Q_2), and a power-output amplifier. The external circuitry of resistor R and capacitor C determines the amount of time delay for the 555 IC timer.

NOTE: A flip-flop is the electronic equivalent of a toggle switch. It has two outputs—one high and one low. When one is high, the other is low, and vice versa.

The comparators compare the input voltages to internal reference voltages that are created by the voltage divider, which consists of R1, R2, and R3. Since the resistors are of equal value, the reference voltage provided by two resistors is two-third of the supply voltage (V_{CC}). The other resistor provides one-third of V_{CC}. The value of V may change 9 volts, 12 volts, 15 volts, and so on, from chip to chip. However, the two-thirds/one-third ratio always remains the same. When the input voltage to either one of the comparators is higher than the reference voltage, the comparator goes into saturation and produces a signal that will trigger the flip-flop. In this IC circuit, the flip-flop has two inputs: S and R.

NOTE: The two comparators are feeding signals into the flip-flop. Comparator 1 is called the threshhold comparator and comparator 2 is called the trigger comparator. Comparator 1 is connected to the S input of the flip-flop and comparator 2 is connected to the R input of the flip-flop.

Whenever the voltage at S is positive and the voltage at R is zero, the output of the flip-flop is high. Whenever the voltage at S is zero and the voltage at R is positive, the output of the flip-flop is low. The output from the flip-flop at point Q is then applied to transistors Q1 and Q2 and to the output amplifier simultaneously. If the signal is high, Q1 will turn ON such that pin 7 (the discharge pin) will be grounded through the emitter-collector circuit. Q1 will then be in a position to turn ON pin 7 to ground through the emitter-collector circuit.

NOTE: Pin 7 is called the discharge pin because it is connected to the timing capacitor. When Q1 conducts, pin 7 is grounded and the capacitor can be discharged.

Referring back to Figure 11-36, the flip-flop signal is also applied to Q2. A signal to pin 4 can be used to reset the flip-flop. Pin 4 can be activated when a low-level voltage signal is applied. Once applied, this signal will override the output signal from the flip-flop. The reset pin (pin 4) will force the output of the flip-flop to go low, no matter what state the other inputs to the flip-flop are in.

The flip-flop signal is also applied to the power output amplifier. The power output amplifier boosts the signal and the 555 timer delivers up to 200 milliamps of current when operated at 15 volts. The output can be used to drive other transistor circuits, and even a small audio speaker. The output of the power output will always be an inverted signal compared to the input. If the input to the power-out-put amplifier is high, the output will be low. If the input is low, the output will be high.

DIGITAL ICs

The two basic types of electronic signals are the analog signal and the digital signal. These voltages and currents vary smoothly or continuously. A digital signal, on the other hand, is a series of pulses that changes levels between either the OFF state or ON state.

The analog and digital processes can be seen in a simple comparison between the light dimmer and light switch. A light dimmer varies the intensity of light from fully OFF to fully ON over a range of brightnesses. This is an example of an analog process. The standard light switch, on the other hand, has only two positions. It is either fully OFF or fully ON. This is an example of digital process. Electronic circuits that process these quickly changing pulses are digital or logic circuits.

When a technician becomes involved with a digital IC, the nature of electronics changes dramatically. The technician is leaving the "analog world" and is entering a "new world" controlled by *digital logic*. This section will not give the technician a complete understanding of digital electronics. Instead, it will give an orientation concerning the kinds of chips that are found in digital electronics. The four most common gates used in digital electronics are the *AND gate, OR gate, NAND gate,* and *NOR gate.*

AND Gate

The quad AND gate is one type of IC chip. See Figure 11-37. The manufacturer places four AND gates in one package—hence, the quad AND gate. By using the numbering system on the chip, any one, or all four AND gates may be utilized. In this case, voltage is applied to the circuit at pins 14 and 7.

The AND gate is a device with an output that is high only when both of its inputs are high. To connect to the AND gate, pins 1, 2, and 3 of the quad AND gate chip could be used. Pins 1 and 2 are the input and pin 3 is the output.

A practical application of an AND gate is in an elevator control circuit. See Figure 11-38. The

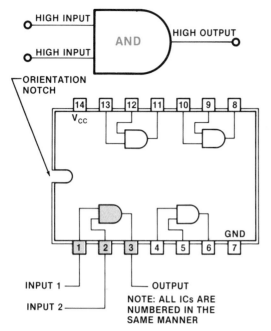

Figure 11-37 The AND gate is a device with an output that is high only when both of its inputs are high. Pins 1, 2, and 3 on the quad AND gate chip could be used to make connections to the AND gate.

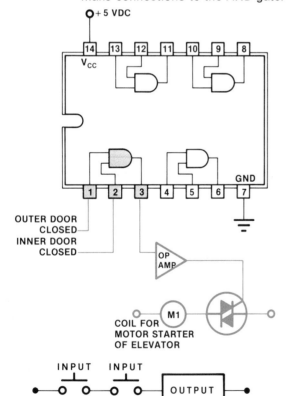

Figure 11-38 A practical application of an AND gate is in an elevator control circuit. The electrical equivalent of an AND gate is two electrical pushbuttons in series.

elevator cannot move unless the inner and outer doors are closed. Once both doors are closed, the output of the AND gate could be fed to an OP amp, which in turn, fires a triac that starts the elevator motor.

OR Gate

The OR gate is a device with an output that is high if either or both inputs are high. See Figure 11-39.

A practical application of an OR gate is in a burglar alarm circuit. If the front door or the back door is open, a signal is sent to the burglar alarm circuit. The electrical equivalent of an OR gate is two pushbuttons in parallel.

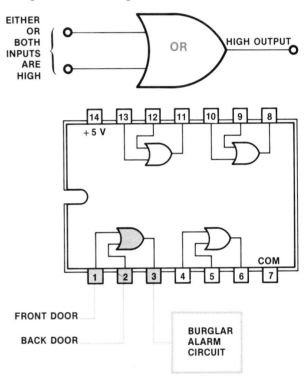

Figure 11-39 The OR gate is a device with an output that is high if either or both inputs are high. A practical application of an OR gate is in a burglar alarm circuit.

NOR Gate

The NOR (NOT-OR) gate is the same as an inverted OR function. Thus, the NOR gate provides a low output if either or both inputs are high. The NOR gate is represented by the OR gate symbol, followed by a small circle indicating an inversion of the output. See Figure 11-40. The NOR gate is a universal building block of digital logic. It is usually used in conjunction with other elements to implement more complex logic functions. NOR gates are also available in quad IC packaging.

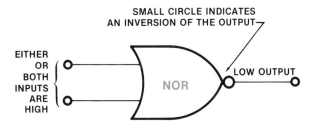

Figure 11-40 The NOR (NOT-OR) gate is the same as an inverted OR function. Thus, the NOR gate provides a low output if either or both inputs are high.

NAND Gate

The NAND (NOT-AND) gate is an inverted AND function. Each low output of the AND function is made high, and each high output is made low. The output is low only if both inputs are high. The NAND gate is represented by the AND symbol followed by a small circle indicating an inversion of the output. See Figure 11-41.

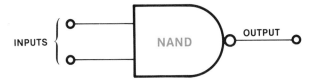

Figure 11-41 The NAND (NOT-AND) gate is an inverted AND function. Each low output of the AND function is made high and each high output is made low. The output is low if and only if both inputs are high.

The NAND gate, like the NOR gate, is a universal building block of digital logic. It is usually used in conjunction with other elements to implement more complex logic functions. NAND gates are also available in quad IC packaging.

FIBER OPTICS

Fiber optics is a technology that uses a thin flexible glass or plastic optical fiber to transmit light. Fiber optics is most commonly used as a transmission link. As a link, it connects two electronic circuits consisting of a transmitter and a receiver. See Figure 11-42. The central part of the transmitter is its source. The source consists of a light emitting diode (LED), infrared emitting diode (IRED), or laser diode, which changes electrical signals into light signals. The receiver usually contains a photodiode that converts light back into electrical signals. The receiver output circuit also amplifies the signal and produces the desired results, such as voice transmission or video signals.

Some advantages of fiber optics are

1. Large bandwidth.
2. Low loss (attenuation).
3. Electromagnetic Interference (EMI) Immunity.
4. Small size.
5. Light weight.
6. Security.

OPTICAL FIBER

Figure 11-43 shows the typical construction of an optical fiber. The *core* is the actual path for light. Although the core occasionally is constructed of plastic, it is typically made of glass. A *cladding* layer, usually of glass or plastic, is bonded to the core. The cladding is enclosed in a jacket for protection.

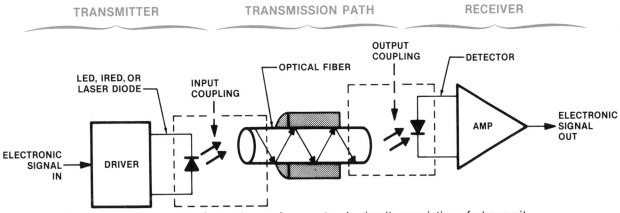

Figure 11-42 Fiber optics are used as a transmission link. The link connects two electronic circuits consisting of a transmitter and a receiver.

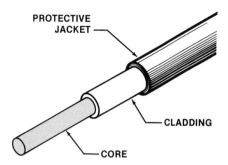

Figure 11-43 Optical fibers consist of a core, cladding, and protective jacket.

LIGHT SOURCE

For a fiber optic cable to operate effectively the light source feeding the cable must be properly matched to the light-activated device. The source must also be of sufficient intensity to drive the light-activated device.

LIGHT EMITTING DEVICES

Light Emitting Diode (LED)

The light emitting diode (LED) is a PN junction diode that emits light when forward biased. The light emitted can be either invisible (infrared), or it can be light in the visible spectrum. LEDs for electronic applications, due to the spectral response of silicon and efficiency considerations, are usually infrared emitting diodes (IRED).

Laser Diode

The laser diode is different from an LED in that it has an *optical cavity,* which is required for lasing production (emitting coherent light). The optical cavity is formed by coating opposite sides of a chip to create two highly reflective surfaces. See Figure 11-44.

FIBER COUPLING

The ideal interconnection of one fiber to another is an interconnection that has two fibers that are optically and physically identical. These two fibers are held together by a connector or splice that squarely aligns them on their center axes. The joining of the fibers is so nearly perfect that the interface between them has no influence on light propagation. Such a perfect connection will always be limited by two factors:

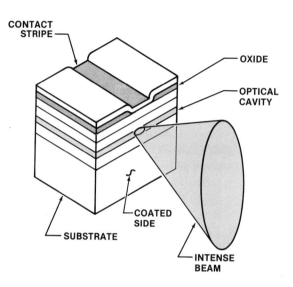

Figure 11-44 Laser action begins at threshhold current level, and the chip emits coherent, or nearly coherent, light. This light is in the form of an intense beam.

1. Variations in fibers.
2. Tolerances required in the connector or splice make their manufacture impractical.

These two factors affect cost and ease of use.

Fiber Coupling Hardware

For an electrician, there is little that can be done about the design of coupling materials available. However, proper installation procedures should be understood and followed. Splices and fiber interconnections are often more of a negative factor than poor quality materials are because of alignment problems that can arise. Figure 11-45 shows three common errors encountered in coupling. The elimination of these problems can be accomplished through proper installation of fiber splices, connectors, and couplers.

LIGHT-ACTIVATED DEVICES

Once light rays have passed through the optical fiber, they must be detected and converted back into electrical signals. The detection and conversion is accomplished with *light-activated devices,* such as PIN photodiodes, phototransistors, photo SCRs, and phototriacs.

PIN Photodiode

Figure 11-50 shows a typical PIN photodiode. (PIN stands for P-type material, insulator, and N-type material.) The operation of a PIN photodiode is

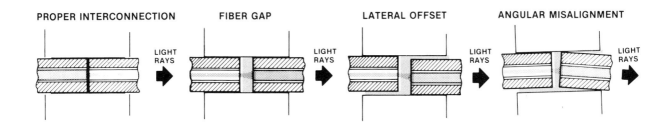

Figure 11-45 Improper interconnection of fiber optic cable can result in improper transmission. Three major causes of transmission loss are fiber gap, lateral offset, and angular misalignment.

based on the principle that light radiation, when exposed to a PN junction, momentarily disturbs the structure of the PN junction. The disturbance is due to a hole created when a high-energy photon strikes the PN junction and causes an electron to be ejected from the junction. Thus, light creates electron-hole pairs, which act as current carriers.

Phototransistor

A phototransistor combines the effect of the photodiode and the switching capability of a transistor. Schematically, the phototransistor may be represented by either one of the symbols shown in Figure 11-46. Electrically, the phototransistor, when connected in a circuit, is placed in series with the bias voltage so that it is forward biased.

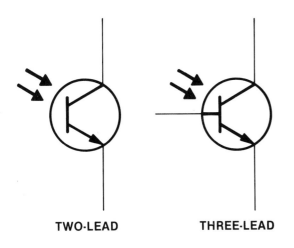

TWO-LEAD   **THREE-LEAD**

Figure 11-46 A phototransistor combines the effect of the photodiode and the switching capability of a transistor.

With a two-lead phototransistor, the base lead is replaced by a clear covering. This covering allows light to fall on the base region. Light falling on the base region causes current to flow between emitter and collector. The collector-base junction is enlarged and works as a reverse-biased photodiode controlling the phototransistor. The phototransistor conducts more or less current, depending upon the light intensity. If light intensity increases, resistance decreases, and more emitter-to-base current is created. Although the base current is relatively small, the amplifying capability of the small base current is used to control the larger emitter-to-collector current. The collector current depends on the light intensity and the DC current gain of the phototransistor. In darkness, the phototransistor is switched OFF with the remaining leakage current. This remaining leakage current is called collector dark current.

Light Activated SCR (LASCR)

The schematic of a light activated SCR (LASCR) is identical to the schematic of a regular SCR. The only difference is that arrows are added in the LASCR schematic to indicate a light-sensitive device. See Figure 11-50.

Like the photodiode, current is of a very low level in an LASCR. Even the largest LASCRs are limited to a maximum of a few amps. When larger current requirements are necessary, the LASCR can be used as a trigger circuit for a conventional SCR.

The primary advantage of the LASCR over the SCR is its ability to provide isolation. Since the LASCR is triggered by light, the LASCR provides complete isolation between the input signal and the output load current.

Phototriac

The gate of the phototriac is light-sensitive. It triggers the triac at a specified light intensity. See Figure 11-47. In darkness, the triac is not triggered. The remaining leakage current is called peak blocking current. The phototriac is bilateral and is designed to switch AC signals.

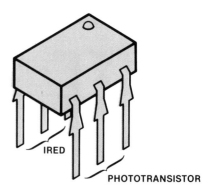

Figure 11-48 An optocoupler or optoisolator is usually constructed as a dual in-line package (DIP).

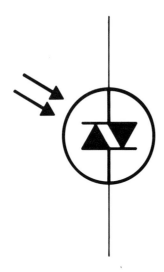

Figure 11-47 The phototriac, like the SCR, can be triggered by light radiation of a certain density.

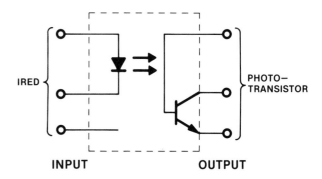

Figure 11-49 Internally, an optocoupler consists of an infrared emitting diode (IRED) as the input stage and a silicon NPN phototransistor as an output stage.

Optocoupler

Externally, an optocoupler is usually constructed as a dual in-line plastic package. See Figure 11-48. An optocoupler consists of an infrared emitting diode (IRED) as the input stage, and a silicon phototransistor as the output stage. See Figure 11-49. Internally, the optocoupler uses a glass dielectric sandwich to separate input from output. The coupling medium between the IRED and sensor is the infrared transmitting glass. This provides one-way transfer of electrical signals from the IRED to the photodetector (phototransistor) without electrical connection between the circuitry containing the devices.

Photons emitted from the IRED (emitter) have wavelengths of about 900 nanometers. The detector (transistor) responds effectively to photons with this same wavelength. Thus, input and output devices are always spectrally matched for maximum transfer characteristics. The signal cannot go back in the opposite direction because the emitters and detectors cannot reverse their operating functions.

SCHEMATIC	PHYSICAL APPEARANCE	TYPICAL APPLICATIONS	EXAMPLE
TRIAC 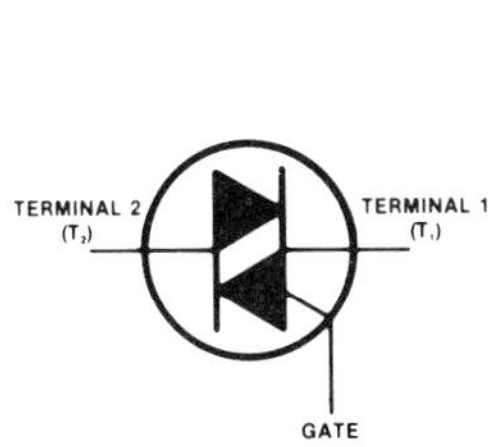	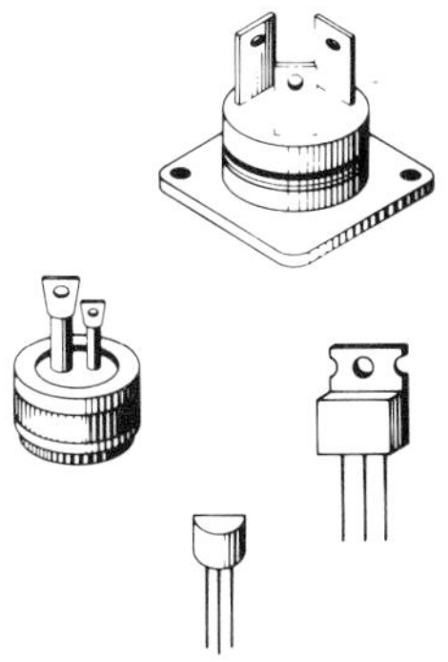	Light Dimmer Speed Motor Controller AC Power Control Interface Devices	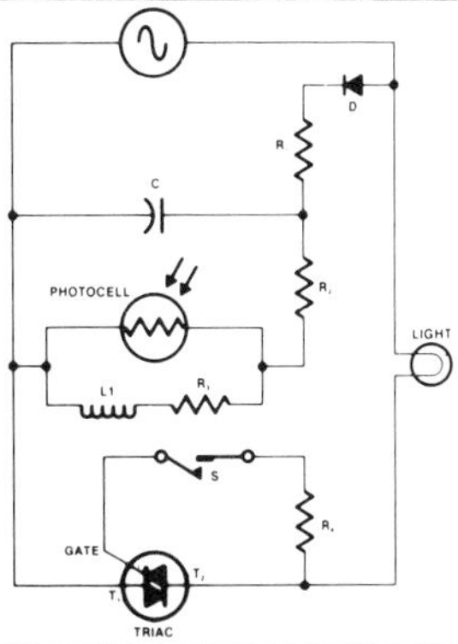
DIAC 	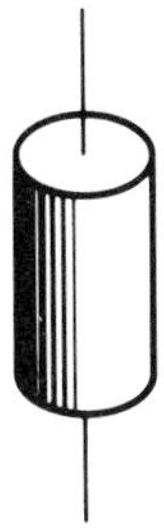	Firing Circuits Limiters Protection Circuits Squarer	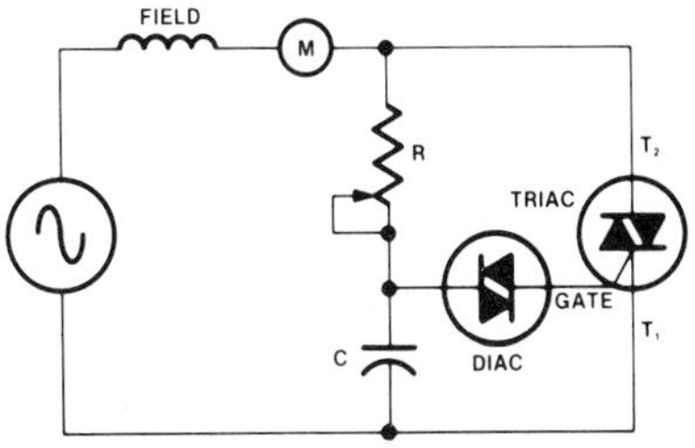
UNIJUNCTION TRANSISTOR 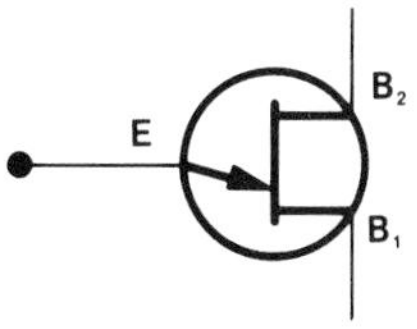	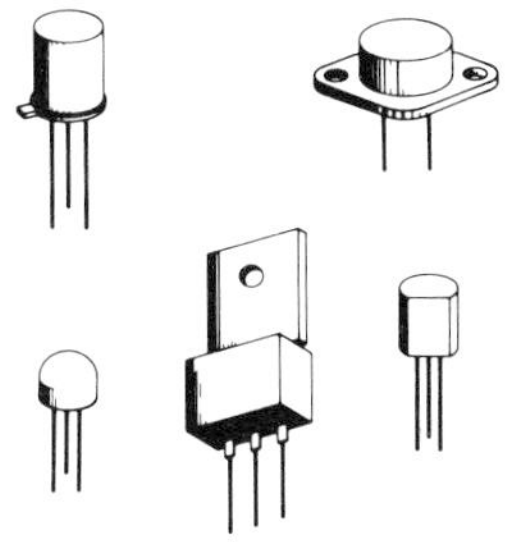	Ramp Circuits Oscillators Timing Circuits Trigger Circuits	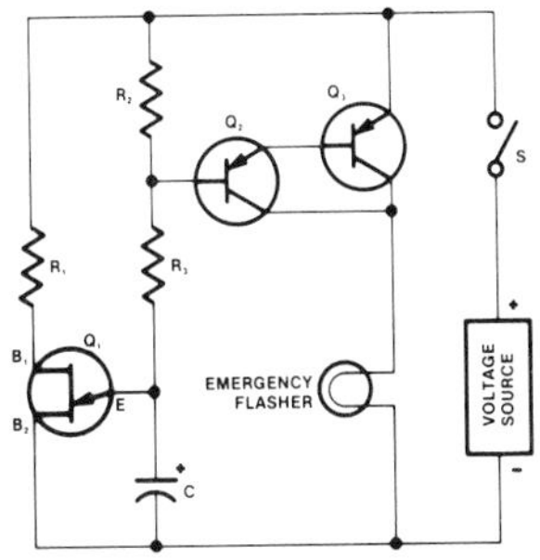
INTEGRATED CIRCUITS 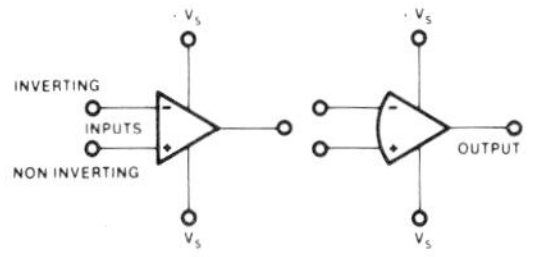	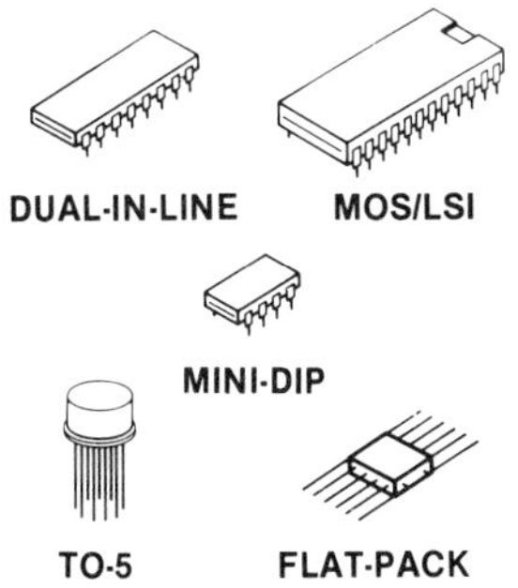 DUAL-IN-LINE MOS/LSI MINI-DIP TO-5 FLAT-PACK	Integrated Circuit Applications Amplifier Voltage Follower Inverter Differentiator Integrator Adder Subtractor Phase Shifter	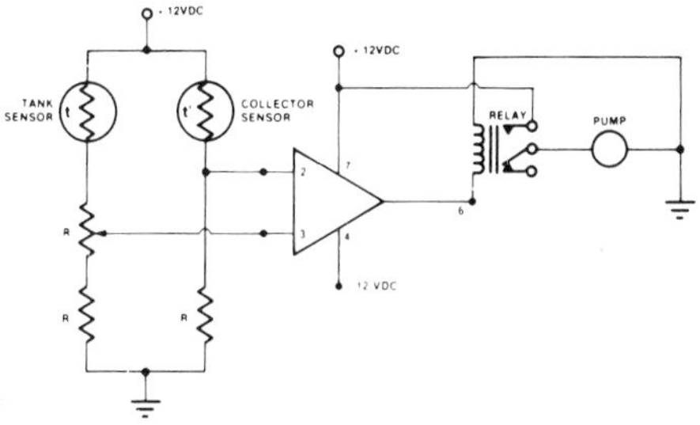
PIN PHOTODIODE 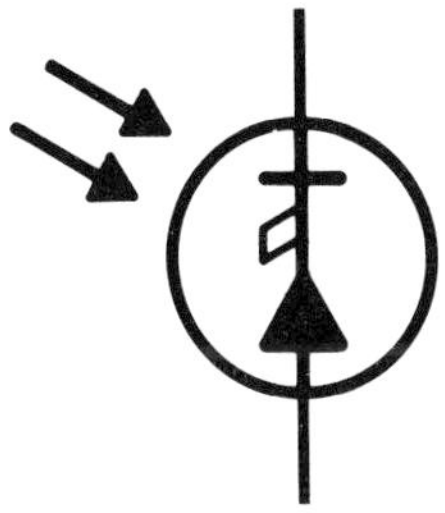	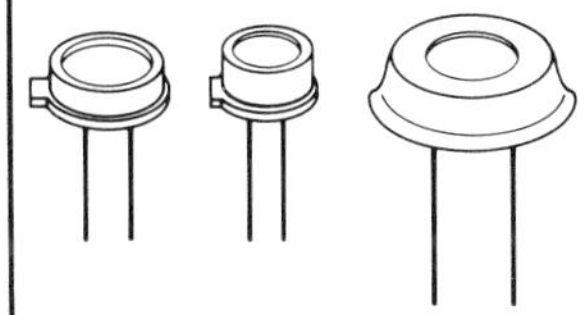	Gas Detector Spectrometer Gas Analyzer	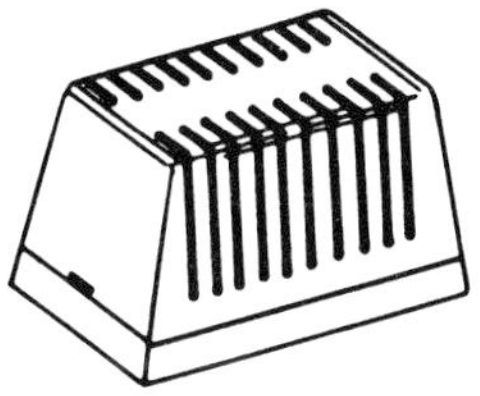

Figure 11-50 Solid state control devices and applications.

SCHEMATIC	PHYSICAL APPEARANCE	TYPICAL APPLICATIONS	EXAMPLE
DIODE CATHODE ANODE	1000A 160A 15A 80A	**Diode Applications** **Rectification** **Light Dimming** **Clipping Circuits** **Clamping Circuits** **Polarity Protection**	METER D. D,
THERMISTOR THERMISTOR ELEMENT HEATER ELEMENT DIRECTLY HEATED THERMISTOR ELEMENT INDIRECTLY HEATED		**Thermistor Applications** **Voltage Regulators** **Vacuum Gauges** **Electronic Timer-Delay** **Precision Temperature** **Measurements** **Temperature** **Compensation**	ALARM NTC THERMISTOR FIRE VOLTAGE SOURCE
PHOTOCONDUCTIVE CELL		**Photoconductive Applications** **On-Off Devices** **Door Openers** **Alarm Systems** **Flame Detectors** **Gas Analyzers** **Spectrometers**	CONTROL CIRCUIT TO MAIN BURNER GAS VALVE PHOTOCONDUCTOR CELL PILOT FLAME PILOT BURNER NOZZLE
PHOTOVOLTAIC CELL (SOLAR CELL)		**Photovoltaic Applications** **Power Source for Satellite** **Power Source for Remote Transmitter** **Light Intensity Meter**	METER MOVEMENT PHOTOVOLTAIC CELL
PHOTOCONDUCTIVE DIODE	LENS	**Photoconductive Diode Applications** **Movie Equipment** **Punched Tape Readers** **Optical Encoders** **Gas Detectors**	LIGHT SOURCE PUNCHED TAPE PHOTODIODES COMPUTER AMPLIFIER

SCHEMATIC	PHYSICAL APPEARANCE	TYPICAL APPLICATIONS	EXAMPLE
HALL SENSOR	HALL EFFECT SENSOR	**Hall Effect Sensor Applications** Computers Serving Machines Automobiles Aircraft Machine Tools Medical Equipment	
PRESSURE SENSOR	PRESSURE SENSING ELEMENT	**Solid State Pressure Sensor Applications** Pressure **Measurements** Relief Valve Activation Compression Tester	AIR COMPRESSOR
LED	METAL PLASTIC	**Light Emitting Diode Applications** Status Indicator Seven-Segment Display Optical Encoders Bar Code Readers	MINI-PROCESOR MODULE
PNP TRANSISTOR **NPN TRANSISTOR**		**Switching** **Output Stage of Amplifier** **High Current Circuits** **Oscillator**	
SCR ANODE GATE CATHODE	5A 1A 100A 2,400A	**Variable Speed Motor Control** Triggering Circuits Temperature Control Light Dimmers	

1. What is a PC board?
2. What determines whether an element will be an insulator, a conductor, or a semiconductor?
3. How is the crystal structure in most semiconductors altered?
4. What is meant by forward bias and reverse bias in semiconductors?
5. How is AC converted to DC using a single-phase rectifier?
6. How does a zener diode operate in a circuit?
7. What is a thermistor?
8. How does a photoconductive cell differ from a photovoltaic cell?
9. Explain the Hall effect principle.
10. What is a solid state pressure sensor?
11. How does a light emitting diode (LED) function?
12. What is a transistor?
13. How is the shape of a transistor identified?
14. How is a transistor used as a switching device?
15. What is meant by gain?
16. What are the three types of transistor amplifiers?
17. What is meant by forward breakover voltage?
18. What is a triac?
19. How does a triac function?
20. How does a diac function?
21. What is an integrated circuit (IC)?
22. How are pins on an IC identified?
23. What is an operational amplifier (OP amp)?
24. What are the two basic types of electronic signals?
25. What are the four most common gates in digital electronics?
26. List one use for each of the four most common gates.
27. What are the basic parts of optical fiber?
28. What is the function of the optical cavity in the laser diode?
29. What is a phototransistor?
30. What is the advantage of an LASCR over an SCR?

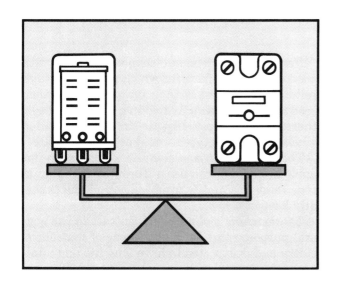

12 Electro-mechanical and Solid State Relays

Relays are used extensively in machine tool control, industrial assembly lines, and commercial equipment. Some applications of relays include switching starting coils and turning on small devices such as pilot lights and audible alarms.

Relays are used primarily as switching devices in a circuit. Depending upon the design, relays generally are not used to control power-consuming devices, except for small motors and solenoids that draw less than two amperes.

There are two major types of relays: the electromechanical relays (EMR) and the solid state relay (SSR). Each type has advantages and limitations as a control device.

INTRODUCTION TO RELAYS

In previous units it was shown how contactors and starters are used to switch heavy currents in the power circuit. A relay, in contrast, is primarily used to switch smaller currents in a control circuit. Depending upon design, relays generally do not control power consuming devices (except for small motors and solenoids which draw less than two amperes). In other words, the relay is primarily used as a switching element in the control circuit. Relays are used extensively in machine tool control, industrial assembly lines and commercial equipment. Relays are used to switch starting coils, heating elements, pilot lights, audible alarms and, in certain cases, small motors.

The function of a relay can be compared to an amplifier. This comparison is appropriate in certain cases as a small voltage applied to a relay can result in a larger voltage being switched. From the circuit of Figure 12-1, we see that applying 24 volts to the relay coils may operate a set of contacts that is controlling a 230 volt or 460 volt circuit. Since relay coils require only a very low current or voltage to switch, but can energize larger currents or voltages, the relay acts as an amplifier of the voltage or current in the control circuit.

Another example of a relay providing an amplifying effect is illustrated in Figure 12-2. Here a single input to the relay results in several other circuits being energized. Since certain mechanical relays can provide eight or more sets of contacts controlled from any one input, the input may be considered to have been amplified.

Although the comparison of an amplifier to a relay can be taken to extremes in its simplest format, this comparison is helpful in distinguishing the relay from other types of control devices.

TYPES OF RELAYS

Relays to date can be placed into two major categories: electromechanical and solid state. *Electromechanical relays* consist of devices which have sets of contacts which are closed by some type of magnetic effect. Pure *solid state relays,* by contrast, have no contacts and switch entirely by electronic devices. A third category that is sometimes recognized is a type of solid state relay called the hybrid relay. *Hybrid relays* are a combination of electromechanical and solid state technology used to overcome unique problems which cannot be resolved by one or the other devices. Each relay category will be treated in this unit as to advantages, limitations and applications.

ELECTROMECHANICAL RELAYS

Electromechanical relays which are common to commercial and industrial applications can be generally subdivided into three classifications: reed relays, general purpose relays and machine control relays. The major difference between these three types of relays is their intended use in a circuit, cost and the life expectancy of the device.

Reed relays are small compact devices with good mechanical features which provide for high reliability. Because of their unique construction, reed relays may be activated in a variety of ways, therefore allowing design circuit application when other relay types would be inappropriate.

The *general purpose relay* is a good relay for applications that can use a "throw away," plug-in type relay to simplify troubleshooting and generally keep cost low.

The *machine tool relay,* in contrast to the general purpose relay, is the backbone of control circuitry and is expected to have long life and minimum problems. Machine tool relays provide easy access for contact maintenance and usually provide additional features like time delay and convertible contacts for maximum circuitry flexibility. Machine tool relays are, as expected, more expensive.

The Reed Relay

Figure 12-3 illustrates a reed relay. The reed relay is a fast operating, single-pole, single-throw switch with normally open (NO) contacts hermetically sealed in a glass envelope. During the sealing operation, dry nitrogen is forced into the tube,

Figure 12-1 Relays may be compared to an amplifier in that a small voltage may result in a large voltage output.

Figure 12-2 Relays may also be compared to an amplifier in that a single input may result in multiple outputs.

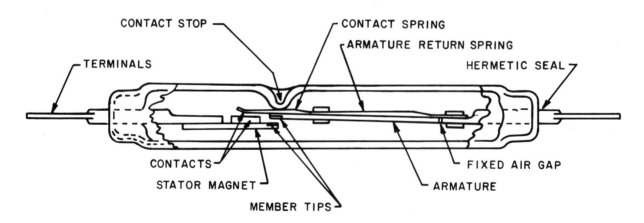

Figure 12-3 The reed relay is a fast operating single-pole, single-throw switch that is activated by a magnetic field.

creating a clean inner atmosphere for the contacts. Because the contacts are sealed, they are unaffected by dust, humidity and fumes—thus their life expectancy is quite high.

Reed relays are designed to be actuated by an external moveable permanent magnet or DC electromagnet. When a magnetic field is brought close to the two flattened reeds sealed in the glass tube, the ferromagnetic (easily magnetized) ends assume opposite magnetic polarity. If the magnetic field is strong enough, the attracting force of the opposing poles overcomes the stiffness of the reed drawing the contacts together. Removing the magnetizing force allows the contacts to spring open.

AC electrogmagnets are not suitable for reed relays since the reed relay switches so fast that it would energize and de-energize on alternate half cycles of a standard 60Hz line.

The Reed Contacts. To obtain a low and consistent contact resistance, the overlapping ends of the contacts may be plated with gold, rhodium, silver alloy or other low resistance metals. Contact resistance is often under 0.1 ohm on closing, yet reed contacts have an open contact resistance of several million ohms.

Most reed contacts are capable of direct switching of industrial solenoids, contacts and starters. When interpreting specifications for reed relays, it should be noted that the contact rating indicates the maximum value of current, voltage and volts/amperes rated. Under no circumstances should these values be exceeded.

Actuation of Reed Relays. As mentioned earlier, a permanent magnet is the most common actuator for a reed relay. Permanent magnet actuation can be arranged in several ways dependent upon the switching requirement. Typically, the most often used arrangements are proximity motion, rotation, shielding and biasing method.

Proximity Motion Actuation. The most obvious method of actuation of a reed relay is proximity motion. Since the reed relay contacts are closed by the presence of a magnetic field, the magnet and relay must be brought within a specific proximity (close distance) of each other to obtain results. The distance for activating any given relay will be dependent upon the sensitivity of the relay and the strength of the magnet. Obviously a more sensitive relay or stronger magnet will need less distance for actuation. Figure 12-4 illustrates examples of proximity motion operation of reed relays.

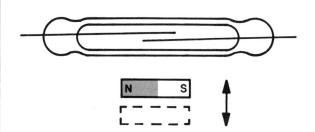

Perpendicular Motion—Provides only one switch closure with maximum magnet movement.

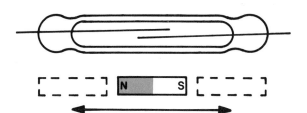

Parallel Motion—Provides as many as three closures with maximum magnet travel. Allows one closure with minimum magnet travel.

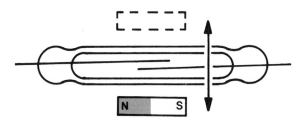

Front-to-Back Motion—Somewhat similar to parallel motion, except magnet motion is at right angles to switch and provides only one switch closure with maximum magnet travel.

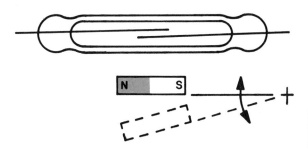

Pivoted Motion—Large angular magnet travel necessary to achieve one switch closure.

Figure 12-4 A reed relay may be activated in several ways by proximity motion. In each situation, movement on the relay or magnet will generate a response.

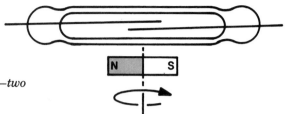

Figure 12-5 A reed relay may be activated by rotary motion. With this sequence, the relay will be activated twice in every 360° rotation.

Rotary Motion—Magnetic polarity is reversed—two switch closures for each complete revolution.

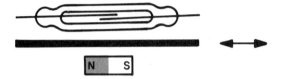

Shielding—Switching by removal of shield. The shield short circuits the magnetic flux, preventing formation of a field at the switch.

Figure 12-6 A reed relay may be activated in several ways utilizing the shielding technique. When the shield between the reed and the magnet is removed, the reed will be activated.

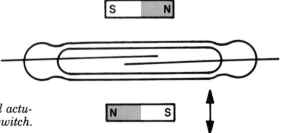

Figure 12-7 A reed relay may be activated through the biasing method. The bias magnet holds the switch closed until the actuating magnet cancels the magnetic flux and opens the switch.

Biasing—Bias magnet holds switch closed until actuating magnet cancels magnetic flux and opens switch.

Figure 12-8 There are several different styles of general purpose relays. Note that most of these relays have a plug-in feature that makes for quick replacement and simple troubleshooting. (Magnecraft Electric Company)

Methods of operation are the pivoted motion, perpendicular motion, parallel motion and front-to-back motion. In each of these methods, it makes no difference if the magnet or relay is moved with the other stationary. In some applications, both the magnet and relay may be in motion. The contacts will operate very quickly with snap action and little wear. In selecting one of these methods, application and switching requirements will determine the best method.

Rotation Operation. Another method of operating a reed relay is illustrated in Figure 12-5. Here revolving the magnet or relay will result in relay contact operation every 180°, or two operations every 360°. When the magnet and relay are parallel, the contacts are closed. When the magnet and relay are perpendicular, the contacts are opened. Although the magnetic poles will reverse every 180°, they still induce the magnetic field of opposite poles on the relay and close the contacts.

Shielding Operation. Another method of operating a reed relay is illustrated in Figure 12-6. In this type of operation, the magnet and relay are permanently fixed so that the relay's contacts are held closed. As a ferromagnetic (iron based) material is passed between the magnet and relay, the contacts are opened. The ferromagnetic material acts like a short circuit or shunt for the magnetic field and eliminates the magnetic field holding the contacts. As the shield is removed, the contacts are closed. It makes no difference at what

angle the shield is passed between the magnet and relay. This method may be used to signal that a protective shield, such as a cover on a high voltage box, has been removed in applications.

Biasing Operation. The last type of operation is called the biasing operation and is illustrated in Figure 12-7. In this operation, one magnet and the relay are arranged so that the relay's contacts are held closed. A second magnet that approaches the first magnet with opposite polarity will cancel the magnetic field of the first magnet and open the relay's contacts. The relay's contacts will only open if the two magnetic lines of force cancel each other. For this reason, strength of the magnetic field and distance must be considered. This application may be used in detecting magnetic polarity or other similar applications.

The General Purpose Relay

Figure 12-8 illustrates several different styles of general purpose relays. These relays are designed for commercial and industrial application where economy and fast replacement are high priorities. Most general purpose relays have a plug-in feature that makes for quick replacement and simple troubleshooting.

Regardless of its many designs, the general purpose relay is basically a mechanical switch operated by a magnetic coil similar to that of Figure 12-9. Since we covered the operation of solenoids

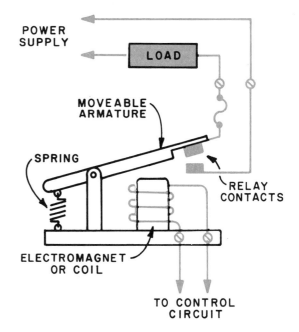

Figure 12-9 A general purpose relay is a mechanical switch operated by a mechanical coil.

and contactors in previous units, we will not elaborate on the relay operation at this time. Rather, we will stress the operating characteristics of the device.

The general purpose relay is available in both AC and DC designs. These relays are available with coils that can open or close the contacts from

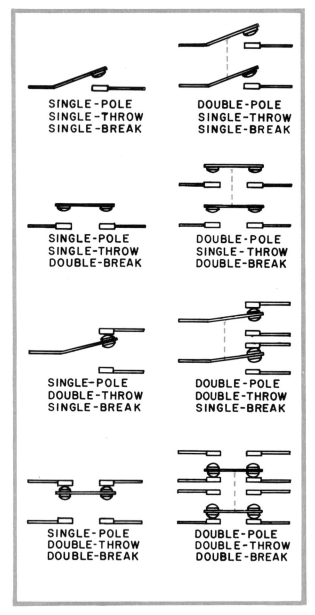

SINGLE-POLE
SINGLE-THROW
SINGLE-BREAK

DOUBLE-POLE
SINGLE-THROW
SINGLE-BREAK

SINGLE-POLE
SINGLE-THROW
DOUBLE-BREAK

DOUBLE-POLE
SINGLE-THROW
DOUBLE-BREAK

SINGLE-POLE
DOUBLE-THROW
SINGLE-BREAK

DOUBLE-POLE
DOUBLE-THROW
SINGLE-BREAK

SINGLE-POLE
DOUBLE-THROW
DOUBLE-BREAK

DOUBLE-POLE
DOUBLE-THROW
DOUBLE-BREAK

TYPES OF CONTACTS

SP - Single pole   NO - Normally open
DP - Double pole   NC - Normally closed
ST - Single throw   SB - Single break
DT - Double throw   DB - Double break

Figure 12-10 The arrangement and types of relay contacts depend on the application of the relay.

millivolts to the several hundred volt range. Relays with a 6, 12, 24, 48, 115 and 230 volt design are the most common. Today designs offer a number of general purpose relays that require as little as 4 milliamperes at 5VDC, or 22 milliamperes at 12VDC, making them IC compatible to TTL and CMOS logic gates. These relays are available in a wide range of switching configurations.

Contact Arrangements. Up to this unit, we have looked at contacts only in their most simple terms. We have seen double-break contacts in contactors and starters, and we have basically covered NO and NC contacts. In relays, however, the use of contacts becomes much more complex. To help understand the terminology and complexity of relays, the following information must be understood.

When describing contact arrangements and types, three words are used: "poles," "throws" and "breaks" (Figure 12-10). *Pole* describes the number of completely isolated circuits that can pass through the switch at one time. *Isolated* means that maximum rated voltage of the same polarity may be applied to each pole of a given switch without danger of shorting between poles or contacts. A double-pole switch can carry current through two circuits simultaneously, with each circuit iso-

CONTACT CODE AND NARM DESIGNATOR	
1 - SPST-NO	16 - 4PST-NC
2 - SPST-NC	17 - 4PDT
3 - SPST-NO-DM	18 - 5PST-NO
4 - SPST-NC-DB	19 - 5PST-NC
5 - SPDT	20 - 5PDT
6 - SPDT-DB	21 - 6PST-NO
7 - DPST-NO	22 - 6PST-NC
8 - DPST-NC	23 - 6PDT
9 - DPST-NO-DB	24 - 7PST-NO
10 - DPST-NC-DB	25 - 7PST-NC
11 - DPDT	26 - 7PDT
12 - 3PST-NO	27 - 8PST-NO
13 - 3PST-NC	28 - 8PST-NC
14 - 3PDT	29 - 8PDT
15 - 4PST-NO	

Figure 12-11 Switching arrangements are identified by NARM (The National Association of Relay Manufacturers).

lated from each other. With double-pole switches, the two circuits are mechanically connected so that they open or close at the same time, while still being electrically insulated from each other. This mechanical connection is represented in the symbol by a dashed line connecting the poles together.

Throws are the number of different *closed* contact positions per pole that are available on the switch. In other words, *throw* denotes the total number of different circuits that each individual *pole* is capable of controlling. The number of throws are independent of the number of poles. It is possible to have a single-throw switch with one or two (or more) poles, as shown in Figure 12-10.

Break is the number of separate contacts the switch uses to open or close each individual circuit. If the switch breaks the electrical circuit in one place, then it is a single-break switch. If the switch breaks the electrical circuit in two places, then it is a double-break switch.

To simplify the listing of contact and switching arrangements, general purpose relays carry National Association of Relay Manufacturers (NARM) code numbers. The numerals are used as abbreviations of the switching arrangements and are listed in Figure 12-11. Figure 12-12 illustrates the basic contact forms that are available. These forms are also designated by the National Association of Relay Manufacturers.

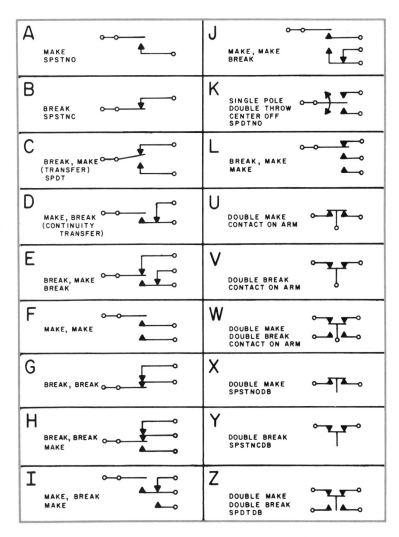

Figure 12-12 A listing of the basic contact forms accepted by NARM (National Association of Relay Manufacturers).

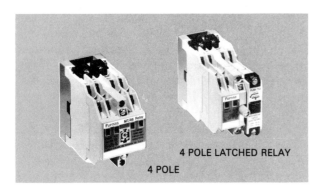

4 POLE LATCHED RELAY

4 POLE

Figure 12-13 The main features of these different styles of machine control relays are quality, reliability and extreme flexibility. (Furnas Electric Company)

The Machine Control Relay

Figure 12-13 illustrates a typical machine control relay. The machine control relay, like the general purpose relay, is merely a mechanical switch controlled by a magnetic coil. This type of relay derives its name from the fact that it is used extensively in machine tools for direct switching of solenoids, contactors and starters. They may also be referred to as heavy duty or industrial control relays.

The popularity of this relay stems from its good quality and reliability, along with its extreme flexibility. In a machine control relay, each contact is a separate removable unit that may be installed to obtain any combination of NO and NC

switching. These contacts are also convertible from NO to NC, and vice versa, as illustrated in Figure 12-14. By merely changing the terminal screws and rotating the unit 180°, the electrician can use the unit as either an NO or NC contact. Relays of one to twelve contact poles are readily assembled from stock parts.

The control coils for machine relays are easily changed from one control voltage to another and are available in AC or DC standard ratings. Further, machine relays have available a large number of accessories that may be added to the relay unit. These include indicating lights to monitor the state of the relay contacts, transient suppression to prevent electrical "noise", latching and time controls, to name a few.

BASIC OPERATION OF SOLID STATE RELAYS

In the recent decade, the industrial control market has been subject to a massive revolution based on solid state electronics. Due to their declining cost, high reliability and immense capability, solid state devices have begun replacing many devices which operated on mechanical and electromechanical principles.

As with anything new, so with solid state electronics, good judgment must be used when considering its acceptance. Just because solid state is new and appears to have certain advantages, you

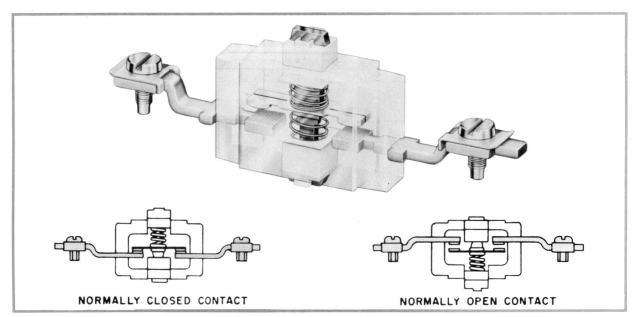

NORMALLY CLOSED CONTACT

NORMALLY OPEN CONTACT

Figure 12-14 Contacts on a machine control relay may be converted from NO to NC, or vice versa, merely by rotating the terminals.

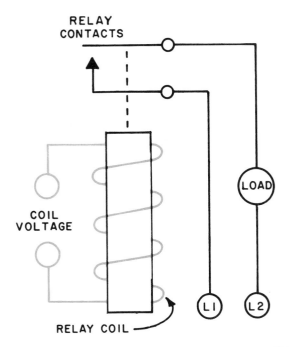

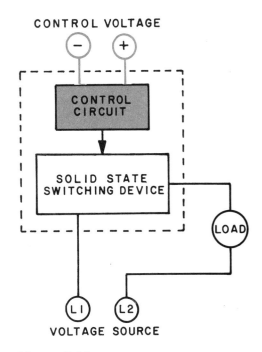

Figure 12-15 The EMR circuit provides switching using electromagnetic devices. The SSR circuit uses SCR's and triacs to switch without contacts.

cannot assume it is best for all applications. Although it can be said that solid state devices will give superior performance in some applications, it is also true that in other applications an electromechanical device will perform better.

In making a choice between solid state and electromechanical, you must compare the electrical, mechanical, and sometimes financial characteristic of each device with the application in which it is to be used. In this portion of the unit, we are going to examine the capabilities of solid state relays versus electromechanical relays and explain the difference between these two devices.

Comparison of Electromechanical Relays to Solid State Relays

Although both electromechanical relays (EMR) and solid state relays (SSR) are designed to provide a common switching function, each accomplishes the final results in different ways.

Basically, the EMR provides switching through the use of electromagnetic devices and sets of contacts, while the SSR depends upon electronic devices such as silicon controlled rectifiers (SCR) and Triacs to switch without contacts. Figure 12-15 illustrates graphically a simple example of both an EMR and SSR with input circuit and load circuit. Figure 12-16 illustrates physically the difference between an EMR and an SSR.

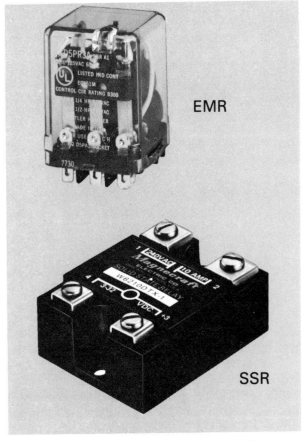

Figure 12-16 Physical features and operating characteristics are different in the EMR and SSR.

Comparable or Equivalent Terminology
for Electromechanical and Solid State Relays

Note: This chart is for comparison use only. Refer to text for complete and detailed information.

ELECTROMECHANICAL RELAYS (EMR)

1. *Coil Voltage:* The minimum voltage necessary to energize or operate the relay. This value is also referred to as the "pick up" voltage.

2. *Coil Current:* In conjunction with the coil voltage, the amount of current necessary to energize or operate the relay.

3. *Hold Current:* The minimum current required to keep a relay energized or operating.

4. *Drop Out Voltage:* The maximum voltage at which the relay is no longer energized.

5. *Pull-in Time:* The amount of time required to operate (open or close) the relay contacts after the coil voltage is applied.

6. *Drop Out Time:* The amount of time required for the relay contacts to return to their normal unoperated position after the coil voltage is removed.

7. *Contact Voltage Rating:* Maximum voltage rating contacts of relay are capable of switching safely.

8. *Contact Current Rating:* Maximum current rating contacts of relay are capable of switching safely.

9. *Surge Current:* Maximum peak current which the contacts on a relay can withstand for short periods of time without damage.

10. *Contact Voltage Drop:* Voltage drop across relay contacts when relay is operating (usually quite low).

11. *Insulation Resistance:* Amount of resistance measured across relay contacts in open position.

12. *No Equivalent or Comparison*

13. *No Equivalent or Comparison*

14. *No Equivalent or Comparison*

SOLID STATE RELAYS (SSR)

1. *Control Voltage:* The minimum voltage required to gate or activate the control circuit of the solid state relay. Generally a maximum value is also specified.

2. *Control Current:* The minimum current required to turn on the solid state control circuit. Generally a maximum value is also specified.

3. *Control Current*

4. *Control Voltage*

5. *Turn-on Time:* The elapsed time between the application of the control voltage and the application of the voltage to the load circuit.

6. *Turn-off Time:* The elapsed time between the removal of the control voltage and the removal of the voltage from the load circuit.

7. *Load Voltage:* The maximum output voltage handling capability of a solid state relay.

8. *Load Current:* The maximum output current handling capability of a solid state relay.

9. *Surge Current:* Maximum peak current which a solid state relay can withstand for short periods of time without damage.

10. *Switch On Voltage Drop:* Voltage drop across solid state relay when operating.

11. *Switch Off Resistance:* Amount of resistance measured across a solid state relay when turned off.

12. *Off State Current Leakage:* Amount of current leakage through solid state relay when turned off but still connected to load voltage.

13. *Zero Current Turn-off:* Turn-off at essentially the zero crossing of the load current that flows through an SSR. A thyristor will turn off only when the current falls below the minimum holding current. If input control is removed when the current is a higher value, turn-off will be delayed until the next zero current crossing.

14. *Zero Voltage Turn-on:* Initial turn-on occurs at a point near zero crossing of the AC line voltage. If input control is applied when the line voltage is at a higher value, initial turn-on will be delayed until the next zero crossing.

Figure 12-17 A comparison chart for EMR and SSR types allows for the comparison terminology used with each type of relay.

Since the basic operating principles and physical structures of an EMR and SSR are so radically different, it should not be surprising to find that a direct comparison of the two devices is somewhat difficult. Differences arise almost immediately both in the terminology used to describe the devices and in their overall ability to perform certain functions. As an aid in developing a comparison between an EMR and SSR, the charts of Figures 12-17 and 12-18 have been developed. The chart of Figure 12-17 allows us to compare equivalent or comparable terminology between EMR and SSR while the chart of Figure 12-18 helps us determine each device's ability to perform certain overall functions.

Input Signals. You will recall from previous discussions dealing with electromagnetic devices that applying a voltage to the input coil of a device created an electromagnet which was capable of pulling in an armature with a set of contacts attached to control a load circuit.

You may also recall that it took more voltage and current to pull in the coil than to hold it in due to the initial air gap between the magnetic coil and the armature. Because of these operating characteristics, we found it necessary to use four different specifications to describe the energizing and de-energizing process of an electromagnetic device; namely, coil voltage, coil current, hold current and drop out voltage (Figure 12-17).

By contrast, the solid state relay has no coil nor contacts and requires only minimum values of voltage and current to turn it on and turn it off. Thus we can see from our chart of Figure 12-17 that only two specifications are needed to describe the input signal for an SSR; namely control voltage and control current.

It is worth noting at this point that the electronic nature of the SSR and its input circuit lends itself well to be directly compatible with digitally controlled logic circuits. Many solid state relays are available with minimum control voltages of three volts and control currents as low as one milliamp, making them ideal for a variety of current state-of-the-art logic circuits.

Response Time. One of the significant advantages of the SSR over the EMR is its response time, or ability to "turn on" and "turn off". Where the EMR may be able to respond hundreds of times per minute, the SSR is capable of switching in the thousands of times per minute with no chattering or bounce.

DC switching times for an SSR are in the micro-second range, while AC switching time, with the use of zero-voltage turn on, is less than nine milliseconds. The reason for this advantage is that the SSR may be "turned on" and "turned off" electronically much more rapidly than a relay may be electromagnetically "pulled in" and "dropped out".

The higher speed of SS relays has become increasingly more important as industry demands

Advantages and Disadvantages of Electromechanical and Solid State Relays

Note: Plusses indicate advantages; minuses indicate disadvantages.

GENERAL CHARACTERISTICS	EMR	SSR
1. Arcless switching of the load	−	+
2. Electronic (IC, etc) Compatibility for interfacing	−	+
3. Effects of temperature	+	−
4. Shock and vibration resistant	−	+
5. Immunity to improper functioning because of transients	+	−
6. Radio frequency switching	+	−
7. Zero voltage turn-on	−	+
8. Acoustic noise	−	+
9. Selection of multipole, multithrow switching capability	+	−
10. No contact bouncing	−	+
11. Ability to stand surge currents	+	−
12. Response time	−	+
13. Voltage drop in load circuit	+	−
14. AC & DC switching with same contacts	+	−
15. Zero current turn-off	−	+
16. Leakage current	+	−
17. Minimum current turn-on	+	−
18. Life expectancy	−	+
19. Costs directly	+	−
20. Real cost-lifetime	−	+

Figure 12-18 The performance of EMR's and SSR's determine the application.

higher productivity from processing equipment. The more rapidly the equipment can process or cycle its output, the greater the productivity of the machine.

Voltage and Current Ratings. When comparing the voltage and current ratings of an EMR and an SSR in the load circuit, you are in effect comparing the maximum safe switching capability of a set of contacts to that of an electronic device such as an SCR or Triac. In each case, the terminology is not *equivalent*, but is somewhat *comparable*. Each device has certain limitations which will determine how much current and voltage it can safely handle. Since this will vary from device to device and from manufacturer to manufacturer, data sheets must be used to determine if a given device will safely switch a given load. Further consideration will be given to data sheets later in this unit.

The obvious advantages of solid state relays in this case, however, are that they have a capacity for arcless switching, have no moving parts to wear out, and are totally enclosed, thus being able to be operated in potentially explosive environments without special enclosures.

The advantage of the EMR is the possibility for replacement of contacts when the device receives an excessive surge current. Although surge current has basically the same meaning for both an SSR and EMR, the net result is quite different. With an EMR, the sets of contacts may be replaced; but with an SSR, the entire device must be replaced.

Voltage Drop. When a set of contacts on an EMR close, the contact resistance will usually be quite low unless the contacts are pitted or corroded. The SSR, however, being constructed of semiconductor materials, opens and closes a circuit by increasing or decreasing its ability to conduct. Even at full conduction, the device presents some residual resistance which can create a voltage drop of up to approximately 1.5 volts in the load circuit. Since this voltage drop is quite small in relation to the load voltage, it is usually considered insignificant and, in a majority of cases, presents no problems. When load voltages are quite small, however, this unique feature may have to be taken into consideration.

Insulation and Leakage. When a set of contacts is opened, the air gap between them provides for an almost infinite resistance through which no current will flow. Solid State devices, again, because of their unique construction, provide a very high but measureable resistance when turned off. Thus an SSR will have a "switched off" resistance not found on EMRs.

Since some conductance is still possible through an SSR even though it is turned off, it is possible for some small amounts of current to pass through the device. This is referred to as off-state current leakage and is also not found on EMRs.

The rating of off-state current leakage in an SSR is usually determined at 200 VDC across the output and usually should not exceed more than 200 milliamps at this voltage. This leakage current normally presents no problem unless the load device is affected by low values of leakage current. For example, in the case of small neon indicator lights and some programmable controllers, they cannot be switched off and will remain on because of the leakage.

Zero Current Turn-Off. Another unique characteristic of SS relays, is the capacity for zero current turn off. Semiconductors by nature will automatically turn-off relay when the AC load current sine wave is crossing the zero axis. This property of an SSR is especially important when switching inductive loads and will be discussed later in the unit.

Zero Voltage Turn-On. Zero Voltage turn-on is not necessarily available on SSRs, but is one which can be added to an SSR to provide certain features which will greatly extend the life of some types of loads. Benefits of zero voltage turn-on will also be discussed later in this unit.

Circuit Capabilities of SSR

The solid state relay can be used to control most of the same circuits that the electromechanical relay is used to control. Because the solid state relay differs from the electromechanical relay in function, the control circuits for solid state relays will differ from those of the electromechanical. This difference will be in how the relay is connected into the circuit and not on the application of the circuit. The solid state relay will perform the same circuit requirements as the electromechanical only with a slightly different control circuit. Following are six basic control circuits that are typical of many circuit requirements.

Two-Wire Control. Figure 12-19 illustrates how a solid state relay is used to control a load using a momentary control such as a pushbutton. In this circuit, the pushbutton when pressed will signal the solid state relay which will turn on the load. To keep the load turned on, the pushbutton must be held down. When the pushbutton is released, the load is turned off. This circuit is identical in operation to the standard two-wire control circuit used with electromechanical relays, magnetic motor starters and contactors. For this reason, the pushbutton used in Figure 12-19 could be changed to any manual, mechanical or automatic control device for simple on/off operation. If, for example, the pushbutton was replaced with a float switch, the same circuit could be used for liquid level control.

Three-Wire Memory Control. Figure 12-20 illustrates how a solid state relay is used with a

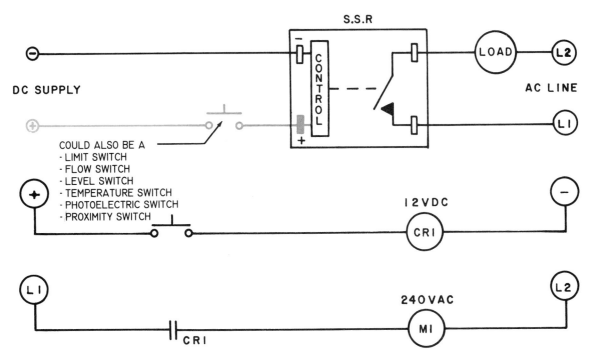

Figure 12-19 A solid state relay can be used to control a load using a momentary control such as a pushbutton.

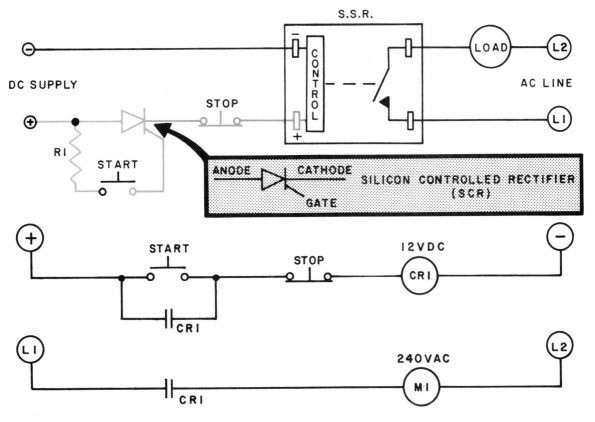

Figure 12-20 A solid state relay with an SCR circuit can be used to latch on the circuit similar to an EMR that would provide MEMORY or three-wire control.

SCR (silicon controlled rectifier) for latching the load on. This circuit is identical in operation to the standard three-wire or MEMORY circuit. To add memory after the start pushbutton is pressed, an SCR is used. An SCR acts as a current operated off-to-on switch. The SCR, as illustrated in Figure 12-20, will not allow the DC control current to pass through until a current is applied to its gate. Merely applying a voltage to an SCR will not cause it to switch. There must be the flow of a definite minimum current to turn the SCR on. This is accomplished by pressing the start pushbutton. Once the gate of the SCR has voltage applied to it, the SCR is latched in the on condition and allows the DC control voltage to pass through even after the start pushbutton is released. Resistor (R1) is used as a current limiting resistor for the gate and is determined by gate current and supply voltage. To stop the anode-to-cathode-flow of DC current to the SCR, the circuit must be opened. This is accomplished by pressing the stop button. If additional starts are required in this circuit, they will be added in parallel with the start pushbutton. Additional stops may be added to the circuit by placing them in series with the stop pushbutton. These additional start/stops may be any manual, mechanical, or automatic control.

Equivalent NC Contacts. Figure 12-21 illustrates how a solid state relay can be used to simulate an equivalent normally closed contact con-

dition. Since most solid state relays have the equivalence of a normally open contact, a normally closed contact must be electrically made. This is accomplished by allowing the DC control voltage to be connected to the solid state relay through a current limiting relay (R). Since the control voltage is present on the solid state, the load is held in the turned on condition. To turn off the load, the pushbutton is pressed, allowing the DC control voltage to take the path of least resistance and electrically removing the control voltage from the relay. This, then, would also turn off the load until the pushbutton is released.

Transistor Control. The control circuits so far have illustrated the control of solid state relays using control devices with mechanical switching contacts (pushbutton, etc.). Solid state relays are also capable of being controlled from electronic control signals from integrated circuits and transistors. This is illustrated in Figure 12-22. In this circuit, a solid state relay is controlled through an NPN transistor which receives its signal (IC logic gates, for example). The two resistors (R) are used as current limiting resistors.

Series and Parallel Control of SSR. Solid state relays can be connected in series or parallel to obtain multicontacts that are controlled by one input device. This is illustrated in both Figures 12-23 and 12-24. Figure 12-23 illustrates three solid state relay control inputs connected in par-

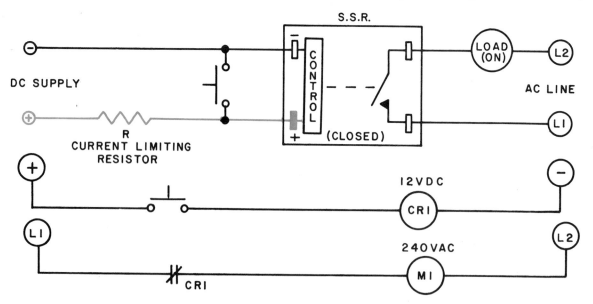

Figure 12-21 A solid state relay can be used to simulate an equivalent normally closed contact condition. A normally closed contact condition can be electrically made since a solid state relay is usually normally open.

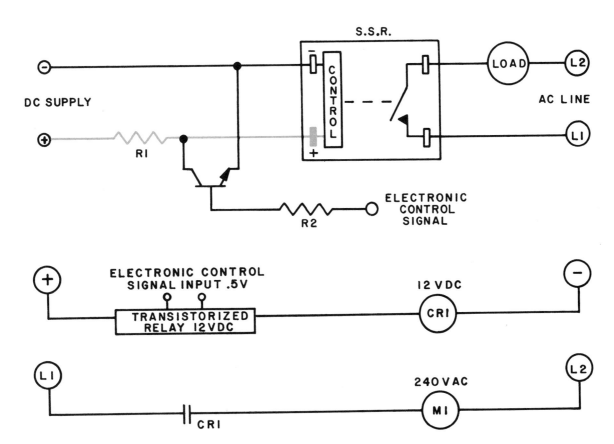

Figure 12-22 A solid state relay can be interfaced through a transistor control circuit to allow signals from integrated circuits and other transistors to activate the SSR.

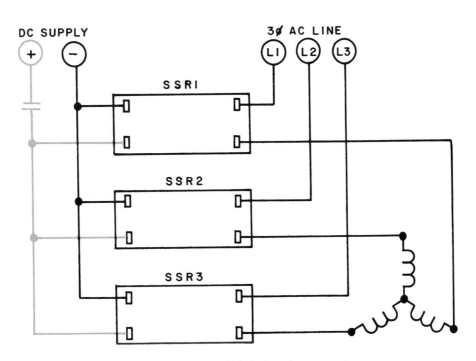

Figure 12-23 Three solid state relays may be parallel wired to control a three-phase circuit. In this arrangement the loss of one relay will not affect the others.

allel so that when the switch is closed all three will be actuated. This then will control the three-phase circuit illustrated in the circuit. In this particular application, the DC control voltage across each solid state relay would equal the DC supply voltage. This is because they are connected in parallel.

Figure 12-24 illustrates how to connect the solid state relays in series to control the same three-phase circuit. Here the DC supply voltage will divide across the three solid state relays when the switch is closed. For this reason, the DC supply voltage must be at least three times greater than the minimum operating voltage of each individual relay.

BASIC SOLID STATE RELAY DESIGNS

There are basically four solid state designs which are predominating the control market today: direct control, transformer isolation, optical (LED) isolation and hybrid solid state relays. Because each offers similar but distinctly different operating characteristics, each will be discussed.

Direct Control

Figure 12-25 shows in block diagram form a direct control or contact closure type of relay for switching AC loads. In this SSR, a set of external switch contacts, connected to the same AC voltage source

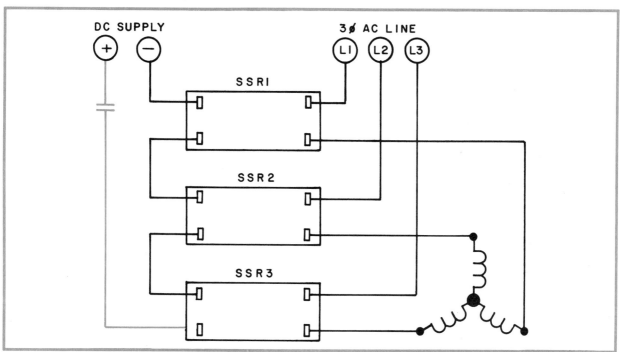

Figure 12-24 Three solid state relays may be series wired to control a three-phase circuit. In this arrangement the loss of one relay will usually drop out the other two relays due to the series wiring.

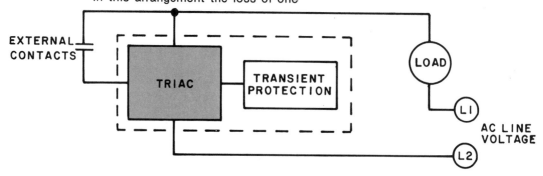

Figure 12-25 A solid state relay with triacs can form a direct control or contact closure type of relay for switching AC loads.

as the load to be controlled, is used as the control circuit. A triode AC semiconductor (triac), or a pair of back-to-back silicon controlled rectifiers (SCRs), may be used as the load switching device.

When the switch contacts are closed, the triac conducts and applies AC source voltage to the load. Opening the external contacts turns off the triac and thus removes the AC source voltage from the load. To protect the triac from undesired turn-on due to transient voltage surges, a transient protection network is included.

When direct control of a DC load is desired, the solid state relay configuration shown in Figure 12-26 is employed. In this circuit, a DC power transistor is used as the electronic switching device. As in the circuit diagram for controlling an AC load (Figure 10-26), external switch contacts are used to control the turn-on and turn-off of the power transistor. Alternatively, a second source voltage may be used in place of the external contacts to control the operation of the power transistor. It should be noted that when external contacts are used to control the operation of an SSR, the source voltage appears at the external

control contacts. Hence, they must be suitably protected to ensure the safety of the user.

Optically (LED) Isolated SSR

An optically isolated solid state relay is the equivalent of a form A (SPST-NO) standard relay. Isolation is provided optically through the use of a light emitting diode and photo-detector illustrated in Figure 12-27. The LED accepts the relay control voltage and through the LED converts this power to light energy. This light is collected by a photo-detector which controls the thyristor gate-firing circuit. When the control voltage specified is reached, because of sufficient light energy being transmitted to the photo detector, the gate circuit fires. Removal or reduction of control voltage reduces light output and stops triggering the circuit. The DC voltage required to operate the LED may be a specific voltage of, say, 5VDC, or may fall in a range typically 3 to 32 volts. The characteristics of the LED permit control circuit design that accepts a wide range of input voltages.

Input isolation from output for this type of relay may reach 10 billion ohms. The breakdown volt-

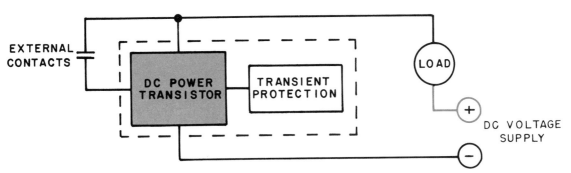

Figure 12-26 A solid state relay with DC power transistors forms a direct control or contact closure type of relay for switching DC loads.

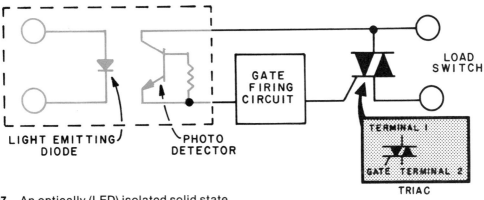

Figure 12-27 An optically (LED) isolated solid state relay can be used for most applications.

age is typically about 1500V rms 50/60 Hz. This isolation can be provided only up to a certain point, which is determined by the ratings of the components used. These ratings can be found in the data sheets of most manufacturers. Once these ratings are exceeded, transients can be introduced into the control circuits.

Based on a 10% reduction in light output, the life expectancy of an opto-coupler is in excess of 50,000 hours. It is actuated in microseconds, is not affected by shock or vibration, has no bounce and can be driven directly by either MOS or TTL gates.

The on/off state of the photo detector controls the state of the logic that permits gating of the output triac. Optically coupled designs usually feature zero voltage turn-on of the triac. This means that no matter when the input control voltage is applied, the triac does not turn on until the source voltage is of the order of 15V. This reduces electromagnetic interference at turn-on to less than one-hundredth that of an EMR and approximately one-fifth that of an SSR without zero voltage turn on.

After initial turn-on, successive half cycle turn on for SSRs require 5 to 10V across the triac, depending on the load being switched. Generally 5 to 16 milliamps (mA) are required to properly operate the coupler. Currents in excess of 20 to 25 mA may start deterioration of the LED coupler, particularly at elevated temperatures.

Transformer Isolation

In many applications, it is desirable or necessary to provide electrical isolation of the control circuit from the load circuit. Isolation is especially necessary when the control circuit is interfaced with low-level logic, due to its susceptibility to transient pulses.

One method of achieving electrical isolation is to use a transformer, as illustrated in Figure 12-28. In this circuit, a DC control voltage is

employed to activate the relay. The control voltage is transformed into an AC signal by a solid state oscillator circuit, the output of which provides primary current to a transformer. Oscillator frequency ranges from 50 kHz to 500 kHz. The transformer's output controls the thyristor gate-firing circuit. Thus, the magnetic coupling of the transformer serves to isolate the control voltage source from the load circuit. An advantage of transformer coupling is that it permits a wide variety of DC control voltage levels to be used. Another advantage is the very low control current required to activate the load.

Input-output isolation resistance and breakdown voltage is the same as that of an opto-isolator; however, the performance of a transformer coupler does not degrade noticeably over the life of the relay. In addition, the transformer coupled SSR has fewer total components than the opto-isolated SSR and is less temperature sensitive. However, it does not have the zero voltage turn-on feature. A transformer coupled SSR does radiate some EMI (electromagnetic interference) from the oscillator circuit, but it is minimal and usually is not a problem. A higher level of EMI may be produced by the triac turning on and off each successive half cycle.

The transient immunity of the transformer coupled SSR is somewhat less than that of the optical coupled SSR.

Hybrid Solid State Relays

Another form of SSR is the hybrid solid state relay. Although not a "true" SSR because it incorporates a mechanical component (a reed relay), it essentially operates as a solid state device.

The use of a reed relay does result in some trade-offs in performance, such as slower switching speed, less immunity to shock and vibration, and reduced life because of its mechanical contacts in the input or control stage. The hybrid,

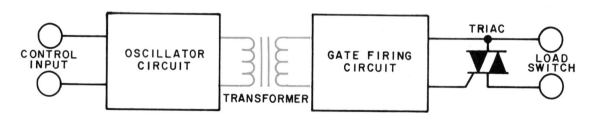

Figure 12-28 DC control voltage is used to activate the relay in a transformer isolated solid state relay.

however, is compatible with certain TTL logic when sufficient output current is available. Although it requires higher control current values than a true SSR, in addition to having a somewhat longer turn-on time, this hybrid is able to handle higher transient input voltages. A final minor limitation is that the hybrid is more vulnerable to shock and vibration because it incorporates a mechanical component.

Figure 12-29 represents a typical hybrid solid state relay incorporating a reed relay in the control circuit. Here the DC control voltage is used to energize a coil which draws the reed relay contacts together. This closes the control circuit, which in turn triggers the triac controller.

Electrical isolation in this circuit is provided by the magnetic coupling between the reed relay coil and the reed relay contacts. A zero voltage turn-

on feature can also be incorporated in this hybrid, as illustrated in Figure 12-30. As with optical isolation, a sensing circuit monitors the AC load voltage and permits the control circuit to trigger the triac only when the sine wave is crossing the zero axis.

When a reed relay is used as the control portion of a hybrid solid state relay, it is normally operated at a low voltage and current values to provide a relatively long life for the control circuit. On an inductive load application, it is normally the high voltage side of an electromechanical relay that will deteriorate first. A hybrid solid state relay can alleviate that problem by using a thyristor load controller to provide long life. A second type of hybrid solid state relay employs a solid state device control circuit to energize the coil of a reed only, the contacts of which switch the load

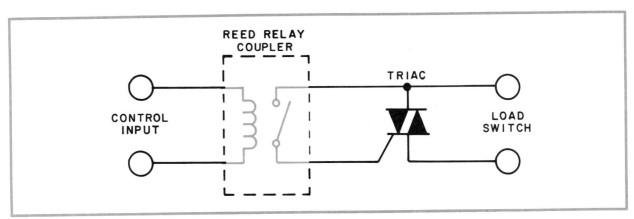

Figure 12-29 A hybrid solid state relay uses a coil to draw the reed contacts together.

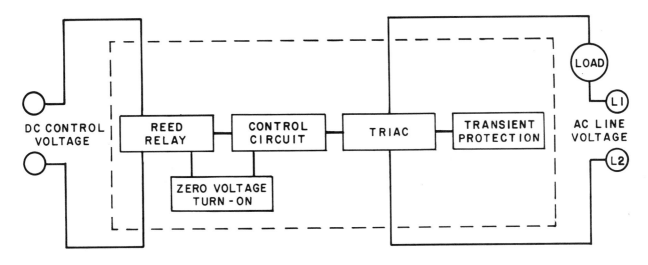

Figure 12-30 A solid state relay that is reed relay isolated can incorporate zero voltage turn on.

circuit. Essentially, this hybrid form is a conventional electromechanical relay with a solid state driver circuit. As such it is subject to many of the same disadvantages as conventional electromechanical relays.

GENERAL ENGINEERING SPECIFICATIONS

When the technician begins the selection process for an SSR either for replacement purposes or new applications, the engineering specifications for the device can be overwhelming. The reason for this confusion is usually based on the versatility or wide range of applications for the device. To simplify this process, we will try to show you specifications and selection procedures for typical circuit applications using SSR.

When selecting SSR, it is necessary to know the application requirements so that an SSR can be selected that will operate within the specified limits and ratings. Failure of solid state relays usually results from subjecting an SSR to an application for which it was not intended.

The following guidelines should be used in determining the basic operational charactrics of the SSR you choose.

Types of Solid State Relay Switching

The three types of solid state relays are the *instant on (IO)*, the *zero switch (ZS)*, and *universal (US)*. Each type is designed to turn on at a different point of time on the AC sine wave and offers certain advantages, depending on the load to be switched. This is illustrated in Figure 12-31.

The instant on relay turns on immediately when the control input is switched. Thus, the instant on relay can turn on anywhere on the outputs AC sine wave. The turn off of the instant on relay is always at zero current, on the point in the sine wave where current crosses zero.

A zero-voltage turn on sensing circuit as the one in Figure 12-32 can be used to overcome the problems of turn-on surges. This is a circuit that monitors the load voltage sine wave and prevents the control circuit from triggering the load controller until the AC voltage sine wave is crossing the zero axis. Figures 12-33 and 12-34 depict in graphic form the zero voltage points of a typical AC source and the delay that can occur between the time the control voltage is applied and the time that the sensing circuit permits the relay to be turned on.

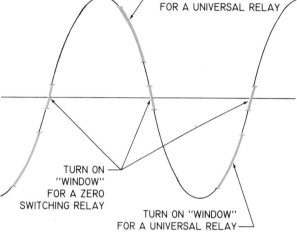

Figure 12-31 Each of the three types of solid state relays are designed to turn on at different points on the AC sine wave. The type of load to be controlled will determine the best relay type to use.

The zero switching relay always turns on at a zero voltage point at the output or when the voltage crosses zero (within a few volts of zero) in the AC sine wave. The turn off of this zero switching relay is always at zero current, just as in the instant on.

The universal relay turns on within a given window on the AC sine wave as illustrated. The universal relay does not turn on at the zero point or will not turn on at the peak of the AC sine wave. This relay type can turn on only inside the given window. The universal relay turns off at zero current, just as in the other two basic types.

Understanding Loads

The first consideration in selecting the proper relay is the type of load to be switched and the demands it will place on the circuit. These load characteristics may include high inrush currents, surge currents, high transient voltages and high induced counter electromotive forces (CEMF). Each of these conditions may erode or even destroy a relay's switching capability. In order to understand how the load affects the relay and what can be done about it, we will examine typical loads encountered with switching relays. The type of load to be switched by a solid state relay will determine the type of SSR to use. Each relay type is designed for a specific load.

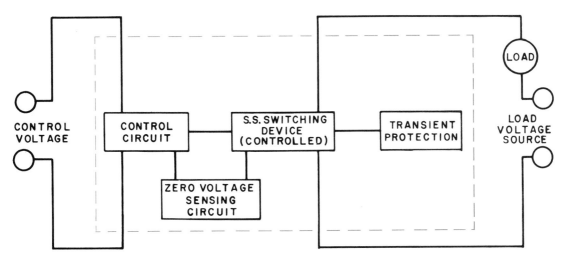

Figure 12-32 A solid state relay may be equipped with a zero voltage turn on sensing cir- cuit to overcome the problem of inductive surges.

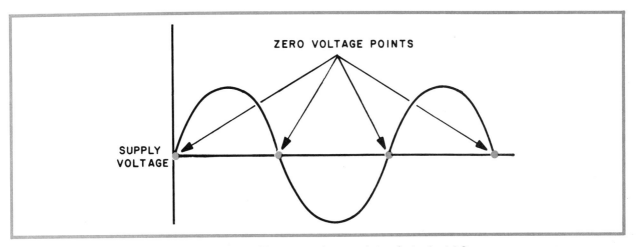

Figure 12-33 The zero voltage points of a typical AC source occur when crossing the zero axis.

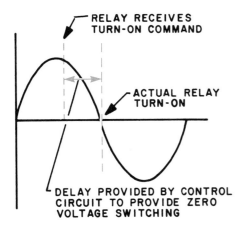

Figure 12-34 A delay can occur between the time the control voltage is applied and the time the sensing circuit permits the relay to be turned on.

Resistive Loads. In resistive loads the current and voltage are in phase, or together. This means if the load is turned on at zero voltage, it is also turned on at zero current. The benefits of turning a load at zero current are numerous. Devices with a very low "cold" resistance, such as light bulbs, are allowed to warm up before they reach the peak line voltage. This will greatly extend the life of this type of load. Figure 12-35 illustrates typical loads that can benefit from a zero voltage turn on. This also includes flashing warning lights, traffic signal lamps, photo lamps and heating elements. Also when switching at zero current, noise spikes are almost completely eliminated. It is for these

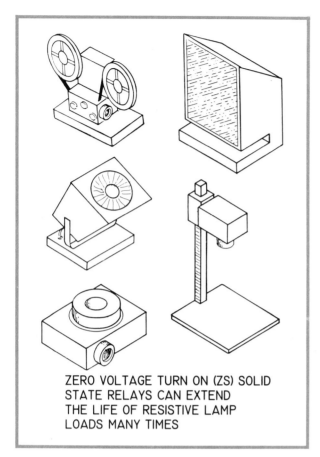

ZERO VOLTAGE TURN ON (ZS) SOLID
STATE RELAYS CAN EXTEND
THE LIFE OF RESISTIVE LAMP
LOADS MANY TIMES

Figure 12-35 Solid state relays can extend the life of resistive type loads, such as many lamp loads.

reasons that a zero switching relay should always be used when switching resistive type loads.

Inductive Loads. At first it may appear that Zero switching is the best for all applications, but this is not true. There are applications where zero switching is a distinct disadvantage, and in fact may damage the relay. In an inductive load such as a solenoid, transformer or motor, the current and voltage are out of phase as illustrated in Figure 12-36. When voltage and current are close to 90 degrees out of phase the load draws maximum current just when the voltage is at zero crossing. This high current can cause the magnetic material used in the load to become magnetically saturated, thus leaving only the pure OHMIC resistance of the load, which being very low will further increase the current draw. Because of this it is recommended to use the Instant On relay type for inductive loads as switching at any point other than the zero point of the voltage cycle is preferable.

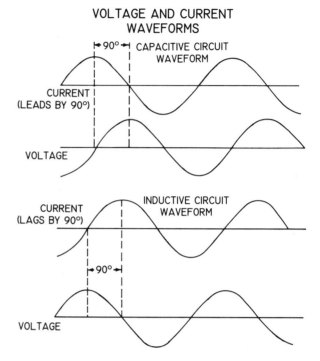

VOLTAGE AND CURRENT WAVEFORMS

90° ← CAPACITIVE CIRCUIT WAVEFORM

CURRENT (LEADS BY 90°)

VOLTAGE

CURRENT (LAGS BY 90°) — INDUCTIVE CIRCUIT WAVEFORM

90°

VOLTAGE

Figure 12-36 Voltage and current are out of phase in circuits with inductive and capacitive loads. In an inductive circuit current lags voltage. In a capacitive circuit current leads voltage. Inductive loads are the most common in industry.

Combination Loads. Often loads are not purely resistive nor purely inductive, but a combination of both. It is for this reason that the universal switching relay is available. The universal switching relay switches neither at the zero point nor the peak, and therefore is often used as a general purpose solid state relay for combination loads.

Lamp Loads. Lamp loads are probably the least understood types of loads that are switched. The inrush current can be extremely high. The cold resistance of a tungsten filament of the typical incandescent lamp is extremely low when cold, resulting in inrush currents of as much as 15 times the steady state current. Such high inrush currents can destroy relays.

Figure 12-37 illustrates a typical graph for the turn on of an incandescent lamp. Tungsten lamps will have high inrush currents which decay over several cycles, but may peak at 10 times their normal current level. Quartz iodide lamps may have inrush current of 15 to 20 times their normal current level. Most relays should be able to handle this peak inrush current and still maintain control

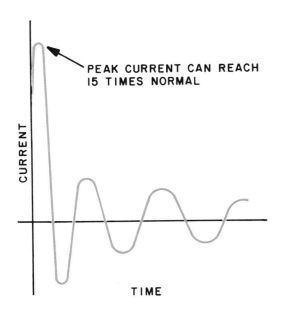

Figure 12-37 High inrush current can be expected when some lamp loads are switched on.

that silicone grease be used between the base of the SSR and the heat sink. This allows the heat to be properly conducted from the relay to the heat sink.

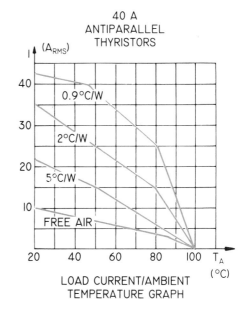

LOAD CURRENT/AMBIENT
TEMPERATURE GRAPH

HEAT SINK

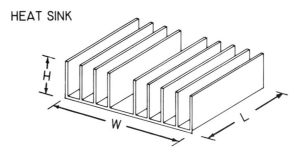

ELECTROMATIC TYPE	H × W × L (MM)	R_{TH} (°C/W)
RHS 01	15 × 79 × 100	2.5
RHS 02	15 × 100 × 100	2.0
RHS 03	25 × 97 × 100	1.5
RHS 04	37 × 120 × 100	0.9
RHS 05	40 × 160 × 150	0.5
RHS 06	40 × 200 × 150	0.4

HEAT SINK SELECTION CHART

Figure 12-38 Selection of the proper heat sink is important to the performance of a solid state relay. The manufacturer's chart can be used to find °C/W generated by the SSR and thus the correct heat sink can be chosen. (Electromatic Controls Corp.)

because start-up time is very fast. Checking the relay's specifications is important and some type of protection circuit may have to be added to extend relay life.

Use of Heat Sinks

The current consumed by the connected load heating of the triac junction and the ambient temperature, require the use of a heat sink to remove heat which may cause damage to the relay.

By determining your load current and assumed ambient temperature of the area where the SSR will be mounted, you can select the proper size heat sink using heat sink charts as illustrated in Figure 12-38. After determining the °C/W generated by the SSR, the correct size heat sink can be chosen.

For example, if a solid state relay is used to switch a 25 amp load at 50 °C as per the chart, the R_{TH} (°C/W) would equal 2. This means a heat sink equal to this number (or a smaller number) must be used. In this example a 15 × 100 × 100mm (RHS 02) would be the minimum correct size. However, the best size would be 25 × 97 × 100mm (RHS 03) to allow for a possible increase in current or ambient temperature.

Whenever a heat sink is required, it is essential

Ambient Temperature. The performance of solid state relays is greatly affected by the ambient temperature. Ambient temperature is the combination of several factors. A major factor is the temperature of the air surrounding the relay. This is determined by the location of the relay and the type of enclosure used. If the enclosure does not provide ventilation, the temperature inside the enclosure will be higher than outside air. In addition, if the enclosure is next to a heat source or in the sun, the inside temperature will also increase. Electronic control circuits and solid state relays will themselves produce heat. In some applications, forced air cooling may be required.

Voltage Drop Dissipation. In a mechanical contact system, the voltage drop across the switching element is negligible. In solid state relays, however, a voltage drop in the output semiconductor is unavoidable. This voltage drop generates heat. By choosing the right heat sink, this heat is dissipated so that temperatures in the relay are within safe limits.

TROUBLESHOOTING SOLID STATE RELAYS

Troubleshooting a solid state relay that appears to have failed is fairly straightforward. This is because there are generally only three conditions under which solid state relays fail. These three conditions are (1) the relay does not turn off when signaled to; (2) the relay does not turn on when signaled to; and (3) the relay performs erratically. Care should be taken to learn the reason the solid state relay has failed and not just replace the inoperative SSR with a new one. Replacing one bad relay with one good relay does not indicate what caused the failure in the first place and does not guarantee that the new relay will not fail.

Relay Fails to Turn Off
If the SSR does not turn off, the first thing to do is disconnect the SSR input. If the SSR still remains on, the problem is usually one of the following.

Leakage Current (Output). In all solid state relays, as in all solid state switches, there is a small amount of current that flows through a non-conducting (open) switch and load circuit. This small current flow is called *leakage* or *residual current.*

When controlling low impedance loads such as contactors, solenoids, and motor starters, leakage current is not a problem. However, when controlling high impedance loads such as small control relays, programmable controller inputs, or other solid state devices, leakage current can affect the load circuit. This current may be enough to keep the load in the on state even if the relay is in the off state. This problem can be taken care of by connecting a resistor in parallel with the load. This resistor is called a loading resistor. A general rule of thumb is to use a 5K-5W resistor.

Load Current. The load to which the SSR is switching is drawing more current than the SSR is rated for. This can cause a permanent short in the SSR. An ammeter can be used to find the current drawn and this can be compared to the rating of the SSR. If necessary, the SSR should be replaced with a higher current rated SSR. Remember, turn on current may be many times higher than steady state current. An SSR can take a high current but only for a short period of time (check specifications).

Heat Sinking. The heat sinking designed to protect the SSR may be insufficient for the current being carried at the ambient temperature that the SSR is subjected to. This would cause permanent failure of the SSR and a larger or more efficient heat sink should be used.

Transient Voltages. If high transient voltages on the line are present and are higher than the rating of the SSR, erratic operation or permanent failure of the SSR may result and extra transient protective circuitry must be provided.

Line Voltage. If the line voltage applied to the SSR is greater than the rated voltage, a permanent short may be caused in the SSR's output. This can be checked with a voltmeter. If the voltage is found to be too high, a higher rated voltage SSR should be used, or a voltage dividing resistor can be connected in series with the SSR. Also check to make sure polarity of the output circuit is correct.

Leakage Current (Input). Solid state relays require very little current to turn them on. The required turn on current for an SSR can be found by checking the specifications for the input impedance. The input impedance, along with the applied voltage, will determine the turn on current. This current may be as small as 2 mA in some solid state relays. It is for this reason that care must be taken to make sure leakage current in the input circuit is reduced to a minimum. If, for example, a solid state proximity switch with a higher leakage current than the turn on current of the SSR is used, a load resistor must be used to prevent the relay from staying on.

Relay Fails to Turn On

If the SSR fails to turn on, measure the input voltage when the relay should be turned on. If this voltage is less than the turn on voltage rating of the SSR, then there is a problem with the input circuit. Correct the problem in the input circuit and the SSR will turn on when the rated voltage is applied.

If the measured input voltage is greater than the turn on voltage rating of the SSR, check the input circuit and supply voltages. In new or modified installations, polarity may not be proper, so this must be checked. If the voltage is present with the proper polarity, check for current flow with an ammeter. If there is no current flow, the SSR's input is open, and the relay is defective and must be replaced. If there is current flow, there is usually a circuit problem in the input circuit which must be corrected. The problem is usually caused by a high resistance somewhere in the input circuit. A quick check can be made by directly supplying the SSR's input with voltage of the proper current capacity and voltage level.

Relay Performs Erratically

This problem may be one of the hardest to find and correct. There are some ways, however, to isolate this problem. Unless a particular problem is suspected, the first thing to do is to replace the SSR. This will take care of the problem if the problem is in the SSR. If not, the circuit will again perform erratically and the control circuit and output circuit must be checked. Check for loose or improper connections. Check to see that input and output wires are not bundled together, as this may

cause indirect coupling and erratic turn on through induction. Erratic turn on's can also be caused by electric noise such as a solenoid or electromechanical contact chattering. In this case, a snubber network on the output usually will help. Check to see that the heat generated by the normal operation of the SSR or other components in the enclosure is not causing the ambient temperature to rise above the allowable temperature.

Suppression Techniques

Switching Off. When the stored energy in an inductive circuit is dissipated across the opening contacts of an electromechanical relay (EMR), contact erosion is possible as well as the creation of unwanted transients. To help ease the adverse effects of this counter electromotive forces (CEMF), certain contact protection circuits have been designed to add on to the relay contacts for protection.

When considering arc suppression techniques, remember that manufacturers have many different types commercially available. In selecting them, remember that the rating of the components used for suppression must take into account relay specifications and load requirements.

EMR/Suppression Techniques

Diode (DC). Figure 12-39 illustrates a diode used for suppression. In this circuit, a diode is con-

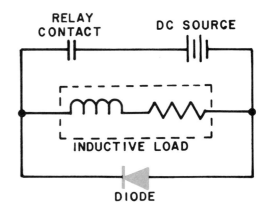

Figure 12-39 A diode may be used for arc suppression. The diode blocks the applied DC voltage at contact closure, but allows a path for stored inductive energy in the load when the contacts open.

nected across the load. The diode across the load blocks the applied voltage (DC) at contact closure, but allows a path for stored inductive energy in the load when the contacts open.

Varistor (DC/AC). In some circuits, a varistor is substituted for the diode (Figure 12-40). The resistance of the varistor is very high at applied voltage but decreases to a small value at high transient peaks. If the varistor carries 10% load current, it will limit the transient voltage to about twice the supply voltage. This suppression method is also used in some AC circuits.

Capacitor/Diode/Resistor Network. Figure 12-41 illustrates another suppression method for AC circuits. In AC circuits, this network circuit can be connected either across the load or across the contacts. For 115 VAC supply voltages, the peak inverse voltage (PIV) rating of the diodes should be 400 volts. The working voltage rating of the capacitors should be 200 VAC. The resistor is usually about 100K ohms.

Other methods of arc suppression not illustrated are: connecting a resistor, two back-to-back zener diodes, a diode and zener diode, or a diode-resistor combination directly across the inductive load. In each of these methods, an additional path for arc suppression is provided. Some suppression circuits may also use a resistor-capacitor (R-C) network across the relay contacts.

SSR/Suppression Techniques

SSR-Zero Current Turn Off. Another feature of SS relays is inherent in the switching device that controls the load. Semiconductors automatically turn-off the relay when the load current sine wave is crossing the zero axis.

This property of SSRs is especially important when switching induction loads. On resistive loads, the current and voltage are in phase, and the zero turn-off points of the voltage and current sine waves are simultaneous. With inductive loads, however, the current and voltage sine waves are out of phase. The solid state relay will turn off when the current sine wave is passing through zero, thereby leaving some minimal voltage to decay to zero.

Figure 12-42 illustrates an example of transient pulses which occur randomly in an AC source voltage and ride the voltage sine wave. These transients can originate in nearby load circuits, in other electrical environments, or in AC power lines feeding the area. Power line pulses

can often have amplitudes exceeding 1000 volts. If these pulses are not suppressed, they may trigger the relay erratically or exceed the surge capability to a point of destruction.

Snubber Networks. Not all SSR manufacturers incorporate transient protection in their products, in which case a snubber network needs to be provided. A snubber network consists of a resistor and a capacitor connected across the SSR output.

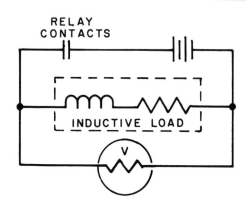

Figure 12-40 A varistor may be substituted for a diode to provide arc suppression. The resistance of the varistor is very high at applied voltage but changes to a small value at high transient peaks.

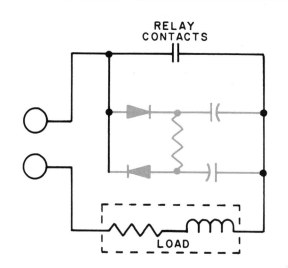

Figure 12-41 A capacitor-diode-resistor network could be used for AC suppression.

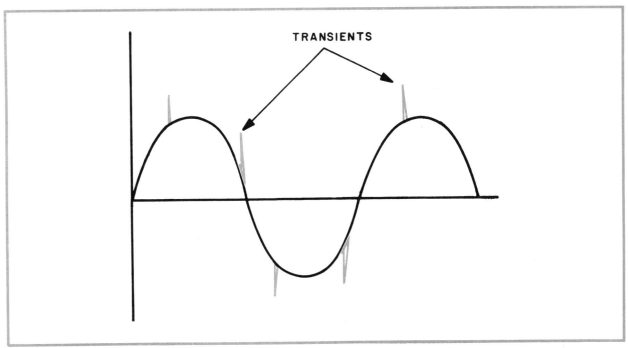

Figure 12-42 Transient pulses in an AC source voltage occur randomly and ride the voltage sine wave.

When transient protection is incorporated, it is specified as a maximum voltage developed during a particular rise time. For example, it may be stated as 400 volts maximum with 0.1 microsecond rise time. This specification defines the absolute maximum transient peak (400 volts) that can be safely handled by the relay when the rise time is 0.1 microseconds. Thus, it defines the rate of change or sharpness of the pulse that the relay can handle. Transients exceeding this specification can turn on the relay for a half cycle of operation; if substantially higher, can destroy the output switching device. It is, therefore, important to know and understand what transient conditions exist in a given application. When selecting a solid state relay for inductive load switching, be sure to look for one with high transient protection.

The amplitude versus time relationship for an SSR with transient protection is very nearly linear, as shown in Figure 12-43. For example, a transient 800 volt pulse with a rise time of 0.2 microseconds has the same rate of change as a 400 volt pulse with 0.1 microsecond rise time. This does not mean, however, that the relay can handle the 800 volt pulse with 0.2 microsecond rise time. Rather, the relay can handle a pulse changing at a rate of 800 volts in 0.2 microseconds, but only if it does not exceed the 400 volt maximum rating.

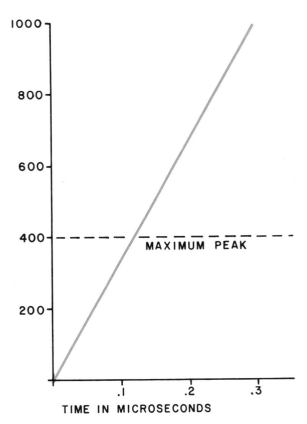

Figure 12-43 The amplitude versus time relationship for an SSR with transient protection is very nearly linear.

Using the preceding example of an SSR with transient protection rating of 400 volts in 0.1 microsecond, the chart in Figure 12-44 illustrates some possible voltages and rise times. Rise time is defined as the time required for a transient voltage to rise from 10% to 90% of its maximum value. Pulse one changes from 0 to 400 volts in 0.1 microseconds and decays in the same time. Since it falls within the specification, pulse one will not turn on the output switching device of the SSR.

Pulse two has the same rate of change, but its amplitude exceeds 400 volts and will, therefore, turn on the relay. Pulse three has a slower rise time of 0.17 microseconds and will not turn on the relay since its rate of change is less than 400 volts in 0.1 microseconds. Pulse four, which has a rise time of 0.05 microseconds, will also turn on the relay since its rate of change exceeds the specifications.

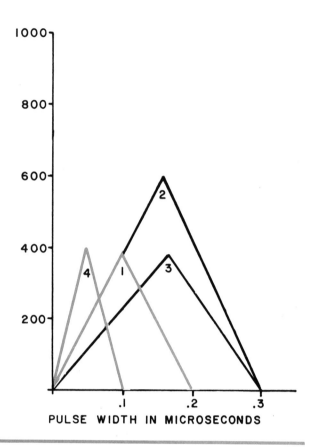

Figure 12-44 Pulses with various response times react to a snubber network with known specifications.

PULSE WIDTH IN MICROSECONDS

QUESTIONS ON UNIT 12
1. What is the major difference between relays and contactors?
2. How may a relay be compared to an amplifier?
3. What are the three major categories of relays?
4. Electromechanical relays can be subdivided into what three classifications?
5. What is a reed relay?
6. On what principle does a reed relay operate?
7. Name six methods of actuating a reed relay.
8. What is a general purpose relay?
9. What does the term "pole" describe in switching?
10. What does the term "throw" describe in switching?
11. What does the term "break" describe in switching?
12. What is a machine control relay?
13. What is a solid state relay?
14. What are some of the advantages of an EMR over an SSR?
15. What are some of the advantages of an SSR over an EMR?
16. What is a hybrid solid state relay?
17. What are some of the characteristics of inductive loads that must be considered when switching?
18. What are some of the characteristics of resistive loads that must be considered when switching?
19. What is a heat sink used for?
20. What are some of the problems that cause SSR's to fail?
21. What is "zero-voltage" turn on?
22. What are some of the methods used in arc suppression?
23. What is meant by "zero-current" turn off?
24. What is a "transient pulse"?
25. What are some of the characteristics of combination loads that must be considered when switching?

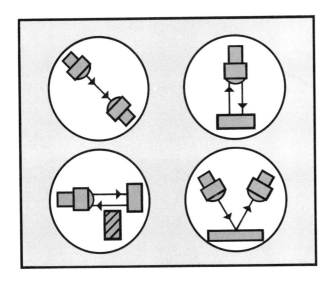

13 Photoelectric and Proximity Control and Applications

Photoelectric and proximity sensors provide the sight, hearing, and touch that link a control with a microcomputer. Microcomputers using sensors have had a tremendous impact on the industrial control industry. The growth in the use of microcomputers has necessitated better sensors. Proximity control using microcomputers are used to control functions for energy efficiency and other industrial processes. Installation and operating characteristics of proximity control devices vary, depending upon the application.

PHOTOELECTRIC CONTROL

Photoelectric controls are becoming increasingly popular in a wide range of industrial applications because they are versatile, safe, reliable and can be used on applications where other mechanical type controls cannot. Their no-touch concept is often the only reliable way to satisfy a switching need.

Photoelectric controls can detect targets moving into the sensing field from any direction and in any sequence. They can be used to detect the presence or absence of practically anything, without touching it and even at considerable distance. Because photoelectric controls are solid state, they combine high reliability, long life and high speed operation with inherent design and application freedom. Figure 13-1 illustrates some typical applications of photoelectric control.

How the Photoelectric Control Is Used in the Circuit

In previous units you have learned that all control circuits are organized into three basic elements. These basic elements are the signal, the decision, and the action parts of a circuit. Figure 13-2 reviews each of these three basic elements in the control circuit. The signal can be supplied by a pushbutton, limit switch, promixity switch, photoelectric switch, or any combination of these control devices.

Once generated, the signal(s) is passed on to the decision making part of the circuit which may be AND, OR, NOT, NAND, NOR, MEMORY, TIME DELAY or a combination of logic functions.

With the decision made, some type of action will usually result through the activation of solenoids, relays, starters or other direct acting loads to accomplish a given task.

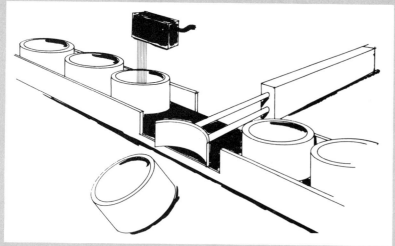

Dark caps are checked for white liners by a photoelectric scanner. The scanner activates a mechanism that rejects caps without the liners.

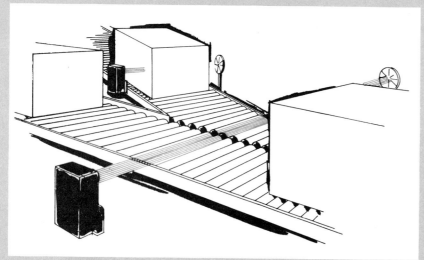

To prevent collisions where two conveyors merge, each conveyor is monitored by a control that powers the other conveyor when its own is cleared.

Using logic for one-shot pulse output, a photoelectric control slows a conveyor and fills the carton which has interrupted the light beam.

Figure 13-1 Applications of photoelectric control. (MICRO SWITCH, a Honeywell Division, Freeport, Illinois)

A tubular light source and photoreceiver in a specially designed bracket detect registration marks to initiate any related operation, such as printing, cutoff, or folding.

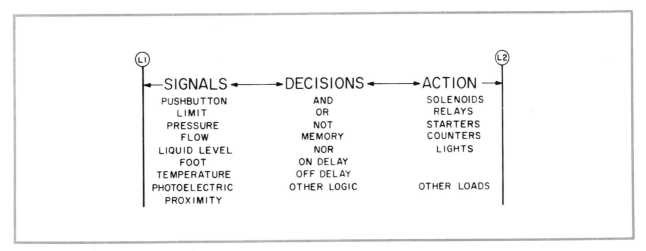

Figure 13-2 Photoelectric and proximity controls are located in the signal portion of a circuit. Based on the signals from these devices, various types of logic may be introduced to perform a variety of actions.

In this unit emphasis will be placed on how photo and proximity sensors produce signals, the available types, and the correct connection into a control circuit. Circuit logic and required interfacing will also be covered.

Photosensors

The term photoelectric cell (or photocell) is a general term used to describe a variety of light sensitive devices. From the electrician's standpoint, this usually means a variety of pre-packaged devices designed to accomplish specific control functions. In most cases the electrician will not need to know the internal working of each device, but rather whether the device is working or not and whether it should be repaired or replaced. Since most photosensors are completely sealed anyway, repair of internal parts is impossible without destroying the device—replacement is the only answer.

With these facts in mind, we will not dwell heavily on photosensor circuity but rather will discuss some of the basic operating principles involved in generating signals through the use of photosensors.

Photovoltaic Cells. A type of photoelectric cell that converts light energy directly to electrical energy is a photovoltaic cell. This type of sensor is sometimes called a solar cell; when several of

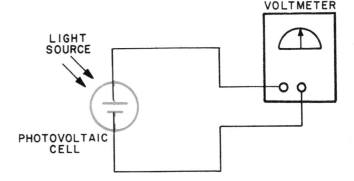

Figure 13-3 Schematic diagram containing a photovoltaic cell.

these sensors are connected they form sources of energy to drive small electrical loads in remote locations. Figure 13-3 illustrates a schematic diagram containing a photovoltaic cell. Notice that no source voltage is required. In this circuit the cell will produce its own voltage according to the amount of light that strikes it. The greater the light, the greater the output voltage, up to the limit of the cell.

Photoemissive Cells. A photoemissive cell has a photosensitive cathode that emits electrons when struck by light. These electrons are gathered by an anode surrounding the cathode and result in current flow. Since the device is sealed in a vacuum tube, current flows readily. These cells were used earlier in industrial equipment but are for the most part obsolete and have been replaced by solid state devices. Figure 13-4 illustrates an earlier model photocell.

Photoresistive Cells. A cell whose resistance changes with light intensity is called photoresistive or photoconductive. When light strikes the cell, the resistance of the cell decreases and current may pass through it more readily. Figure 13-5 illustrates a practical circuit using this principle. With this circuit we can control contactor C1 by the amount of light striking the photoresistive cell. As the light intensity increases, the current through C1 is increased, causing it to energize. Once energized, the NC contacts of line two open and the light is turned off. Thus we have created an automatic light sensitive control to turn lights on when it is dark and off when it is light. To make this system less sensitive to light flashes such as those from lightning and passing cars, a time delay circuit could be added to respond more uniformly to light changes.

Photodiodes/Phototransistor. When a diode is reverse biased, it becomes sensitive to light energy and can be used as a photodiode. When used in this manner, the case of the diode must be designed with a transparent area to allow light to enter. As light intensity increases, diode current increases, making it ideal as a control device. Figure 13-6 illustrates a simple circuit using a silicon diode as a photoelectric sensor. Since the sensitivity of this device is quite low (less than 50 microamperes), photodiodes are usually coupled with different transistor configurations to form more practical control sensors called phototransistors. The big advantage of photodiodes and phototransistors is the response time. Photodiode response times are in fractions of microseconds as compared to several milliseconds for other photosensitive devices.

SCANNING TECHNIQUES (SIGNALS)

The signal input of a photoelectric control is accomplished by scanning. Scanning is the initial phase of photoelectric control. In scanning, a light source and photosensor are used to detect the external object or condition that triggers the control. In other words, the photosensor is made to recognize a change in light. This condition is illustrated in Figure 13-7, where a photoelectric control is used to detect a truck entering a loading bay and signal the driver when to stop.

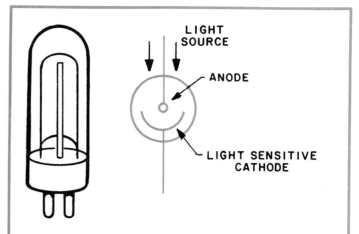

Figure 13-4 Photoemissive cells have been replaced by solid state devices.

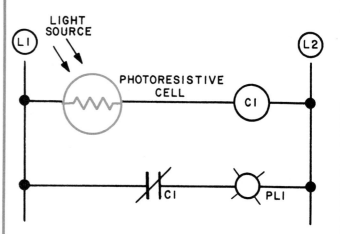

Figure 13-5 Resistance changes with light intensity with a photoresistive cell.

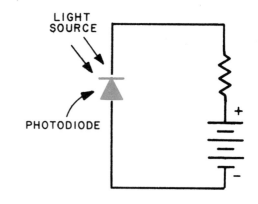

Figure 13-6 In a simple photodiode circuit, as light strikes the photodiode, current increases in the circuit, making it ideal as a control device.

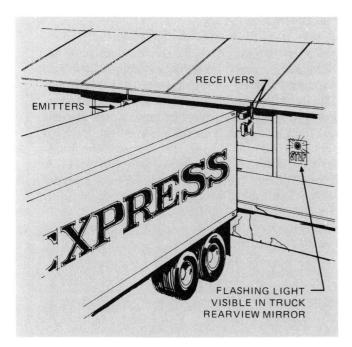

Figure 13-7 A signal generated from a photoelectric control can be used to monitor the height of an incoming truck. (MICRO SWITCH, a Honeywell Division, Freeport, Illinois)

There are several scan techniques or ways in which to set up the light source and photoreceiver to detect objects. The best technique to use is the one that yields the highest signal ratio for the particular object to be detected, subject to scanning distance and mounting restrictions.

Characteristics of the objects to be detected that have a bearing on scan technique include whether the objects are: (1) opaque or translucent; (2) highly or only slightly reflective; (3) in the same position or randomly positioned as they pass the sensor. Detecting change in color is a special consideration.

Direct Scan

Figure 13-8 illustrates direct scanning. In direct scan, the light source and photoreceiver are posi-

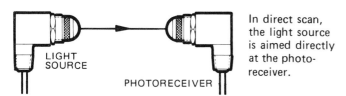

In direct scan, the light source is aimed directly at the photoreceiver.

Figure 13-8 In direct scan the light source is aimed directly at the photoreceiver. (MICRO SWITCH, a Honeywell Division, Freeport, Illinois)

tioned opposite each other, so light from the source shines directly at the sensor. The object to be detected passes between the two. If the object is opaque, direct scan will usually yield the highest signal ratio and is usually the best choice (Figure 13-9).

As long as an object blocks enough light as it interrupts the light beam, it may be skewed or tipped in any manner. As a rule of thumb, object size should be at least 50% of the diameter of the photoreceiver lens. To block enough light when detecting small objects, special converging lenses

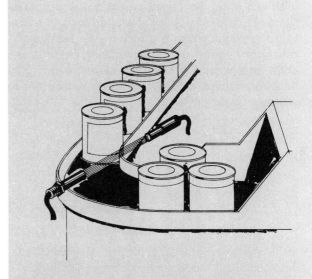

Counting products is a common application of photoelectric controls. Counting batches or groups of cans or other items prior to packaging or group processing is also common.

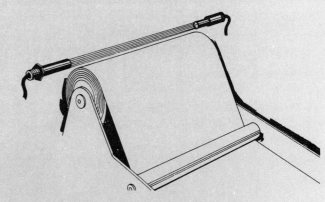

The size of a paper or fabric roll can be controlled by positioning a light source and a photoreceiver so the roll diameter blocks the beam.

Figure 13-9 In direct scan, the highest signal ratio will occur with opaque objects. (MICRO SWITCH, a Honeywell Division, Freeport, Illinois)

for the light source and photoreceiver can be used to focus the light in a small bright spot (where the object should be made to pass), thereby eliminating the need for the object to be half the lens diameter. An alternative is to place an aperture, like that found in a camera, over the photoreceiver lens in order to reduce its diameter. Detecting small objects typically requires direct scan.

Because direct scan does not rely on the reflectiveness of the object (or on a permanent reflector) to be detected, no light is lost at the photoreceiver. Therefore, the direct scan technique lets you scan farther than reflective scanning.

Direct scan, however, is not without limitations. Alignment is critical and difficult to maintain where vibration is a factor. Also, with separate light source and photoreciever, there is the additional wiring of the light source, which may be inconvenient if the application is difficult to reach.

Reflective Scan

In reflective scan, the light source and photoreceiver are placed on the same side of the object to be detected (Figure 13-10). Limited space or mounting restrictions may prevent aiming the light source directly at the photoreceiver, so the light beam is reflected either from a permanent reflective target or surface, or from the object to be detected, back to the photoreceiver. There are five types of reflective scan: retroreflective scan, polarized scan, convergent beam specular scan and diffuse scan.

Retroreflective Scan

Figure 13-11 illustrates an application with retroreflective scanning. With retroreflective scan, light source and photosensor occupy a common housing referred to as a retroreflective scanner. The light beam is directed at a retroreflective target (reflective acrylic disc or tape) which returns the light along the same path it was sent. Perhaps the most commonly used retro target is the familiar bicycle-type reflector shown in Figure 13-11. A larger reflector returns more light to the photosensor and thus allows you to scan farther. With

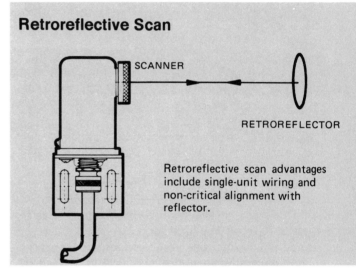

Retroreflective Scan

SCANNER

RETROREFLECTOR

Retroreflective scan advantages include single-unit wiring and non-critical alignment with reflector.

Figure 13-10 Retroreflective scan advantages include single unit wiring and non-critical alignment with reflector. (MICRO SWITCH, a Honeywell Division, Freeport, Illinois)

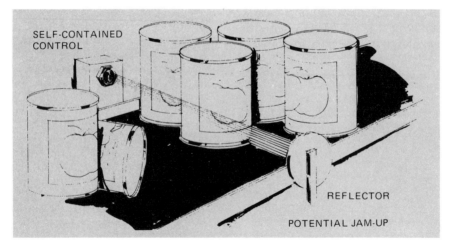

SELF-CONTAINED CONTROL

REFLECTOR

POTENTIAL JAM-UP

Figure 13-11 The familiar bicycle type reflector can be used on retroreflective scan to stop potential jam-ups. (MICRO SWITCH, a Honeywell Division, Freeport, Illinois)

retro-targets, alignment is not so critical. The light source/photosensor can be as much as 15° to either side of the perpendicular to the target. Also, since alignment need not be exact, retroreflective scan is an excellent way to counteract vibration.

Retroreflection from a stationary target normally provides a high signal ratio as long as the object passing between scanner and target is not highly reflective and passes very near the scanner. Retroreflective scan is a preferred technique for detecting translucent objects and assures a higher signal ratio than is obtainable with direct scan. With direct scan, the "dark" signal may not register very dark at the photosensor, because some light will pass through the object. With retroreflective scan, however, any light that passes through the translucent object on the way to the reflector is diminished again as it returns from the reflector.

Another way to use retroreflective scan is to apply retroreflective tape coding to cartons or other items that must be sorted. This is illustrated in Figure 13-12.

Retroreflective scan can normally be used at distances up to 40 feet in clean air conditions. As the distance to the target increases, a larger retro target should be used to intercept and return as much light as possible.

Single-unit wiring and maintenance are secondary advantages of retroreflective scanning. This is because there is no separate light source to wire and all of the working components are housed in one common unit.

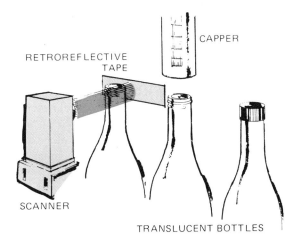

Figure 13-12 Retroreflective tape can be used with a scanner to determine when a bottle should be capped. (MICRO SWITCH, a Honeywell Division, Freeport, Illinois)

Polarized Scan

Figure 13-13 illustrates polarized scan. As with retroreflective scan the light source (emitter) and photoreceiver are placed on the same side of the object to be detected. Polarized controls use a special lens that filters the emitter's beam of light so that it is projected in one plane only. The receiver responds only to the de-polarized reflected light from corner cube-type reflectors or polarized sensitive reflective tape. It ignores the light reflected from most varieties of shrink wrap materials, shiny luggage, aluminum cans or common reflective objects.

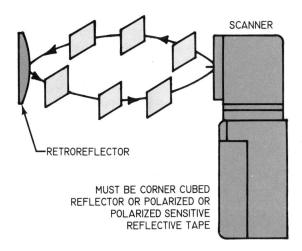

Figure 13-13 Polarized controls use a special lens which filters the emitter's beam of light so that it is projected in one plane only. The receiver responds only to de-polarized reflected light. (MICRO SWITCH, a Honeywell Division, Freeport, Illinois)

Convergent Beam

Figure 13-14 illustrates convergent beam scanning. Convergent beam scanning is used to detect products that are only inches away from another reflective surface. It is the first choice for edge-guiding or positioning clear or translucent materials. Because the beam is well-defined, it is also a good second choice for position sensing of opaque materials.

Convergent beam scanning is a special variation of the diffuse mode. The control's optical system is the key to its operation. It simultaneously focuses and converges the light beams to a fixed focal point in front of the control. The control is essentially blind a short distance before and beyond the focal point. Operation is even possible

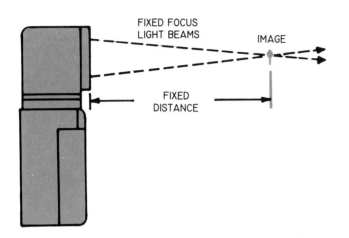

FIXED FOCUS
LIGHT BEAMS

IMAGE

FIXED
DISTANCE

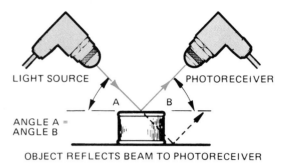

LIGHT SOURCE   PHOTORECEIVER

ANGLE A =
ANGLE B

OBJECT REFLECTS BEAM TO PHOTORECEIVER

Figure 13-14 The depth of field is determined by the angle between light source and photoreceiver. A narrower angle yields greater depth of field, for objects subject to height variation. If the aim is to keep height variation or fill level within precise limits, a wider angle is used. (MICRO SWITCH, a Honeywell Division, Freeport, Illinois)

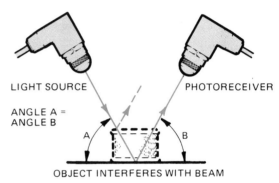

LIGHT SOURCE   PHOTORECEIVER

ANGLE A =
ANGLE B

OBJECT INTERFERES WITH BEAM

Figure 13-15 Specular scan is used with objects with a shiny surface.

when highly reflective backgrounds are present. Like diffuse scanning, convergent beam scanning senses light reflected back directly from an object. Convergent beam scanning can be a very useful mode for detecting the presence or absence of small objects while ignoring nearby background surfaces.

Parts can be sensed on a conveyor from above while ignoring the conveyor belt. Or they can be sensed from the side without detecting guides or rails directly in back of the object. Convergent beam scanning can detect the presence of fine wire, resistor leads, needles, bottle caps, pencils, the stack height of material, fill level of clear liquids, and discriminate the product from its background. It is also capable of sensing black code marks against a contrasting background.

Specular Scan

Figure 13-15 illustrates specular scanning. The specular scan technique employs a very shiny surface, such as rolled or polished metal, shiny plastic, or a mirror to reflect light to the photosensor. With a shiny surface, the angle at which light strikes the reflecting surface equals the angle at which it reflects from the surface—just as in billiards, where a ball leaves the cushion at an angle equal to the angle it struck the cushion. If billiard balls followed the principle of retroreflection, they would rebound back to the cue. With this in mind you can see that the positioning of the light source

and photoreceiver must be precise, and the distance of the reflecting surface from the light source and photoreceiver must be consistently controlled. If they are not, much of the light will be lost because of an improper angle.

The size of the angle between light source and photoreceiver determines the depth of scanning field. This depth of the scanning field is illustrated in Figure 13-16. With a narrower angle, there is more depth of field. With a wider angle, there is less depth of field. In a fill level detection application, for example, this means that a wider angle between light source and photoreceiver enables the detection of a fill level more precisely.

Specular scan also gives a good signal ratio when required to distinguish between shiny and non-shiny (matte) surfaces, or when using depth of field to reflect selectively off shiny surfaces of a certain height. When monitoring a non-flat shiny surface with high and/or low points that fall outside the depth of field, the photosensor will see the points as dark signals.

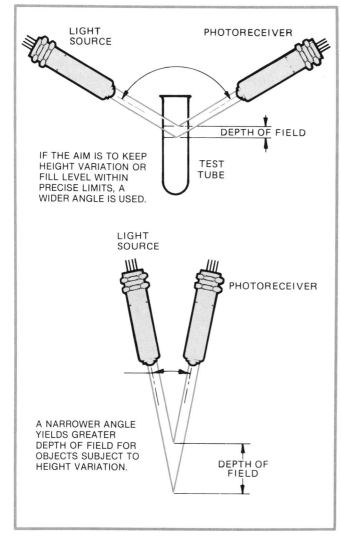

Figure 13-16 The depth of field is determined by the angle between light source and photoreceiver. (MICRO SWITCH, a Honeywell Division, Freeport, Illinois)

Figure 13-16 labels:

LIGHT SOURCE

PHOTORECEIVER

DEPTH OF FIELD

IF THE AIM IS TO KEEP HEIGHT VARIATION OR FILL LEVEL WITHIN PRECISE LIMITS, A WIDER ANGLE IS USED.

TEST TUBE

LIGHT SOURCE

PHOTORECEIVER

A NARROWER ANGLE YIELDS GREATER DEPTH OF FIELD FOR OBJECTS SUBJECT TO HEIGHT VARIATION.

DEPTH OF FIELD

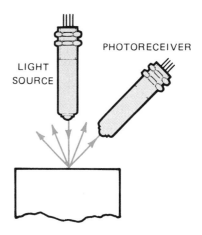

Figure 13-17 Diffuse scanning is used on objects with a non-shiny surface.

Figure 13-17 labels:

LIGHT SOURCE

PHOTORECEIVER

Diffuse Scan

Figure 13-17 illustrates diffuse scanning. Diffuse scanning is the use of light to detect the proximity of an object. Non-shiny (matte) surfaces such as kraft paper, rubber, cork and wooden pallets absorb most direct light and reflect only a small amount. Light is reflected or scattered nearly equally in all directions. In diffuse scan, the light source is positioned perpendicular to a dull surface and the photoreceiver placed equally close and off to the side (usually 45°) to pick up some of the diffuse (scattered) reflection. Because the light is scattered, only a small percentage of it reaches the photosensor. Therefore, scanning dis-

tance is usually limited to six inches maximum (except with some high-intensity modulated LED controls), even with very bright light sources. It is often difficult to get a sufficient signal ratio with diffuse scanning when the surface to be detected is almost the same distance from the sensor as is another surface—for instance, a nearly flat or low profile cork liner moving along a conveyor belt. When the sensor is trying to distinguish between the reflection from two unlike surfaces, signal ratio can be improved considerably where the unlike surfaces are contrasting colors.

Diffuse scan is used in registration control, and to detect material (corrugated metal, for example) with a slight vertical flutter which might present a consistent signal with specular scan. Alignment is not critical in picking up diffuse reflection.

Fiber Optics

Fiber optics is not a scanning technique, but a method of controlling or transmitting the signal (light beam) from or to the control. Fiber optics use transparent fibers of glass or plastic to conduct and guide light energy. They are used in photoelectric controls as light pipes.

The control's beam is transmitted through a cable. It returns through a separate cable either combined in the same cable assembly (*bifurcated*) or within a separate cable assembly (*thru scan*) to the receiver.

Scanning options depend on the type of cable selected. Figure 13-18 illustrates three types of fiber optic scan techniques. Retroreflective and diffuse scan use a bifurcated cable and thru scan uses two separate cables (emitter and receiver). Scan distances vary, depending on type of scan, from typically 0.4 to 54 inches. An optical lens accessory that attaches to some cable ends significantly increases scan distances. Figure 13-19 illustrates typical fiber optic controls.

Combining the optic cables with photoelectric controls has many advantages. Small parts detec-

tion and usage in limited mounting space is obvious. High temperature, high vibration or high electrical noise levels at any control can cause false triggering. With fiber optics, the light emitting and receiving components are located remotely at the control's housing and only passive light-transmission fibers need be exposed to the severe environmental or hazardous areas.

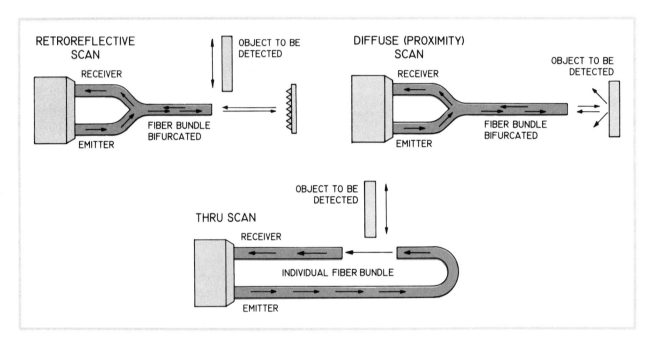

Figure 13-18 Fiber optics use transparent fibers of glass or plastic to conduct and guide light energy. Fiber optic control can be connected for each type of scan technique. (MICRO SWITCH, a Honeywell Division, Freeport, Illinois)

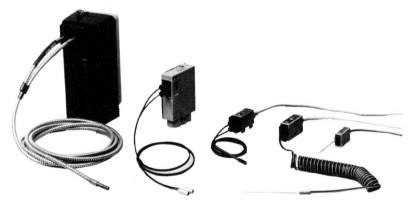

Figure 13-19 Fiber optic controllers are available in different sizes and configurations. They often use the same basic housing as standard photoelectric controls. (MICRO SWITCH, a Honeywell Division, Freeport, Illinois)

Selection of Scanning Techniques

Selecting the correct scanning technique is an important part of applying photoelectric controls. In many applications several methods of scanning will work, but usually one method is the best. Figure 13-20 illustrates a summary of the basic scanning techniques, listing major advantages and disadvantages of each type.

LOGIC MODULES (THE DECISION)

The second phase of photoelectric control involves amplifying the weak signal from the photosensor and adding logic (ON/OFF, TIME DELAY, COUNTING, etc.). This second phase occurs in the amplifier/logic circuitry of a separate control base or is a part of the self-contained control. This amplifier/logic circuit is usually a plug-in component, making changing circuit logic easy. In selecting a control base and amplifier/logic circuit, the electrician must consider the specific application and environment of the circuit, for these determine the final selection.

Color Differentiation (Registration Control)

Contrast is the key to distinguishing color. High contrast (dark on light or vice versa) provides the best signal ratio and control reliability.

When the background is clear (transparent) the best method of detecting any color mark is thru scan. When the background is a second color, contrasts such as black-on-white usually assure sufficient signal ratio (difference between light and dark) to be handled with diffuse scan. Red (or a color that contains red pigment, such as yellow, orange or brown) on a white or light background is a special case.

Because white light contains red, photosensors that are more sensitive to the red-infrared region of the spectrum do not register enough of a signal ratio between white and red. CdSe (cadmium selenide) photocells and all phototransistors fall in this range. CdS (cadmium sulphide) cells, however, peak in the green-yellow portion of the spectrum

SCANNING FEATURES	DIRECT SCAN	RETRO-REFLECTIVE SCAN	SPECULAR SCAN	DIFFUSED SCAN
Configuration	A transmitter on one side sends a signal to a receiver on the other side. The object to be detected passes between the two.	The transmitter and receiver are housed in one package and are placed on the same side of the object to be detected. The signal from the transmitter is reflected back to the receiver by a retroreflector.	The transmitter sends a signal to the receiver by reflecting the signal of the object to be detected. The transmitter and receiver are not housed in the same package and the receiver must be positioned precisely to receive the reflected light.	The transmitter and receiver are housed in one package. It is the object being detected that reflects the signal back to the detector. No retro-reflector is used.
Major Advantages	Reliable performance in contaminated areas; Long range scanning; detection of small parts; and the most well-defined effective beam of all scanning techniques.	Ease of installation in that wiring on only one side is required; alignment need not be exact, more tolerant to vibration.	Good for detecting a shiny versus a dull surface; depth of field can be changed by changing transmitter, receiver angle.	Ease of installation in that wiring on only one side is required since detected object returns the signal; alignment is not critical; the best scanning technique for transparent on translucent materials.
Major Disadvantages	Wiring and alignment required for both the transmitter and receiver; higher installation cost.	Sensitive to contamination since light source must travel to the retro-reflector and back; hard to detect transparent or translucent materials; not good for small part detection.	Wiring required for both the transmitter and receiver even though they are in the same housing; alignment is important.	Limited range since object is used to reflect transmitted light back; performance changes from one type of object to be detected to another type.

Figure 13-20 Selecting the correct scanning technique is an important part of applying photoelectric control.

and are not sensitive to red light. Also, photoreceivers with a CdS cell include a blue-green filter (transmits only blue and green) to block red wavelengths. A photoreceiver with a CdS cell makes red appear dark on a light background.

As illustrated in Figure 13-21 a retroreflective control with a short focal length lens (but without a retro target) can be used to detect registration marks. It is placed near the mark and is actually used in the diffuse scan technique. If a retro scanner is used to detect marks on a shiny surface, it should be cocked somewhat off the perpendicular to make certain only diffuse reflections are picked up. Otherwise, the shiny surface of the mark could mirror-reflect so brightly it could overcome the dark signal a CdS cell normally gets from red. This would mean a light signal from both background and mark. In detecting colors, you should use diffuse (weakened) rather than specular (mirror) reflection.

Either a cadmium selenide photocell or a phototransistor can be used to detect colored marks—black, blue, green—which contain little or no red pigment. (Red as a background presents no difficulty, since a black or blue mark against a red background would appear dark on light.) The phototransistor, because of its longer life, faster response, and temperature stability, is usually preferable.

Figure 13-21 A retroreflective control with a short focal length lens can be used to detect registration marks. This can be useful in application requiring positioning or identification. (MICRO SWITCH, a Honeywell Division, Freeport, Illinois)

MODULATED AND UNMODULATED CONTROLS

In photoelectric controls the light source is either modulated or unmodulated, as illustrated in Figure 13-22. Most manufacturers offer both modulated and unmodulated controls.

In an unmodulated light source, the light beam is constant and not turned on and off. The advantage to unmodulated light is that it is faster and less expensive. Use of unmodulated light should be considered when scanning range is very short and when dirt accumulation, ambient light and penetrating power is not a problem. Unmodulated light sources can sometimes be used for high speed counting since the beam is continually transmitted.

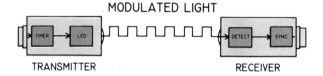

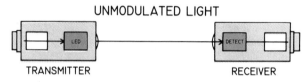

MODULATED CONTROL

ADVANTAGES	APPLICATION CONSIDERATIONS
1. Ambient light immunity.	1. Not as fast as non-modulated controls.

2. High excess gain.
  a. For penetration power through dust, dirt, mist, steam, etc.
  b. For long distance scanning (up to 2500 feet).	

3. For applications with high vibration	
4. For diffuse scanning of objects with low reflectivity.	

UNMODULATED CONTROL

ADVANTAGES	APPLICATION CONSIDERATIONS
1. Best for registration control when an incandescent light source and photocell are used.	1. Not recommended when stray ambient light is present.
2. Fastest operating speed (less than 1 msec. response).	2. Not recommended where high excess gain is needed.

Figure 13-22 In the modulated controller the infrared light source is turned on and off at a very high frequency. In a unmodulated light source, the light beam is constant and not turned on and off.

In the modulated controller the infrared light source is turned on and off at a very high frequency. The modulated controller then responds to modulating frequency rather than light intensity. Because the receiver circuitry is tuned to the emitter modulating frequency, the control does not respond to ambient light. This also helps achieve rejection of noise.

Modulated controls are the answer to many troublesome problems with high ambient light. A modulated controller should always be used outdoors, in low visibility scanning conditions, high vibration areas, and any other place where environmental challenges are severe. They are good for penetrating dirt, dust and mist, as well as long range scanning. Some controls are modulated at different frequencies than others. This prohibits using the emitter of one control with the receiver of others. This may be desirable in applications where more than one control is needed in close proximity. As a general rule, a modulated light source should always be the first choice when using a photoelectric control.

EXCESS GAIN

Excess gain is the ratio of the amount of transmitted light falling on the receiver over and above the minimum amount of transmitted light required just to operate the photoelectric control. The signal that is just barely adequate for operation is called the *threshold signal*. Thus, in equation form:

$$\text{Excess Gain} = \frac{\text{Received Signal}}{\text{Threshold Signal}}$$

In a clean environment an excess gain equal to or greater than 1 is sufficient. However, if an environment in which 50% of the light energy is attenuated, a minimum excess gain of 2 is required. Attenuation is the reduction of the light energy due to dust, dirt, mist, lint, fog, smoke, etc., either in the air or on the lens of the photoelectric control.

Attenuation through absorption and scattering reduces the available light transmitted to the receiver. In photoelectric applications where one end of the control system will be in a dirtier environment than the other, always place the transmitter at the cleaner end. This will minimize the scattering effect of dirt. Figure 13-23 illustrates an environment in which high excess gain is required.

For minimum operating performance of a photoelectric control the following guidelines can be used.

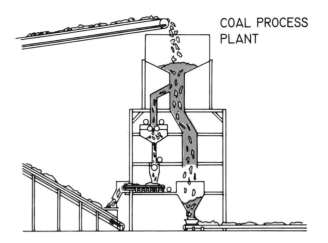

Figure 13-23 In a high dust and dirt environment high excess gain of 50 or more may be required.

1. Allow an excess gain of 1.5 for clean air, clean lens conditions.

2. Allow an excess gain of 4 to 6 for slightly dirty applications such as slight amounts of dust, dirt, and mist in the air or on the photo lens.

3. Allow an excess gain of 10 to 20 for an obviously dirty area.

4. An excess gain of 50 or more will be required in areas with heavy dirt, dust, smoke or lint.

5. Always keep the lens clean on the photoelectric control to reduce attenuation. An additional cover may be required to help prevent lens contamination.

RESPONSE TIME

The response time of a photoelectric control is important when the object to be detected moves past the beam at a very high speed, or when the object to be detected is not much bigger than the effective beam of the controller. This response time of the controller will determine how many pulses (objects) per second the controller can detect. This information is listed in the specification sheet of the photoelectric control. For example, a photoelectric control may have an activating frequency of 10 pulses per second.

As illustrated in Figure 13-24 the beam must be totally blocked before the receiver will shut off. The receiver will turn on when the object just uncovers the edge of the beam. This has the effect of shortening the size of the object to be detected as seen by the photoelectric control.

The length of time that an object breaks the beam

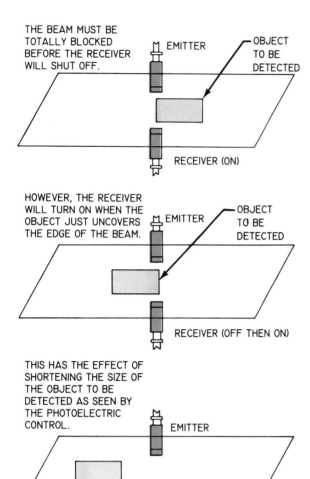

THE BEAM MUST BE TOTALLY BLOCKED BEFORE THE RECEIVER WILL SHUT OFF.

EMITTER

OBJECT TO BE DETECTED

RECEIVER (ON)

HOWEVER, THE RECEIVER WILL TURN ON WHEN THE OBJECT JUST UNCOVERS THE EDGE OF THE BEAM.

EMITTER

OBJECT TO BE DETECTED

RECEIVER (OFF THEN ON)

THIS HAS THE EFFECT OF SHORTENING THE SIZE OF THE OBJECT TO BE DETECTED AS SEEN BY THE PHOTOELECTRIC CONTROL.

EMITTER

OBJECT TO BE DETECTED

RECEIVER (ON)

Figure 13-24 Response time of a photoelectric control is important when the object to be detected moves past the beam at a very high speed, or when the object to be detected is not much larger than the effective beam of the controller.

can be found by the following formula:

$$T = \frac{W - D}{S}$$

Where:

T = Time that the object takes to break the beam
W = Width of object moving through the beam in inches
D = Effective beam diameter in inches
S = Speed of object in inches per second

Sensitivity Adjustment

The sensitivity adjustment on photoelectric controls determines the operating point or the level of light intensity that triggers the output. An adjustment is provided for setting the sensitivity adjustment between a minimum and maximum range.

The goal of sensitivity adjustment is to determine the minimum and maximum setting at which the unit will operate, and then to set the adjustment halfway between for best operational stability. A lower sensitivity setting may be necessary under the following conditions: (1) in high ambient light; (2) where electrical interference can be expected; (3) when detecting translucent objects.

The lower setting, however, should be kept as high above the threshold as possible for operational stability.

LIGHT OPERATED/DARK OPERATED

Thru Scan

In a thru scan system, dark operated means that the output energizes when a target is present, breaking the beam. The light operated condition occurs when the target is missing, as illustrated in Figure 13-25.

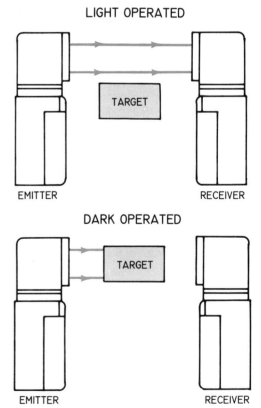

LIGHT OPERATED

TARGET

EMITTER
RECEIVER

DARK OPERATED

TARGET

EMITTER
RECEIVER

Figure 13-25 The photoelectric control can be light operated and dark operated when used with thru scan control. (MICRO SWITCH, a Honeywell Division, Freeport, Illinois)

Retroreflective Scan

In a retroreflective scan system the conditions are the same as for thru scan. The dark operated condition occurs when the target is present and the receiver sees light when the target is gone, as illustrated in Figure 13-26.

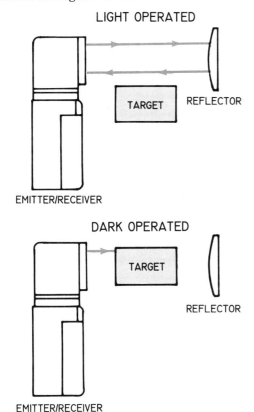

Figure 13-26 The difference between light operated and dark operated, as used with retroreflective scan control. The photoelectric control can be set for the required operation. (MICRO SWITCH, a Honeywell Division, Freeport, Illinois)

Combining Light/Dark Operations

Combining a control's receiver condition, light or dark operated with logic terms AND/OR, can result in logic gate interfacing with multiple control hookups.

Light-operated OR or *light-OR* means that the output will energize when any one (or more) of the receivers sees light. *Dark-operated OR* or *dark-OR* means that the output will energize when any one

(or more) of the receivers sees dark. *Dark-operated AND* or *dark-AND* means that no output will occur until all receivers are seeing dark. *Light-operated AND* or *light-AND* means that no output will occur until all receivers are seeing light.

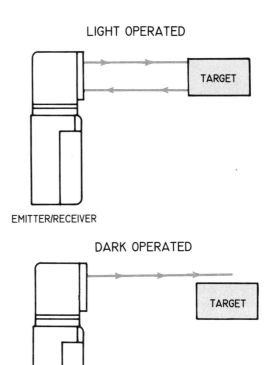

Figure 13-27 The difference between light operated and dark operated, as used with diffuse (specular/convergent) scan control. (MICRO SWITCH, a Honeywell Division, Freeport, Illinois)

The four logic conditions of light-OR, dark-OR, light-AND, and dark-AND cover most multiple control hookup situations. The combination of this logic is shown in the table at the top of page 354.

Diffuse Scan

In diffuse, specular, or convergent beam scanning the conditions reverse from thru scan. The light operated condition occurs when the target is present, making the beam. When the target is missing, no light is returned, as illustrated in Figure 13-27.

Light-OR Logic	Receiver A	Receiver B	Load
	Light	Light	Energized
	Light	Dark	Energized
	Dark	Light	Energized
	Dark	Dark	De-energized
Dark-OR Logic	**Receiver A**	**Receiver B**	**Load**
	Light	Light	De-energized
	Light	Dark	Energized
	Dark	Light	Energized
	Dark	Dark	Energized
Dark-AND Logic	**Receiver A**	**Receiver B**	**Load**
	Light	Light	De-energized
	Light	Dark	De-energized
	Dark	Light	De-energized
	Dark	Dark	Energized
Light-AND Logic	**Receiver A**	**Receiver B**	**Load**
	Light	Light	Energized
	Light	Dark	De-energized
	Dark	Light	De-energized
	Dark	Dark	De-energized

ENVIRONMENTAL INFLUENCES IN PHOTOELECTRIC CONTROL

Ambient Light

Use any photoreceiver in fluorescent or moderate incandescent light. In high incandescent light, use a bright light source. If possible, shade the photoreceiver or face it away from the ambient light. Always use a modulated light source outdoors to combat sunlight. Common sources of light problems are dirt accumulation on lenses, aging light sources, gradual drift in alignment between light source and photoreceiver, and poor color contrast between light and dark signals. Any of these factors can degrade the signal ratio, which means that the control sensitivity must be set very near the trip point, where it is more susceptible to triggering from a slight variation in light.

Vibration

Use any light source and photoreceiver where there is little vibration. To compensate for moderate vibration, use a bright light source and focus it to present a large image or use retroreflective scanning. A retroreflective target returns the beam to the scanner even if, due to vibration, the two are no longer precisely aligned. For high vibration, use an LED light source or self-contained modulated light source control.

Temperature

Temperature can affect photoelectric control to the point where it no longer functions. For this reason temperature may dictate the choice of the photoreceivers used. Use a phototransistor type receiver for temperatures below 0°C (32°F) or above 60°C (122°F). Any photoreceiver is suitable between 0°C and 60°C. For extremely high temperatures up to 150°C (302°F), specify a high temperature phototransistor type receiver. In high temperature areas it is a good idea to cool the photoreceiver with air.

Moisture

In areas of indirect splash or high humidity, use a splash-proof light source and photoreceiver. With a high pressure or hosedown water area, use a photoreceiver and light source that meet NEMA 4 standards (water tight and dust tight, indoor and outdoor). In areas of high-pressure detergent stream, use a painted (not anodized finish) photoreceiver and light source. This helps combat the effect of humidity.

Oil, Paint, Dust

Avoid direct exposure to spray. Mount the light source and photoreceiver outside the spray area. Use a bright light source and set the control sensitivity well above the operating point. If possible, direct clean air across the light source and photoreceiver lenses and avoid retroreflective scanning. It may become necessary to clean the lenses regularly in some applications.

Explosive Atmosphere

Use explosion-proof light source and photoreceivers, or mount standard units in explosion-proof housings. Explosion-proof light sources and photoreceivers can be used in *National Electrical Code®* Class 1, Division 1, Group A, B, C and D locations. (Refer to the current *National Electrical Code®* for details.)

Line Voltage Variation

Voltage variations affect the light source. If the variance is 10 to 30 volts, use a bright light source and set the control sensitivity well above the minimum operating point. Use a voltage regulator where variance exceeds 30 volts.

Electrical Noise

Electrical noise can interfere with the photoelectric's circuit performance. Electrical noise can originate in the vicinity of photoreceiver leads, lines, or control output device. This type of interference is caused by contact bounce, power line surges, inductive loads being turned on or off, rotating industrial machinery, and contact arcing. Precaution must be taken to eliminate potential sources of electrical noise from the photoelectric circuit. To help accomplish this, ground all photoreceiver shields and control cases. If necessary, some filtering may have to be added.

INTERFACING

When interfacing a photoelectric control to a DC output load, either a current sinking (open collector) output or a current sourcing output is used.

Current Sinking Output

Figure 13-28, A and B, represents the output stage of a typical current sinking output photoelectric control. In this circuit configuration, the load is generally connected between the supply voltage and output terminal (collector) of the control. When the control is actuated (turned ON), current flows through the load, into the output transistor to ground. The supply voltage of the control (VS) need not be the same value as the load supply (VLS); however, it is usually convenient to use a single supply. The control output voltage is measured between the output terminal (collector) and ground (−). When the control is not actuated, current will

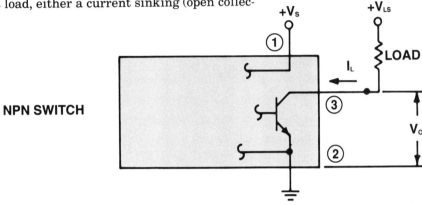

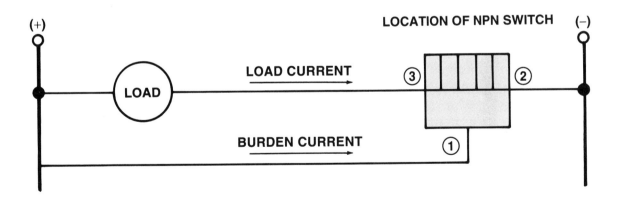

Figure 13-28 In a typical current sinking output circuit the load is connected between the supply voltage (+) and output terminal of the control.

not flow through the output transistor (except for a small leakage current). The output voltage, in this condition, will be equal to V_{LS} (neglecting the leakage current). When the control is actuated, the output voltage will drop to ground potential if the saturation voltage of the output transistor is neglected. The output voltage of a current sinking photoelectric control is normally ON (V_L high).

Current sinking derives its name from the fact that it "sinks current from a load." The current flows from the load into the control. Like a mechanical switch, the control allows current to flow when turned ON and blocks current flow when turned OFF. Unlike an ideal switch, the control has a voltage drop when turned ON and a small current (leakage) when turned OFF.

A current sinking output (and a current sourcing output) can generally control loads of only a few hundred milliamperes. If higher current switching is required an interface will be needed. The solid state relay is the most common interface used.

Current Sourcing Output

Figure 13-29, A and B, represents the output stage of a typical current sourcing output photoelectric control. In this circuit configuration, the load is generally connected between the output terminal (emitter) of the control and ground. When the control is actuated, current flows from the output transistor into the load to ground. The control output voltage is measured between the output terminal and ground ($-$) and is equal to the voltage across the load. When the control is not actuated, current will not flow through the output transistor (except for the leakage current). The output voltage will equal zero (neglecting the leakage current). When the control is actuated, the output voltage will rise to V_S less the collector-to-emitter voltage drop of the output transistor. The output voltage of a current sourcing control is normally OFF (V_O low). When interfacing to solid state DC inputs it is best to check input requirements of the system before selecting the control.

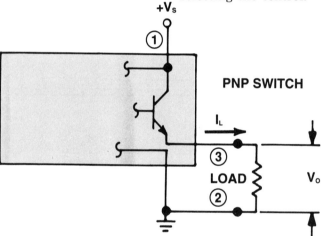

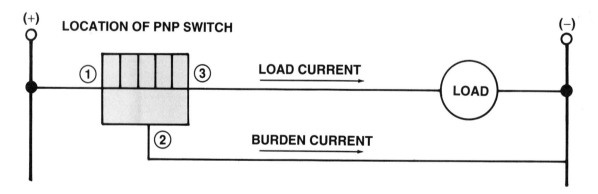

Figure 13-29 In a typical current sourcing output circuit the load is connected between the supply voltage ($-$) and output terminal of the control.

DC Sinking Series Wiring

Most current sinking output photoelectric controls can be wired in series. For 3-wire DC controls, the output current rating and voltage drop are the most important parameters.

Figure 13-30 illustrates how 3-wire DC current sinking output controls can be connected in series to provide AND logic control. In this example all controls must be ON (conducting) in order for the load to operate.

but must be taken into account. Each control in series must carry the load current plus the burden current for all downstream controls. In this application control number three must have an output circuit capable of handling both the load and current consumption of the control numbers one and two.

+ = TO POSITIVE OF POWER SUPPLY

- = TO NEGATIVE OF POWER SUPPLY

0 = OUTPUT OF SWITCH

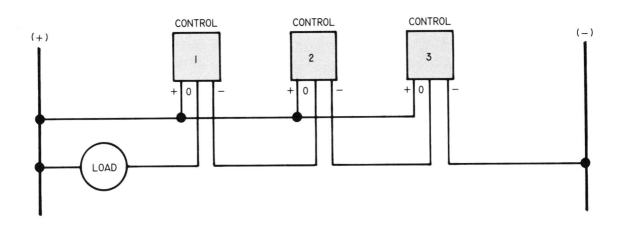

Figure 13-30 When wiring DC sinking in series for AND logic control, all controls must be activated in order for the load to be energized. (MICRO SWITCH, a Honeywell Division, Freeport, Illinois)

The maximum number of DC photoelectric controls that can be wired in series is limited by:

Supply voltage
Control's current consumption
Control's output current
Minimum operating voltage of the control
Minimum voltage of the load
Load current

Figure 13-31 illustrates an example of three current sinking output controls wired in series. Each sensor, when conducting, will have a voltage drop across it, thus reducing the available load voltage. This voltage drop is usually small for each control

DC Sinking Parallel Wiring

Figure 13-32 illustrates how 3-wire DC current sinking output controls can be connected in parallel to provide OR logic control. In this example any DC photoelectric control can be ON (conducting) to operate the load.

When 3-wire DC current sinking output controls are wired in parallel, leakage currents are important. The leakage currents of all sensors are added together for the total leakage current of the system. In most cases, the leakage currents are small enough that adding small numbers of them together will not cause a problem.

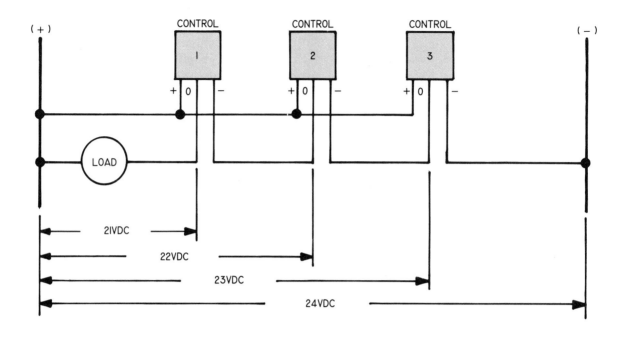

Figure 13-31 Voltage will drop when connecting sinking controls in series. Each sensor, when activated, will have a voltage drop across it, thus reducing the available load voltage. (MICRO SWITCH, a Honeywell Division, Freeport, Illinois)

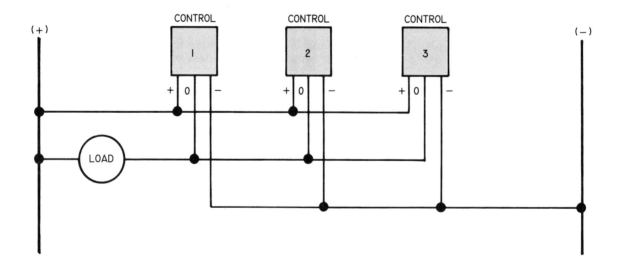

Figure 13-32 When wiring DC sinking in parallel for OR logic control, any control that is activated will energize the output. (MICRO SWITCH, a Honeywell Division, Freeport, Illinois)

Some manufacturers recommend using a blocking diode to prevent damage to the non-conducting sensor that can be caused when reverse polarity is applied to the sensor.

DC Sourcing Series Wiring
Figure 13-33 illustrates how 3-wire DC current sourcing output controls can be connected in series to provide AND logic control. In this example all controls must be ON (conducting) in order to operate the load.

DC Sourcing Parallel Wiring
Figure 13-34 illustrates how 3-wire DC current sourcing output controls can be connected in parallel to provide OR logic control. In this example any photoelectric control can be ON (conducting) to operate the load.

Interfacing DC Controls to AC Loads
Interfacing DC controls to AC loads always requires a relay as the interface. A solid state relay is the most common relay used. Factors that must

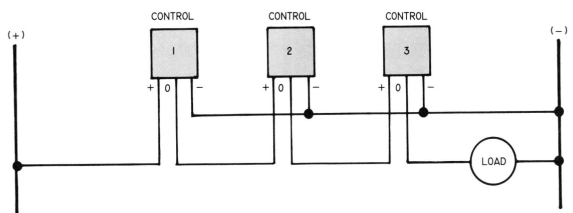

Figure 13-33 When wiring DC sourcing in series for AND logic control all controls must be activated in order for the load to be energized. (MICRO SWITCH, a Honeywell Division, Freeport, Illinois)

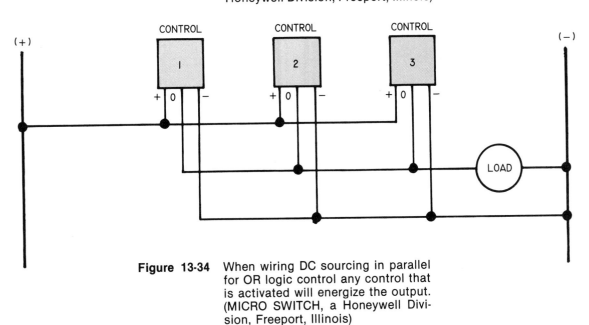

Figure 13-34 When wiring DC sourcing in parallel for OR logic control any control that is activated will energize the output. (MICRO SWITCH, a Honeywell Division, Freeport, Illinois)

be taken into consideration when selecting the SSR are load rating, switching speed and type of load.

AC Series Wiring

Figure 13-35 illustrates how a 2-wire AC photoelectric control can connected in series to provide AND logic control. In this example, all photoelectric outputs must be ON (conducting) to operate the load.

The maximum number of AC photoelectric controls that can be wired in series is limited by:

Supply voltage
Control's voltage drop
Minimum operating voltage of the control
Minimum operating voltage of the load

Figure 13-36 illustrates five controls connected in series. Typically a 2-wire control will have a voltage drop of about 7 volts across it. As illustrated this will leave the load with only about 80 volts, which is not recommended.

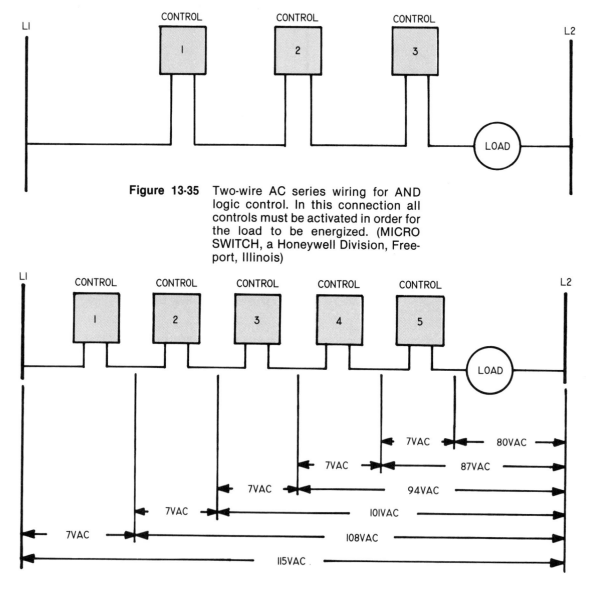

Figure 13-35 Two-wire AC series wiring for AND logic control. In this connection all controls must be activated in order for the load to be energized. (MICRO SWITCH, a Honeywell Division, Freeport, Illinois)

Figure 13-36 Each control will have a voltage drop across it when connecting two-wire AC in series, reducing the total voltage to the load. (MICRO SWITCH, a Honeywell Division, Freeport, Illinois)

AC Parallel Wiring

Figure 13-37 illustrates how 2-wire AC photoelectric controls can be connected in parallel to provide OR logic control. In this example any photoelectric control output can be ON (conducting) to operate the load.

The maximum number of AC photoelectric controls that can be wired in parallel is limited by:

Control's leakage current

Maximum OFF state load current

When connecting AC photoelectric controls in parallel, care must be taken to ensure that the sum of the leakage current for each control is less than the maximum OFF state load current. Also, because of false pulse protection, if one photoelectric control is operated then released, all others in the parallel circuit will not be able to respond to a target for up to 100 milliseconds. Parallel operation for 2-wire photoelectric controls should only be used when there is no danger that any two controls will be operated within 100 milliseconds of each other.

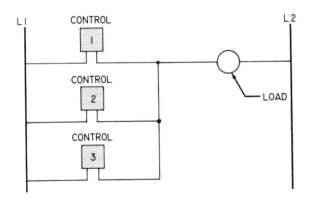

Figure 13-37 In two-wire AC parallel wiring for OR logic control, any control that is activated will energize the output. (MICRO SWITCH, a Honeywell Division, Freeport, Illinois).

PHOTOELECTRIC APPLICATIONS

Height and Distance Monitoring

Figure 13-38 A and B illustrates the line diagram of the height monitoring circuit and distance monitoring circuit for a truck loading bay. This application monitors a truck in a loading bay for necessary clearance and distance. (The dimensions given are arbitrary and will vary, depending upon particular needs.)

Any truck 16 feet - 8 inches high or larger must be unloaded at another bay. In the control circuit

photoelectric control one is used to turn on an alarm in line two if the truck is too high. Photoelectric control two is used to start a recycler timer that will flash a yellow light on and off at a distance of 2 feet from the dock. Photoelectric control three is used to start a recycler timer that will turn on a red light at a distance of 6 inches from the dock.

In this application direct scan using modulated controls will work the best. The photoelectric control will be connected for a dark operation of the controller, allowing for operation only when the beam is blocked by the truck.

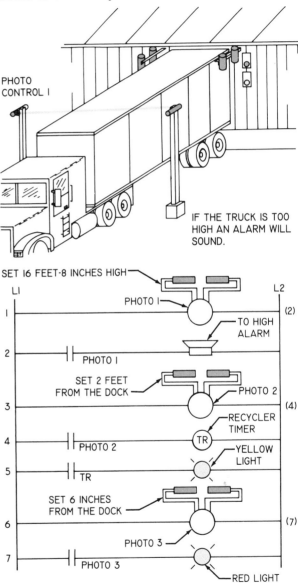

Figure 13-38 Three photoelectric controls monitor a truck loading bay for clearance and distance. When the truck is 2' from the bay, a yellow light will flash; at 6" away, a red light will turn on.

Product Monitoring

Figure 13-39, A and B, illustrates the use of photoelectric controls to detect a backup of a product on a line. In this application three photoelectric controls are used to turn on a warning light and turn off the line if required.

Photoelectric control one is used to turn on a warning light, indicating a product is at the end of the conveyor line. At this time an operator may remove the product or wait until more products are on the line. This will allow for best utilization of a worker's time. If the products back up to the se-

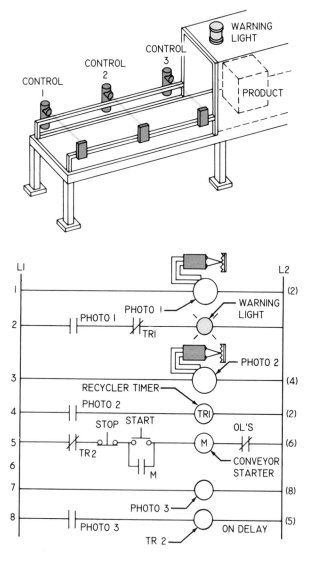

Figure 13-39 Photoelectric controls can be used to detect a backup of a product on a line. Three controls are used to turn on a warning light and turn off the line if required.

cond photoelectric control, this control will start a recycler timer, starting a flashing of the warning light. At this time the operator should unload the conveyor. If the conveyor is not unloaded and the products back up to the third photoelectric control, this control will stop the line, preventing a problem.

In this application, retroreflective scan is used for ease of installation and because of vibration of the conveyor line. In an application like this one, if a conveyor is turned off, all upstream conveyors and machines must also be turned off to prevent a jam-up.

INTRODUCTION TO PROXIMITY CONTROLS

Proximity controls are becoming increasingly popular in a wide range of industrial applications because they are versatile, safe, reliable and can be used on applications where mechanical type controls cannot. Proximity controls can be used to replace mechanical limit switches. Proximity switches can be used to detect the presence or absence of practically anything, without touching the target. Because proximity controls are solid state, they combine high reliability, long life and high speed operation with inherent design and application freedom.

As the use of solid state sensors becomes more prevalent, manufacturers and electricians will be faced with two choices: whether or not to use them and how to properly install them. The purpose of this portion of the unit is to provide an insight into solid state electronic switching and provide a working knowledge of proximity capabilities and limitations.

TYPES OF PROXIMITY SENSORS

Basically there are three types of proximity sensors that are available in a variety of packages to meet the many application requirements possible. Three types are *inductive, capacitive,* and *Hall effect* proximity controls.

Inductive

Inductive proximity sensors detect metallic objects only. Typical applications include positioning, fan blade detection, drill bit breakage, and solid state replacement of mechanical limit switches.

Most all-metal-responsive sensors are RF (radio frequency) inductive devices operated on the Eddy Current Killed Oscillator (ECKO) principle. The oscillator consists of an LC tank circuit (tuned circuit) and an amplifier circuit with positive feedback. The oscillator frequency is determined by the inductance and capacitance of the LC network. The inductance (L) portion of the tuned circuit is formed by the sensor coil and ferrite core. The oscillator circuit has just enough positive feedback to keep it oscillating, generating an AC wave shape (sine wave) which varies in amplitude depending on whether or not a target is present.

When a metal target is placed in front of the sensor coil, the RF field generates eddy currents in the surface of the target material. These eddy currents upset the AC inductance of the tuned circuit, causing the oscillations to die. The level detector (Schmitt trigger), connected to the tank circuit, produces a digital (square wave) output which is either ON or OFF, in the manner of a mechanical switch. Ten percent hysteresis (differential travel) is built into the sensor to avoid output chatter when the target is situated right at the operate point. An integrated circuit incorporates most of the circuitry needed to produce an ECKO sensor.

An Eddy Current Killed Oscillator proximity sensor detects any metal placed in front of the sensor face, and metals beside the sensor if it is not shielded. Typically, it will detect ferrous metals more efficiently than non-ferrous materials. Sensitivity to aluminum, copper, and brass is generally one-third that of sensitivity to steel. A peculiar phenomenon of ECKO sensors is that they will detect thin aluminum foil almost as efficiently as they detect ferrous metals. This is known as the "skin effect." The "skin effect" causes as great an AC resistance change as do ferrous materials. Therefore, these foils have a sensitivity equal to ferrous materials. As the

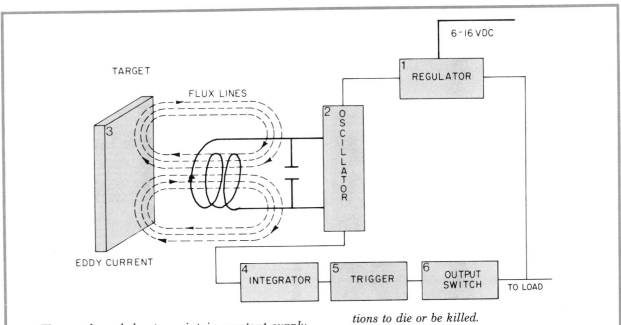

1. The regulator helps to maintain constant supply voltage despite supply voltage variations.

2. The oscillator at center is used to generate an rf (radio frequency) field at the face of the sensor. The oscillator and sensor together form a tuned circuit, which oscillates at a pre-determined frequency.

3. When metal is brought near the sensor, Eddy currents are established in the metal causing the oscillations to die or be killed.

4. The integrator converts the sine wave signal generated by the oscillator into a DC signal.

5. The DC signal, which varies in amplitude with the amplitude of oscillation is sensed by the Schmitt trigger and is converted to a digital signal.

6. The digital signal from the Schmitt trigger is then used to power the output transistor.

Figure 13-40 An eddy current killed oscillator proximity sensor detects any metal placed in front of the sensor face.

thickness increases, the "skin effect" disappears, and the characteristics return to normal.

Eddy Current Killed Oscillator (Three-Wire)

A simple diagram of the integrated circuit contained in a three-wire DC version of the ECKO sensor is shown in Figure 13-40. An explanation of the diagram follows. (This explanation is repeated underneath the diagram in Figure 13-40.)

1. The regulator helps to maintain constant supply voltage despite supply voltage variations.

2. The oscillator is used to generate an RF (radio frequency) field at the face of the sensor. The oscillator and sensor together form a tuned circuit which oscillates at a pre-determined frequency.

3. When metal is brought near the sensor, eddy currents are established in the metal, causing the oscillations to die, or be "killed."

4. The integrator converts the sine wave signal generated by the oscillator into a DC signal.

5. The DC signal, which varies in amplitude with the amplitude of oscillation, is sensed by the Schmitt trigger and is converted to a digital signal.

6. The digital signal from the Schmitt trigger is then used to power the output transistor.

Two-Wire AC Proximity Sensor (ECKO)

Another type of ECKO sensor is the two-wire control shown in Figure 13-41. The two wire control has the following features:

1. The two-wire line voltage proximity sensors operate on the ECKO principle, the same as the three-wire types.

2. The AC to DC conversion circuitry and output circuitry is additional.

3. When no target is present, the SCR is off. However, there is still a leakage current flowing through the load from the noise suppression circuitry, and other parts of the circuit. This leakage current should be considered when applying the sensors.

4. When a target is present, the circuit is completed through the bridge rectifier, the SCR, and when included, the LED indicator. This results in a 9VAC drop across the sensor, such that the voltage will not be available to the load.

5. This metal oxide varistor, and resistor and capacitor, as shown, serve as the noise suppression circuit.

6. The leakage current when OFF, and the voltage drop when ON, must be considered, especially when wiring sensors in series and parallel.

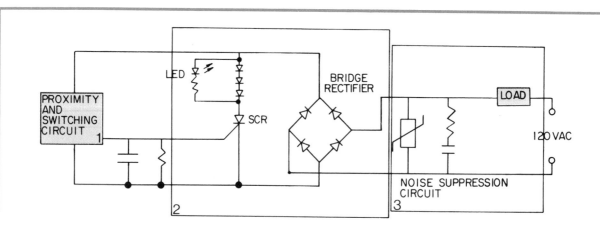

1. The two-wire line voltage proximity switches operate on the ECKO principle, the same as the three-wire types.

2. When a target is present, the circuit is completed through the bridge rectifier, the SCR, and, when included, the LED indicator. This results in a 9 VAC drop across the switch which obviously will not be available to the load.

3. This metal oxide varistor, and resistor and capacitor, as shown, serve as the noise suppression circuit.

NOTE: The leakage current when off, and the voltage drop when on, must be taken into consideration, especially when wiring switches in series and parallel.

Figure 13-41 When no target is present, the SCR is off in a two-wire eddy current killed oscillator.

Capacitive

These sensors detect most solid, fluid, or granulated substances being conductive or non-conductive. When an object approaches the sensor, an oscillator circuit causes oscillation, which then activates the output. Nominal sensing distances typically range from 3mm–15mm. The maximum sensing distance depends on the physical and electrical characteristics (dielectric) of the object to be detected. The larger the dielectric number, the easier it is to detect the material with a capacitive sensor. Figure 13-42 illustrates the dielectric characteristics of typical substances.

CAPACITIVE PROXIMITY

Capacitive sensors work on the dielectric of the material to be sensed. Generally, any material rated *greater than 1.2* can be sensed.

MATERIAL	DIELECTRIC CONSTANT	MATERIAL	DIELECTRIC CONSTANT
Acetone	19.5	Perapex	3.5
Acrylic Resin	2.7–4.5	Petroleum	2.0–2.2
Air	1.000264	Phenol Resin	4–12
Ammonia	15–24	Polyacetal	3.6–3.7
Aniline	6.9	Polyester Resin	2.8–8.1
Aqueous Solutions	50–80	Polypropylene	2.0–2.2
Benzene	2.3	Polyvinyl Chloride Resin	2.8–3.1
Carbon Dioxide	1.000985	Porcelain	5–7
Carbon Tetrachloride	2.2	Powdered Milk	3.5–4
Cement Powder	4	Press Board	2–5
Cereal	3–5	Rubber	2.5–35
Chlorine Liquid	2.0	Salt	6
Ebonite	2.7–2.9	Sand	3–5
Epoxy Resin	2.5–6	Shellac	2.5–4.7
Ethanol	24	Shell Lime	1.2
Ethylene Glycol	38.7	Silicon Varnish	2.8–3.3
Fired Ash	1.5–1.7	Soybean Oil	2.9–3.5
Flour	2.5–3.0	Styrene Resin	2.3–3.4
Freon R22 & 502 (liquid)	6.11	Sugar	3.0
Gasoline	2.2	Sulphur	3.4
Glass	3.7–10	Tetrafluoroethylene Resin	2.0
Glycerine	47	Toluene	2.3
Marble	8.5	Turpentine	2.2
Melamine Resin	4.7–10.2	Urea Resin	5–8
Mica	5.7–6.7	Vaseline	2.2–2.9
Motrpbemzome	36	Water	80
Nylon	4–5	Wood, Dry	2–6
Parafin	1.9–2.5	Wood, Wet	10–30
Paper	1.6–2.6		

Figure 13-42 Capacitive sensors work on the dielectric of the material to be sensed. Listed is the dielectric constant of typical materials. The larger the dielectric number, the easier it is to detect the material with a capacitive sensor.

Hall Effect

The Hall effect sensor detects the proximity of a magnetic field. The Hall effect principle upon which the sensor operates is rather well known. In fact, it was discovered in 1879 by Edward H. Hall at Johns Hopkins University. Mr. Hall found that when a magnet was placed in a position where its field was perpendicular to one face of a thin rectangle of gold through which current was flowing, a difference of potential appeared at the opposite edges. He found this voltage in turn was proportional to the current flowing through the conductor and the flux density or magnetic induction perpendicular to the conductor.

Today, semiconductors rather than gold are used for the Hall element. It was discovered that Hall voltages obtained with semiconductors are much higher than those obtained with gold. They are also less expensive.

Theory of Operation. Figure 13-43 illustrates the Hall effect principle. Shown is a thin strip of semiconductor material (Hall Generator) through which a constant control current is passed. When a magnet is brought near, its field directed at right angles to the face of the semiconductor, a small voltage appears at the contacts placed across the narrow dimension of the strip. As the magnet is removed, the Hall voltage reduces to zero. Thus the Hall voltage is dependent on the presence of the magnetic field and on the current flowing in the element. If either the current or the magnetic field is removed, the output of the Hall generator is zero.

In most Hall effect devices, the control current is held essentially constant and the flux density is changed by movement of a permanent magnet.

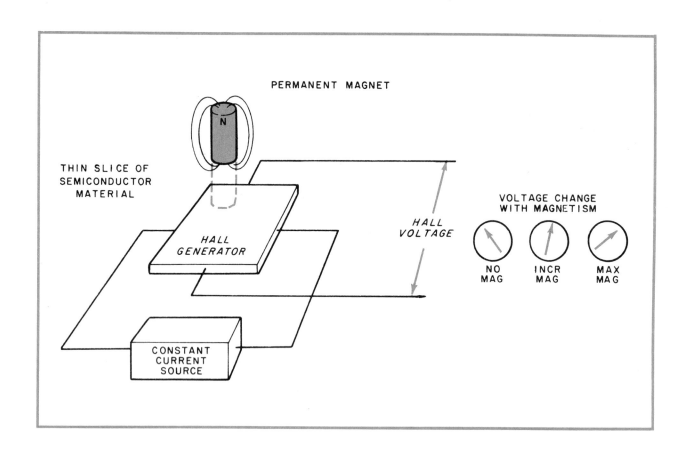

Figure 13-43 In the Hall effect principle, voltage is affected by the magnetic field.

Types of Hall Effect Sensors

Figure 13-44 illustrates several ways in which Hall effect devices can be packaged. In addition, a brief description follows of what each of these sensors can do.

Cylindrical. Magnetically operated Hall sensor. This product consists of an integrated circuit sealed in a threaded aluminum bushing. Many different types of integrated circuits can be used in this package. These sensors are available with digital current sourcing and current sinking outputs and an analog (linear) output.

Vane. This easy-to-use package consists of a plastic package with a Hall sensor on one side and a magnet on the other. When a ferrous vane is passed through the opening, the magnetic field is shunted, producing a signal change.

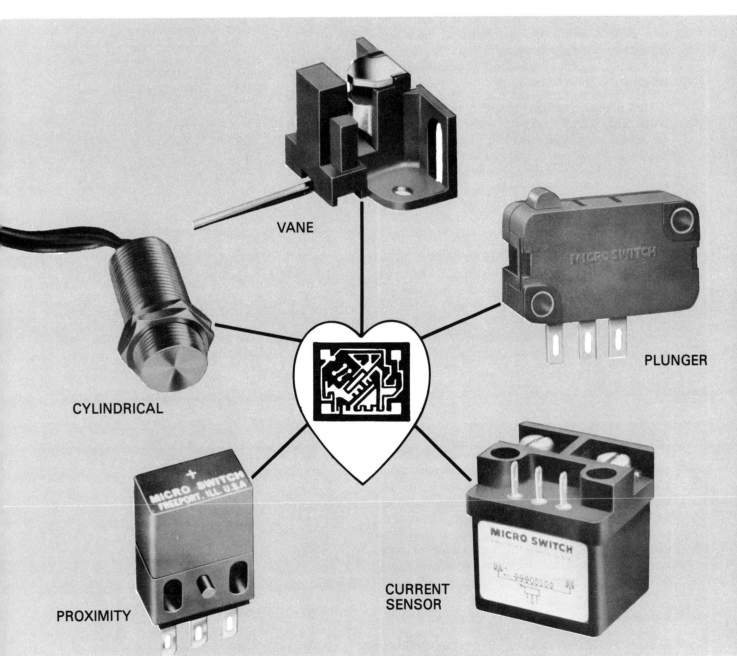

Figure 13-44 Hall effect devices can be packaged in several ways. (MICRO SWITCH, a Honeywell Division, Freeport, Illinois)

Plunger operated solid state sensor. This mechanically (plunger) operated solid state sensor is a marriage of mechanical switch mounting convenience and solid state reliability. It is available with either current sinking or current sourcing outputs and can be used directly with most electronic circuits.

Current sensor. The current sensor is a Hall device which is activated when current passes through a coil of wire. This device is designed to act as a superfast switching device or to detect DC current in a line. A unique feature of these sensors is the speed of operation. A signal is produced in a fraction of a millisecond after the current reaches the operate level.

Hall position sensor. This easy-to-mount plastic package is supplied with solder-quick-connect terminals, elongated mounting holes, and a locating boss to aid in gang-mounting the sensors. These reliable devices are operated by the magnetic field from a permanent magnet or electromagnet. Current sinking and current sourcing digital outputs are available.

Actuation (Active): Head-On Operation

Figure 13-45 shows a head-on form of Hall sensor actuation. The magnet is oriented perpendicularly to the surface of the device and is usually centered over the point of maximum sensitivity. The direction of movement is directly toward and away from the Hall sensor. The actuator and Hall sensor are positioned so the S pole of the magnet will approach the sensitive face of the position sensor.

Actuation (Active): Slide-By Operation

Figure 13-46 illustrates the slide-by form of Hall sensor actuation. Instead of the magnet being moved "head-on" to the sensor, it is moved across the face of the position sensor at some constant gap distance. The primary advantage of the slide-by actuation over the head-on type is that less actuator travel is needed to produce a signal large enough to cycle the device between operate and release.

Actuation (Active): Pendulum Operation

The pendulum mode of operation (Figure 13-47) is a combination of the head-on and the slide-by modes. Two methods (single pole and multipole) are available. With this type of operation, single or multiple signals may be generated by one actuator.

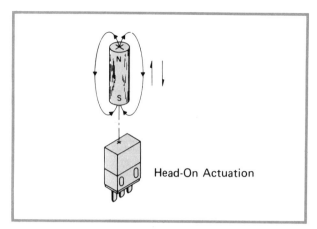

Figure 13-45 A head-on form of Hall sensor actuation. (MICRO SWITCH, a Honeywell Division, Freeport, Illinois)

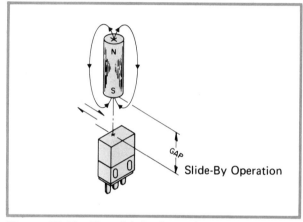

Figure 13-46 The slide-by form of Hall sensor actuation. (MICRO SWITCH, a Honeywell Division, Freeport, Illinois)

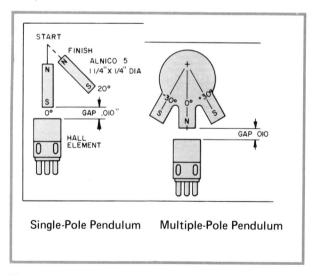

Figure 13-47 The pendulum form of Hall sensor actuation. (MICRO SWITCH, a Honeywell Division, Freeport, Illinois)

Actuation (Active): Rotary Operation

A multipolar ring magnet or collection of magnets can be used to produce an alternating magnetic pattern which is called rotary actuation. Figure 13-48 illustrates a typical rotary mode of operation and the induction pattern (+ and −) seen by the sensor located in the threaded tubular housing. Ring magnets with up to 60 or more poles are available for multiple output signals.

Actuation (Passive): Vane Operation

A passive target incorporates a magnetic source that is not part of the actuator. Figure 13-49 illustrates a ferrous (iron) vane which will shunt or redirect the magnetic field in the air gap away from the sensor. When the iron blade (vane) is moved through the air gap between the Hall-effect sensor and magnet, the sensor can be made to turn on and off sequentially at nearly any speed due to the shunting effect. The same effect can be achieved in the rotary mode of operation as illustrated by Figure 13-50.

Actuation (Passive): Ferrous Proximity Shunting (Open)

Instead of introducing a ferrous vane between the magnet and Hall sensor, it is possible to shunt the flux around the sensor with an arrangement illustrated in Figure 13-51. The gear causes the magnetic induction to be shunted from the sensor when a tooth is present and pass through the sensor when the tooth is absent. This variable flux concentration is what causes the Hall sensor to be activated, producing an output signal.

Actuation (Passive): Electromagnetic Operation

Electromagnetism may also be used to activate Hall sensors. With an electromagnet, we must consider the strength of the magnetic field in terms of current in the turns of the coil as depicted in Figure 13-52. The more current and the more turns, the stronger the magnetic field. The coil serves the same purpose as the bar magnet with poles at the ends.

Since electromagnets dissipate heat, attention must be given to thermal consideration. Care must be exercised to see that the current—and thus temperature—limits are not exceeded.

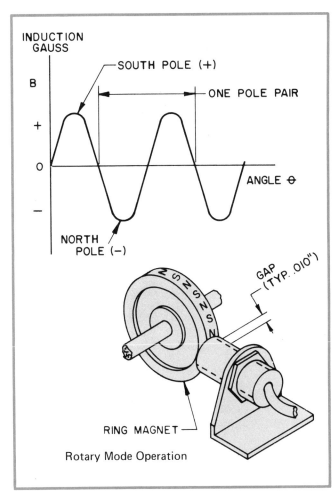

Figure 13-48 A typical rotary form of Hall sensor actuation. (MICRO SWITCH, a Honeywell Division, Freeport, Illinois)

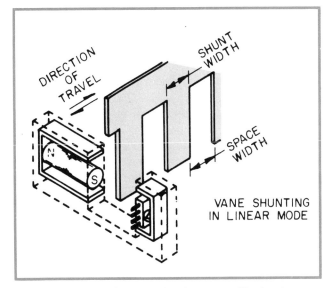

Figure 13-49 A ferrous (iron) vane will shunt or redirect the magnetic field in the air gap away from the sensor. (MICRO SWITCH, a Honeywell Division, Freeport, Illinois)

TYPICAL APPLICATIONS FOR HALL EFFECT SENSORS

Conveyor Belt Application

Figure 13-53 illustrates a simple solution for monitoring a remote conveyor operation. A cylindrical Hall effect solid state sensor is mounted to the frame of the conveyor. A magnet mounted on the tail pully revolves past the sensor to cause an intermittent visual or audible signal at a remote location to assure that all is well. Any shutdown of the conveyor will interfer with the normal signal and alert operators of trouble. With no physical contact, levers or linkages, the sensor can be installed and forgotten.

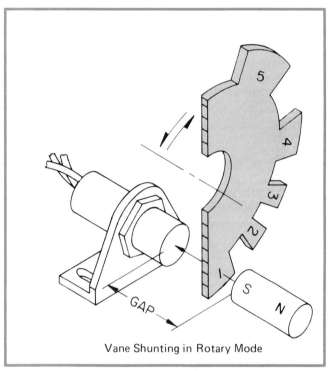

Vane Shunting in Rotary Mode

Figure 13-50 A rotary ferrous vane can be used to shunt magnetic fields. (MICRO SWITCH, a Honeywell Division, Freeport, Illinois)

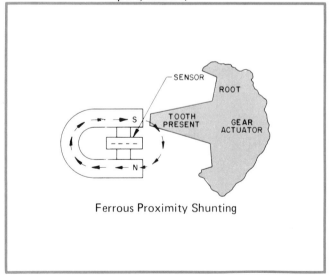

Ferrous Proximity Shunting

Figure 13-51 A gear type shunt. (MICRO SWITCH, a Honeywell Division, Freeport, Illinois)

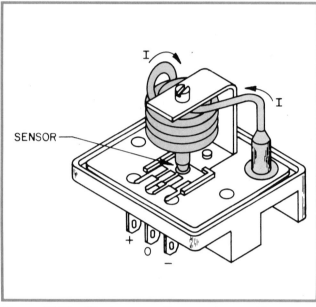

Figure 13-52 Electromagnetism can be used to activate a Hall sensor. (MICRO SWITCH, a Honeywell Division, Freeport, Illinois)

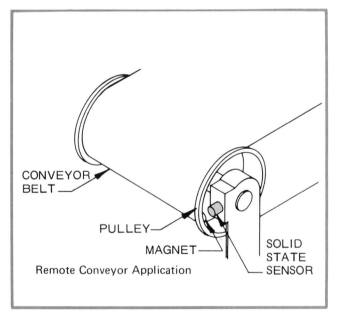

Remote Conveyor Application

Figure 13-53 A simple way of monitoring a remote conveyor operation. (MICRO SWITCH, a Honeywell Division, Freeport, Illinois)

Limit Switch For Production Machinery

A Hall effect sensor application in high speed production machinery is illustrated in Figure 13-54. Four cylindrical Hall sensors that are threaded into a aluminum housing are actuated by four small rod magnets pressed into an aluminum spacing bar. Functionally, one magnet passing four sensors would generate event signals as needed, but the diameter of the sensors in this case precludes their being spaced closely enough to measure the small movements involved.

In use, event signals are generated by the sensors and represent distances measured from a mechanical reference surface on the machine. These signals define the acceptable dimensional limits between which the item under test must generate electrical pulses. In a known application, each of the Hall sensors has accummulated at least eight million operate/release cycles per month and is still operating, without replacement or maintenance, after 35 million operations.

Current Sensor Application

A fast acting, automatically resetting circuit can be made by using a Hall effect current sensor. The internal construction is illustrated in Figure 13-55.

An overload signal could change state from low to high, or vice versa, when the current exceeded the design trip point. This signal could be used to trigger some warning alarm or to control the current directly, by electronic means. These current sensors have been built in sizes ranging from the diameter of a dime to about half the size of a deck of playing cards, and indicating overload currents from a few milliamperes to several amperes.

Different reset characteristics can be achieved by tailoring the electromagnetic design to the sensor's requirements. In addition, use of a linear Hall sensor in place of the digital Hall sensor can provide an output signal proportional to the input current, but electrically isolated from it.

Speed Sensing

The use of a ring magnet with a bipolar actuated sensor can provide a valuable alternative to coil pickup or variable reluctance methods. Techniques which depend upon inducing a voltage in a coil, either by passing magnets near it or by changing the air gap reluctance of a fixed magnet, have the disadvantage that both the frequency and amplitude of the voltage waveform change

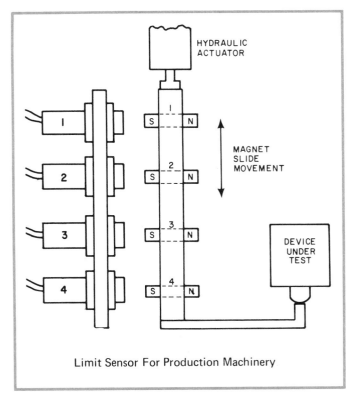

Limit Sensor For Production Machinery

Figure 13-54 A Hall effect sensor can be used in high speed production. (MICRO SWITCH, a Honeywell Division, Freeport, Illinois)

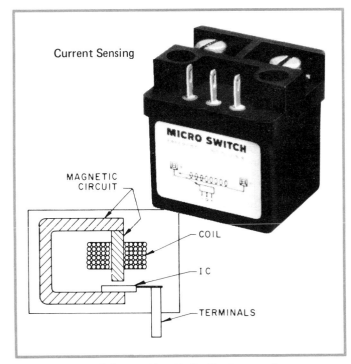

Figure 13-55 A fast acting Hall effect current sensor. (MICRO SWITCH, a Honeywell Division, Freeport, Illinois)

with speed. Hall sensors have no minimum speed of operation.

A typical speed sensing application of the Hall sensor is illustrated in Figure 13-56. The sensor's output provides an electrical waveform whose repetition frequency varies directly with shaft speed, but whose amplitude does not.

Hall Effect Switch for Instrumentation

Figure 13-57 illustrates a cylindrical Hall effect sensor and ring magnet combination used in an instrumentation application. This basic mechanism shows a Hall sensor actuated by a rotating magnet to initiate resistance measurements of electrical circuits being life tested (not shown). A second sensor actuated 180° after the first sensor checks that the circuits are open. The no-touch actuation and exceptionally long life are ideal in instruments and apparatus designs.

Indexing of Rotary Table

A pair of Hall effect sensors (hidden) and eight magnets are used in Figure 13-58 to verify the proper angular indexing (rotation) of a circular plate with work stations around its perimeter.

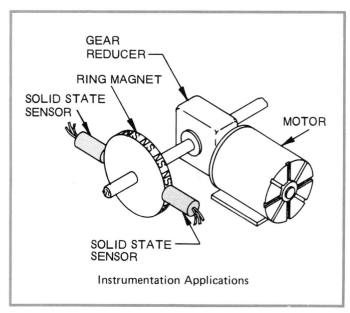

Figure **13-57**   A Hall sensor can be used in testing instrumentation. (MICRO SWITCH, a Honeywell Division, Freeport, Illinois)

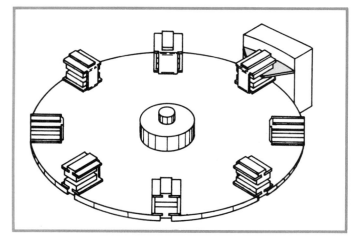

Figure **13-58**   A Hall sensor is being used to verify the proper angular indexing of a circular plate. (MICRO SWITCH, a Honeywell Division, Freeport, Illinois)

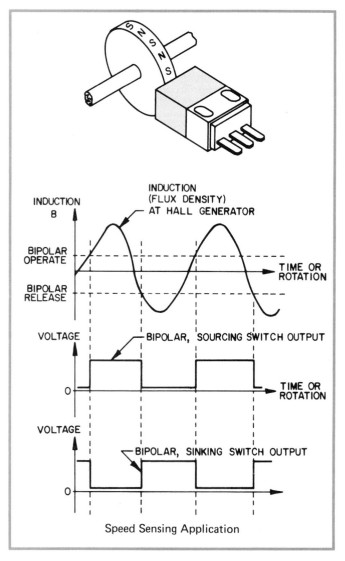

Figure **13-56**   A Hall sensor can be used for speed sensing. (MICRO SWITCH, a Honeywell Division, Freeport, Illinois)

By electronically sensing the sequence in which the peripheral magnets actuate the sensors during indexing, the sensor makes it possible to detect reverse rotation, lack of rotation, and improper forward rotation. In the drawing, the plate containing the actuator magnets is visible.

Beverage Gun Application

Figure 13-59 illustrates a typical application for Hall effect sensors in a beverage gun. Some unique features offered are:

1. Small size. There are nine sensors in the hand-held device.

2. Sealed construction. The "gun" cannot be contaminated by syrups, etc. It is completely submersible in water for easy cleaning.

3. Easy logic interface. No buffering circuitry is necessary between the "gun" and automatic totalizer cash register.

4. Improved reliability. Less maintenance.

Sequencing Operation

Sequence of a number of operations can be achieved by clamping a number of metal discs to a common shaft as illustrated in Figure 13-60. The metal discs are rotated in the gaps of Hall effect vane sensors. When a disc is rotating in tandem with its mates, it can be used to create a binary code which will establish a sequence of operation. Programs can be altered by replacing the discs shown with others having a different air-to-ferrous cam ratio.

Length Measurement

Measurement of a strip of product can be made by mounting a disc with two notches on the extension of a drive motor shaft as illustrated in Figure 13-61. A Hall effect vane sensor is then mounted so that the disc will pass through the gap. Each notch represents a fixed length of material and can be used to measure tape, fabric, wire, rope, thread, aluminum foil, plastic bags, etc.

Shaft Encoding

Figure 13-62 illustrates how a cylindrical Hall effect sensor can be used for shaft encoding. A ring magnet is mounted on the motor shaft. Each pair of N-S poles actuates a Hall effect sensor. Each pulse represents angular movement.

Level . . . Degree of Tilt Measurement

Cylindrical Hall effect sensors can be installed in

the base of a machine with magnets installed in a pendulum fashion, as illustrated in Figure 13-63. As long as the magnet remains directly over the sensor, the machine is level. A change in state of output (when a magnet swings away from a sensor) is indication that the machine is not level.

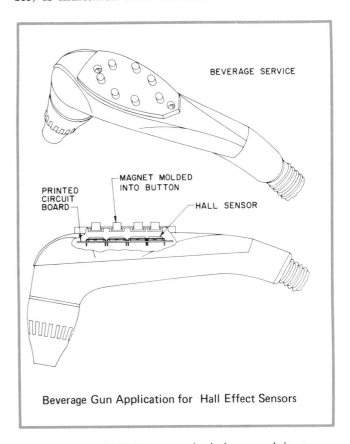

Beverage Gun Application for Hall Effect Sensors

Figure 13-59 A Hall sensor is being used in a beverage gun. (MICRO SWITCH, a Honeywell Division, Freeport, Illinois)

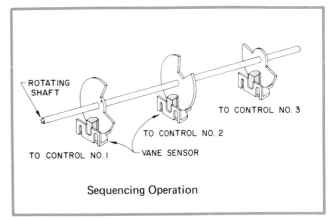

Sequencing Operation

Figure 13-60 A Hall sensor is being used for sequence operations. (MICRO SWITCH, a Honeywell Division, Freeport, Illinois)

The sensor/magnet combination may also be installed in such a manner as to indicate the degree of tilt.

Mini "Joy Stick"

Figure 13-64 illustrates how a Hall effect digital sensor can be used in a "joy stick" application. The Hall sensor inside the joy stick housing is actuated by a magnet on the toggle. The proximity of the magnet to the sensor controls position for wire bonder, aircraft or crane operators, wheel chairs, etc. Use of an analog device also achieves degree of movement measurements such as speed.

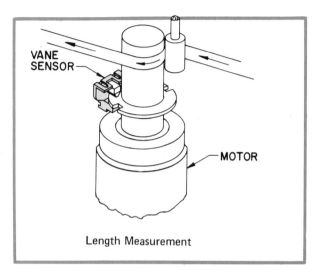

Length Measurement

Figure 13-61 A Hall sensor is being used for length measurement. (MICRO SWITCH, a Honeywell Division, Freeport, Illinois)

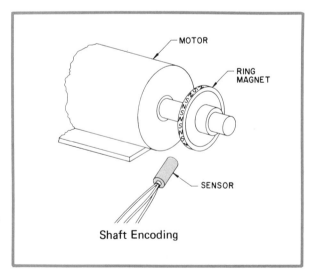

Shaft Encoding

Figure 13-62 A Hall sensor is being used for shaft encoding. (MICRO SWITCH, a Honeywell Division, Freeport, Illinois)

Cam Actuator

Figure 13-65 illustrates how a Hall effect vane sensor can be used in conjunction with a cam actu-

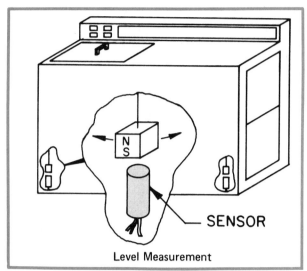

Level Measurement

Figure 13-63 A Hall sensor is being used for setting level or degree of tilt. (MICRO SWITCH, a Honeywell Division, Freeport, Illinois)

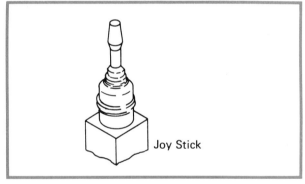

Joy Stick

Figure 13-64 A Hall sensor is being used with a mini joy stick operator. (MICRO SWITCH, a Honeywell Division, Freeport, Illinois)

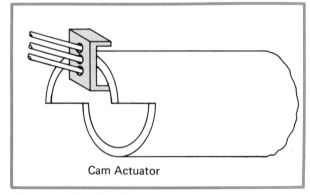

Cam Actuator

Figure 13-65 A Hall sensor used with a cam actuator. (MICRO SWITCH, a Honeywell Division, Freeport, Illinois)

ator. Some unique features of this combination are:

1. Fast response time—High RPM applications.

2. Sealed construction—dust or dirt will not contaminate the switch.

3. Vane device saves assembly time—there is no need to mount a separate magnet.

4. Logic interface—no buffer circuitry is necessary since the vane switch directly interfaces with logic circuitry such as mini-computers

5. Product life is increased—the Hall devices have essentially unlimited life. There are no mechanical parts to wear out.

Paper Detector

Figure 13-66 illustrates how a Hall effect plunger type sensor can be used in copy machines to detect paper "flow." The advantages of a magnetically operated Hall effect sensor over mechanical switches are:

1. No contamination by toner—there are no contacts to become gummy or corroded.

2. Long-life—the extremely long life of the sensor means fewer service calls and lower maintenance costs.

3. Digital ON/OFF output—direct interface with logic circuitry.

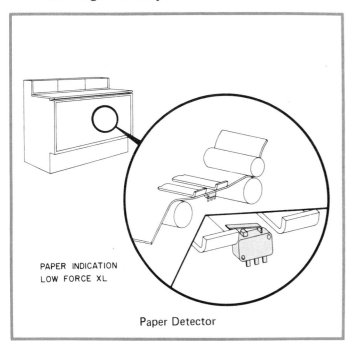

PAPER INDICATION
LOW FORCE XL

Paper Detector

Figure 13-66 A Hall effect plunger type sensor can be used in copy machines to detect paper flow. (MICRO SWITCH, a Honeywell Division, Freeport, Illinois)

INSTALLING SOLID STATE SENSORS

The following application suggestions are based on applying a Hall effect sensor with a digital output, but the suggestions will work for many analog applications as well.

1. Mount a sensor and a magnet according to your application.

2. Adjust either the sensor and/or magnet to make the sensor operate.

3. Measure this distance. This is the operating point.

4. To provide for slight shifts in operating characteristics which result principally from temperature changes, the final magnet to sensor distance should be reduced somewhat from the measured operating point. You cannot damage the sensor by magnet overdrive.

5. Adjust the sensor-to-magnet distance as short as possible. Avoid any chance of the magnet physically hitting the switch. If travel is somewhat limited, the gap should be narrowed to a dimension about one-half the originally measured sensor-to-magnet distance when the output changes state (operating point).

6. If the magnet movement of the other extreme is adequate, at least as far beyond the release point (the distance between switch and magnet at the point where the output state returns to the unactuated condition) as the forward stroke is ahead of the operating point, the applied sensor and actuator pair will work properly.

Several more sensor and magnet combinations should be tested in the previously described manner. A pattern should begin to appear.

Determine by the final system checkout that there aren't any marginal sensors before completing the check. After a fairly large number of sensors have been successfully installed, installation should become routine for the remaining sensors in that delivery lot.

TROUBLESHOOTING

During final equipment checkout, the sensor's operation should be checked at its temperature extremes. Should a sensor fail to work at some point in the checkout, the amount of overtravel (magnet movement past the operating points) and/or the release travel (magnet travel away from the sensor after the release point was reached initially) will have to be increased.

1. If the sensor does not actuate, the magnet-to-sensor distance must be shortened.

2. If the sensor does not release, the sensor-to-magnet distance must be opened up by a small amount.

3. If within the maximum magnet travel and sensor position, the sensor will not release, a lower strength magnet should be substituted.

4. If the sensor does not operate and reducing the magnet-to-sensor spacing to the minimum practical does not result in proper sensor function, a stronger magnet is needed.

5. If other magnets from the same group work, the wrong magnetic pole may be facing the sensor. Turn the magnet around.

6. In the event that the above procedure does not result in proper sensor operation, consult manufacturers representative for additional design information.

INTERFACING RELAY OPERATION WITH A HALL EFFECT SENSOR

Figure 13-67 is a typical example of how a Hall effect sensor can operate an AC relay when the sensor's output is too low.

The alternative is to use a triac with the relay. Functionally, a triac is like a relay because it can conduct large AC loads, yet it is controlled by a small voltage and current that are supplied to the gate terminal. The gate terminal is similar to the coil of a relay except that one terminal is tied internally on the triac.

For this application (Figure 13-68) the gate voltage and current required to trigger the triac are to be supplied by a Hall sensor. The triac selected for this particular application requires 10 milliamperes, 1.5 volts at the gate terminal to cause it to operate.

A six volt DC supply voltage is available for the Hall sensor. The application requires the triac to remain unoperated when the magnet is not in proximity of the Hall sensor. Therefore, a normally low output is needed. To get this, a current sourcing sensor should be used.

In the operated condition, the output voltage of the Hall sensor should be:

^{V} supply = 1.5V (typical drop across a current sourcing Hall sensor in the operated condition) = 6 − 1.5.

^{V}output = 4.5 volts

Thus, the output voltage of the Hall sensor will satisfy the voltage requirements of the triac, providing the proper circuit resistance is used.

Resistor R1 (Figure 13-68) is a current limiting device which must be supplied by the customer.

Its purpose is to limit current from the Hall sensor to 10 milliamperes. The value of R1 is determined as follows:

Voltage drop across R1 = ^{V}output (voltage of sensor) − ^{V}gate (voltage required at triac) = 4.5 − 1.5 = 3 volts

Applying Ohms Law we find that:

$$\frac{V(R1)}{1(R1)} = \frac{3 \text{ volts}}{.010 \text{ amps}} = 300 \text{ ohms} = R1$$

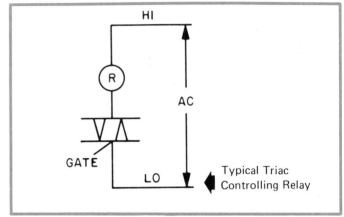

Figure 13-67 A typical triac like this one can be added to a Hall sensor. (MICRO SWITCH, a Honeywell Division, Freeport, Illinois)

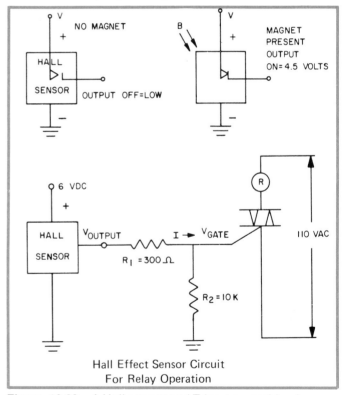

Hall Effect Sensor Circuit For Relay Operation

Figure 13-68 A Hall sensor and Triac are combined to control a relay. (MICRO SWITCH, a Honeywell Division, Freeport, Illinois)

The function of resistor R2 (a pull-down resistor) is to pull the voltage at the gate down to ground when the Hall effect sensor is in the off condition. Factors that must be considered in determining the resistor's value and selection include:

1. Ability to absorb leakage current.
2. Ability to absorb electrical noise generated in the output circuit when the sensor is unoperated.

3. Ability to conduct a small percentage of load current when the sensor is operated.

A 10K resistor is selected for R2 in this application based on the design objectives.

With this application you now have a circuit which results in the relay being normally open and which is capable of handling large currents for the price of two resistors and one triac. This is considerably less expensive than other methods of switching.

RESISTOR COLOR CODE				
COLOR	1st NUMBER	2nd NUMBER	MULTIPLIER	TOLERANCE (%)
Black	0	0	1	0
Brown	1	1	10	
Red	2	2	100	
Orange	3	3	1000	
Yellow	4	4	10,000	
Green	5	5	100,000	
Blue	6	6	1,000,000	
Violet	7	7	10,000,000	
Gray	8	8	100,000,000	
White	9	9	1,000,000,000	
Gold			0.1	5
Silver			0.01	10
None			0	20
	BAND 1	BAND 2	BAND 3	BAND 4

QUESTIONS ON UNIT 13

1. Name four types of photosensors.
2. What are three characteristics of an object that must be considered when selecting a scanning technique?
3. What is meant by direct scan?
4. What is meant by reflective scan?
5. What is meant by retroflective scan?
6. What is meant by polarized scan?
7. What is meant by convergent beam scan?
8. What is meant by specular scan?
9. What is meant by diffused scan?
10. What is an advantage of using a fiber optics scan method?
11. What is the difference between modulated and unmodulated controls?
12. When is an excess gain important?
13. When is knowing the response time important?
14. What are five environmental influences that must be considered when installing a photoelectric control?
15. What is meant by a current sink output?
16. What is meant by a current source output?
17. What are the three basic types of proximity sensors?
18. What is the basic operating principle of the Hall effect sensor?
19. What are two advantages in using a Hall effect sensor?
20. What are the five basic types of Hall effect sensors?
21. Name four methods of actuating a Hall effect sensor.

SPEED CONVERSION TABLE FOR USE IN PHOTOELECTRIC APPLICATIONS			
Ft/Min	In/Min	In/Sec	Sec/In
1	12	.2	5.000
2	24	.4	2.500
3	36	.6	1.667
4	48	.8	1.250
5	60	1.0	1.000
6	72	1.2	0.833
7	84	1.4	.714
8	96	1.6	.625
9	108	1.8	.556
10	120	2.0	.500
11	132	2.2	.455
12	144	2.4	.417
13	156	2.6	.385
14	168	2.8	.357
15	180	3.0	.333
16	192	3.2	.313
17	204	3.4	.294
18	216	3.6	.278
19	228	3.8	.263
20	240	4.0	.250
21	252	4.2	.238
22	264	4.4	.227
23	276	4.6	.217
24	288	4.8	.208
25	300	5.0	.200
30	360	6	.167
40	480	8	.125
50	600	10	.100
60	720	12	.083
70	840	14	.071
80	960	16	.063
90	1080	18	.056
100	1200	20	.050
125	1500	25	.040
150	1800	30	.033
175	2100	35	.028
200	2400	40	.025
225	2700	45	.022
250	3000	50	.020
275	3300	55	.018
300	3600	60	.017
325	3900	65	.015
350	4200	70	.014
375	4500	75	.013
400	4800	80	.013
450	5400	90	.011
500	6000	100	.010
600	7200	120	.0083
700	8400	140	.0071
800	9600	160	.0063
900	10800	180	.0056
1000	12000	200	.0050
1250	15000	250	.0040
1500	18000	300	.0033
1750	21000	350	.0029
2000	24000	400	.0025
2500	30000	500	.0020
5000	60000	1000	.0010

14 Programmable Controllers

A programmable controller (PC) is a control system consisting of four basic sections: power supply section, input/output interface section, processing section, and programming section.

The power supply section provides all the voltages required for the internal operation of the PC. The input/output sections of the PC interface low-level incoming signals from sensors into higher voltage and current control signals for load devices. The processing section organizes all control activity by receiving inputs, performing logical decisions in relation to the stored program, and controlling outputs. The programming section of the PC allows programs to be entered or changed in the programming section of the PC.

PROGRAMMABLE CONTROLLERS

Programmable controllers are electrical devices designed to control machines and industrial processes automatically. Programmable controllers are capable of many industrial functions and applications and are now widely used in automated industrial applications.

The automotive industry was the first to recognize the advantages of programmable controllers. Annual model changes required constant modifications of production equipment controlled by relay circuitry. In some cases entire control panels had to be scrapped and new ones designed and built with new components, which resulted in increased production costs.

The automotive industry was looking for equipment that could reduce changeover costs required by model changes. In addition, the equipment had to be able to operate in a harsh factory environment of dirty air, vibration, electrical noise, and a wide range of temperature and humidity.

To meet this need, a ruggedly constructed computer-like control, which became the programmable controller, was developed. The programmable controller could easily accommodate constant circuit change using a keyboard to introduce new operation instructions. In 1968, the first programmable controller was delivered to General Motors (GM) in Detroit by Modicon (now Gould).

The first programmable controllers were large and costly. Their initial use was in large systems with the equivalent of 100 or more relays. Today programmable controllers come in all sizes, from micro, which is cost effective down to the equivalent to as few as 10 relays, to the large units with the equivalent of thousands of inputs and outputs.

Programmable controllers have become popular because they could be programmed and reprogrammed using line (ladder) diagrams that plant personnel understood. The way a machine or process was supposed to operate could be programmed and read as a line diagram showing open and closed contacts. This was the same approach used for years to describe relay logic circuits. This allowed the use of a computer-like device without learning a computer language. Additional advantages of using a programmable controller include:

1. A reduction in hard wiring and wiring cost.

2. Reduced space requirements, due to a much smaller size as compared to using standard relays, timers, counters, and other control components.

3. Flexible control because all operations are programmable.

4. High reliability using solid-state components.

5. Microprocessor-based memory allows storage of large programs and data.

6. Improved on-line monitoring and troubleshooting by monitoring and diagnosing its own failures as well as the machines and process it controls.

7. Eliminates the need to stop a controlled process to change set parameters.

8. Can provide for analog, digital, and voltage inputs as well as discrete inputs such as pushbuttons and limit switches.

9. Modular design allows components to be added, substituted, and rearranged as requirements change.

10. Types of programming languages used are familiar and follow industrial standards, such as line diagrams.

Although the first programmable controllers were designed to replace relays, today's programmable controllers are now used to achieve factory automation, often interfacing with robots, numerical control equipment, CAD/CAM systems, and general-purpose computers. Programmable controllers are used in virtually every segment of industry where automation is required. Figure 14-1 illustrates today's typical programmable controller.

The Programmable Controller Market

The programmable controller market is one of the fastest growing segments in the motor control industry. In 1976 annual sales totaled $30 million. By 1981 sales had grown to $400 million and today sales are in the billions.

The market for programmable controllers are divided into two market areas: discrete parts manufacturing and process manufacturing.

Discrete Parts Manufacturing. The discrete parts manufacturing market area represents durable goods such as automobiles, washers, refrigerators, machines, and tractors. Discrete parts manufacturing is done primarily by stand-alone machines that bend, drill, punch, grind and shear metals. All of these machines can be

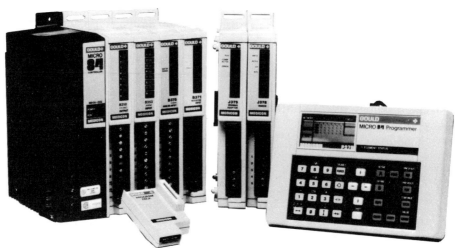

Figure 14-1 Programmable controllers can be programmed using conventional line diagrams. (Gould Inc., Modicon Programmable Control Division)

automated with programmable controllers. Figure 14-2 illustrates a typical discrete parts machine.

The programmable controller allows each machine to have its own unique capability using standard hardware. When the functional requirements for the machine change, the programmable controller allows each modification of the controls. Using modular replacement of programmable controllers reduces downtime of the machine.

Specialty machine builders called original equipment manufacturers (OEM) were among the first to use programmable controllers. Since every system produced by an OEM was unique, the programmable controller allowed flexibility in the machine function. Using the programmable controller helped reduce start-up and de-bug time, and also allowed the manufacturer to incorporate additional user requirements for changes in machine operations after start-up.

Today the programmable controller has become the standard for machine builders. Increased capabilities in a reduced size allow today's programmable controllers to control one machine or link up to many machines in any network configuration.

Process Manufacturing. The process manufacturing market area produces consumables such as food, gas, paint, pharmaceuticals, and chemicals. Most of these processes require systems to blend, cook, dry, separate or mix ingredients. Figure 14-3 illustrates a typical process manufacturing system.

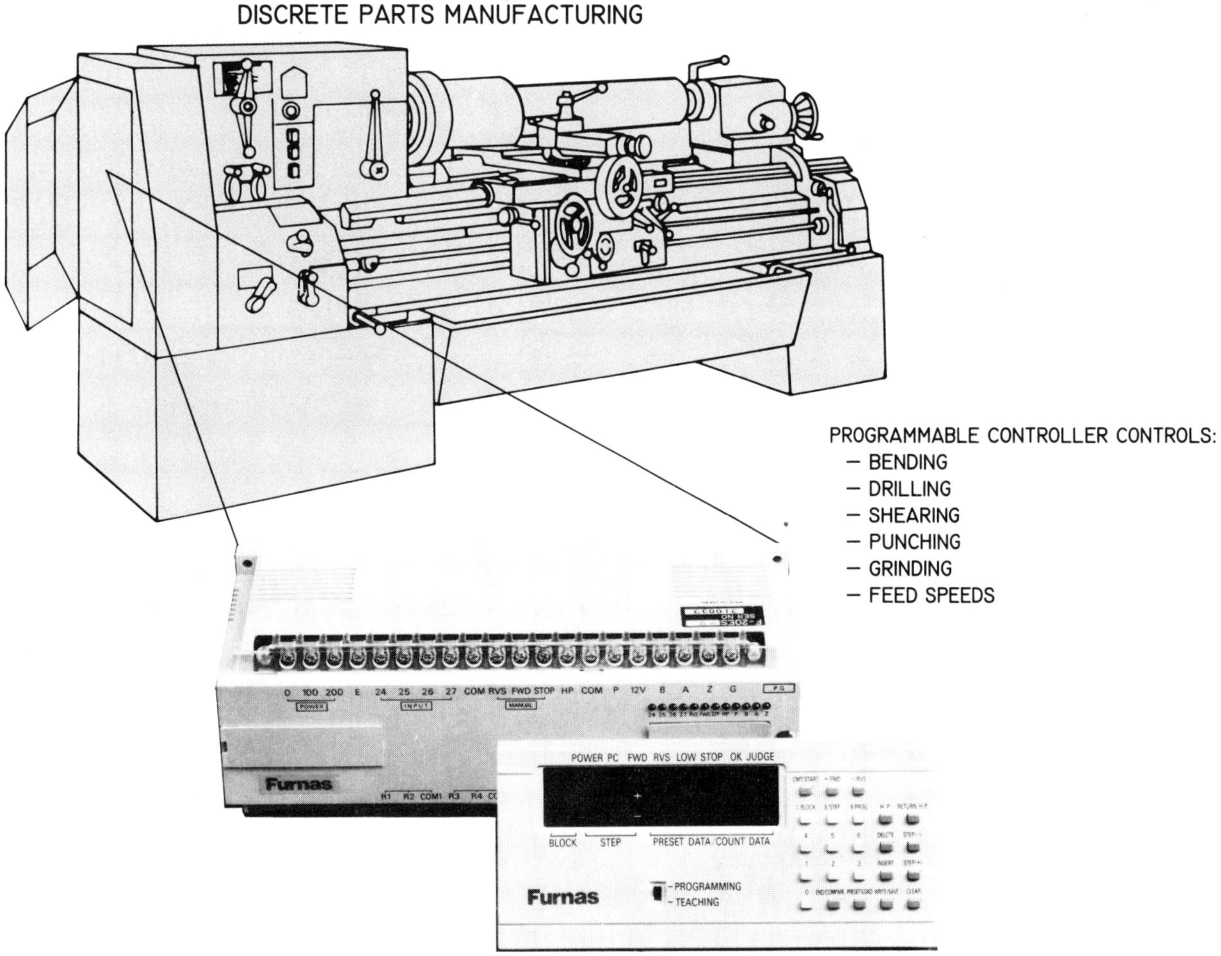

DISCRETE PARTS MANUFACTURING

PROGRAMMABLE CONTROLLER CONTROLS:
− BENDING
− DRILLING
− SHEARING
− PUNCHING
− GRINDING
− FEED SPEEDS

Figure 14-2 The programmable controller can be used to control all electrical functions on a machine used in discrete parts manufacturing. (Furnas Electric Company)

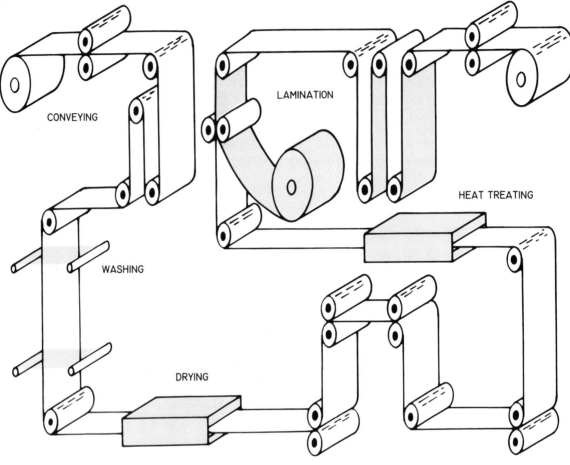

PROGRAMMABLE CONTROLLER CONTROLS:
- CONVEYING
- PALLETIZING
- STORAGE
- CONDITIONING
- TREATMENT
- ALARMS
- INTERLOCKS
- PREVENTIVE MAINTENANCE

Figure 14-3 Programmable controllers are used in the automation of process manufacturing.

Automation of this type of process is required for opening and closing valves, and controlling motors in the proper sequence and at the correct time. The programmable controller is ideal for this type of operation. If the time, temperature, or flow requirements of the product changes, the programmable controller allows for easy modifications to the system.

Today's programmable controllers control process manufacturing such as the conveying of the product, palletizing, storage, treatment, alarms, in-

terlocks, and preventive maintenance functions for the system. The programmable controller can also generate reports that are used to determine production efficiency.

Manufacturers of programmable controllers offer a variety of programmable controllers from micro through very large units as illustrated in Figure 14-4. For machines and processes that have limited capability and little potential for future expansion, the micro or small programmable controller is the best choice. A medium or large con-

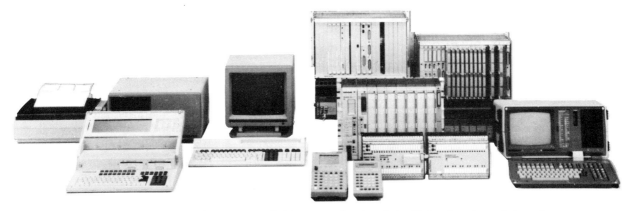

Figure 14-4 Programmable controllers are available for different applications from various manufacturers. Capabilities of programmable controllers range from a simple mini with ten inputs and six outputs to a powerful controller equivalent to a process computer. (Siemens Energy And Automation, Inc.)

troller is recommended for more complex control requirements and future expansion.

PROGRAMMABLE CONTROLLERS VERSUS MICROCOMPUTERS

Programmable controllers have grown in popularity in applications which were once performed exclusively by the microcomputer. Microcomputers feature fast number manipulation and powerful text-handling capabilities. However, programmable controllers offer several advantages for industrial control applications that microcomputers do not.

The first difference between the programmable controller and the microcomputer is that the programmable controller is designed to communicate directly with inputs from the process and controlling outputs. The programmable controller recognizes these inputs and outputs (I/O's) as part of its internally programmed system. Inputs include limit switches, pushbuttons, temperature controls, photoelectric controls, analog signals, ASCII (American National Standard Code for Information Interconnections), serial data, and other inputs. The outputs include voltage or current levels that drive end devices such as solenoids, motor starters, relays, and lights. Other outputs are to analog devices, digital binary coded decimal (BCD) displays, ASCII compatible devices, and other programmable controllers and computers.

The second difference between programmable controllers and microcomputers is the ease in programming. The programmable controller uses simple programming techniques that are easily learned and understood. Simple relay line (ladder) diagram programming does not require knowledge of FORTRAN, BASIC, or other computer languages. The programmable controller can be programmed and reprogrammed "on-line" while the process is running, without hardware modifications.

The third difference between programmable controllers and microcomputers is that the programmable controller is designed specifically for use in an industrial environment (Figure 14-5). Variations in levels of noise, vibration, temperature, and

Figure 14-5 Programmable controllers are designed to withstand fluctuations in noise, vibration, temperature, and humidity in the industrial environment. (Allen-Bradley, A Rockwell International Company)

humidity will not adversely affect operations. A microcomputer cannot withstand the typical industrial environment.

PARTS OF A PROGRAMMABLE CONTROLLER

All programmable controllers have the same basic parts and characteristics. The four basic parts of the programmable controller are the power supply, input/output interface sections, processor section, and programming section. These basic parts are illustrated in Figure 14-6.

Programs are stored and retrieved from memory as required. Sections of the programmable controller are interconnected and work together to allow the programmable controller to accept inputs from a variety of sensors, make a logical decision as programmed, and control outputs such as motor starters, solenoids, valves, and drives.

The Power Supply
The power supply provides all the necessary voltage levels required for the programmable controllers' internal operations. In addition, the power

supply may provide power for the input/output modules. The power supply can be a separate unit or built into the processing section. Its function is to take the incoming voltage (usually 120 or 240 VAC) and change this voltage as required (usually 5 to 32 VDC).

The power supply must provide constant output voltage free of transient voltage spikes and other electrical noise. The power supply also charges an internal battery in programmable controllers to prevent loss of memory when external power is removed. Memory retention time may vary from hours up to 10 years on many programmable controllers.

The Input/Output Interface Section
The input and output interface section functions as the eyes, ears, and hands of the programmable controller. The input section is designed to receive information from pushbuttons, temperature switches, pressure switches, photoelectric and proximity switches, and other sensors. The output section is designed to deliver the output voltage required to control alarms, lights, solenoids, starters and other supports.

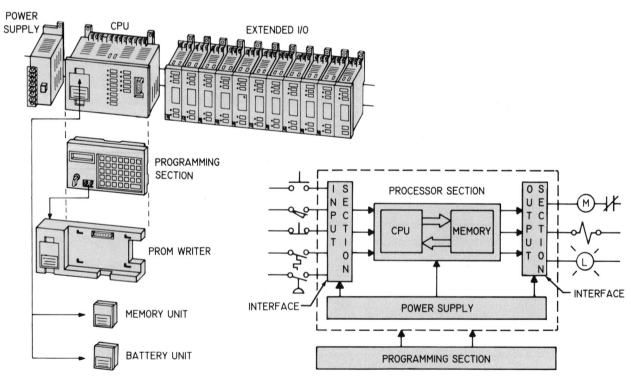

Figure 14-6 The four basic parts of a programmable controller include the power supply, input/output interface section, processor section and programming section. (Omron Company; Gould Inc., Modicon Programmable Control Division)

The input section receives incoming signals (usually at a high voltage level) and interfaces the signal to the low power digital processor section. The processor can then register and compare the incoming signals to the program.

The output section receives low power digital signals from the processor and converts them into high power signals. These high power signals can drive industrial loads than can light, move, grip, rotate, extend, release, heat, and perform other functions.

Discrete I/O. The most common type of inputs and outputs are the discrete type. This type of I/O uses bits, with each bit representing a signal that is separate and distinct, such as ON/OFF, open/closed or energized/de-energized. The processor reads this as the presence or absence of power.

Examples of discrete inputs are pushbuttons, selector switches, joy sticks, relay contacts, starter contacts, temperature switches, pressure switches, level switches, flow switches, limit switches, photoelectric switches, and proximity switches. Discrete outputs include lights, relays, solenoids, starters, alarms, valves, heating elements, and motors.

Data I/O. In many applications, more complex information is required than the simple discrete I/O is capable of. For example, measuring temperature (72 °F) may be required as an input into the programmable controller and numerical data (001) may be required as an output.

These types of inputs and outputs are called *data I/O*. They may be an analog type, which allows for monitoring and control of analog voltages and currents, or they may be digital, such as BCD inputs and outputs.

When an analog signal (such as voltage or current) is input into an analog input card, the signal is converted from analog to digital by an analog to digital (A to D) converter. The converted value, which is proportional to the analog signal, is sent to the processor section. After the processor has processed the information according to the program, the processor outputs the information to a digital to analog (D to A) converter. The converted signal can provide an analog voltage or current output that can be used or displayed on an instrument in a variety of processes and applications.

Examples of data inputs are potentiometers, rheostats, temperature transducers, level transducers, pressure transducers, humidity transducers, encoders, bar code readers, and wind speed transducers. Examples of data outputs are analog meters, digital meters, stepping motor (signals), variable voltage outputs, and variable current outputs.

I/O Capacity. A factor that determines the size of a programmable controller is the controller's input/output and capacity. Typical input/output capacities of different size programmable controllers are listed.

1. Mini/Micro—usually 32 or less I/O but may have up to 64
2. Small—usually have 64 to 128 I/O, but may have up to 256
3. Medium—usually have 256 to 512 I/O, but may have up to 1023
4. Large—usually have 1024 to 2048 I/O, but may have many thousands more on very large units.

The inputs and outputs may be directly connected to the programmable controller or may be in a remote location. I/Os in a remote location from the processor section can be hard wired back to the controller, multiplexed over a pair of wires, or sent by a fiber optic cable. In any case, the remote I/O is still under the control of the central processing section. Typical remote I/Os are of the 16, 32, 64, 128, and 256 size.

Fiber optic communications modules (Figure 14-7) route signals to and from I/Os to the processor

Figure 14-7 I/O modules connected with fiberoptic cable provide transmission of data unaffected by noise interference. (Allen-Bradley, A Rockwell International Company)

section. Fiber optics communications modules are unaffected by noise interference and are commonly used for process applications in the food industry, petrochemicals, and hazardous locations.

Processor Section

The processor section is the brain of the programmable controller. This section organizes all control activity by receiving inputs, performing logical decisions according to the program, and controlling the outputs.

The processor section evaluates all input signals and levels. This data is then compared to the memory in the programmable controller, which contains the logic of how the inputs are interconnected in the circuit. The interconnections are programmed into the processor by the programming section. Based upon the input conditions and program the processor section then controls the outputs. The processor continuously examines the status of the inputs and outputs and updates them according to the program. Figure 14-8 illustrates some of the many functions the processor performs.

The process of evaluating the input/output status, executing the program, and updating the system is called *scan*. The time it takes a programmable controller to make a sweep of the program

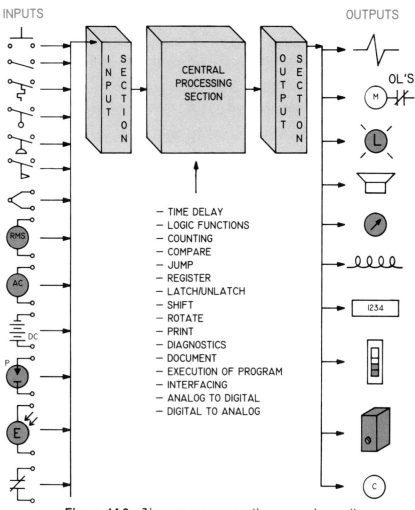

Figure 14-8 The processor section organizes all control activity by receiving inputs, performing logical decisions as programmed and controlling the outputs.

is called the *scan time*. Scan time is usually given as the time per 1k byte of memory and typically runs in the 1 to 25 millisecond range. Scanning is a continuous and sequential process of checking the status of inputs, evaluating the logic, and updating the outputs.

The Programming Section

The programming section of the programmable controller allows input into the programmable controller through a keyboard. Even though the programmable controller has a brain (the processor section) it still must be told what to do. The processor must be given exact, step-by-step directions. This includes communicating to the processor such things as load, set, reset, clear, enter in, move, and start timing.

Programming a programmable controller involves two components. The first component is the programming device that allows access to the processor. The second component is the programming language that allows the operator to communicate with the processor section.

Programming Devices. Programming devices vary in size, capability and function. Programming devices are available as simpler, small, handheld units or complex color CRTs with monitoring and graphics capabilities (Figure 14-9).

A programming device may be connected permanently to the programmable controller or connected only while the program is being entered. Once a program is entered, the programming device is no longer needed, except to make changes in the program or for monitoring functions. Some programmable controllers are designed to use an existing personal computer, such as an IBM ® for programming. Using a personal computer to program is called *off line programming*. This permits the computer to be used for other purposes when not being used with the programmable controller.

Language of Programmable Controllers. The first programmable controllers used a language that was compatible with industry—the line (ladder) diagram. Line diagrams are still commonly used as a language for programmable controllers throughout the world. Other languages used are Boolean, Functional blocks, and English statement. Line diagrams and Boolean are basic programmable controller languages. Functional blocks and English statement are higher level languages required to execute more powerful

HANDHELD UNIT

CRT MONITOR WITH KEYBOARD

Figure 14-9 A hand-held programming unit can be used to input information into the processor. Larger programming units include a CRT for illustrating and monitoring the program. (Furnas Electric Company; Allen-Bradley, A Rockwell International Company)

operations, such as data manipulations, diagnostics, and report generation.

The line diagram is drawn in a series of rungs. Each rung contains one or more inputs and the output (or outputs) controlled by the inputs. The rung relates to the machine or process controls, and the programming instructions relate the desired logic to the processor. Figure 14-10 illustrates basic circuit logic functions and the typical program required to enter the circuit's logical operation into the processor section. The program is entered into the controller through the keyboard.

Programming a programmable controller is not difficult since the program follows a logical process. Inputs and outputs are entered into the controller

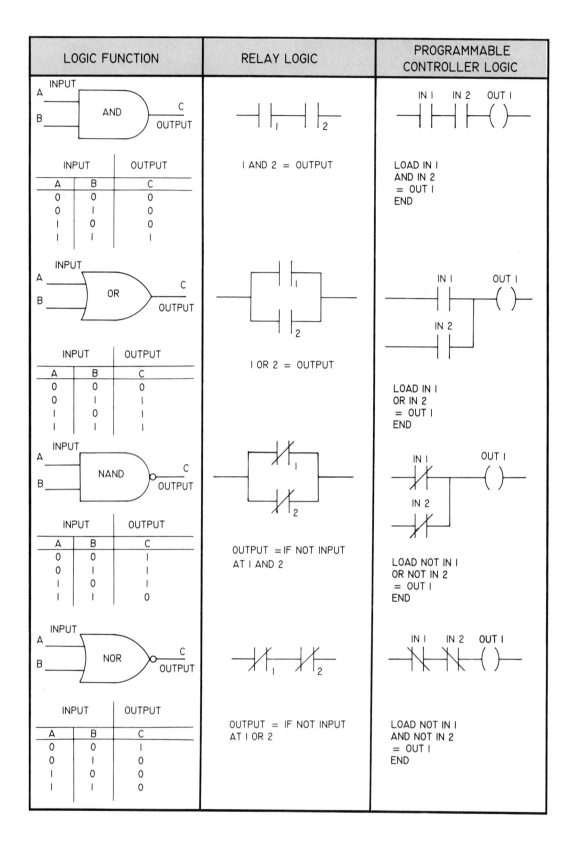

Figure 14-10 Basic circuit logic functions are used to enter the circuits logical operation into the processor.

in the same manner as if connecting them by "hard wiring." The difference in programming is that although a circuit is the same, each manufacturer has a different method of entering that circuit. However, there are more similarities than differences from manufacturer to manufacturer. Figure 14-11 illustrates a typical keyboard for a pocket programmer. This device has both uppercase and lowercase functions on a key. The uppercase functions (FRC OFF, FRC ON, A, etc.) can be obtained by first pressing the shift key.

Developing a Typical Program. Before a program can be entered into a programmable controller, several steps must be taken.

The first step is to develop the logic required of the circuit into a line diagram. Figure 14-12 illustrates a typical line diagram for a control circuit. In this circuit when control switches 1 and 2 are closed, motor starter 1 and the timer are energized. After the set time delay, motor starter 2 is energized. An overload on either motor will de-energize the starter coil. In this circuit, control switches 1 and 2 could be a manual, mechanical or automatic input.

The second step is to take the line diagram and convert it into a programming diagram. A programming diagram is a line diagram that better matches the programmable controller's language. With a programmable controller all contacts are listed as inputs on the left and the loads are listed as outputs on the right. As illustrated, the overload contacts would then be moved to the left on this diagram.

The third step is to enter the desired logic of the circuit into the controller. Every manufacturer will have a slightly different set of steps and functions

FRC OFF	FRC ON	A	B	C
0	1	2	3	4
D	E	F	EVENT	TIME
5	6	7	8	9
MODE	PRT UNPRT	REMOVE	INSERT	LAST
SEARCH	-(SQO)- -(CTU)-	-(SQI)- -(CTD)-	-(ZCL)- -(MCR)-	NEXT
RUNG	-(L)-	-(U)-	-(RTF)- -(RTO)-	SHIFT
CANCEL CMD	-| |-	-(RST)- -|*|-	-| |-	ENTER

Pocket programmer keyboard.

Abbreviations and Symbols

FRC OFF	Force OFF	-(RTF)-	Retentive Timer Off-Delay
FRC ON	Force ON	-(RTO)-	Retentive Timer On-Delay
PRT	Protect	CANCEL CMD	Cancel Command
UNPRT	Not Protect	-(RST)-	Reset
-(SQO)-	Sequencer Output	Ͳ	Branch Open
-(SQI)-	Sequencer Input	Ⅎ	Branch Close
-(CTU)-	Up Counter	-| |-	Examine ON
-(CTD)-	Down Counter	-|*|-	Examine OFF
-(ZCL)-	Zone Control Last State	-| |-	Output Energize
-(MCR)-	Master Control Reset		
-(L)-	Latch		
-(U)-	Unlatch		

Figure 14-11 A handheld programmer keyboard is used to input circuit logic functions. (Allen-Bradley, A Rockwell International Company)

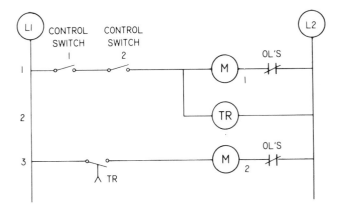

Figure 14-12 In this line diagram, control switches 1 and 2 must be closed to energize starter 1 and the timer. After the set time, starter 2 will be energized.

to enter the program into the programmable controller. Figure 14-13 illustrates the programming (line) diagram and the programming sheet required to enter the logic of Figure 14-12 into a typical programmable controller. This program matches the keys that must be pressed when programming. (Figure 14-14).

The fourth step is to take the written program and enter it into the programmable controller. Once the program is entered, it can be tested.

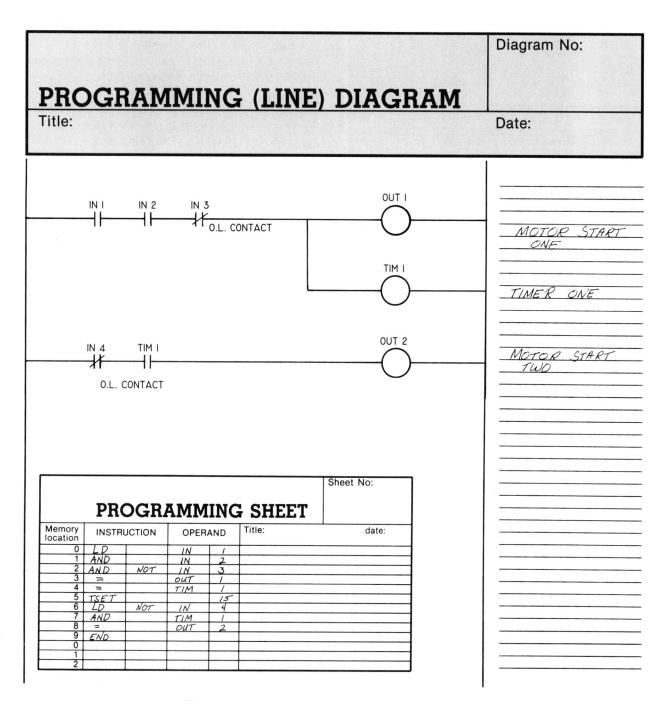

PROGRAMMING (LINE) DIAGRAM

Diagram No:

Title:

Date:

IN 1	IN 2	IN 3 O.L. CONTACT		OUT 1
IN 4	TIM 1	O.L. CONTACT		TIM 1
				OUT 2

MOTOR START
ONE

TIMER ONE

MOTOR START
TWO

PROGRAMMING SHEET

Sheet No:

Memory location	INSTRUCTION		OPERAND		Title:	date:
0	LD		IN	1		
1	AND		IN	2		
2	AND	NOT	IN	3		
3	=		OUT	1		
4	=		TIM	1		
5	TSET			15		
6	LD	NOT	IN	4		
7	AND		TIM	1		
8	=		OUT	2		
9	END					
0						
1						
2						

Figure 14-13 A line diagram indicates the logic of the circuit. The programming sheet lists the keys that must be pressed when entering the program.

Figure 14-14 Keys are pressed on the programmer as indicated on the programming sheet to enter the program. (Electromatic Controls Corporation)

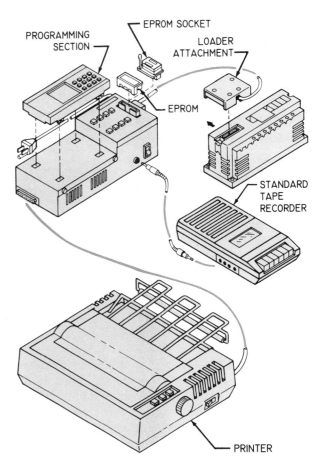

Figure 14-15 The program can be stored on a cassette for later use. A copy of the program can be made by connecting the controller to a printer. (Furnas Electric Company)

Storing and Documentation. Once a program has been developed it may be necessary to store the program outside of the controller or document the program by printing it out (Figure 14-15). This allows for a means of storing and retrieving control programs, which makes for fast changes in a process or operation. Storage of a program is commonly achieved using a cassette tape recorder. For example, one tape may have the program for controlling 8-ounce bottles and a second tape may have the program for controlling 16-ounce bottles. When a change from one size bottle to another is required, one need only load the programmable controller with the correct tape to start the line for all the proper control settings.

Even if the programmable controller is not likely to ever have its program changed, the program should be stored on a tape. This ensures the safety of the program in the event of a problem.

Another method of storing a program is to use a read-only memory (ROM), random access memory (RAM), or erasable programmable read only memory (EPROM) type memory chip. This allows a permanently stored program (EPROM) to provide the memory storage. These chips allow for a program to be stored on them. Many original equipment manufacturers (OEM) use this type of storage to store the machine's program after it has been developed. They can be mass produced and placed in the machine as needed.

Once a program has been entered into the programmable controller, a copy of the program can

be made by connecting the controller to a printer. The printout can be used as a hard copy of the program for documentation and future reference.

INTERFACING SOLID STATE CONTROLS

Programmable controllers can have many types of inputs, including pushbuttons, level switches, temperature controls, and photoelectric controls. Inputs such as pushbuttons and temperature controls are usually easy to input. However, more complex solid state control inputs such as proximity and photoelectric inputs require special consideration because of their function.

Solid state proximity and photoelectric controls are an important part of the automated system. Figure 14-16 illustrates a typical process using

PHOTOELECTRIC CONTROLS CAN BE
INPUT INTO PROGRAMMABLE
CONTROLLERS FOR—
• DETECTION
• INSPECTION
• MONITORING
• COUNTING
• DOCUMENTATION

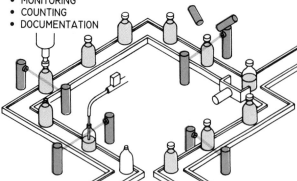

Figure 14-16 Photoelectric and proximity controls commonly have a solid state output and are ideal for use with programmable controllers.

photoelectric controls. These controls usually have a solid state output and are ideal for inputting to programmable controllers. Available outputs include two- and three-wire types with thyristor and transistor outputs that can be connected individually or in series/parallel combinations.

Two-wire Thyristor Output Type Sensors

Two-wire thyristor output sensors are available in a supply voltage range of 20–270 VAC at about 180 to 500 mA ranges in either normally open (NO) or normally closed (NC) versions. Two-wire thyristor output sensors have only two wires and are wired

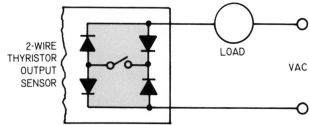

Figure 14-17 The power to operate a two-wire thyristor output type sensor for control of AC loads is received through the load. (Electromatic Controls Corporation)

in series with the load like a mechanical switch, as illustrated in Figure 14-17. The power to operate these sensors is received through the load when the load is not being operated. As with any thyristor output device, some consideration must be given to "off state leakage current" and "minimum load current." Unlike a mechanical switch, there is current consumed by the proximity sensor in the inactivated mode. However, the current is small enough that most industrial loads are not affected.

In the case of high impedance loads and some programmable controllers, this leakage current may be enough to activate the load. This problem can be corrected by placing a load resistor across the load as illustrated in Figure 14-18. The resistor value should be chosen to ensure that minimum load current is exceeded and the effective load impedance is reduced, preventing off state leakage current turn on. This resistance value is typically in the range of 4.5K to 7.5K ohms. A general rule is to use a 5K-5W resistor for most applications.

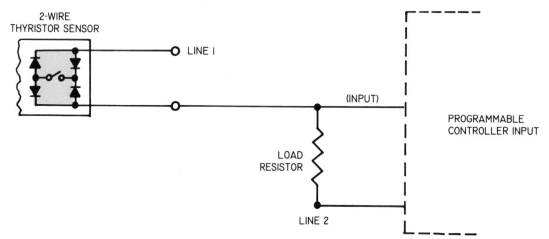

Figure 14-18 A load resistor may be required when connecting a sensor to the programmable controller to prevent leakage current of the sensor from inputting into the controller. (Electromatic Controls Corporation)

Surge Protection for Inductive AC Loads

Inductive loads such as solenoids, motor starters, and motors all hold stored energy in their coils that is dissipated when the loads are switched off. To prevent the dissipated energy (called *transients*) from damaging the programmable controller, it can be controlled by using some type of surge suppression method.

Figure 14-19 illustrates common methods of surge protection. These surge protection circuits are connected directly across the output load in most applications. By using surge protection on inductive AC loads, the output section of the programmable controller is protected. Surge protection should be used on all heavy inductive loads, including programmable controllers with internal protection. Figure 14-20 illustrates the connection of external surge protection in addition to internal protection. Surge protection is more important with a programmable controller compared to separate relays. Unlike individual relays, in a programmable controller the complete output module often must be replaced if one output does not function.

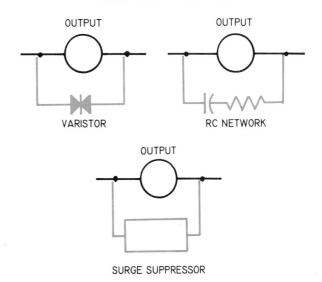

Figure 14-19 Surge protection protects the programmable controller from transients. (Allen-Bradley, A Rockwell International Company)

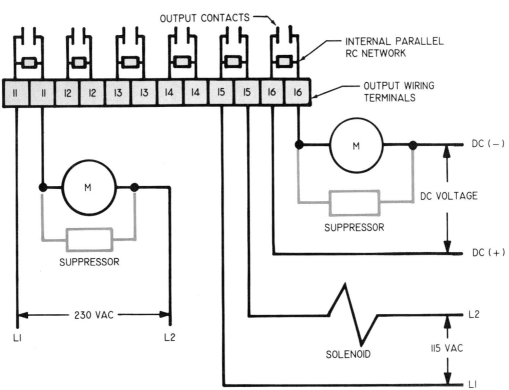

Figure 14-20 In addition to internal surge protection, external surge protection is recommended for heavy inductive loads. (Allen-Bradley, A Rockwell International Company)

Three-wire Transistor Output Sensors

Three-wire transistor output sensors are available in a supply range of 10–40 VDC at about 200 mA range. This type of sensor is easily interfaced with other electronic circuitry and programmable controllers. Output sensor types consist of either an open collector NPN or PNP transistor type. Both NO or NC versions are available.

These sensors receive their power to operate through two of the leads (plus and minus, respectively) from the power source. The third lead is used to switch power to the load either using the same source of power as the proximity switch or an independent source of power as illustrated in Figure 14-21. When an independent source is used, one lead of that source is common with one lead of the source used to power the sensor. When using an independent power source for the load, the voltage level must be within the specifications of the sensor used.

APPLICATIONS OF PROGRAMMABLE CONTROLLERS

Programmable controllers are useful in increasing production, and improving overall plant efficiency. Programmable controllers can control individual machines and link the machines together into a system. The flexibility provided by a programmable controller has resulted in many applications in manufacturing and process control. Process control has gone through many changes in

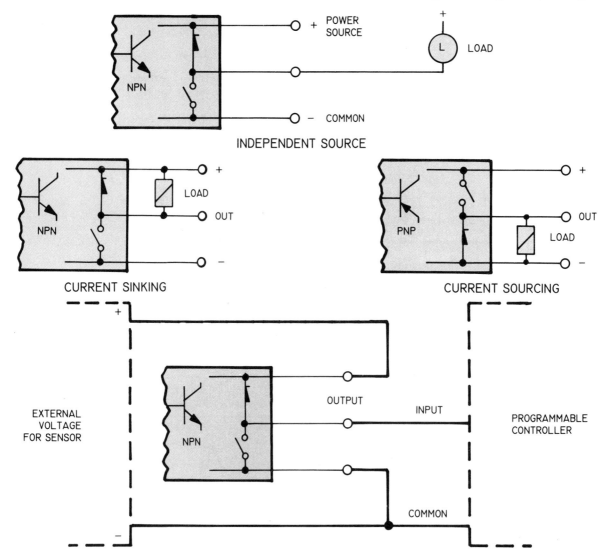

Figure 14-21 Three-wire transistor output sensors use either an NPN or a PNP transistor to control the load. Power is received through two of the leads with the third lead used to switch the load. (Electromatic Controls Corporation)

the past few years. In the past, process control was mostly accomplished by manual control. Flow, temperature, level, pressure and other control functions were monitored and controlled at each stage by production workers. Today, using programmable controllers, an entire process can automatically be monitored and controlled with little or no workers involved at all. Following is a list of a few process applications in which programmable controllers have been used.

1. Grain operations involving storage, handling, and bagging.

2. Syrup refinery involving product storage tanks, pumping, filtration, clarification, evaporators, and all fluid distribution systems.

3. Fats and oils processing involving filtration units, cookers, separators, and all charging and discharging functions.

4. Dairy plant operations involving all process control from raw milk delivered to finished dairy products.

5. Oil and gas production and refinement from the well pumps in the fields to finished product delivered to the customer.

6. Bakery applications from raw material to finished product.

7. Beer and wine processing, including the required quality control and documentation procedures.

Each of these processes involve much more control then can be listed here.

Welding

In manufacturing of discrete parts, welding is a major part of the system. Figure 14-22 illustrates a typical welding operation that has been automated. In this application the programmable controller can control the length of the weld and the power required to produce the correct weld. The programmable controller is programmed to allow the weld to occur only if all inputs and conditions are correct. This includes determining

1. if the parts are present and in the correct position;

2. the correct weld cycle speed and power setting;

3. the correct rate of speed on the line for the given application; and

4. if all interlocks and safety features are functioning.

In addition, the programmable controller can be used to determine if parts are running low and be set to automatically turn on and off the line as re-

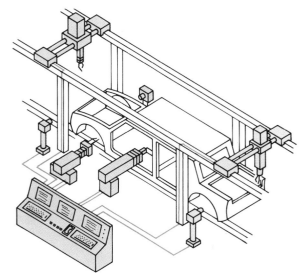

Figure 14-22 Programmable controllers can be used to control and automate industrial welding processes.

quired. Documentation of production efficiency can be generated for quality control and inventory requirements.

Although this application illustrates only one welding station the programmable controller can be used to control and interlock many welders. Welders at one station may require more power than is available if all the welders are on simultaneously. In this case, a large power draw can cause poor quality welds. The requirements in a system using many welders is to limit the amount of power being consumed at any one point in time. This is accomplished by time sharing the power feed to each welder.

The programmable controller can be programmed for a maximum power draw. When a welder requires power, the controller can determine if power is available. If the correct power level is available, the weld will take place. If not, the controller will remember the request and when power is available, it will permit the welder to proceed with the weld cycle. The programmable controller can also be programmed to determine which welder has priority.

Machine Control

When machines are linked together to form an automated system, controls must be synchronized.

Figure 14-23 illustrates a typical manufacturing application in which one machine receives a part from another machine and sends it on to yet another. In this application each machine may be controlled by a programmable controller, with another controller synchronizing the operation. This is likely if the machines are purchased from different manufacturers. In this case, each machine may include a programmable controller to control all the functions on that machine only. If the machines are purchased by one manufacturer or designed in-plant, it is possible to use one large programmable controller to control each individual machine and synchronize the process.

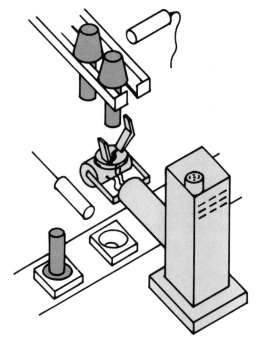

PROGRAMMABLE CONTROLLERS ARE USED
TO CONTROL INDUSTRIAL ROBOTS

Figure 14-24 Programmable controllers can be used to control the operations of an industrial robot.

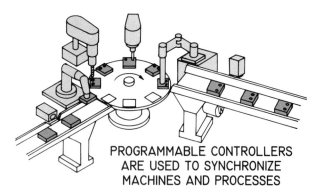

PROGRAMMABLE CONTROLLERS
ARE USED TO SYNCHRONIZE
MACHINES AND PROCESSES

Figure 14-23 Programmable controllers can be used to control individual machines and synchronize operation with other machines.

Control of Industrial Robots

Programmable controllers are ideal devices for controlling any industrial robot. Figure 14-24 illustrates a programmable controller used to control a robot. The programmable controller can be used to control all operations such as rotate, grip, withdraw, extend, and lift. Since most industrial robots operate in an industrial environment, a programmable controller is recommended.

Fluid Power Control

When a linear movement is required in an automated application, fluid power cylinders are usually chosen. Pneumatic cylinders are common because they are easy to install and most plants have compressed air. Pneumatics works well for most robot grippers and drives, positioning cylinders, machine loading and unloading and tool working applications. When a manufacturing process requires high forces, hydraulic cylinders are used. Hydraulic systems of several thousand psi are often used to punch, bend, form and move components.

Figure 14-25 illustrates a typical industrial fluid power circuit. This circuit includes both linear and rotary actuators. This system, as with any fluid power circuit, is ideal for control by a programmable controller. The controller's output module would be connected to control the four solenoids. The function of each solenoid is as follows:

Solenoid	Equals
A	cylinder in
B	cylinder out
C	rotate forward
D	rotate reverse

The programmable controller is used to control the energizing or deenergizing of the solenoids. Solenoids control the directional control valves, which would control the actuators.

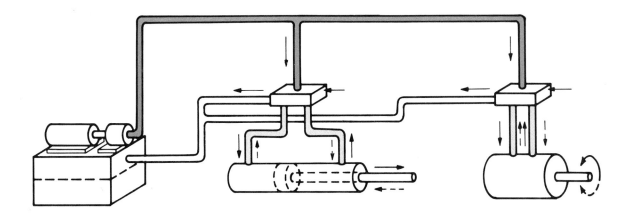

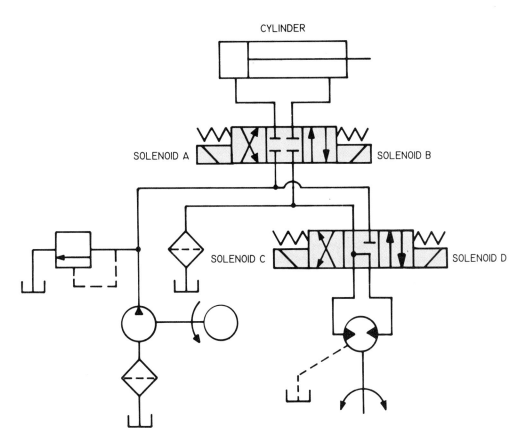

CYLINDER

SOLENOID A

SOLENOID B

SOLENOID C

SOLENOID D

PROGRAMMABLE CONTROLLERS ARE USED
TO CONTROL FLUID POWER APPLICATIONS

Figure 14-25 Programmable controllers can be
used to control linear and rotary actuators in an industrial fluid power
circuit.

Control of Industrial Drives

Typically, motors have been connected directly to the power lines and operated at a set speed. As systems have become more automated, there are many applications that require a variable motor speed. Adjustable speed controls are available to control the speed of standard AC and DC motors. These controls are usually manually set for the desired speed, but many allow for automatic control of the set speed. Figure 14-26 shows a programmable controller used to control AC drives, which control a given application. The drives can accept frequency and direction commands in a BCD format that the programmable controller can provide with a BCD output module.

The Pulp and Paper Industries. Figure 14-27 illustrates a typical pulp and paper mill operation. This type of operation covers a large area. The control of this type of operation is ideal for a programmable controller since most of the control logic is of the start/stop, time delay, count sequential and interlock type functions. The programmable controller allows for required I/O, which when multiplexed, can transmit multiple signals over a single pair of wires.

The basic operation of a paper mill is to receive raw material such as logs, pulp wood or chips, and process, size, store and deliver the material. This includes a large conveyor system that has diverter gates, over travel switches, speed control and interlocking. A break in any part of the system can shut down the entire system. Since the system covers a large area, finding a fault can be time-consuming. To solve this problem a programmable controller with fault diagnostics can be used to analyze the system and give an alarm and printout of where the problem exists with suggested solutions.

Batch Process Control System. Batch processing blends sequential, step-by-step functions with continuous closed-loop control. Process control is systems control, and systems are made up of many parts. This means that individual programmable controllers can be used to control each part and step of the process with additional programmable controllers and computers supervising the total operation.

Figure 14-28 illustrates the major components in a programmable controlled batch process control system. Note that as part of the system an operator interface is added. This may be in the form of an

PROCESS THAT REQUIRES MULTI-SPEED CONTROL

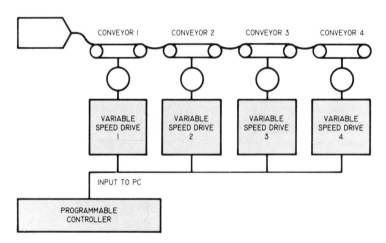

PROGRAMMABLE CONTROLLERS ARE USED
TO CONTROL INDUSTRIAL DRIVES

Figure 14-26 Programmable controllers can be used to control and synchronize the speed of conveyors on the assembly line. One programmable controller can control all drives or individual drives. (Allen-Bradley, A Rockwell International Company)

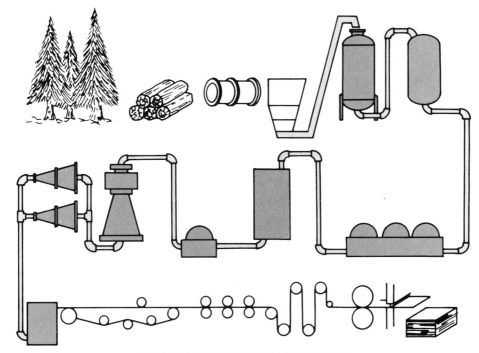

Figure 14-27 In a paper mill, the programmable controller can be used to control each operation and diagnose a problem in the system.

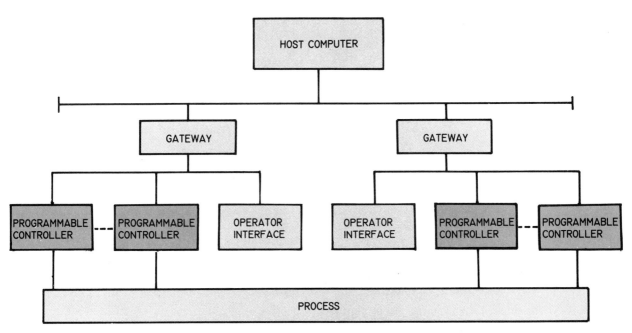

Figure 14-28 In a batch process control system, the operator interface is used for instrumentation or other monitoring functions. (Gould Inc., Programmable Control Division)

instrumentation and process control station, a CRT terminal, or any other type of interface. To add in interfacing and monitoring a programmable based system, an access panel can be connected as required. Figure 14-29 illustrates a data access panel that will give message displays.

Figure 14-29 The data access panel is used to monitor the system by providing message displays. (Gould Inc., Programmable Control Division)

MULTIPLEXING

Multiplexing is a method of transmitting more than one signal over a single transmission system. As the distance increases between any transmitting and receiving point, the cost of multiconductor cable with separate wires for each signal becomes very expensive through installation, maintenance and replacements. With multiplexing, a single-wire pair can serve multiple transmitters and receivers. A multiplexing system is ideal when used with programmable controllers, as all inputs and outputs can be connected with just one pair of wires. A multiplexing system is also called a *two-wire system.*

Many advantages exist in using a multiplexing system for control. One of the main advantages is the elimination of costly hard wiring. Figure 14-30 illustrates how eight control switches are hard wired to control eight loads. A pair of wires connected through conduit is required for each control switch. This means that time and money would be wasted for even the shortest distance. As the distance between the control switches and loads increases, the cost of time and materials for the hard wired circuit increases.

This same circuit can be connected using a multiplexing system. Only one pair of wires is required between the eight control switches and eight loads. Additional control switches to be added require no additional transmission wires. Additional transmitters, receivers, displays, or programmable controllers can all be connected to the same pair of wires.

As a control circuit increases in size and function, wiring it becomes more difficult. Figure 14-31 illustrates how a multiplexing system can send back a signal to indicate that the load is energized. The multiplexing system is much simpler than hard wiring and can be expanded to almost any number of inputs and outputs, all controlled by a programmable controller. The programmable controller controls all inputs and outputs and makes timing, counting, and sequencing decisions and any other required logic decisions. Figure 14-32 illustrates how a programmable controller could be connected to the system.

The multiplexing system can be used to transmit both analog and digital signals on the same two-wire system. This makes the system ideal for any instrumentation application, including the transmission and control of temperatures, BCD signals, rpm, voltage and current levels, and counts.

In addition, a 24-hour clock and printer can be added to the system for documentation. This addition would make it possible to print out the time of day when a certain event has taken place on the multiplexing system.

Security System

A multiplexing system can be used in a security system for a plant or building. Each door and window can be connected to the two-wire system. A display is located in a central control location to

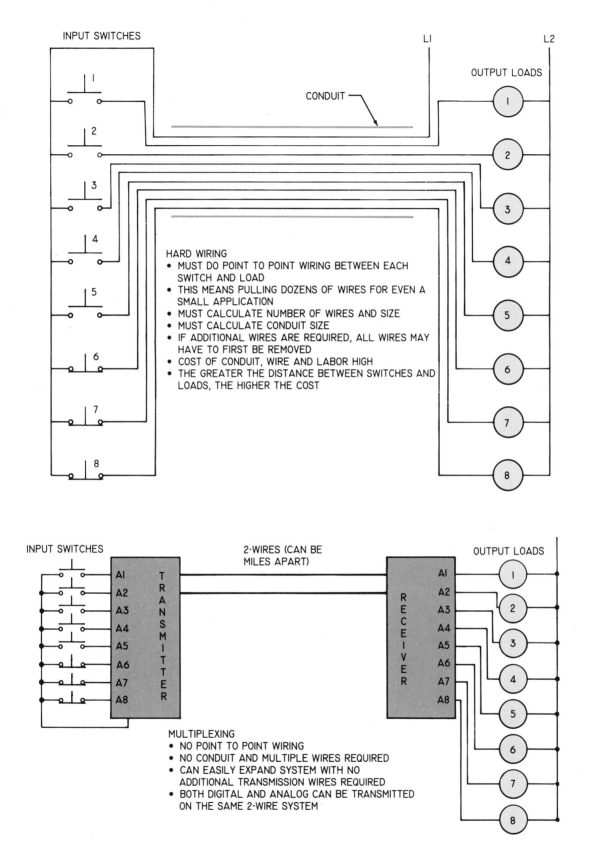

INPUT SWITCHES

L1

L2

OUTPUT LOADS

CONDUIT

HARD WIRING
• MUST DO POINT TO POINT WIRING BETWEEN EACH SWITCH AND LOAD
• THIS MEANS PULLING DOZENS OF WIRES FOR EVEN A SMALL APPLICATION
• MUST CALCULATE NUMBER OF WIRES AND SIZE
• MUST CALCULATE CONDUIT SIZE
• IF ADDITIONAL WIRES ARE REQUIRED, ALL WIRES MAY HAVE TO FIRST BE REMOVED
• COST OF CONDUIT, WIRE AND LABOR HIGH
• THE GREATER THE DISTANCE BETWEEN SWITCHES AND LOADS, THE HIGHER THE COST

INPUT SWITCHES

2-WIRES (CAN BE MILES APART)

OUTPUT LOADS

TRANSMITTER

RECEIVER

A1
A2
A3
A4
A5
A6
A7
A8

MULTIPLEXING
• NO POINT TO POINT WIRING
• NO CONDUIT AND MULTIPLE WIRES REQUIRED
• CAN EASILY EXPAND SYSTEM WITH NO ADDITIONAL TRANSMISSION WIRES REQUIRED
• BOTH DIGITAL AND ANALOG CAN BE TRANSMITTED ON THE SAME 2-WIRE SYSTEM

Figure 14-30 Multiplexing eliminates the need for costly hard wiring in a system.

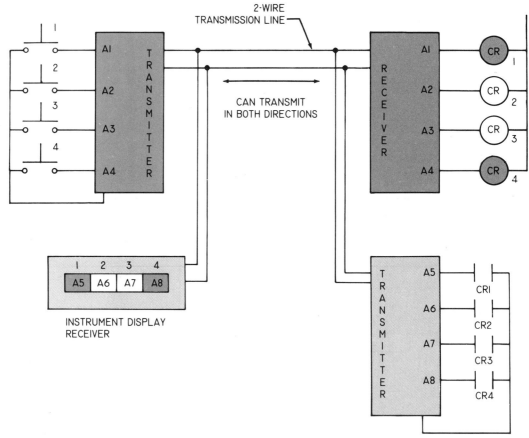

Figure 14-31 In this multiplexed system, four control switches to control four loads use signals sent back to indicate the loads are energized. (Electromatic Controls Corp.)

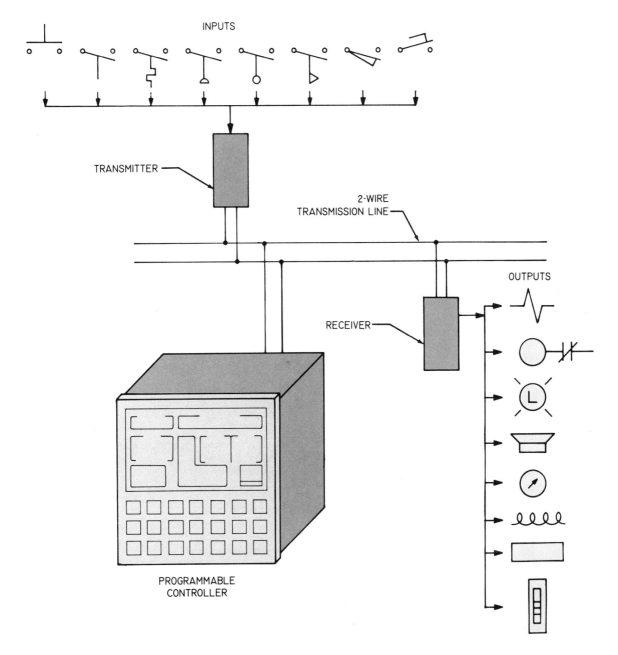

Figure 14-32 The programmable controller makes decisions based on signals provided by the inputs and outputs.

monitor the total system. Figure 14-33 illustrates a basic security system used to monitor a building.

A clock and printer can be added to record the time each door or window is opened and closed. A programmable controller could be added to control all required circuit logic.

The programmable controller's controlling functions on the multiplexing system can be expanded as necessary. For example, if a security guard is

to patrol a building, the controller can be programmed to monitor the guard as well as the building. As the guard moves through the building, the controller would monitor the movement by recording when a door is open and/or when the guard activates an assigned switch. The controller would know how long it should take the guard to move from station to station. If something happens to the guard, the controller would detect this and

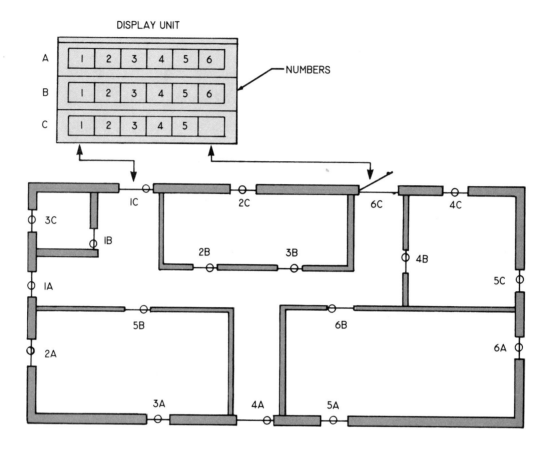

Figure 14-33 A two-wire multiplexing system can be
used for a building security system.

Conveyor System

Conveyor systems are commonly used in industry for movement of materials. As industrial systems become more automated, additional control is required. Additional control means additional wires to be connected from machine to machine. Again, multiplexing can be used to reduce the total wires required.

As in any assembly line application, a fault or breakdown at one station requires that all upstream machines be turned off to prevent a jam-up. Because this system may cover miles in many applications, multiplexing could be used to link the system together. Figure 14-34 shows a number of machines connected together by conveyors.

As illustrated, a sensor could be used to detect a fault at one location and send a signal over the two-wire system to stop all upstream machines and conveyors. This system could also be connected for total control of all functions using the multiplexing system and a programmable controller.

take corrective action, such as alarming a central control station.

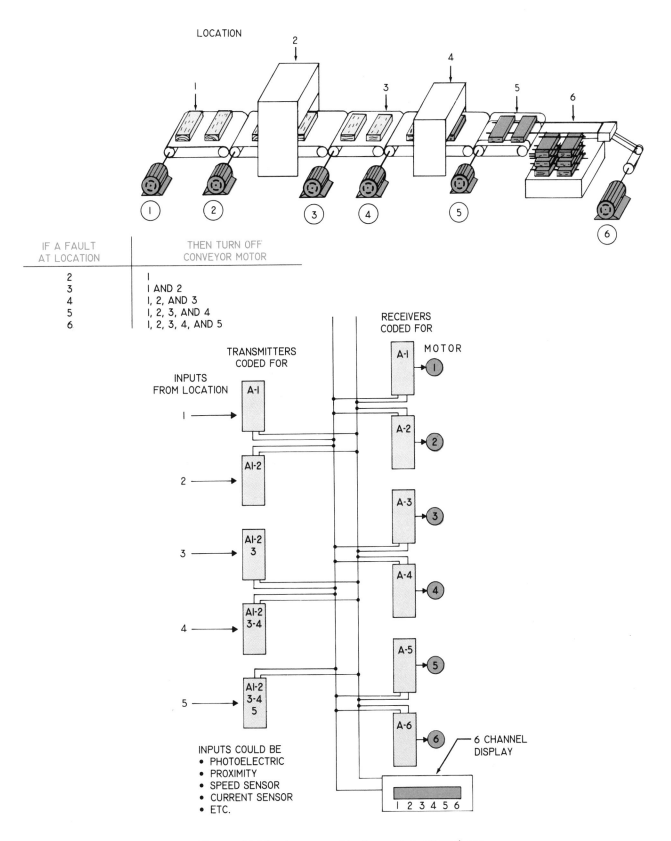

IF A FAULT AT LOCATION	THEN TURN OFF CONVEYOR MOTOR
2	I
3	I AND 2
4	I, 2, AND 3
5	I, 2, 3, AND 4
6	I, 2, 3, 4, AND 5

INPUTS COULD BE
• PHOTOELECTRIC
• PROXIMITY
• SPEED SENSOR
• CURRENT SENSOR
• ETC.

Figure 14-34 An assembly line using several conveyors can be controlled by a two-wire multiplexing system.

1. What are five major advantages of using programmable controllers in an industrial application?
2. What are two major market areas that programmable controllers are used in?
3. How does a programmable controller differ from a computer?
4. What are the four basic parts of a programmable controller?
5. What is the function of the power supply section?
6. What is the function of the input/output section?
7. What is the function of the processor section?
8. What is the function of the programming section?
9. What is an example of a discrete input?
10. What is an example of a discrete output?
11. What is an example of a data input?
12. What is an example of a data output?
13. What are five typical functions performed by the central processor?
14. What is the most popular language used with programmable controllers?
15. What are the steps that must first take place before entering a program into the programmable controller?
16. What are two methods of storing a program once it has been developed?
17. What is the value of the load register used to reduce off state leakage current when using a two-wire sensor as an input into a programmable controller?
18. What type of loads may require surge protection to be added into the controller?
19. What is one of the main advantages of using a multiplexing system?
20. How many wires are required to transmit and receive signals over a multiplexing system?

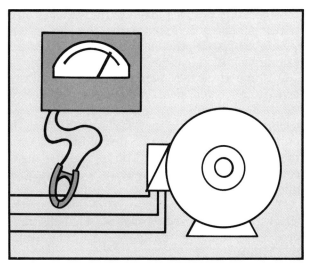

15 AC Reduced Voltage Starters

The five basic types of reduced voltage starters are primary resistance, autotransformers, part-winding, wye-delta and solid state. In primary resistance starting a resistor controlled by a timing circuit is connected to the limit current in each motor line. Autotransformer starting uses a tapped autotransformer to limit in-rush current. Part-winding starting uses a sequence of connections to connect parts of the winding until all parts are connected. Wye-delta starting utilizes a different connection for starting (wye) than for running (delta) to limit in-rush current. Solid state starting utilizes semiconductor electronic control devices to limit in-rush current.

REASONS FOR REDUCED VOLTAGE STARTING

Until this unit, we have discussed only the theory and applications of full voltage starters. For many applications, especially those involving small horsepower motors, this is the least expensive and most efficient means of starting a motor.

There are, however, many other applications involving large horsepower DC motors and AC motors where full voltage starting may be objectionable because it creates interference with other systems.

Reduced voltage starting eliminates many of these problems by reducing interference in these major areas: the power source, the load, and the electrical environment surrounding the motor. How reduced voltage starting minimizes the problems in each area will be discussed in detail in the following paragraphs.

Power Source

One of the most predominant reasons for using reduced voltage starting is to reduce the large current drawn from the power company lines by the across-the-line start of a large motor. When an induction motor is started, it acts much like a short circuit would in the secondary of a transformer. The current drawn by the motor will be from 2.5 to 10 times the current rating found on the motor nameplate. This sudden demand for large current can reflect back into the power lines and create problems.

Electric utilities usually limit the inrush current drawn from their lines to a given maximum amount for a specified period of time. Such limitations are necessary for smooth, steady power regulation and eliminating objectionable voltage disturbance such as annoying light flicker. In these cases, the utility company is not limiting the total maximum amount of current that can be drawn, but rather dividing the amount of current into steps. This permits an incremental start that allows the utility voltage regulators sufficient time to compensate for the large current drawn. This maximum current permitted by the utility in any one step of an increment start is called the *increment current*. (This increment current can be found by checking with your local utility company).

Reduced voltage starting in this case resolves the problem of large current surges by providing incremental current draw over a longer period of time.

Torque and Starting Requirements of a Load

In several industries, especially those dealing with paper and other delicate fabrics, care must be exercised to avoid sudden high starting torque (turning force), which could stretch or tear the product. To prevent this type of product damage or damage to gears, belts, and chain drives, it becomes necessary to limit starting torque surges.

Reduced voltage starting can be used to overcome excessive starting torque by providing a gentle start and smooth acceleration. To understand how torque is reduced requires an understanding of the relationship between voltage, current and torque.

We will recall from Ohm's law that as voltage is reduced, current is reduced. We should also remember that as current is reduced, torque is reduced. Thus a reduction in voltage reduces current which in turn reduces torque to provide a gentle start. As voltage is increased, current and torque increase, providing smooth acceleration.

It is important to remember at this point that reduced voltage starting is *not* a type of speed controller. Rather, reduced voltage starting acts as a buffer or shock absorber to the load when it is starting. Speed controllers and their applications will be covered in detail in Unit 16, "Accelerating and Decelerating Methods and Circuits."

A second important point to keep in mind is that reduced voltage starting should not be considered for use on loads which are difficult to start. If the load is difficult to start at full voltage, it positively will not start at a reduced voltage.

Electrical Environment

When an electrical system requiring large amounts of power is being designed or installed, consideration must be given to the system's electrical neighbors. In other words, the new system should not create problems for the systems which are already installed and working properly.

It is important to remember that electrical current surges, even small ones, can cause disruptions. For example, a current surge may cause timers to reset or relay and starters to drop out. In buildings which are totally air conditioned, compressor motors have been known to cause major computers and microcomputers to malfunction due to current surges.

Since reduced voltage starting reduces current surges, it may be used to solve internal electrical problems even when not required by the utility companies.

REDUCED VOLTAGE STARTING AND THE DC MOTOR

The DC Motor is used to convert electrical energy into mechanical energy. Although both the AC and DC motors operate on the same fundamental principles of magnetism, they each differ in the way the conversion of the electrical power to mechanical power is accomplished. This difference gives each motor type its own operating characteristics. The two fundamental operating characteristics of DC motors that make them the choice for some applications are high torque outputs and good speed control.

Another factor in using DC motors is the available source of power, AC or DC. For applications such as the automobile starter, the DC motor is compatible to the power source—a battery which is capable of delivering only DC. This application of DC motors run by batteries is also true in industrial applications using portable power equipment—fork lift trucks, dollies, and small locomotives used to move material and supplies. However, in applications where the motor is to be connected to a power source other than a battery, the available power source may be either AC or DC. Rectifying AC to DC was covered in Unit 10, "Power Distribution Systems."

DC Motors

As we have learned in previous units, any conductor carrying current has a magnetic field around it. In a DC motor, a magnetic field, caused by current flow in a conductor, interacts with another magnetic field. This interaction causes the conductor to move. In the alternator studied in Unit 10, a voltage was induced into a conductor as it was moved through the magnetic field. In Unit 10 mechanical motion was converted into electricity.

Before describing the operation of DC motors, let's review some of the facts about current flow, magnetic fields, and induced motion. Motor operation is based on these principles.

DC Motor Action

When a current-carrying conductor is placed between the poles of a magnet, the conductor moves. The direction of the induced motor depends upon the directions of both the current and the magnetic field. The relationship between the factors involved is called the right-hand, or motor, rule.

Figure 15-1 illustrates this rule, with arrows indicating the various directions. The index finger points in the direction of the magnetic field (N to S); the middle finger points in the direction of electron current flow in the conductor movement.

The line carries a current at right angles to the lines of the magnetic field. The force felt by the conductor is at right angles to both the current and the magnetic field. The amount of this force is

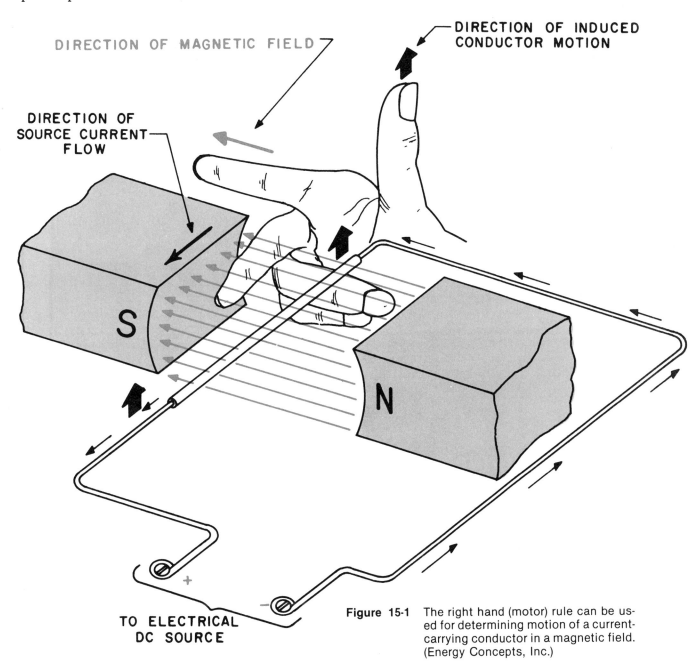

Figure 15-1 The right hand (motor) rule can be used for determining motion of a current-carrying conductor in a magnetic field. (Energy Concepts, Inc.)

dependent upon three factors: (1) The intensity of the magnetic field; (2) The current through the conductor; (3) The length of the conductor.

By increasing either of these three factors, the amount of force can be increased. Generally, it is the intensity of the field and the amount of current which are changed to increase force.

In the illustration of Figure 15-2 we have changed the single conductor into a simple coil or loop of wire. You will note in this diagram that the conductor current is at a right angle to the magnetic field. This is a requirement for induced motion, since no force is felt by a conductor if the current and the field direction are the same (parallel).

From the illustration of Figure 15-2 we can see that both sections of the loop AB and CD have a force exerted on them, since the direction of current flow in these segments is at right angles to the magnetic lines of force. Note, however, that the exertions of force on AB and CD are opposite in direction since the current flow is opposite in each section.

The result of the two magnetic fields intersecting creates a turning force, or torque, on the loop. The torque tends to rotate the loop in a counterclockwise direction, as indicated by the larger arrow. Figure 15-3 illustrates a cross-sectional view of the induced loop motion. The small circle with the sign (X) in it represents the AB loop seg-

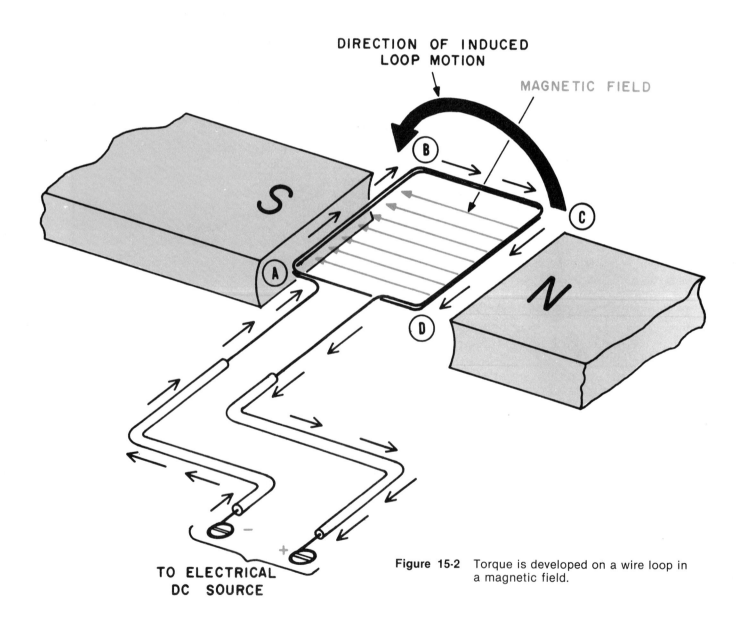

Figure 15-2 Torque is developed on a wire loop in a magnetic field.

ment. The sign (X) indicates that current is flowing away from the view, or into the page. The small circle with the dot sign (·) in it represents the DC loop segment. The dot sign (·) indicates that current is flowing toward the viewer, or out of the page.

As indicated, the interaction between the two magnetic fields causes a bending of the lines of force. When the lines tend to straighten out, they cause the loop to undergo the rotating motion. As indicated by the arrows, the left conductor is forced downward, and the right conductor is forced upward, causing a counterclockwise rotation.

Simple Motor Construction

Now that we understand the underlying principles of a DC motor, it is necessary to determine how this can be applied to a simple device. Figure 15-4 illustrates the construction and operation of a simple DC motor coil. From this illustration, we can see that the four main parts of a DC motor are its field, armature, brushes and commutator.

Since we have seen how the field and armature interact, we will concentrate on the purpose of the brushes and commutator.

When the coil in Figure 15-4 is positioned so that the plane of the loop is parallel to the magnetic flux and the loop sides are at right angles to the magnetic field with a current flowing through the coil as shown in position "A," a turning force is exerted. If the conductor coil was moved and stopped in position B, however, no further movement would take place. This position is called the neutral plane minimum flux position, or zero axis. In the neutral plane, no further torque (turning force) is produced because the forces acting on the conductor are upward on the top coil side and downward on the lower coil side, thus producing *no* torque.

The conductor, at this point, probably would not stop because of inertia; it would continue forward for a short distance. As this movement progresses, a problem arises. The magnetic field in the conductor is opposite that of the field, and this will tend to push the conductor back toward where it came, stopping the rotating motion.

To solve the problem regarding rotation, some method must be employed to reverse the current in the conductor every one-half rotation so that the magnetic fields will work together to maintain a positive rotation. The addition of brushes and a commutator are the solution to the problem.

Figure 15-4 illustrates how the commutator is used to reverse the direction of current flow in the armature coils. The commutator is split into two sections with each section connected to one side of the armature coil winding. The split ring commutator is supplied voltage through the brushes.

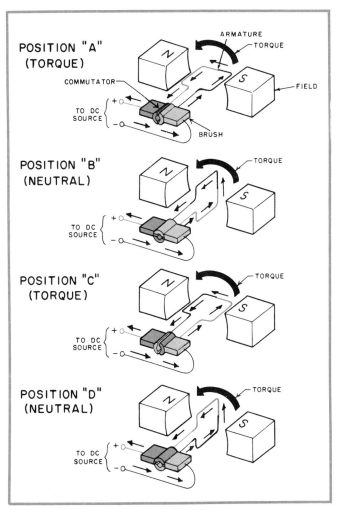

Figure 15-4 A two segment commutator is used to reverse the direction of current flow in the armature coils. (Energy Concepts, Inc.)

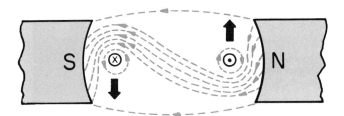

Figure 15-3 The distortion of magnetic field causes loop motion. (Energy Concepts, Inc.)

Each brush supplies a constant current from the power supply and does not change polarity. For purposes of illustration, one brush is in black and the other in red. Also for purposes of illustration, one-half of the armature coil winding is illustrated in black and the other half in red.

Starting with position A, we can see that the black brush is in contact with the black half of the coil winding and the red brush is in contact with the red half. As the coil is rotated 90° through the magnetic field, position B, we can see that the black brush will break contact with the black half of the coil and make contact with the red half of the coil. Likewise, the red brush will break contact with the red half of the coil and make contact with the black half. Since the flow of current remains the same polarity on the brushes at all times, the flow of current through the coil winding is reversed. This allows the coil winding to rotate another 180° in the same direction, as shown in positions C & D. After the additional 180° rotation, the black brush will break contact with the red half of the coil winding and make contact with the black half. Likewise, the red brush will break contact with the black half of the coil winding and make contact with the red half. This will again reverse the direction of current in the coil windings and allow for another 180° of rotation. As long as the coil winding is supplied with current and there is a magnetic field, the armature will continue to rotate.

The simple DC motor just described has some shortcomings. Each time the armature is in a neutral position, no torque is produced. In positions A and C, maximum torque is produced. The change in the amount of torque is shown graphically in Figure 15-5. The speed of the motor also changes because of the changes of torque. This is shown by the dashed lines in the graph of Figure 15-5. Most practical devices require a motor to turn at a uniform speed, so the simple DC motor just described would not be suitable.

Another problem is that simple DC motors do not start easily. This is particularly true if the armature is in or near a neutral position. The armature must be moved out of the neutral position in order to start the motor. In a practical DC motor, the armature is never in a neutral position, and the torque is always maximum. Maximum torque is accomplished by using an armature with more than one loop. A four loop armature is shown in Figure 15-6.

Each loop of the armature is connected to a pair of commutator segments as shown in Figure 15-6,

top. A single pair of brushes makes contact with the commutator segments. The armature acts like two series circuits connected in parallel, as shown schematically in Figure 15-7.

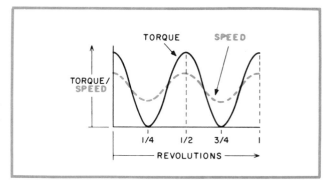

Figure 15-5 Torque and speed vary as the armature rotates. (Energy Concepts, Inc.)

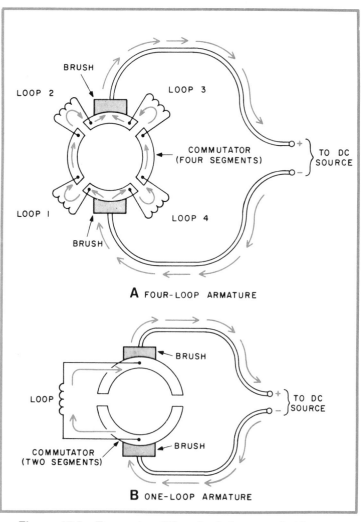

Figure 15-6 Torque in a DC motor is increased with additional loops. (Energy Concepts, Inc.)

When current flows through the brushes, all four loops act together and each loop adds to the total torque at all times. There is no neutral armature position where torque is absent. Notice that the brushes are larger than the gaps between the commutator segments. This means that contact with the commutator is maintained at every instant of rotation of the armature. A DC motor of this type has uniform torque, both for running and for starting. It is a definite improvement over the simple DC motor previously described.

Practical DC Motor

Field Coils. Figure 15-8 illustrates a disassembled view of a typical four-pole DC motor with a closeup view of the brush assembly and armature with commutator segments. Note from these diagrams that many turns or windings are involved in the field circuits. The larger the number of coils used in a DC motor, the smoother the motor will run. Also note the presence of field poles used to concentrate the magnetic lines of force created by the field windings. The number of field poles used must always be an even number, with each set consisting of a North and South pole.

Armature. The armature is constructed of steel laminations and is suspended at each end of the motor by bearings set in the motor frame. The commutator on the armature, along with the brushes, is used to supply the coil windings with current and reverse the current flow as needed. The typical commutator is constructed of drawn copper commutator bars which are insulated, one from the other. It is to each of the copper commutator bars that the armature coils are connected.

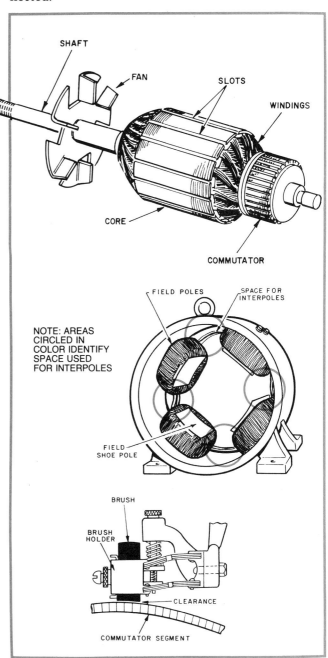

Figure 15-8 A DC motor will run smoother with a greater number of coils.

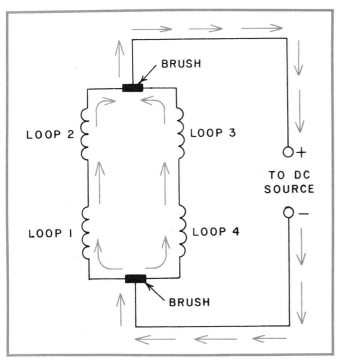

Figure 15-7 A four-loop armature provides uniform torque for running and starting. (Energy Concepts, Inc.)

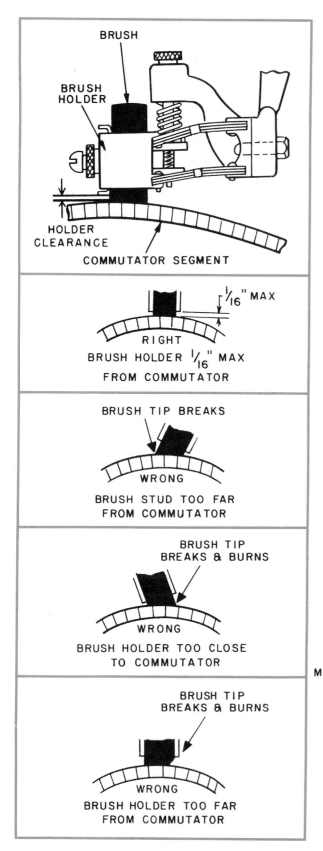

Figure 15-9 The brush must be positioned correctly for proper contact with the commutator.

The Brushes. The brushes of the motor are used to provide the contact between the external power circuit and the commutator. The current is supplied to the commutator by the brush, which rides on the commutator, making contact as it is turned. The brushes are held in a stationary position by brush holders and are usually made from various grades of carbon. A flexible braided-copper conductor, called the pigtail, is used to connect the brush to the external circuit.

Figure 15-9 illustrates how the brush is used to make contact with the commutator. Each brush is free to move up and down on the brush holder; this freedom allows the brush to follow irregularities in the surface of the commutator. A spring placed behind each brush forces the brush to make contact on the commutator. The spring pressure is usually adjustable, as is the entire brush holder assembly, allowing shifting of the position of the brushes on the commutator.

Figure 15-9, top, illustrates the proper way the brush should make contact with the commutator. The brush must be pressed on the commutator with about 1.5 to 2 pounds of pressure for each square inch of brush surface making contact with the armature. For this pressure to be applied, the spring must be allowed to move the brush up and down freely. Also there must be as small a space as possible between the surface of the brush making contact with the commutator and the brush holder. Too much space will cause the brushes to chatter and break all or part of the brush. A brush may become wedged in the brush holder or by broken brush parts. If this happens, the brush will not make good contact with the commutator and an open circuit will exist.

Interpoles. Figure 15-10 illustrates the location and connection of interpoles used in a DC

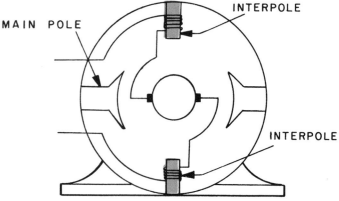

Figure 15-10 Interpoles reduce sparking at the brushes in larger DC motors.

motor. Interpoles are auxiliary poles that are placed between the main field poles of the motor. The interpoles are made with larger size wire than the main field poles in order to carry armature current. They are smaller in overall size than the main field poles because of less windings. The interpoles are connected in series with the armature windings. Interpoles are also called commutating field poles.

The interpoles are used to reduce sparking at the brushes of larger DC motors. Interpoles are usually used with shunt and compound DC motors of 0.5 H.P. or more. The interpole reduces sparking at the brush by helping to overcome the effects of what is called armature reaction.

Figure 15-11 illustrates armature reaction in the DC motor. Armature reaction is the effect that the armature coil magnetic field has upon the

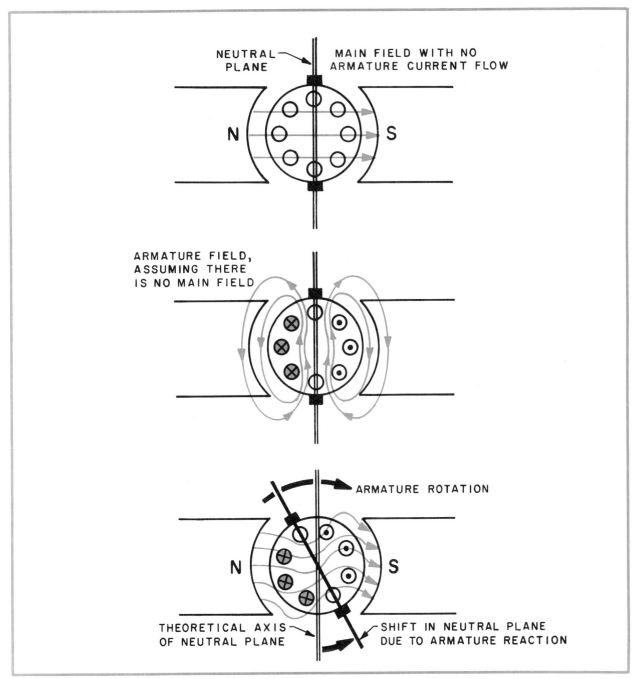

Figure 15-11 Armature reaction in the DC motor is caused by the magnetic field of the armature coil and the main pole windings.

magnetic field of the main pole windings. Figure 15-11, top, shows the effect of the field's magnetic force on the armature without a magnetic field in the armature. As illustrated, the lines of magnetic flux produced by the field poles will be directed through the armature's iron core from left to right. Figure 15-11, middle, shows the effect of the current flowing through the armature. This current creates a magnetic field in the field windings. As illustrated, the lines of magnetic flux produced by the armature coils will be directed through the armature's iron core and field poles in an up-and-down movement. If both of these magnetic fields are combined, as they are in all DC motors, the resultant magnetic field will change the angle of the neutral plane. Figure 15-11, bottom, shows the effect of the two magnetic fields in changing the angle of the neutral plane.

The shifting of the neutral plane, because of armature reaction, will have a direct effect on the commutation of the motor. This is because the brushes are best connected to the commutator at the neutral plane. With the brushes making contact at the neutral plane, the current through the commutated armature coil is minimum. However, with the neutral plane being moved, a current is produced by the magnetic field which allows current to flow through the brushes and commutator at the time when the brushes come in contact with the armature. This current will cause sparking at the brushes and result in burning at the commutator.

To correct this sparking, the brushes can be shifted in a direction opposite the direction of motor rotation. This will move the brushes back to the neutral plane when they make contact with the armature. This is illustrated in Figure 15-12. This method, however, is not the best way to prevent sparking because the neutral plane will change angles with each change in the load connected to the motor. For a load that is constant, the brushes may be shifted to an angle that will produce the least amount of sparking in an attempt to reduce the effect of armature reaction.

However, the best way to reduce sparking at the brushes is by using interpoles. As illustrated in Figure 15-13, the interpole's magnetic field will oppose the distorted magnetic flux caused by armature reaction. The result is that the combined magnetic force will keep the neutral plane at a fixed angle. Since the neutral plane cannot be changed in motors that have interpoles, sparking can be kept to a minimum and the motor will not be influenced by a changing neutral plane.

Reason for Using Reduced Voltage Starting with DC Motors

As previously illustrated, all DC motors are supplied with current directly connected to the armature and field windings. When current is connected directly to the motor, the current is limited only by the resistance of the wire in the coil winding. The larger the motor, the less the resistance and the larger the current. In DC motors this current will be so high that it will damage the motor. It is for this reason that reduced voltage starting must be applied to DC motors generally larger than 1 HP.

Reduced voltage starting of a DC motor, as in AC motors, will reduce the amount of current during starting. As the motor is accelerated, the

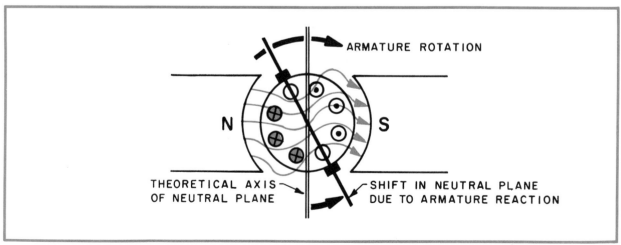

Figure 15-12 Brushes can be repositioned in a DC motor to reduce arcing.

reduced voltage may be removed since the current in the motor will decrease with an increase in motor speed. This decrease in current is a result of the motor generating a voltage that is opposite to the applied voltage as the motor is accelerated. This opposing voltage is called counter electromotive force, or counter EMF. The amount of counter electromotive force is dependent on the speed of the motor. Counter electromotive force is zero at standstill and increases with motor speed.

If, for example, a DC motor with an armature resistance of one ohm were connected to a 200 volt supply, the current of the motor at start would be:

$$I = \frac{E}{R} = \frac{200}{1} =$$

200 amps if no external starting resistance is used

As the motor is accelerated and a counter electromotive force of 100 volts is generated, the current of the motor is reduced to:

$$I = \frac{E}{R} = \frac{200 - 100(\text{CEMF})}{1} = \frac{100}{1} = 100 \text{ amps}$$

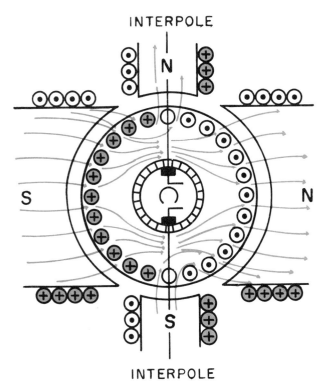

INTERPOLE

INTERPOLE

Figure 15-13 Interpoles may be used to automatically shift the neutral plane and reduce arcing at the brushes.

When the motor is at full speed, a counter electromotive force of 180 volts would allow a running current of:

$$I = \frac{E}{R} = \frac{200 - 180(\text{CEMF})}{1} = \frac{20}{1} = 20 \text{ amps}$$

It is the 200 amps (or starting current of any large DC motor) that the motor must be protected from to prevent damage.

REDUCED VOLTAGE STARTING AND THE SQUIRREL-CAGE MOTOR

Although other types of motors are used with reduced voltage starting, a majority of the industrial applications will usually involve the use of a squirrel-cage motor.

Squirrel-cage motors are usually chosen over other types of motors because of their simplicity, ruggedness and reliability. Because of these unique features, squirrel-cage motors have practically become the accepted standard for alternating current, all purpose constant speed motor applications. Without a doubt, the squirrel-cage motor is truly the workhorse of the industry.

Motor Construction

The squirrel-cage induction motor has certain advantages over the DC motor. There are only two points of mechanical wear on the squirrel cage motor: the two bearings. Since it has no commutator, there are no brushes to wear; therefore, maintenance is minimal. Nor are sparks generated to create a hazard in the presence of flammable material.

Figure 15-14 illustrates a cutaway view of a typical AC squirrel-cage motor, showing the over-

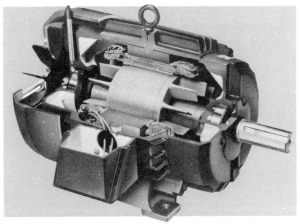

Figure 15-14 The squirrel cage induction motor has no brushes.

all assembly and the simplicity of construction. The motor consists of two main parts connected by the bearings: the fixed frame, called the *stator,* and a rotating member, called the *rotor.* The motor is named for its distinguished rotor, which resembles a squirrel cage, and for the fact that currents flowing in the stator induce AC currents in the rotor like those found in a transformer. Figure 15-15 illustrates electrically the coils involved in producing the rotating field in a three phase induction motor.

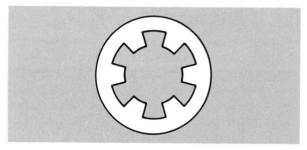

Figure 15-15 The stator portion of a squirrel cage induction motor is the stationary or fixed frame portion of the motor.

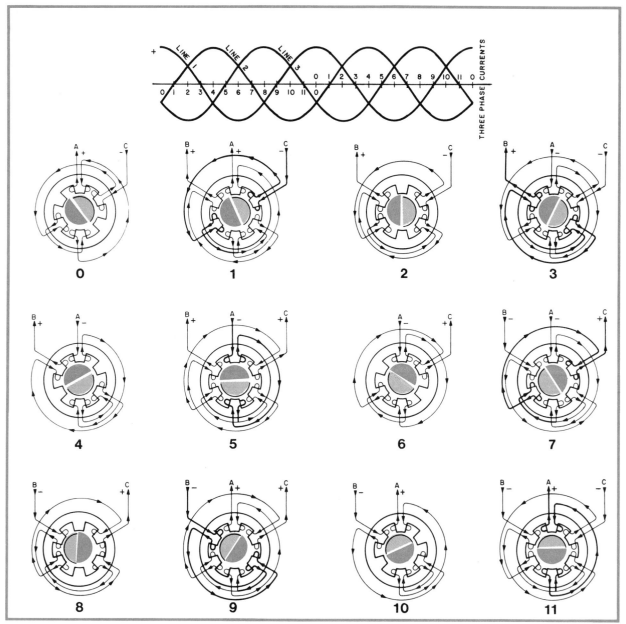

Figure 15-16 A rotating magnetic field can be generated in an induction motor when three-phase power is applied.

AC Motor Operation

In a three-phase induction motor, there are three sets of windings on the stator frame, arranged to produce a revolving magnetic field when connected to a three-phase source. Figure 15-16 illustrates how these fields revolve around the stator as each of the three phases rises and falls with the line frequency.

Notice in this illustration that a permanent bar magnet has been inserted within this rotating field to show how it tracks with the field for a complete 360 degree rotation. The use of a bar magnet is impractical for use in industrial motors, so a rotor similar to the one in Figure 15-17 is substituted in place of the permanent magnet.

The rotor consists of steel laminations mounted rigidly on the motor shaft. Copper or aluminum bars placed or cast in the slots of the laminated steel core form the rotor winding (in contrast to the copper wire coils found in many other motor types). These bars are designed to extend a sufficient distance beyond the end of the core so that they can all be interconnected by what is termed short circuited end rings. In other words, all the conducting bars are connected into a short circuited closed loop.

When the cutting lines of magnetic force originating in the stator cross into these short-circuited bars of the rotor, a voltage is induced into the rotor by transformer action, resulting in a heavy current flow in the rotor. From Unit 9, "Power Distribution Systems", you will recall that any induced voltages from transformer action will be of reverse polarity to the voltage creating it. This, in turn, will result in a magnetic field opposite to that of the stator. The combined electromagnetic effects of the stator and rotor currents and their magnetic fields produce the torque or force to create rotation. Careful study and observation of Figure 15-16 will lead to a thorough understanding of a three-phase induction motor. All three phases—A, B and C—are shown changing in amplitude at the top of the illustration, while the 12-part sequence below shows how the electromagnet interaction is turned into mechanical rotation of the rotor.

It should be pointed out that when the rotor is inserted into the stator, the air gap between the rotor and stator is kept extremely small in order to increase efficiency.

Speed of the Motor

The squirrel-cage motor is basically a constant speed device. It cannot operate for any appreciable period of time at speeds below those shown on the nameplate without danger of burning out. The synchronous speed of an induction motor is found by the following formula:

$$\text{Synchronous RPM} = \frac{120F}{P}$$

RPM = revolutions per minute
F = supply frequency in cycles/second
P = number of poles in the motor winding

Thus, motors designed for 60 cycle use (standard in the US) have synchronous speeds as follows:

Poles:	2	4	6	8
RPM	3600	1800	1200	900
Poles:	10	12	14	16
RPM	720	600	514	450

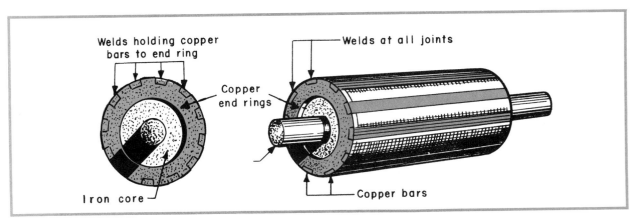

Figure 15-17 The rotor of a squirrel cage is the rotating portion of the motor.

All induction motors will have a full load speed somewhat below the synchronous speed. The percentage reduction in speed below synchronous speed is known as the "percent slip". Thus, a normal motor with, say, 2.8% slip and 1800 rpm synchronous speed would have a slip of 50 rpm and a full load speed of 1750 rpm (1800 − 50 = 1750 rpm). It is this full load speed that will be found on the motor's nameplate.

It is important to remember how the speed of an induction motor is determined. That is by the number of poles and the frequency of the power supply (*not* the supply voltage). It is for this reason that a reduced voltage starter *is not* a speed controller and must not be used as one. The reduced voltage starter is designed for specific reasons, and speed control is not one of the reasons.

Reasons for Reduced Voltage Starting of AC Motors

A heavy current is drawn from the power lines when an induction motor is started. This sudden demand for large current can reflect back into the power lines and create problems.

The revolving field of the stator induces this large current in the short circuited rotor bars. The current will be highest when the rotor is at standstill and decrease as the motor comes up to speed. The current at start is excessive because of a lack of counter EMF (electro-motive-force) at the instant of starting. Once rotation has begun, counter EMF is built up in proportion to speed, and the current decreases.

Figure 15-18 illustrates the behavior of the current drawn by a typical induction motor at various speeds. The percent of full load current is marked on the horizontal scale and the percent of motor speed is marked on the vertical scale. From this graph, two facts stand out: first, the starting current is quite high compared to the running current; second, the starting current remains fairly constant at this high value as the speed of the motor increases, but then drops sharply during the last few revolutions. This means, of course, that the heating rate is quite high during acceleration, since the heating rate is a function of I^2, (I = current). It also means that a motor may be considered to be in the locked condition during nearly all of the accelerating period.

The steady-state current taken from the power line with the rotor locked (stopped and with the voltage applied) is called the *locked rotor current* (LRC). The locked rotor current and the resulting

torque produced in the motor shaft (in addition to load requirements) are factors which determine whether the motor can be connected across the line, or whether the current has to be reduced through a reduced voltage starter.

The locked rotor current is not the current that is given on the motor's nameplate. The current that is given on the motor's nameplate is called the *full load current* (FLC). The full load current is the current required by the motor to produce full load torque at the motor's rated speed. If the motor is not called upon to deliver full torque, then the load current will be less than what is given on the nameplate. This information is important to remember when bench testing a motor (running a motor without a load). Since the motor does not have a load, the only torque it must produce is that which is needed to overcome its own internal friction and winding losses; as a result, the current will be less than that which is given on the nameplate.

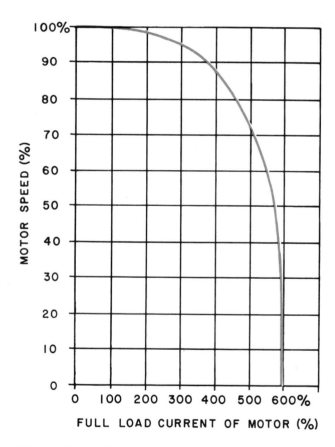

Figure 15-18 The starting current draw is higher when the motor is started.

PRIMARY RESISTOR STARTING

Figure 15-19 illustrates a typical primary resistor reduced voltage starting system. In primary resistor starting, a resistor is connected in each motor line (in one line only in a single-phase starter) to produce a voltage drop due to the motor starting current passing through the resistor.

A timer is provided in the control circuit to short out the resistors after the motor accelerates to a specified point. Thus, the motor is started at reduced voltage but operates at full line voltage.

Primary resistor starters provide extremely smooth starting due to the increasing voltage across the motor terminals as the motor accelerates.

Standard primary resistor starters provide two-point acceleration (one step of resistance) with approximately 70% of line voltage at the motor terminals at the instant of motor starting. When extra smooth starting and acceleration is needed, a multiple-step starting is possible by using additional contacts and resistors. This additional resistance stepping may be required in paper or fabric applications where even a small jolt in starting may tear the paper or snap the thread.

Primary Resistor Starter Circuits

Figure 15-20 illustrates a typical circuit used to provide reduced voltage primary resistance starting. From this circuit, we can see how external

Figure 15-19 With a primary resistor reduced voltage starting system voltage is increased as the motor accelerates. (Furnas Electric Company)

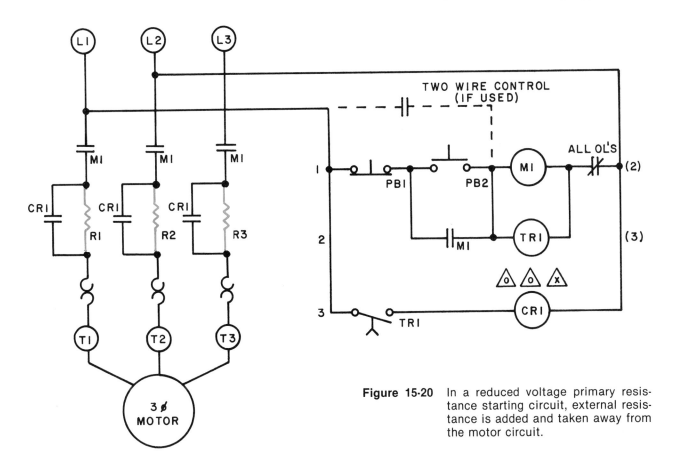

Figure 15-20 In a reduced voltage primary resistance starting circuit, external resistance is added and taken away from the motor circuit.

resistance is added to and taken away from the motor circuit to provide reduced voltage upon starting.

The control circuit of Figure 15-20 consists of the magnetic motor starter coil M1, timer relay coil TR1 and contactor coil C1. Coil M1 controls the magnetic motor starter which energizes the motor and provides overload protection. Coil TR1 energizes the timer to provide on delay logic from the point where coil M1 energizes until contacts C1 close, shorting out resistors R1. Coil C1 energizes the contactor which provides short circuiting across the resistors.

The circuit of Figure 15-20 would operate in the following manner: depressing start button PB2 energizes starter coil M1 and timer coil TR1. Coil M1 causes contacts M1 to close, creating MEMORY, while coil TR1 causes the contacts TR1 to remain open during reset, stay open during timing and close after timed out as indicated by the coding over the load (refer to Unit 7 to review coding method). Once timed out, contacts coil C1 energizes, causing contacts C1 to close and short out the resistors.

The circuit of Figure 15-20 is very typical of those used on reduced voltage starting. It should be noted, however, that changes are often made in the values of resistance and wattage to accommodate motors of different horsepowers.

Figure 15-21 Autotransformers use a tapped three-phase autotransformer to provide reduced voltage starting. (Furnas Electric Company)

AUTOTRANSFORMER STARTING

Figure 15-21 illustrates a photograph of a typical autotransformer reduced voltage starting system. Autotransformer starters provide reduced voltage starting at the motor terminals through the use of a tapped, three-phase autotransformer.

Autotransformer starting is one of the most effective methods of reduced voltage starting. It is preferred over primary resistance starting when the starting current drawn from the line must be held to a minimum value yet the maximum starting torque per line ampere is required.

To fully understand autotransformer starting, some electrical characteristics of motors that are often overlooked must be understood. First, it is important to realize that the motor terminal voltage does not depend upon the load current and that the motor voltage will remain substantially constant during the acceleration period. In other words, the current to the motor may change, because of the motor's changing characteristics; but the voltage to the motor will remain relatively constant.

Second, autotransformer starting may use its turn ratio advantage to provide more current on the load side of the transformer than on the line side. In autotransformer starting, one must distinguish between transformer motor current and line current, which are not equal as they are in primary resistor starting.

To illustrate how we can take advantage of the turn's ratio, we will apply the concept to a typical example. Let us assume that we have a motor with a full voltage starting torque of 120% and a full voltage starting current of 600%. The power company has set a limitation of 400% current draw from the line. Note that the limitation is set only for the line side of the controller. Since the transformer will have a step down ratio, the motor current on the transformer's secondary will be larger than the line current as long as the primary of the transformer does not exceed 400%.

In the example noted, with the line current limited to 400%, one can apply 80% voltage to this motor, have 80% motor current, and still have only $0.8 \times 80\%$, or 64% line current due to the 1/0.8 turns-ratio for the transformer. The advantage is that the starting torque is $80\% \times 80\%$ of 120%, or 77%, instead of the 51% obtained in primary resistor starting. This additional percentage may be sufficient accelerating energy to start a load which may have been otherwise difficult to start.

Two types of autotransformer connections or

control schemes are in common use today. One is designated closed circuit transition and the other is open circuit transition. The basic difference between the systems is that in closed circuit transition the motor is never removed from a source of voltage when moving from one incremental voltage to another. Conversely, open circuit transition means that the motor may be temporarily disconnected when moving from one incremental voltage to another.

Although both systems are used and function on the same principle, we will describe closed circuit transition since it poses the least amount of interference to the electrical environment. The closed circuit transition is, however, the more expensive of the two.

Autotransformer Starting Circuits

Figure 15-22 illustrates a typical circuit used to provide autotransformer reduced voltage start-

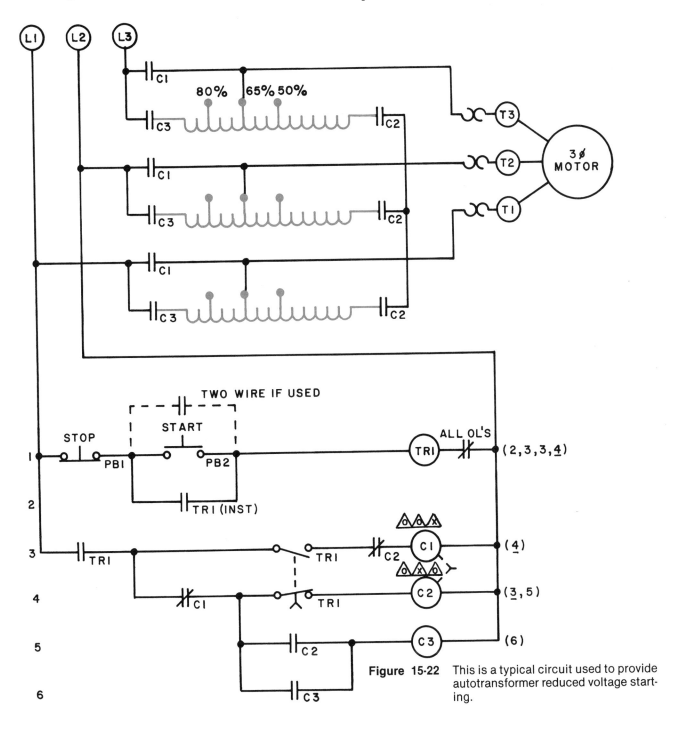

Figure 15-22 This is a typical circuit used to provide autotransformer reduced voltage starting.

ing. From this circuit, we can see how the various windings of the transformer are added to and taken away from the motor circuit to provide reduced voltage upon starting.

The control circuit of Figure 15-22 consists of timer relay coil TR1 and contactor coils C1, C2 and C3. The circuit of Figure 15-22 would operate in the following manner: depressing start button PB2 energizes timer relay coil TR1, causing instantaneous contacts in line two and three to close. Closing the NO contacts in line two provides MEMORY for relay TR1, while closing NO contacts in line three completes an electrical path through line four, energizing contactor coil C2. The energizing of coil C2 in turn causes NO contacts C2 in line five to close, energizing contactor coil C3. NC contacts in line three also provide electrical interlocking for coil C1 so that they may not be energized together. When coil C2 energizes, the NO contacts of contactor C2 close, connecting the ends of the autotransformers together. When coil C3 energizes, the NO contacts of contactor C3 close and connect the motor through the transformer taps to the power line, starting the motor at reduced inrush current and starting torque. MEMORY was also provided to coil C3 by contacts C3 in line six.

After a predetermined time, the on delay timer will time out and the following sequence will take place: NC timer contacts TR1 will open in line four, de-energizing contactor coil C2, and NO timer contacts TR1 will close in line three, energizing coil C1. In addition, NC contacts C1 will provide electrical interlock in line four, and NC contacts C2 in line three will return to their normally closed position. The net result of de-energizing C2 and energizing C1 is to connect the motor to full line voltage. Note that during the transition from starting to full line voltage the motor was not disconnected from the circuit, indicating closed circuit transition.

As long as the motor is running in the full voltage condition, the timer TR1 and contactor C1 will remain energized. Only an overload or depressing the stop button would normally stop the motor and reset the circuit.

In this circuit, as with all circuits, it is important to remember that although pushbuttons are used to control the circuit, any normally open and normally closed device will work. Thus, if this was an air conditioning system, the pushbuttons would be replaced with a temperature switch, and the circuit would be connected for two-wire control.

PART WINDING STARTING

Wye Connected Motor
Figure 15-23 illustrates a photograph of a typical part winding reduced voltage starting system. Part winding starting depends upon the use of a part winding motor. A part winding motor has two sets of identical windings which are intended to be operated in parallel. When these windings are energized in sequence, they produce reduced starting current and reduced starting torque. Most, (but not all) dual voltage 230/460 motors are suitable for part winding starting at 230 volts.

Part winding starters are available in either two or three step construction. The more common two step starter is designed so that when the control circuit is energized, one winding of the motor is connected directly to the line. This winding then draws about 65% of normal locked rotor current and develops approximately 45% of normal motor torque. After about one second, the second winding is connected in parallel with the first winding in such a way that the motor is then electrically complete across the line and develops its normal torque.

From the preceding discussion, it should be pointed out that part winding starting is not truly a reduced voltage means; however, it is usually so classified because of the reduced current and torque resulting.

Delta Connected Motor
When a dual voltage *Delta*-connected motor is operated at 230 volts from a part winding starter having a three-pole starting and a three-pole running contactor, an unequal current division occurs during normal operation, resulting in over-

Figure 15-23 A part-winding reduced voltage starting system is used with a part winding motor. (Cutler Hammer)

loading the starting contactor. To overcome this defect, some part winding starters are furnished with a four-pole starting contactor and a two-pole running contactor. Not only does such an arrangement eliminate the unequal current division obtained with a Delta wound motor, but it also enables Wye-connected part winding motors to be given either a one-half or two-thirds part winding start.

Advantages and Disadvantages of Part Winding Starting

Part winding starting has certain obvious advantages, but it also has several disadvantages. On the positive side, part winding is less expensive than most other methods because it requires no voltage reducing elements such as transformers or resistors; and it uses only two one-half size contactors. Furthermore, its transition is inherently closed circuit.

On the negative side, part winding has poor starting torque because it is fixed. In addition, the starter is almost always an increment start device. Second, not all motors should be part winding started, and it is quite important that the motor manufacturer's specifications be consulted before this type of starting is applied. Some motors are wound sectionally with part winding in mind, but indiscriminate application to just any dual voltage motor (for example, a 230/460 volt motor which is to run at 230 volts) can lead to excessive noise and vibration during start, cause overheating and lead to extremely high transient currents upon switching.

One further word of caution should be expressed concerning fusing a part winding starter. Since the current requirements in part windings allow the use of smaller contactors and proportionally smaller overload elements, the fuses must be proportionally smaller in order to protect these elements. Dual element fuses are, therefore, a virtual necessity.

Part-Winding Starter Circuit

Figure 15-24 illustrates a typical circuit used to provide part-winding reduced voltage starting.

The control circuit of Figure 15-24 consists of magnet motor starter coil M1, timer relay coil TR1 and starter coil M2. The circuit of Figure 15-24 would operate in the following manner: depressing start button PB2 energizes starter coil

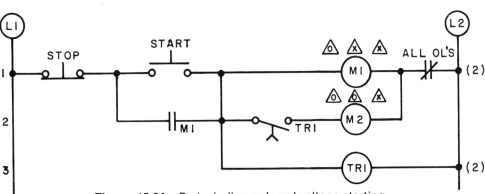

Figure 15-24 Part-winding reduced voltage starting is less expensive than other methods but produces less starting torque.

M1 and timer relay coil TR1. Coil M1 energizes motor starter M1 and closes NO auxiliary contacts M1 in line two to provide MEMORY. With the motor starter M1 energized, L1 is connected to T1, L2 to T2 and L3 to T3, starting the motor at reduced current and torque through one-half of the motor windings.

The on delay NO contacts of timer relay TR1, line two remain open during reset, stay open during timing, and close after timed out, energizing coil M2 as indicated by our code. With the second starter M2 energized, L1 is connected to T7, L2 to T8, and L3 to T9, applying voltage to the second set of Wye windings. The motor now has both sets of windings connected to the supply voltage for full current and torque. The motor may normally be stopped by pressing the stop button PBI or by an overload in any line.

Notice should be taken that each magnetic motor starter need be only half-size, since each one controls only one-half of the winding. Overloads, of course, must be sized accordingly.

WYE-DELTA STARTING

Wye-Delta starting, widespread in Europe, was hardly ever used in the United States until the manufacturers of large centrifugal air conditioning units started using them. Since more and more buildings are becoming centrally air conditioned, this starting method has become more popular in the United States.

Figure 15-25 illustrates a typical Wye-Delta reduced voltage starting system. Wye-Delta starting accomplishes reduced voltage by first connecting the motor leads into a Wye configuration for starting. The motor started in the Wye connection will receive approximately 58% of the normal voltage and develop approximately 33% torque.

Wye-Delta motors are specially wound with six leads brought out to enable the windings to be connected either in Wye or in Delta. When a Wye-Delta starter is energized, two contactors close, with one contactor connecting the windings in a Wye configuration and the second contactor connecting the motor to line voltage. After a time delay, the Wye contactor opens (momentarily de-energizing the motor), and the third contactor closes to reconnect the motor to the lines with the windings connected in Delta. Since the leads of the motor are disconnected and then reconnected, the Wye-Delta starter system is inherently an open transition type.

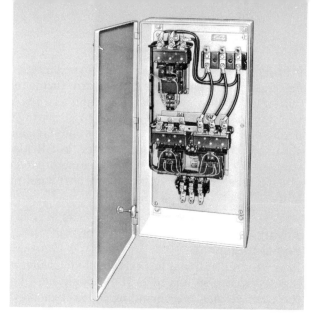

Figure 15-25 Wye-Delta reduced voltage starting systems are commonly used on large centrifugal air conditioning units. (Allen-Bradley Company)

The appeal of this starting method, as with the part winding system, comes from the reduction of any accessory voltage reducing equipment such as resistors and transformers. Wye-Delta gives a higher starting torque per line ampere than a part winding, with noise and vibration considerably less.

Wye-Delta starters have the disadvantage of being open transition; however, closed transition versions are available at additional cost. In closed transition, the motor windings are kept energized for the few cycles required to transfer the motor windings from Wye to Delta. Such starters are provided with one additional contactor plus a resistor bank—hence the additional cost.

Theory of Wye-Delta Motors

Figure 15-26, top, illustrates the Wye-Delta motor windings with the beginning and ends of each of the three starter windings brought outside the motor. There are no internal connections on this motor as there are on standard Wye and standard Delta motors. This allows the electrician to connect the motor leads into a Wye connected motor, as illustrated in Figure 15-26, middle, or to connect the motor leads into a Delta connected motor, as illustrated in Figure 15-26, bottom.

If a Delta connected motor is connected across a 208 volt three-phase power line, each coil winding in the motor will receive 208 volts. This is because each coil winding in the motor is connected directly across two power lines (Figure 15-27).

If a Wye connected motor is connected across a 208 volt, three-phase power line, each coil wind-

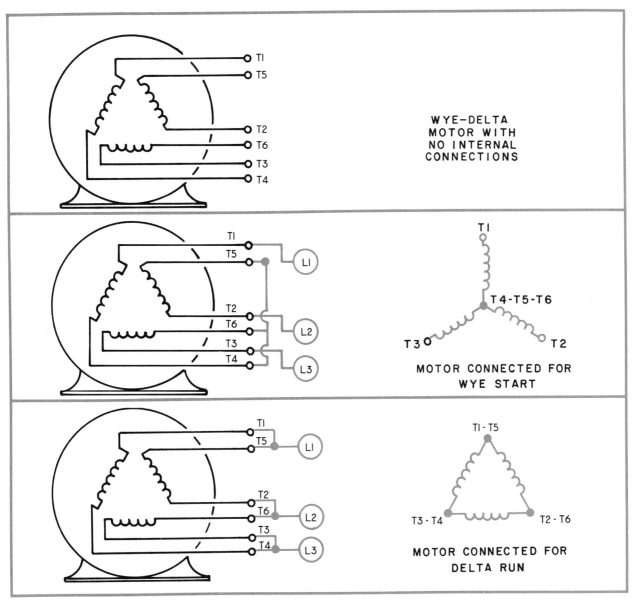

Figure 15-26 Windings on a Wye-Delta motor may be joined to form a Wye or Delta con- figuration.

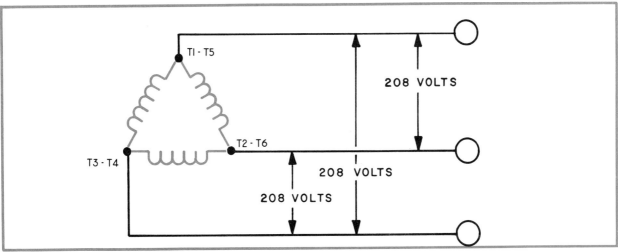

Figure 15-27 A Delta-connected motor has each coil winding directly connected across two power lines so each winding receives the entire source voltage of 208V.

ing in the motor will receive 120 volts. This is because there are now two coils connected in series across any pair of power lines, as illustrated in Figure 15-28.

In figuring the voltage in the coil for a Wye-connected circuit, remember that the voltage is equal to the line voltage divided by the square root of 3 (1.73). Since the line voltage is equal to 208 volts, the coil voltage is equal to 208 ÷ (1.73), or 120 volts. (See Unit 10 on power distribution for further information on figuring voltage and current in a Wye and Delta power distribution system.) This, then, means that if a Wye-Delta motor is connected to a line voltage of 208 volts, the motor will start with 120 volts (Wye) and run with 208 volts (Delta) across the motor windings. This, then, is another way of reduced voltage starting.

Wye-Delta Starter Circuit

Figure 15-29 illustrates a typical circuit used to provide Wye-Delta reduced voltage starting.

The control circuit of Figure 15-29 consists of magnetic motor starter coils M1 and M2, contactor coil C1 and timer relay coil TR1.

The circuit of Figure 15-29 would operate in the following manner: depressing start button PB2 energizes M1, which provides MEMORY in line two and connects the power lines L1 to T1, L2 to T2, and L3 to T3. C1 is energized, providing electrical interlock in line two and connecting motor terminals T4 and T5 to T6, so the motor will start in a Wye configuration. TR1 is also energized and after a preset time the on delay timer will time out, causing the normally open TR1 contacts to close and normally closed TR2 contacts to open. The opening of the NC contacts in line three disconnects the contactor C1, and an instant later the NO contacts in line two energize the second motor starter through Coil M2; the short time delay between M2 and C1 is necessary to prevent a short circuit in the power lines and is provided through the NC auxiliary contacts of C1 in line two. With contactor C1 de-energized, terminals T5, T6, and T4 are connected to the power lines T1, T2, and T3, since L1, L2, and L3 are still connected to run in a Delta configuration.

The circuit, at this point, can normally be stopped only by an overload in any line or by pressing the stop pushbutton.

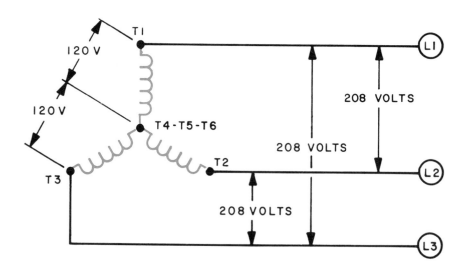

Figure 15-28 A Wye-connected motor has the two power lines distributed across two sets of windings. With this arrangement, the voltage across the coils is equal to the line voltage divided by the square root of 3 (1.73), resulting in 120 volts across the individual coils. Since this is considerably less than 208 volts, the unit is started at reduced voltage.

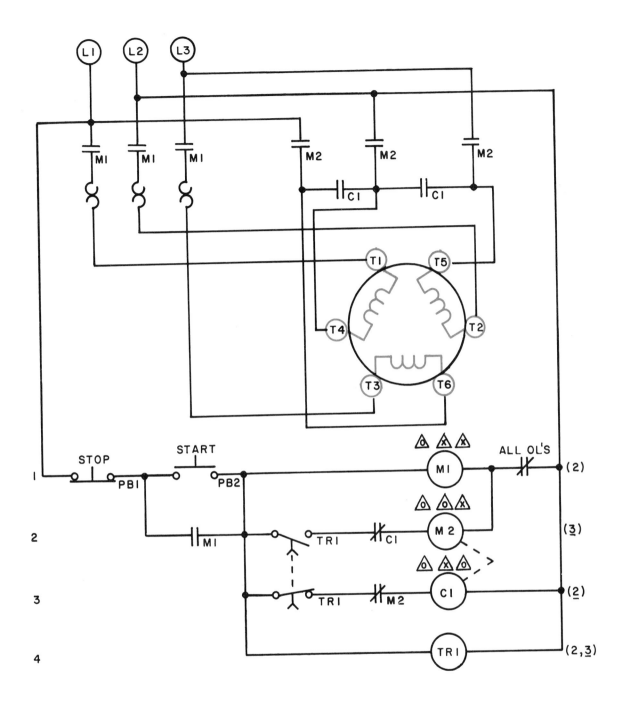

Figure 15-29 Typical circuit used to provide a Wye-Delta reduced voltage starting system.

Figure 15-30 illustrates a typical solid state reduced voltage starting system. The solid state starter is the newest method of reduced voltage starting for standard squirrel-cage motors. The heart of the solid state system is the silicon controlled rectifier (SCR) which controls motor voltage, current and torque during acceleration.

The SCR is a solid state rectifier with the ability to rapidly switch heavy currents. Because of this unique switching capability, this type of starting provides a smooth, stepless acceleration in applications such as starting conveyors, compressors, pumps and a wide range of other industrial applications.

The advantage of SCRs is the fact that they are small in size, are rugged and have no contacts. Unlimited life can be expected when SCRs are operated within specifications.

The major disadvantage of solid state at this point is its relatively high cost in relation to other systems.

Electronic Control Circuitry

Figure 15-30 illustrates the electronic circuitry needed to regulate SCR controls. The top section of the controller determines to what degree the SCRs should be phased on, thereby controlling the voltage, current, and the torque applied to the motor.

The top section of the solid state controller also includes current limiting fuses and current transformers for protection of the unit. The current limiting fuses are used to protect the SCRs from excess current, while the current transformers are used to feed back information to the controller. Heat sinks and thermostat switches are also used to protect the SCRs from high temperatures.

The controller further provides the sequential logic necessary for interfacing other control functions of the starter, such as line loss detection during acceleration. Here if any voltage is lost or too low on any one line, the controller is turned off. This may happen if one line would open or a fuse would blow.

SCR Operation

Figure 15-31 illustrates a typical SCR along with its schematic symbol. SCRs are also sometimes called thyristors. From the schematic, we can see that the SCR is composed of three elements: the anode, the cathode and the gate.

The anode and cathode of the SCR are similar to the anode and cathode of the diode rectifier cov-

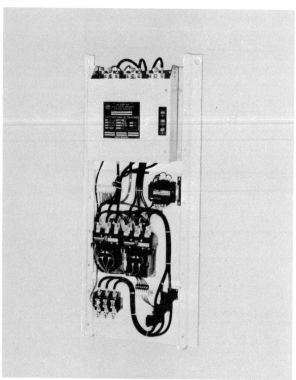

Figure 15-30 A solid state reduced voltage starting system uses an SCR to control motor voltage, current, and torque during acceleration. (Allen-Bradley Company)

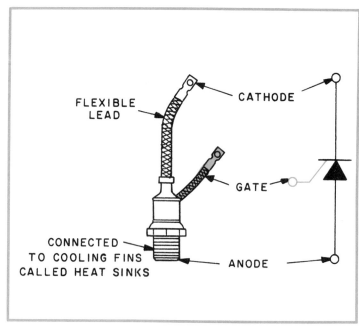

Figure 15-31 An SCR includes an anode, a cathode, and a gate.

ered in Unit 10, "Power Distribution Systems." The addition of the third element, the gate, however, gives the SCR added control not possible with an ordinary diode.

With the gate, an SCR can be made to operate as an off-to-on switch controlled by a voltage signal to the gate. Unlike an ordinary diode, an SCR will not pass current from Cathode to Anode unless an appropriate live signal is applied to the gate. When the signal is applied to the gate, however, the SCR is triggered on and the anode resistance decreases sharply such that the resulting current flow through the SCR is only limited by the resistance of the load. The advantage of this device is its ability to turn on at any point in the half-cycle as illustrated in Figure 15-32. Since the amount of conduction can be varied, the average amount of voltage and current can be reduced or increased by the firing or phasing of the SCRs.

SCRs may be used alone in a circuit to provide one way current control or may be wired in reverse parallel circuits to control AC line current in both directions. Figure 15-33 illustrates an SCR circuit with reverse parallel using an SCR to provide maximum control. More information on SCRs and other solid state control devices will be given in Unit 16, Accelerating and Decelerating Methods and Circuits.

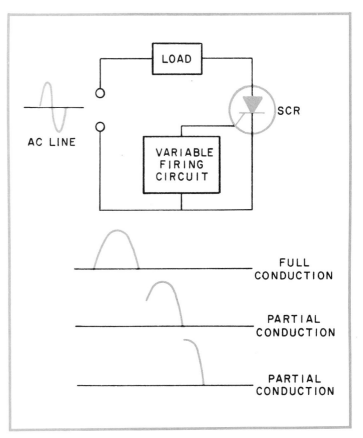

Figure 15-32 The output of an SCR will differ at different firing control settings.

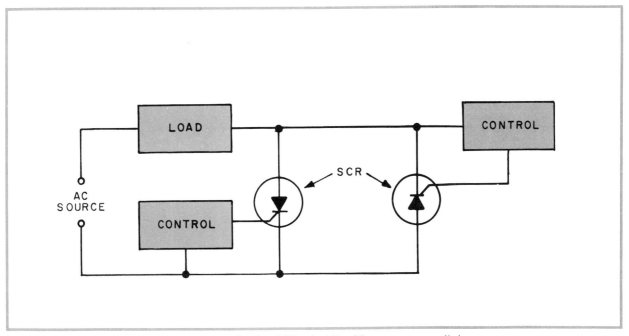

Figure 15-33 An SCR circuit with reverse parallel wiring of SCRs provides maximum control of an AC load.

Solid State Starter Circuits

Figure 15-34 illustrates a typical circuit used in solid state starting. The circuit consists of both start and run contactors C1 and C2. The start contactor C1 contacts are in series with the SCRs and the run contactor C2 contacts are in parallel with the SCRs.

When the starter is energized, the start contacts C1 close and the acceleration of the motor is controlled by phasing on the SCRs. The SCRs then control the motor until it approaches full speed, at which time the run contacts of C2 close, connecting the motor directly across the power line. At this point, the SCRs are turned off and the motor runs with full power applied to the motor terminals.

COMPARING THE DIFFERENT STARTING METHODS

When an industrial application calls for using reduced voltage starting, any one of several methods is available. When selecting a method of start-

ing, one may select from primary resistor, autotransformer, part-winding, Wye-Delta or solid state. The selection is not simply a matter of selecting the starting method that reduces the most amount of current. For whenever the current is reduced, so is the starting torque; and if the starting torque is reduced too much, the motor will not start and the motor overloads will trip. When selecting the method of starting, you must consider three major factors: the amount of reduced current, the amount of reduced torque, and the cost of each starting method.

Figure 15-35 illustrates a general comparison of the amount of reduced current for each type of starting method compared to across-the-line starting. The amount of reduced current is adjustable when using the solid-state or autotransformer starters. The autotransformer starter has taps, so the amount of reduced current is adjustable in part. The solid-state starter is adjustable through its range. Some primary resistance starters are adjustable, others are not. The part-winding and Wye-Delta starters are not adjustable. It is impor-

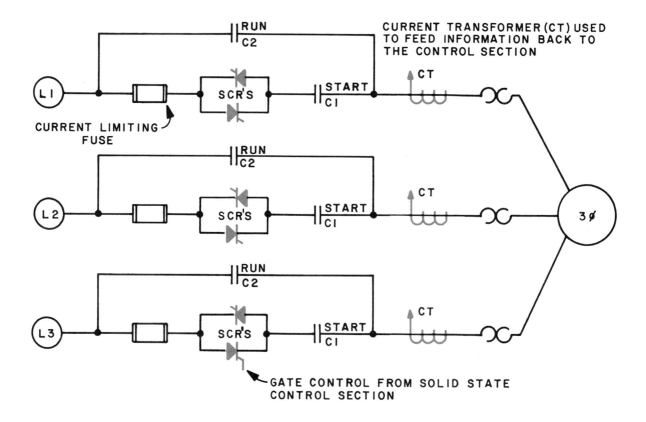

Figure 15-34 Typical circuit used in solid state starting.

tant to remember that as the amount of starting current is reduced, so is the torque of the motor.

Figure 15-36 illustrates a general comparison of the amount of reduced torque for each type of starting method compared to across-the-line starting. The amount of reduced torque is adjustable when using the solid-state or autotransformer starters. The autotransformer starter has taps, so the amount of reduced current is adjustable in part. The solid-state starter is adjustable through its range. Remember that if the load requires more torque than the motor can deliver, then the motor overloads will trip. The torque requirements of the load must be taken into consideration when selecting a starting method.

Figure 15-36 also illustrates a general comparison of the costs for each type of starting method compared to across-the-line starting. Although reducing the amount of starting current or starting torque in comparison to the load or requirements is the primary consideration for selecting a starting method, cost may also have to be considered. As you can see, the costs vary considerably for each starting method. Only when all of these factors are considered can the correct type of starting method be selected.

Following is a brief comparison of the applications, performance, and cost of each type of starting method. Figure 15-37 illustrates a general comparison table for basic types.

Primary Resistor: used when it is necessary to restrict inrush current to predetermined incre-

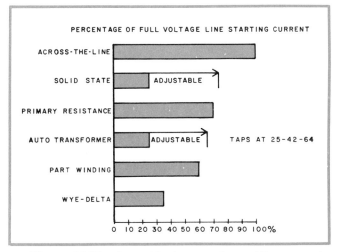

Figure 15-35 A graphical comparison of the amount of reduced current for each type of starting method compared to across-the-line starting.

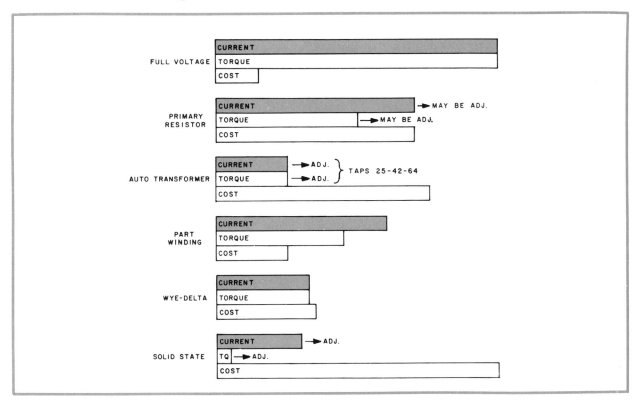

Figure 15-36 Several factors must be considered when selecting reduced voltage starting systems.

ments. Primary resistors can be built to meet almost any current inrush limitation. They also provide smooth acceleration and can be used where it is necessary to control starting torque. (This type of starting can be used with any squirrel cage motor.)

Autotransformer: this starter provides the highest possible starting torque per ampere of line current, and is, thus, the most effective means of motor starting for applications where the inrush current must be reduced with a minimum sacrifice of starting torque. Three taps are provided on the transformers, making it field adjustable. Since the autotransformer is the most expensive type of transformer, cost may have to be considered. (This type of starting can be used with any squirrel cage motor.)

Part-Winding: this starter is simple in construction and economical in cost. It provides a simple method of accelerating fans, blowers, and other loads involving low starting torque. This type of starter requires a nine-lead Wye motor, and this must be considered. The cost is less because no external resistors or transformers are required.

Wye-Delta: this starting method is particularly suitable for applications involving long accelerating times or frequent starts. They are typically used for high inertia loads such as centrifugal air conditioning units, although they are applicable in cases where low starting torque is necessary or where low starting current and low starting torque is permissible. With this type of starter, a special six lead motor is required.

Solid-State: this starter provides smooth, stepless acceleration in applications such as starting of conveyors, compressors and pumps. It uses a solid state controller using silicon controlled rectifiers (SCRs) to control motor voltage, current and torque during acceleration. Although this starting method offers the most control over a wide range, it is also the most expensive at present technology.

MONITORING THE LOAD ON A MOTOR

Whether a motor is starting or running, there is always a phase difference between motor current and voltage. Because a motor is an inductive load, the angle (COS ϕ) always exists, and its change is proportional to the actual motor load. The true working force (torque) of the motor can be measured by monitoring this phase difference.

Figure 15-38 illustrates the use of a power monitoring relay used in a typical control process. This application is for a milling process in which grain must be ground to a certain consistency. The size of the granules (or the consistency of the grain) is determined by the feed of grain. The feed of grain can be controlled by monitoring the load on the motor used to drive the grinder. For example, perhaps the correct feed of grain may be attained by setting the power monitor to 80% of the full motor load of the grinding motor.

This type of monitoring can be used for other applications in addition to grinding, tool braking, and conveying.

TYPE OF STARTER	STARTING CHARACTERISTICS IN PERCENT OF FULL VOLTAGE VALUES			STANDARD MOTOR	TRANSITION	EXTRA ACCELER. STEPS AVAILABLE	COST OF INSTALLATION	ADVANTAGES	DISADVANTAGES	REMARKS	APPLICATIONS
	VOLTAGE AT MOTOR	LINE CURRENT	STARTING TORQUE								
ACROSS-THE-LINE A10	100%	100%	100%	Yes	None	None	Lowest	• Inexpensive • Readily available • Simple to maintain • Maximum starting torque	• High inrush • High starting torque		Many and various
AUTO-TRANS-FORMER A400	80% 65% 50%	64% 42% 25%	64% 42% 25%	Yes	Closed	No	High	• Provides highest torque per ampere of line current • 3 different starting torques available through auto-transformer taps • Suitable for relatively long starting periods • Motor current is greater than line current during starting	• In lower hp ratings is most expensive design • Low power factor • Large physical Size	• Most flexible • Very efficient	Blowers Pumps Compressors Conveyors
PRIMARY RESISTOR A430	65%	65%	42%	Yes	Closed	Yes	High	• Smooth acceleration — motor voltage increases with speed • High power factor during start • Less expensive than autotransformer starter in lower HP's • Available with as many as 5 accelerating points	• Low torque efficiency • Resistors give off heat • Starting time in excess of 5 seconds requires expensive resistors • Difficult to change starting torques under varying conditions	• Can be designed so starting characteristics closely match requirements of load	Belt and gear drives Conveyors Textile machines
PART WINDING A460	100%	65%	48%	❶	Closed	Yes (but very un-common)	Low	• Least expensive reduced voltage starter • Most dual voltage motors can be started part winding on lower voltage • Small physical size	• Unsuited for high inertia, long starting loads • Requires special motor design for voltage higher than 230 • Motor will not start if the torque demanded by the load exceeds that developed by the motor when thet first half of the motor is energized • First step of acceleration must not exceed 5 seconds or else motor will overheat	• Not really a reduced voltage starter. Is considered an increment starter because it achieves objective by reconnecting motor winding.	Reciprocating compressors Pumps Blowers Fans
WYE DELTA A490	100%	33%	33%	No	Open ❷	No	Medium	• Suitable for high inertia, long acceleration, loads • High torque efficiency • Ideal for especially stringent inrush restrictions. • Ideal for frequent starts.	• Requires special motor • Low starting torque • During open transition there is a high momentary inrush when the delta contactor is closed	• Same as part winding (above) • Very efficient	Centrifugal compressors Centrifuges

❶ Standard dual voltage 230/460 volt motor can be used on 230 volt systems.
❷ Closed transition available for about 30% more in price.

Figure 15-37 Motor starter types are designed for specific applications. (Cutler Hammer)

Power Factor Monitor

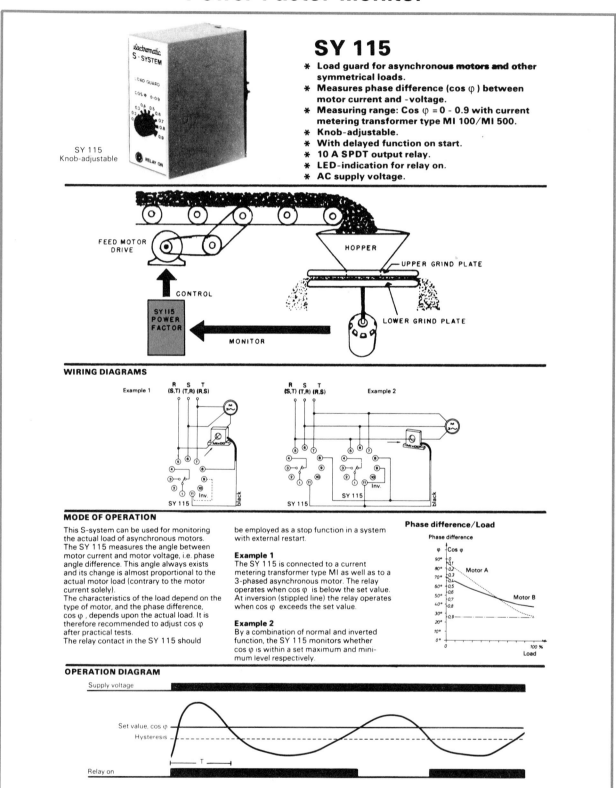

SY 115

* Load guard for asynchronous motors and other symmetrical loads.
* Measures phase difference (cos φ) between motor current and -voltage.
* Measuring range: Cos φ = 0 - 0.9 with current metering transformer type MI 100/MI 500.
* Knob-adjustable.
* With delayed function on start.
* 10 A SPDT output relay.
* LED-indication for relay on.
* AC supply voltage.

SY 115
Knob-adjustable

FEED MOTOR DRIVE

HOPPER

UPPER GRIND PLATE

LOWER GRIND PLATE

CONTROL

SY 115 POWER FACTOR

MONITOR

WIRING DIAGRAMS

Example 1

R S T
(S,T) (T,R) (R,S)

SY 115

Inv.

black

Example 2

R S T
(S,T) (T,R) (R,S)

SY 115

SY 115

Inv.

black

MODE OF OPERATION

This S-system can be used for monitoring the actual load of asynchronous motors. The SY 115 measures the angle between motor current and motor voltage, i.e. phase angle difference. This angle always exists and its change is almost proportional to the actual motor load (contrary to the motor current solely).
The characteristics of the load depend on the type of motor, and the phase difference, cos φ , depends upon the actual load. It is therefore recommended to adjust cos φ after practical tests.
The relay contact in the SY 115 should

be employed as a stop function in a system with external restart.

Example 1
The SY 115 is connected to a current metering transformer type MI as well as to a 3-phased asynchronous motor. The relay operates when cos φ is below the set value. At inversion (stippled line) the relay operates when cos φ exceeds the set value.

Example 2
By a combination of normal and inverted function, the SY 115 monitors whether cos φ is within a set maximum and minimum level respectively.

Phase difference/Load

Phase difference

Motor A

Motor B

Load

OPERATION DIAGRAM

Supply voltage

Set value, cos φ

Hysteresis

T

Relay on

Figure 15-38 A power factor monitor is used to monitor the load on a motor. (Electromatic Controls Corp.)

1. What are some of the reasons for using reduced voltage starting?
2. What is the difference in starting current between a full voltage start and a reduced voltage start?
3. What is the difference in starting torque between a full voltage start and a reduced voltage start?
4. What are the four main parts of a DC motor?
5. Why does a DC motor use many loops in the armature instead of just one loop?
6. In the DC motor, how is the external power delivered to the commutator coils?
7. What are interpoles used for in DC motors?
8. Why is reduced voltage starting used with large DC motors?
9. What are the main parts of the AC squirrel cage motor?
10. What determines the speed of the AC squirrel cage motor?
11. Why is reduced voltage starting used with large AC motors?
12. What is meant by "locked rotor current"?
13. What is meant by "full voltage current"?
14. In primary resistor starting, if an additional reduction in current is needed, how can this be accomplished?
15. In Figure 15-20, what determines the time setting on the timer (TR1)?
16. What is the difference between open transition and closed transition?
17. What are some advantages of autotransformer starting over primary resistor starting?
18. What are some advantages of part winding starting?
19. What are some disadvantages of part winding starting?
20. Why must the overloads be sized to one-half that of a full voltage start when using part winding starting?
21. Why is a special motor with no internal connections needed for Wye-Delta starting?
22. How is the voltage reduced to the motor coils in Wye-Delta starting?
23. Will a six-lead motor connected for Wye run at a different speed when connected for Delta? Why, why not?
24. In a solid state reduced voltage starter, what device is used to reduce the voltage applied to the motor?
25. Is the solid state reduced voltage starter limited to a preset number of steps (as is the Wye-Delta, for example)?

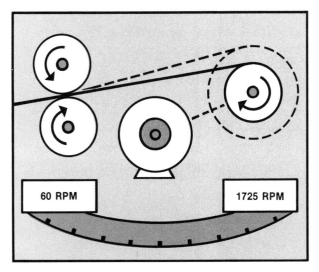

16 Accelerating and Decelerating Methods and Circuits

Speed control and braking are important considerations in the safe operation of a motor. Speed control and braking can be accomplished by mechanical means and electrical/ electronic means. Traditionally, speed control was mechanical and controlled by gears and pulleys with a friction brake. Mechanical speed and braking controls are still in use. However, many new systems use solid state speed controls and electronically controlled dynamic braking.

Because speed control relies heavily on the operating characteristics of the motor, an understanding of the relationship of speed, horsepower and torque is necessary.

BRAKING

When a motor is disconnected from the power supply, the motor will coast to a stop. The time taken by the motor to come to rest will depend on the inertia of the moving parts (both the motor and the motor's load) and friction. When it is necessary to stop a motor more quickly than coasting allows, stopping is accomplished by braking. Braking can be accomplished by many different methods, each method having advantages and disadvantages. The braking method used will depend on the application, available power, circuit requirement, cost and desired results.

It is important to understand each type of braking method in order to choose the best method for a given application. Applications of braking vary greatly. For example, braking may be applied to the motor every time the motor is stopped, or it may be applied to the motor only when there is an emergency. In the first application the braking

action requires a method which will do the job correctly over a long period of time. In the second case the fastest method of stopping the motor may be used with little or no consideration given to the damage braking may do to the motor or motor load. Hazard braking may be required to protect an operator (hand in equipment, etc.) even if braking is not part of the normal stopping method. There are brakes existing today in industry that have never been used and hopefully never will be used.

Friction Brakes

Friction brakes (also called magnetic or mechanical) have been used to stop motors for a long time. They work very much like the brakes on your automobile. Figure 16-1 illustrates a typical friction brake. The friction braking method usually consists of two friction surfaces, called shoes, that come in contact with a wheel mounted onto the motor shaft. Spring tension holds the shoes on the wheel and braking occurs as a result of the friction between the shoes and the wheel.

Solenoid Operation. The friction brake is usually controlled by a solenoid which activates the brake shoes. When the motor is running, the solenoid is energized, keeping the brake shoes from touching the drum mounted on the motor shaft. When the motor is turned off, the solenoid is de-energized and the brake shoes are applied through spring tension.

Figure 16-2 illustrates two methods used to connect the solenoid into the circuit so that it activates the brake whenever the motor is turned on and off. The left circuit is used if the solenoid has a voltage rating equal to the motor voltage rating. The right circuit is used if the solenoid has a voltage rating equal to the voltage between L2 and the neutral. It is a good idea to try to connect the solenoid of the brake directly into the motor circuit and not into the control circuit. This will help eliminate improper activation of the brake.

Brake Shoes. In friction braking, the braking action is applied to the wheel mounted on the shaft of the motor rather than to the shaft directly. The purpose for mounting the wheel is to provide a much larger braking surface than could be obtained from the shaft alone. This permits the use of a larger brake shoe lining and lower shoe pressure. Low shoe pressure, equally distributed over a large area, will result in even wear and braking torque. Braking torque developed is directly proportional to surface area and spring pressure. The spring pressure is adjustable on almost all friction brakes.

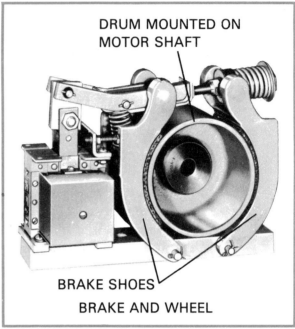

Figure 16-1   A friction brake uses pressure exerted by brake shoes on a drum mounted on the motor shaft. (Cutler Hammer)

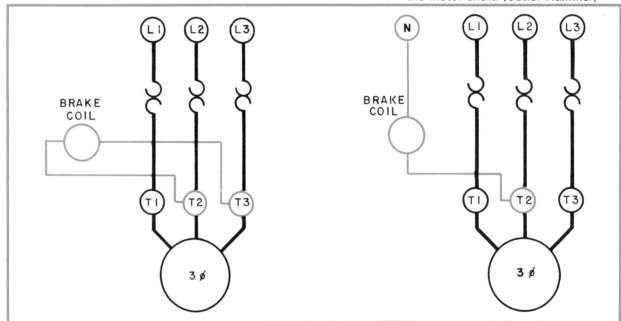

Figure 16-2   A friction brake may be connected to full line voltage or to a voltage equal to that produced between L1 and neutral.

Determining Braking Torque. A method often used to determine the required braking torque is to calculate the full load motor torque. This is calculated using the following formula:

$$T = \frac{5252 \times HP}{RPM}$$

T = full load motor torque in foot-pounds.
HP = Motor horsepower
RPM = speed of shaft on which brake wheel is mounted
5252 = Constant which is used to figure motor torque when HP and RPM are known.

The torque rating of the brake selected should be at least equal to or greater than the full load motor torque. Manufacturers of electric brakes will list the torque rating in foot-pounds for their brakes.

Figure 16-3 illustrates another method for determining braking torque. With the aid of this chart you merely draw a line between the horsepower and the rpm to find the torque. In this case a 50HP motor at 900 rpms would require a braking torque of 300 foot-pounds.

Advantages and Disadvantages of Friction Brakes. One disadvantage of friction brakes is that they require more maintenance than other braking methods. This maintenance consists of replacing the shoes and is dependent on the number of times the motor is stopped. A motor that is stopped often will need more maintenance and a

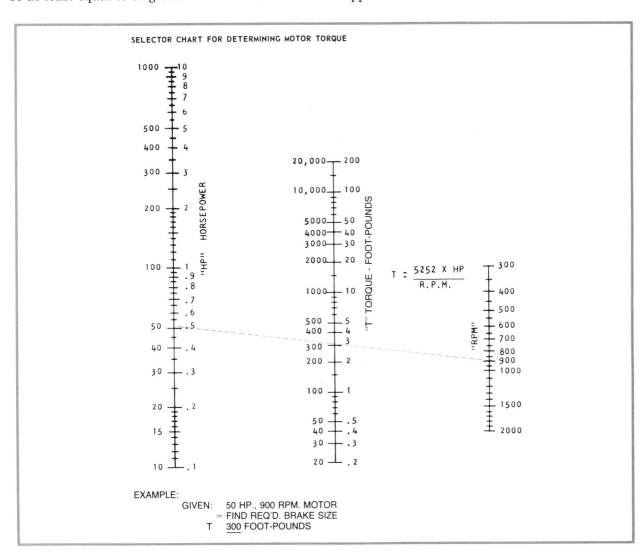

Figure 16-3 Brake torque can be determined using a chart, and drawing a line from the RPM column to the HP column. (Cutler Hammer)

motor that is hardly ever stopped may need no maintenance. The advantage in using friction brakes is initial cost and simplified maintenance. Maintenance is simplified because it is easy to see whether the shoes are worn and the brake is working. (Friction brakes are available in both AC and DC design to meet almost any application.)

Typical applications of friction brakes include printing presses, small cranes, overhead doors, hoisting equipment and machine tool control.

Plugging

Plugging is a way of braking in which the motor connections are reversed so that the motor develops a counter torque which acts as a braking force. This is accomplished by reversing the motor at full speed with the reversed motor torque opposing the forward inertia torque of the motor and its mechanical load. Plugging of the motor allows for a very rapid stop. Although manual and electromechanical controls can be used to reverse the direction of a motor, a plugging switch similar to those in Figure 16-4 is usually used in plugging applications.

The plugging switch is connected mechanically to the shaft of the motor or driven machinery. The rotating motion of the motor is transmitted to the plugging switch contacts either by a centrifugal mechanism or by a magnetic induction arrangement. The contacts on the plugging switch are NO, NC, or both, and will actuate at a given speed. The primary function of the plugging switch is to prevent the reversal of the controlled load once the counter torque action of plugging has brought the load to a standstill. If the plugging switch were not present, the motor and load would simply continue to run in the opposite direction without stopping.

Operation of the Plugging Switch. The plugging switch is designed to open and close sets of contacts as the shaft speed on the switch varies.

As the shaft speed increases, the contacts are set to change at a given rpm. As the shaft speed decreases, the contacts return to their normal condition. As the shaft speed increases, the contact set point (the point at which the contacts operate) will be at a higher rpm than the point at which the contacts reset (return to their normal position) on decreasing speed. The difference in these contact operating values is called the differential speed or rpm.

In plugging, the continuous running speed should be many times the speed at which the contacts are required to operate. This will provide high contact holding force and reduce possible contact chatter or false operation of the switch.

Continuous Plugging. Figure 16-5 illustrates the use of a plugging switch used to plug a motor to a stop. In this circuit the NO contacts of the plugging switch are connected to the reversing starter through an interlock contact. Pushing the start pushbutton energizes the forward starter, starting the motor in forward, and adding MEMORY to the control circuit. As the motor accelerates in speed, the NO plugging contacts close. The closing of the NO plugging contacts will not energize the reversing starter because of the interlocks. Pushing the stop pushbutton drops out the forward starter and interlocks. This allows the reverse starter to immediately energize through the plugging switch and the normally closed forward interlock. As a result, the motor is

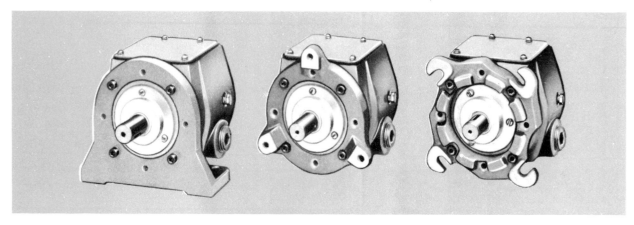

Figure 16-4 Plugging switches prevent the reversal of the controlled load after the load has stopped. (Allen-Bradley Company)

reversed and the motor is braked to a stop. When the motor is stopped, the plugging switch opens to disconnect the reversing starter before the motor is actually reversed. For the actual individual wiring diagrams for each motor type, refer to Unit 9 on reversing circuits.

Plugging for Emergency Stops. In the circuit of Figure 16-5, the motor would be plugged every time the motor was stopped. If plugging was required only in emergency stops, then the circuit of Figure 16-6 could be used. In this circuit the motor is started in the forward direction by pushing the start pushbutton. This starts the motor and adds MEMORY to the control circuit. As the motor accelerates in speed, the NO plugging contacts close. Pushing the stop pushbutton will de-energize the forward starter but not energize the reverse starter. This is because there is no path for the L1 power to reach the reverse starter. The motor will coast to a stop. Pushing the emergency

stop pushbutton will de-energize the forward starter and simultaneously energize the reversing starter. Energizing the reversing starter adds MEMORY in the control circuit and plugs the motor to a stop. When the motor is stopped, the plugging switch opens to disconnect the reversing starter before the motor is actually reversed. The de-energizing of the reversing starter also removes the MEMORY from the circuit.

Limitations of Plugging
Reversing. Most motors can be used for plugging provided they can be connected for reversing at full speed. If a motor cannot be reversed at full speed, as in the case of the shaded pole single-phase motor, it cannot be used for plugging. For other types of single-phase motors full speed reversing is not possible due to the centrifugal switch. At full speed the centrifugal switch holds open part of the circuitry necessary for reversing.

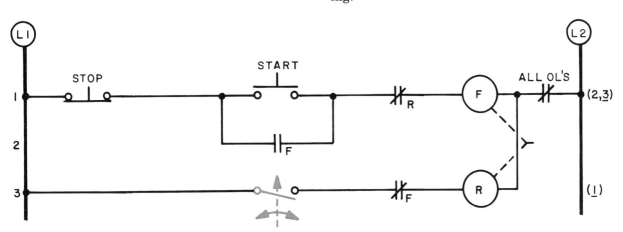

Figure 16-5 In this circuit, the motor will plug to a stop every time the motor comes to a stop.

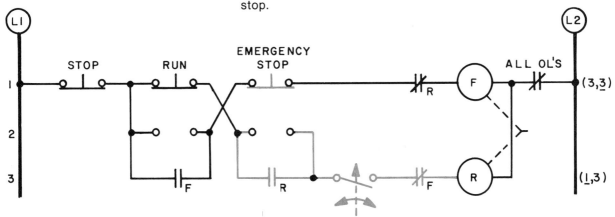

Figure 16-6 In a circuit where plugging is required only in an emergency, the plugging circuit is not used in a normal stop.

Heat. All three-phase motors and most single-phase and DC motors can be used for plugging. However, it is important to understand that high current and heating will result from plugging a motor to a stop. Remember that the motor is not being connected in reverse at standstill but at full opposite speed. For this reason current may be three or more times higher during plugging than during normal starting. That is why a motor designated for plugging or with a high service factor should be used in all cases except emergency stops. The service factor (marked SF on motor nameplate) should be 1.35 or more for plugging applications.

Plugging Using a Timing Relay

Plugging can also be accomplished by using a timing relay. The advantage of using a timing relay is usually in cost since a timer is much less expensive and does not have to be connected mechanically to the motor shaft or driven machine. The disadvantage is that, unlike the plugging switch, the timer does not compensate for a change in the load condition (which affects stop time) once the timer is preset. However, in applications where the time needed to decelerate the motor is constant and known, a timer may be used.

Figure 16-7 illustrates a plugging circuit using an off delay timer. In this circuit, the normally open contacts of the timer are connected into the circuit in the same manner in which the plugging switch was used. The coil of the timer is connected in parallel with the forward starter. When the start pushbutton is pushed, the motor is started and MEMORY is added to the circuit. In addition to energizing the forward starter, the off delay

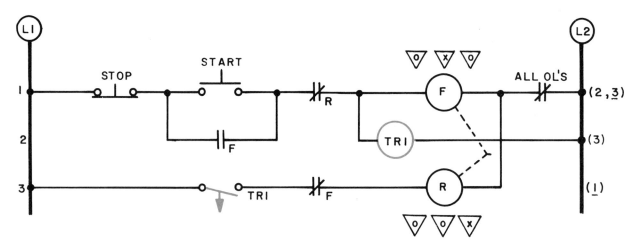

Figure 16-7 The time setting must be less than or equal to the deceleration time of the motor when plugged.

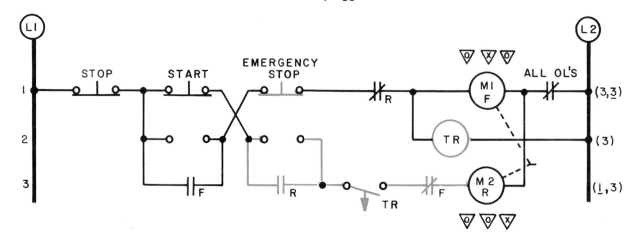

Figure 16-8 An off delay timer is used for plugging a motor to stop during emergency stops.

timer is also energized. The energizing of the off delay timer immediately closes the NO timer contacts. The closing of these contacts, however, will not energize the reverse contacts because of the interlocks. When the stop pushbutton is pushed, the forward starter and timer coil are de-energized. The NO timing contact will remain held closed for the setting of the time. The holding closed of the timing contact energizes the reversing starter for the period of time set on the timer. This plugs the motor to a stop. It is important that the timer's contact re-opens before the motor is actually reversed. If the time setting is too long, the motor will reverse direction.

Figure 16-8 illustrates how an off delay timer can be used for plugging a motor to stop during emergency stops. Here again the timer's contacts are connected in the same manner as the plugging switch was. When the start pushbutton is pressed, the motor is started and MEMORY is added to the circuit. If the stop pushbutton is pressed, the forward starter and timer are de-energized. Although the timer's NO contacts will be held closed for the time period set on the timer, the reversing starter will not be energized. This is because no power is applied to the reversing starter from line one. If the emergency stop pushbutton is pressed, the forward starter and timer are de-energized and the reversing starter is energized. The energizing of the reversing starter adds MEMORY to the circuit and plugs the motor to a stop. The opening of the timing contact de-energizes the reversing starter and removes the MEMORY.

Electric Braking

Electric braking is a method of braking in which direct current is applied to the stationary windings of a motor after the AC voltage is removed. This is an efficient and effective method of braking most AC motors. Electric braking provides a quick and smooth braking action on all types of loads including high speed and high inertia loads. Since there are no parts that come in physical contact during braking, maintenance is kept to a minimum.

Operating Principle of Electric Braking. To understand the operation of electric braking, it will be helpful to quickly review the basic principle behind the operation of motors. Referring to Figure 16-9, you will recall that if a South pole of a rotating magnet is brought near a stationary magnet with its North pole facing the rotating

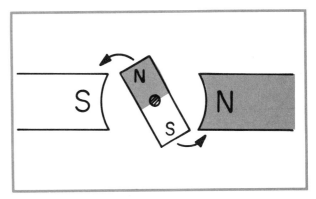

Figure 16-9 Electric braking is achieved by applying DC voltage to the stationary windings once the AC is removed.

magnet, the two unlike poles will attract each other. Likewise, if a North pole of a rotating magnet is brought near a stationary South pole, the two unlike poles will also attract each other. However, if two North or two South poles are brought near each other, they will repel each other. This principle, when applied to both AC and DC motors, is the reason why the motor shaft rotates. It should be noted, however, that the method in which the magnetic fields are created does change from one type of motor to another.

You will recall specifically from Unit 15, "Reduced Voltage Starting," that in the case of an AC induction motor, the opposing magnet fields are induced from the starter windings into the rotor windings by transformer action. As long as the AC voltage is applied, the motor will continue to turn. When the AC voltage is removed, however, the motor will coast to a standstill over a period of time since there is no induced field to keep it rotating.

Since the coasting time may be unacceptable, particularly in an emergency situation, electric braking can be used to provide a more immediate stop. Electric braking can accomplish this by applying a DC voltage to the stationary windings once the AC is removed. This DC voltage creates a magnetic field in the stator that will not change polarity.

This constant magnetic field in the stator in turn creates a magnetic field in the rotor. Since the magnetic field of the stator is not changing in polarity, it will attempt to stop the rotor when the magnetic fields are aligned (N to S and S to N). The only thing that can keep the rotor from stopping with the first alignment is the rotational inertia of the load connected to the motor shaft. However, since the braking action of the stator is present at all times, the motor is braked quickly

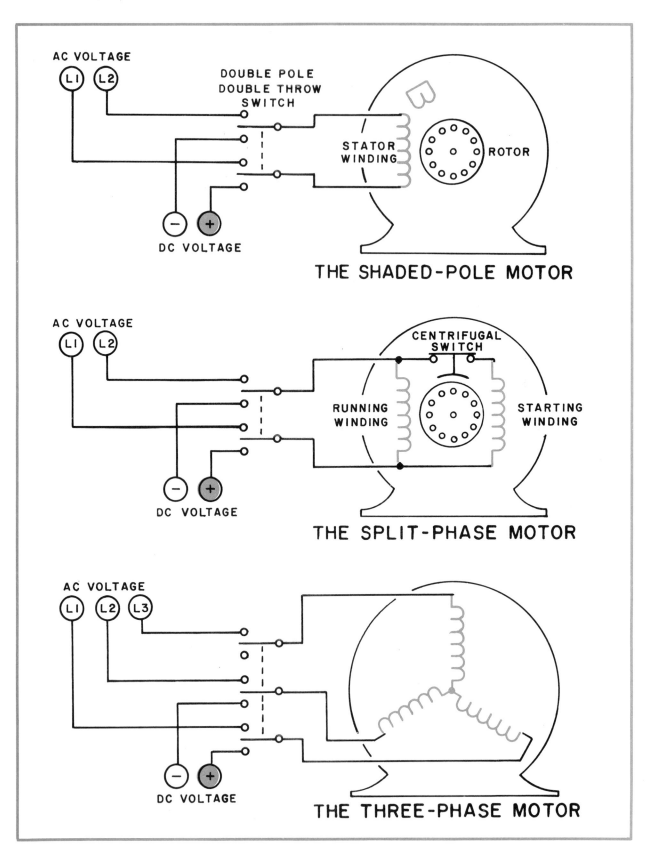

Figure 16-10 DC can be applied to different types of motors to provide electrical braking.

and smoothly to a standstill. Figure 16-10 illustrates how DC can be applied in different motor types for electric braking.

Circuit Used in DC Electric Braking. Figure 16-11 illustrates a typical circuit used to apply DC to a motor after the AC is removed. This circuit, like most DC braking circuits, uses a bridge rectifier circuit to change the AC into DC. For more information on bridge rectifiers, refer back to Unit 10 "Power Distribution."

This circuit illustrates a typical three-phase AC motor that is connected to the three-phase power by the magnetic motor starter. The magnetic motor starter is controlled by a standard stop/start pushbutton station with MEMORY. An off-delay timer is connected in parallel with the magnetic motor starter. This off-delay timer controls a NO contact that is used to apply power to the braking contactor for a short period of time after the stop pushbutton is pressed. This timing contact is adjusted to remain closed until the motor comes to a complete stop.

The braking contactor connects two motor leads to the DC supply. A transformer with tapped windings is used in this circuit to adjust the amount of braking torque applied to the motor; however, current limiting resistors could be used

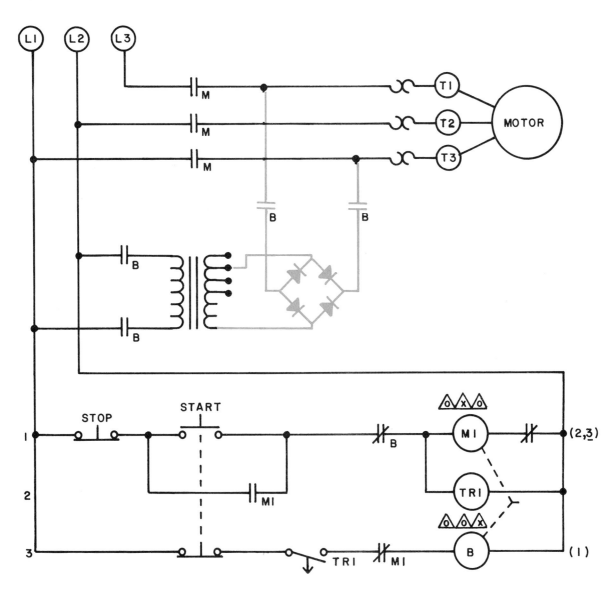

Figure 16-11 DC is applied after the AC is removed to bring the motor to a stop quickly.

for the same purpose. This allows for a low or high braking action depending upon application.

The interlock system in the control circuit prevents the motor starter and braking contactor from being on at the same time. This is required because the AC and DC power supplies must never be connected to the motor at the same time. This is one example of where total interlocking should always be used. Total interlocking is the use of mechanical, electrical and pushbutton interlocking. A standard forward and reversing motor starter can be used in this circuit, as it can with most electric braking circuits.

Dynamic Braking

Another method of braking is achieved if a motor that is running is reconnected to act as a generator immediately after it is turned off. The result of connecting the motor in this way is to make the motor act as a loaded generator that develops a retarding torque which rapidly stops the motor. The generator action converts the mechanical energy of rotation to electrical energy that can be dissipated as heat in a resistor. This method of braking is called dynamic braking.

Dynamic braking is usually applied to DC motors because in order to reconnect the motor to act as a generator, there must be access to the rotor windings. The rotating windings on a DC motor are called the armature. The stationary windings on a DC motor are called the fields, either series or shunt. Access is accomplished through the brushes on DC motors.

Figure 16-12 illustrates how dynamic braking is applied to a DC motor. Dynamic braking of a DC motor may be needed because of the fact that the DC motor is often used for lifting and moving heavy loads that may be difficult to stop.

In this circuit, the armature terminals of the DC motor are disconnected from the power supply and immediately connected across a resistor which acts as a load.

The smaller the resistance of the resistor, the greater the rate of energy dissipation and the faster the motor comes to rest. The field windings of the DC motor are left connected to the power supply. The armature generates a voltage referred to as counter electromotive force (CEMF). This CEMF causes current to flow through the resistor and armature. The current causes heat to be dissipated in the resistance in the form of electrical watts. This removes energy from the system and slows the motor rotation.

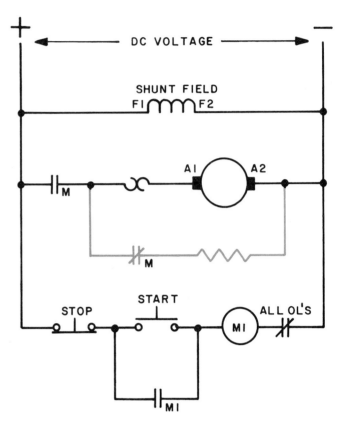

Figure 16-12 Dynamic braking is applied to a DC motor.

The generated CEMF decreases as the speed of the motor decreases. As the motor speed approaches zero, the generated voltage also approaches zero. This means that the braking action lessens as the speed of the motor decreases. As a result, a motor cannot be braked to a complete stop using dynamic braking. Dynamic braking also cannot hold a load once it is stopped because there is no braking action. For this reason electromechanical friction brakes are sometimes used along with dynamic braking in applications that require the load to be held. A combination of dynamic braking and friction braking can also be used in applications where a large heavy load is to be stopped. In this application, the force of the load would wear the friction shoe brakes too fast, so dynamic braking can be used to slow the load before the friction brakes are applied. This is similar to using a parachute to slow a race car before applying the brakes.

SPEED CONTROL

Speed control is essential in many residential, commercial and industrial applications. For example, the motors found around your home are often called upon to turn the load at different speeds. This may be in your washing machine that runs at low speed (gentle cycle), medium speed (normal load) or high speed (spin dry). Commercial furnaces and air conditioners, as well as electrical appliances such as mixers and blenders, will have more than one speed.

Industrial applications of speed control are even more diversified. Typical applications requiring speed control include: mining machines, printing presses, cranes and hoists, elevators, assembly line conveyors, food processing equipment and metalworking or woodworking lathes.

AC and DC Motor Types

In choosing speed control for any given application, the motor type is the fundamental consideration. Some motor types offer excellent speed control through a total range of speeds, while other types may offer only two or three different speeds. Other motor types offer only one speed that cannot be changed except by external means such as gears, pulley drives, or changes of power source frequency.

There are two fundamental types of motors used in speed control applications: AC and DC motors. Each of these motor types will be discussed in terms of range of speed control, application, and cost effectiveness.

Load Requirements

It is important to remember that the loads that are connected to and controlled by motors vary considerably. Each motor type, AC or DC, has its own ability to control different loads at different speeds. For example, certain types of motors will be rated at high starting torque with low running torque, and other types will be rated at poor starting torque with high running torque. In order to select the correct motor type for a given application, it is necessary to understand the load requirements first. This is especially true in applications that require speed control.

To fully appreciate how speed control is obtained for a given load, we need to have a thorough understanding of the requirements the motor must meet in controlling the load. Specifically, we need to understand the concepts of force, work, torque, power and horsepower in relation to speed.

Force and Work. Work is done when a force overcomes a resistance. Work is measured by multiplying the force that must be overcome by the distance through which it acts. Thus the formula used to determine the amount of work done is:

$$Work = Distance \times Force$$

If, for example, you were required to lift a 50 pound bag of groceries (load or force) and haul it from street level vertically to the 8th floor of a building approximately 100 feet, 5000 foot-pounds (ft-lb) of work is done. If the groceries were heavier or the distance longer, more work would be required. In order to perform work, the resistance must be overcome. For example, a building foundation exerts a large force in holding up the building, but since the foundation does not move, no work is accomplished. In the case of electric motors, force is not exerted vertically but rather is applied to a cylindrical shaft. When force is exerted in this manner, it is called torque.

Torque. The turning or twisting effort produced at the shaft of a motor is called the torque. The torque is directly proportional to the force exerted on the rotor (magnetic repulsion) and the radial distance through which the force acts (Figure 16-13).

Torque is usually measured in foot-pounds (ft-lbs). Mathematically, torque (T) is equal to the product of the force (F) trying to produce rotation, times the distance from the center of rotation to the point of application of the force (radius R) such that $T = FR$. Torque can exist even if there is no rotation. If there is no rotation, no work will be performed, yet there can exist a torque trying to produce rotation. With motors, this is called locked rotor torque and it exists when the motor is energized but cannot turn.

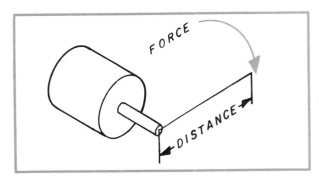

Figure 16-13 Torque is directly proportional to the force exerted on the rotor (magnetic repulsion) and the radial distance through which the force acts.

Power and Horsepower. Up to this point we have discussed how force can be applied vertically and through rotation to produce work. We have, however, not given any consideration to how fast we wish this work to be accomplished.

Power is the *rate* of doing work. The formula for determining power is:

$$\text{Power} = \frac{\text{Work}}{\text{Time}}$$

Time is important in figuring power because it tells us how fast the work is done. If, for example, the 50-pound load that was to be moved 100 feet in our first example was connected to a very small motor, it might take the motor several minutes to move the load. On the other hand, if a larger motor was used, it might move the load in only a few seconds. The reason for this difference is the amount of work that can be delivered in a given amount of time. Obviously a larger motor should be able to deliver more work in a given time than one which is considerably smaller. It is this difference that determines the power rating of the motor.

Motors are rated in horsepower (HP) or fractions of horsepower (1/4, 1/3, 1/2, etc.) One horsepower (1 HP) is equal to 33,000 pounds being moved 1 foot in 1 minute or 33,000 ft-lb/min. Electrical power can also be measured in watts (P=E × I). One horsepower is equal to 746 watts of electrical power.

Using the above information, we will figure the horsepower of our small and large motor used in the example of lifting a 50 pound load 100 feet.

Using this example we will see that if a motor must lift a fixed load at a higher rate of speed, the HP rating of the motor must also be higher. For example, if the small motor took two minutes to lift the 50 pound load 100 feet, then the HP of the motor would be: 0.075 HP (or about 5/64 HP.)

Work = Distance × Force

Work = 100 ft × 50 lb

Work = 5000 ft-lb

$$\text{Power} = \frac{\text{Work}}{\text{Time}}$$

$$\text{Power} = \frac{5000 \text{ ft-lb}}{2 \text{ min}}$$

Power = 2500 ft-lb per min.

And since 33,000 ft-lb per min equals 1 HP,

then

$$\text{HP (of small motor)} = \frac{2500}{33,000}$$

HP = 0.075

If the large motor took only 15 seconds (0.25 of a min) to move the 50 lb load 100 ft, then the HP of the motor would be: 0.6 HP (or about 5/8 HP).

Work = 5000 ft-lb

$$\text{Power} = \frac{5000 \text{ ft-lb}}{.25 \text{ min}}$$

Power = 20,000 ft-lb per min

$$\text{HP} = \frac{20,000}{33,000}$$

HP = 0.6 HP

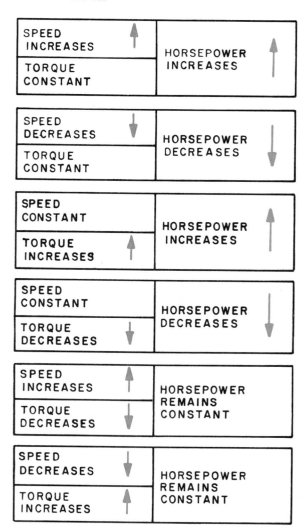

Figure 16-14 The relationship between torque, speed and horsepower.

Relationships Between Torque, Speed and Horsepower

Up to this point we have discussed torque, speed and horsepower as separate concepts. Actually, all three factors are interrelated when turning a load. Basically torque and speed are proportional to horsepower. This means that if torque remains constant and speed increases, horsepower increases. Also, if torque increases and speed remains constant, horsepower increases. Conversely, if either torque or speed decreases while the other remains constant, horsepower will decrease. Finally, if both torque and speed vary simultaneously but in opposite directions, horsepower may remain constant. Figure 16-14 illustrates the relationship between torque, speed and horsepower.

When motors are driving a load, they will be called upon to deliver a constant or variable torque—torque at a constant or variable horsepower. The amount of torque and horsepower needed will depend upon the speed and size of the load(s).

Constant Torque/Variable Horsepower. Figure 16-15 illustrates a typical application for a constant torque motor along with a graph indicating the relationship between torque, speed and horsepower.

This example is typical of machines that have mainly friction type loads—for example, conveyors, gear-type pumps and machines, or load lifting equipment. In this situation the HP required increases when speed increases. The torque requirement does not vary throughout the speed range except for the extra starting torque needed to overcome the breakaway friction. The torque remains constant because the force of the load does not change.

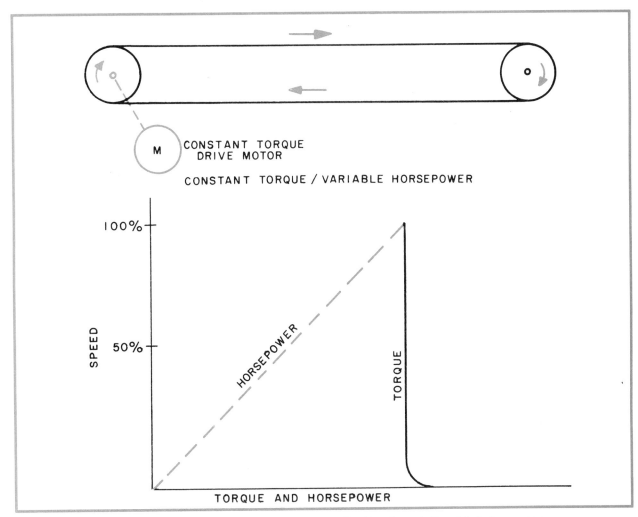

Figure 16-15 Typical application for a constant torque motor along with a graph indicating the relationship between torque, speed and horsepower.

Constant Horsepower/Variable Torque.
Figure 16-16 illustrates a typical constant HP load with its characteristics graph. This application applies to loads that demand high torque at low speeds and low torque at high speeds. Examples of these loads are machines that roll and unroll material such as paper or metal. Since the work is done on a varying diameter with tension and linear speed of the material constant, HP must also be constant. Although in these applications the speed of the material is kept constant, the motor speed is not. This is because the diameter of the material on the roll that is driven by the motor is constantly changing as material is added. At start, the motor must run at high speed to maintain the correct material speed while torque is kept at a minimum. As material is added to the roll, the motor must deliver more torque at a slower speed. In this application both torque and speed are constantly changing while motor HP remains the same.

Variable Torque/Variable Horsepower.
Figure 16-17 illustrates a typical variable torque load with its characteristics graph. This graph applies to loads that have a varying torque and HP at different speeds. Typical examples are fans, blowers, centrifugal pumps, mixers and agitators. In these examples, as the motor speed is increased so is the load output. Since the motor must work harder to deliver more output at a faster speed, both torque and HP are increased with increased speed.

NEMA DESIGN

Since each motor type has its own characteristics of horsepower, torque and speed, different motor types are more suited for particular applications. The basic characteristics of each motor type are determined by the design of the motor and the supply voltage used. These design types are classified and given a letter designation which can be found on the nameplate of some motor types listed

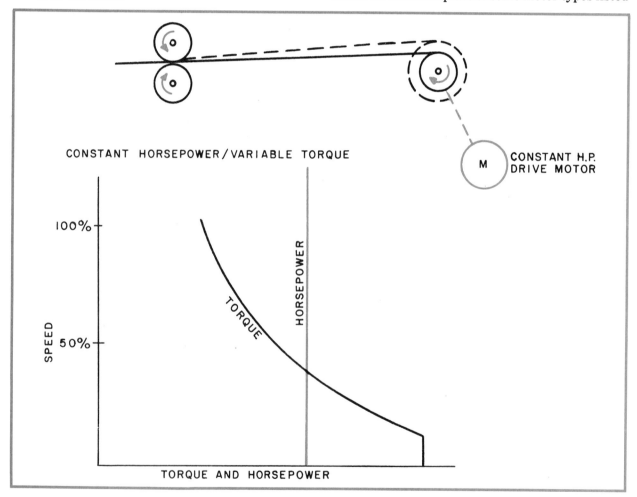

Figure 16-16 In a constant horsepower motor, horsepower remains the same at varying speeds.

as "NEMA Design." The NEMA design, along with basic characteristics and typical uses for AC motors, is illustrated in Figure 16-18.

MULTISPEED MOTORS

When motors are required to run at different speeds, the motor's torque or HP characteristics will change with a change in speed. In these applications, the motor types are broken down into three categories previously discussed:

1. Constant Torque-Variable HP
2. Constant HP-Variable Torque
3. Variable Torque-Variable HP

Each of these types may be shortened to simply:
1. Constant Torque (CT)
2. Constant HP (CHP)
3. Variable Torque (VT)

The choice of motor type to use will depend on the application that the motor must perform. Once this selection is made, the motor will have to be connected into the circuit. Figure 16-19 illustrates several typical motor connection arrangements conforming to NEMA standards. Not all possible arrangements are shown, however, this should be a helpful guide to many common applications.

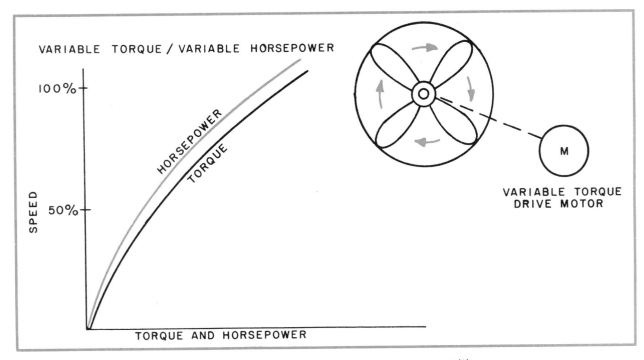

Figure 16-17 Torque and horsepower increase with speed in a variable torque motor.

NEMA DESIGN	STARTING TORQUE	STARTING CURRENT	BREAKDOWN TORQUE	FULL LOAD SLIP	TYPICAL APPLICATIONS
A	NORMAL	NORMAL	HIGH	LOW	MACHINE TOOLS, FANS, CENTRIFUGAL PUMPS
B	NORMAL	LOW	HIGH	LOW	MACHINE TOOLS, FANS, CENTRIFUGAL PUMPS
C	HIGH	LOW	NORMAL	LOW	LOADED COMPRESSOR LOADED CONVEYOR
D	VERY HIGH	LOW	- - -	HIGH	PUNCH PRESSES

Figure 16-18 Motor design types are classified by NEMA.

MULTISPEED MOTOR CONNECTIONS

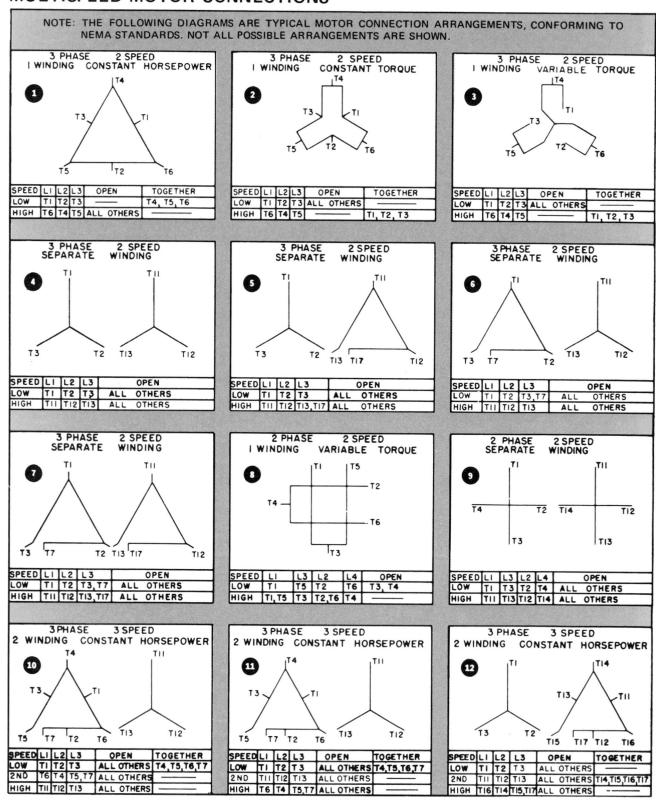

NOTE: THE FOLLOWING DIAGRAMS ARE TYPICAL MOTOR CONNECTION ARRANGEMENTS, CONFORMING TO NEMA STANDARDS. NOT ALL POSSIBLE ARRANGEMENTS ARE SHOWN.

① 3 PHASE 2 SPEED 1 WINDING CONSTANT HORSEPOWER

SPEED	L1	L2	L3	OPEN	TOGETHER
LOW	T1	T2	T3	—	T4, T5, T6
HIGH	T6	T4	T5	ALL OTHERS	

② 3 PHASE 2 SPEED 1 WINDING CONSTANT TORQUE

SPEED	L1	L2	L3	OPEN	TOGETHER
LOW	T1	T2	T3	ALL OTHERS	—
HIGH	T6	T4	T5	—	T1, T2, T3

③ 3 PHASE 2 SPEED 1 WINDING VARIABLE TORQUE

SPEED	L1	L2	L3	OPEN	TOGETHER
LOW	T1	T2	T3	ALL OTHERS	—
HIGH	T6	T4	T5	—	T1, T2, T3

④ 3 PHASE 2 SPEED SEPARATE WINDING

SPEED	L1	L2	L3	OPEN
LOW	T1	T2	T3	ALL OTHERS
HIGH	T11	T12	T13	ALL OTHERS

⑤ 3 PHASE 2 SPEED SEPARATE WINDING

SPEED	L1	L2	L3	OPEN
LOW	T1	T2	T3	ALL OTHERS
HIGH	T11	T12	T13, T17	ALL OTHERS

⑥ 3 PHASE 2 SPEED SEPARATE WINDING

SPEED	L1	L2	L3	OPEN
LOW	T1	T2	T3, T7	ALL OTHERS
HIGH	T11	T12	T13	ALL OTHERS

⑦ 3 PHASE 2 SPEED SEPARATE WINDING

SPEED	L1	L2	L3	OPEN
LOW	T1	T2	T3, T7	ALL OTHERS
HIGH	T11	T12	T13, T17	ALL OTHERS

⑧ 2 PHASE 2 SPEED 1 WINDING VARIABLE TORQUE

SPEED	L1	L3	L2	L4	OPEN
LOW	T1	T5	T2	T6	T3, T4
HIGH	T1, T5	T3	T2, T6	T4	—

⑨ 2 PHASE 2 SPEED SEPARATE WINDING

SPEED	L1	L3	L2	L4	OPEN
LOW	T1	T3	T2	T4	ALL OTHERS
HIGH	T11	T13	T12	T14	ALL OTHERS

⑩ 3 PHASE 3 SPEED 2 WINDING CONSTANT HORSEPOWER

SPEED	L1	L2	L3	OPEN	TOGETHER
LOW	T1	T2	T3	ALL OTHERS	T4, T5, T6, T7
2ND	T6	T4	T5, T7	ALL OTHERS	—
HIGH	T11	T12	T13	ALL OTHERS	—

⑪ 3 PHASE 3 SPEED 2 WINDING CONSTANT HORSEPOWER

SPEED	L1	L2	L3	OPEN	TOGETHER
LOW	T1	T2	T3	ALL OTHERS	T4, T5, T6, T7
2ND	T11	T12	T13	ALL OTHERS	—
HIGH	T6	T4	T5, T7	ALL OTHERS	—

⑫ 3 PHASE 3 SPEED 2 WINDING CONSTANT HORSEPOWER

SPEED	L1	L2	L3	OPEN	TOGETHER
LOW	T1	T2	T3	ALL OTHERS	
2ND	T11	T12	T13	ALL OTHERS	T14, T15, T16, T17
HIGH	T16	T14	T15, T17	ALL OTHERS	—

Figure 16-19 Several typical motor connection arrangements conforming to NEMA standards. (Square D Company)

MULTISPEED MOTOR CONNECTIONS

NOTE: THE FOLLOWING DIAGRAMS ARE TYPICAL MOTOR CONNECTION ARRANGEMENTS, CONFORMING TO NEMA STANDARDS. NOT ALL POSSIBLE ARRANGEMENTS ARE SHOWN.

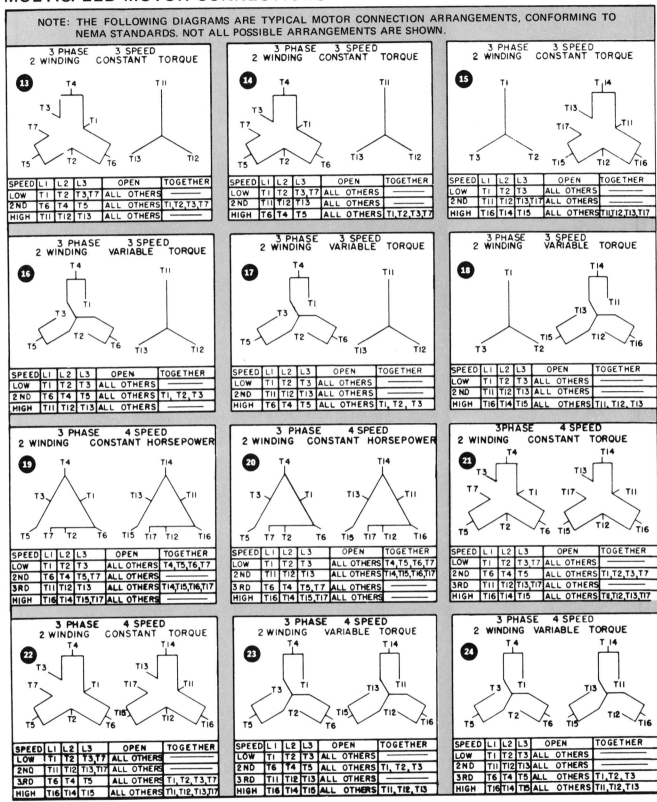

13 — 3 PHASE 3 SPEED 2 WINDING CONSTANT TORQUE

SPEED	L1	L2	L3	OPEN	TOGETHER
LOW	T1	T2	T3,T7	ALL OTHERS	——
2ND	T6	T4	T5	ALL OTHERS	T1,T2,T3,T7
HIGH	T11	T12	T13	ALL OTHERS	——

14 — 3 PHASE 3 SPEED 2 WINDING CONSTANT TORQUE

SPEED	L1	L2	L3	OPEN	TOGETHER
LOW	T1	T2	T3,T7	ALL OTHERS	——
2ND	T11	T12	T13	ALL OTHERS	——
HIGH	T6	T4	T5	ALL OTHERS	T1,T2,T3,T7

15 — 3 PHASE 3 SPEED 2 WINDING CONSTANT TORQUE

SPEED	L1	L2	L3	OPEN	TOGETHER
LOW	T1	T2	T3	ALL OTHERS	——
2ND	T11	T12	T13,T17	ALL OTHERS	——
HIGH	T16	T14	T15	ALL OTHERS	T11,T12,T13,T17

16 — 3 PHASE 3 SPEED 2 WINDING VARIABLE TORQUE

SPEED	L1	L2	L3	OPEN	TOGETHER
LOW	T1	T2	T3	ALL OTHERS	——
2ND	T6	T4	T5	ALL OTHERS	T1, T2, T3
HIGH	T11	T12	T13	ALL OTHERS	——

17 — 3 PHASE 3 SPEED 2 WINDING VARIABLE TORQUE

SPEED	L1	L2	L3	OPEN	TOGETHER
LOW	T1	T2	T3	ALL OTHERS	——
2ND	T11	T12	T13	ALL OTHERS	——
HIGH	T6	T4	T5	ALL OTHERS	T1, T2, T3

18 — 3 PHASE 3 SPEED 2 WINDING VARIABLE TORQUE

SPEED	L1	L2	L3	OPEN	TOGETHER
LOW	T1	T2	T3	ALL OTHERS	——
2ND	T11	T12	T13	ALL OTHERS	——
HIGH	T16	T14	T15	ALL OTHERS	T11, T12, T13

19 — 3 PHASE 4 SPEED 2 WINDING CONSTANT HORSEPOWER

SPEED	L1	L2	L3	OPEN	TOGETHER
LOW	T1	T2	T3	ALL OTHERS	T4,T5,T6,T7
2ND	T6	T4	T5,T7	ALL OTHERS	——
3RD	T11	T12	T13	ALL OTHERS	T14,T15,T16,T17
HIGH	T16	T14	T15,T17	ALL OTHERS	——

20 — 3 PHASE 4 SPEED 2 WINDING CONSTANT HORSEPOWER

SPEED	L1	L2	L3	OPEN	TOGETHER
LOW	T1	T2	T3	ALL OTHERS	T4,T5,T6,T7
2ND	T11	T12	T13	ALL OTHERS	T14,T15,T16,T17
3RD	T6	T4	T5,T7	ALL OTHERS	——
HIGH	T16	T14	T15,T17	ALL OTHERS	——

21 — 3 PHASE 4 SPEED 2 WINDING CONSTANT TORQUE

SPEED	L1	L2	L3	OPEN	TOGETHER
LOW	T1	T2	T3,T7	ALL OTHERS	——
2ND	T6	T4	T5	ALL OTHERS	T1,T2,T3,T7
3RD	T11	T12	T13,T17	ALL OTHERS	——
HIGH	T16	T14	T15	ALL OTHERS	T11,T12,T13,T17

22 — 3 PHASE 4 SPEED 2 WINDING CONSTANT TORQUE

SPEED	L1	L2	L3	OPEN	TOGETHER
LOW	T1	T2	T3,T7	ALL OTHERS	——
2ND	T11	T12	T13,T17	ALL OTHERS	——
3RD	T6	T4	T5	ALL OTHERS	T1,T2,T3,T7
HIGH	T16	T14	T15	ALL OTHERS	T11,T12,T13,T17

23 — 3 PHASE 4 SPEED 2 WINDING VARIABLE TORQUE

SPEED	L1	L2	L3	OPEN	TOGETHER
LOW	T1	T2	T3	ALL OTHERS	——
2ND	T6	T4	T5	ALL OTHERS	T1, T2, T3
3RD	T11	T12	T13	ALL OTHERS	——
HIGH	T16	T14	T15	ALL OTHERS	T11, T12, T13

24 — 3 PHASE 4 SPEED 2 WINDING VARIABLE TORQUE

SPEED	L1	L2	L3	OPEN	TOGETHER
LOW	T1	T2	T3	ALL OTHERS	——
2ND	T11	T12	T13	ALL OTHERS	——
3RD	T6	T4	T5	ALL OTHERS	T1,T2, T3
HIGH	T16	T14	T15	ALL OTHERS	T11,T12, T13

SPEED CONTROL OF DC MOTORS

Direct current motors are used in industrial applications that require variable speed control, high torque, or both. Since the speed of most DC motors can be controlled smoothly and easily from zero to full speed, DC motors are used in many acceleration and deceleration applications.

In addition to having excellent speed control, the DC motor is ideal in applications that call for momentarily higher torque outputs. This is because the DC motor can deliver three to five times its rated torque for short periods of time. Most AC motors will stall with a load that requires twice the rated torque. Good speed control and high torque are the reasons DC motors are used in running large machine tools found in the mining industry, cranes and hoists.

DC Series Motor

Figure 16-20 illustrates the DC series connected motor with the operating characteristics graph. The field coil of the motor is connected in series with the armature. Although speed control is poor, the DC series motor produces very high starting torque and is ideal for applications in which the starting load is large. Examples of such loads are cranes, hoists, electric buses, streetcars, railroads and other heavy traction applications.

The torque that is produced by a motor is dependent on the strength of the magnetic fields within

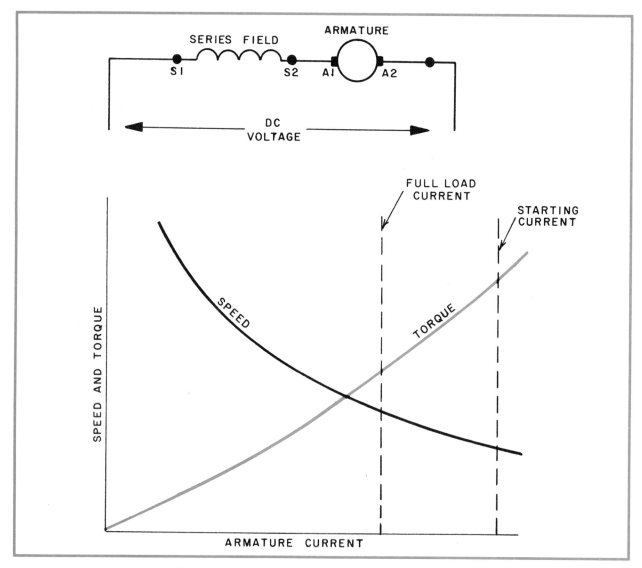

Figure 16-20 The DC series connected motor produces high starting torque.

the motor. The strength of the magnetic field is dependent upon how much current flows through the coil winding. The amount of current that flows through a motor is dependent upon the size of the load. The larger the load, the greater the current flow. Since the armature and field are connected in series, any increase in load will increase current in both the armature and field. This increased current flow in both the armature and field is what gives the series motor a high torque output.

In the DC series motor, speed changes rapidly with torque changes. This is illustrated in the motor characteristics graph. When torque is high, speed is low, and when speed is high, torque is low. This is because as the increased current (created by the load) flows through the series field, there will be a larger flux increase. This increased flux will produce a large counter electromotive force which will greatly decrease the speed of the motor. As the load is removed, the motor will rapidly speed up. In fact, without a load, the motor would gain speed and tend to run away. In certain cases, the speed may become great enough to damage the motor. It is for this reason that the DC series motor should always be connected directly (not through belts, chains, etc.) to the load.

The speed of a DC series motor is controlled by varying the applied voltage. Speed control of a series motor is poor when compared to that of the shunt motor. Although the speed regulation of the series motor is not as good as that of the shunt motor, remember that not all applications require good speed regulation. The advantage of a high torque output outweighs good speed regulation in certain applications—for example, the starter motor in your automobile.

DC Shunt Motor

Figure 16-21 illustrates the DC shunt motor with the motors characteristics chart. In the DC shunt motor, the field coil is connected in parallel (shunt) with the armature. The DC shunt motor has good speed regulation and is used in applications that require a constant speed at any control setting, even with a changing load. The DC shunt motor is considered a constant speed motor for all reasonable loads.

Speed is determined by one of two factors: the voltage across the armature and the strength of the field. If the voltage to the armature is reduced, the speed is also reduced. If the strength of the magnetic field is reduced, the motor will speed up. The reason that the motor will speed up with a reduction in field strength is that with less field strength, there is less counter electromotive force (CEMF) developed in the armature. When the counter electromotive force is lowered, the armature current increases, producing increased

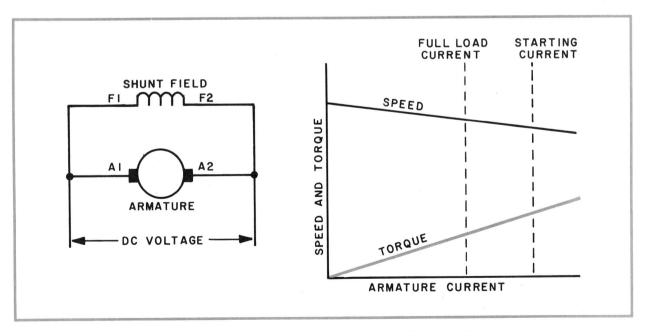

Figure 16-21 The field coil is connected in parallel (shunt) with the armature in the DC shunt motor.

torque and speed. Thus, to control the speed of a DC shunt motor, the voltage to the armature is varied or the field current is varied.

Figure 16-22 illustrates how a field rheostat or armature rheostat could be used to adjust the speed of a DC shunt motor. The rheostat is used to increase or decrease the strength of the field or armature. Once the strength of the field is set, it will remain constant regardless of changes in armature current. As the load is increased on the armature, the armature current and torque of the motor will increase. This will tend to slow down the armature, but the reduction of counter electromotive force will simultaneously allow a further increase in armature current and thus tend to return the motor to set speed. Thus the motor will run at a fairly constant speed at any control setting.

The DC shunt motor has fairly high torque at any speed. As illustrated in the motor characteristics graph, the motor torque is directly proportional to the armature current. As armature current is increased, so is motor torque, with only a slight drop in motor speed.

DC Compound Motor

Figure 16-23 illustrates the DC compound motor with the motor's characteristics chart. In the DC compound motor both a series and shunt field are connected in relationship to the armature. The compound motor combines the operating characteristics of both the series and shunt motor. This produces the higher starting torque characteristic of the series motor and better speed control characteristic of the shunt motor. However, the compound motor will not have as high a starting torque as the series motor and not as good a speed regulation as the shunt motor. Compound motors are used in applications where the load changes and precise speed control is not required. Applications include elevators, hoists, cranes and conveyors.

Speed control is obtained in the compound motor by changing the shunt field current strength or changing the voltage applied to the armature. This can be accomplished by using a controller that uses resistors to reduce the applied voltage or by using a variable voltage supply.

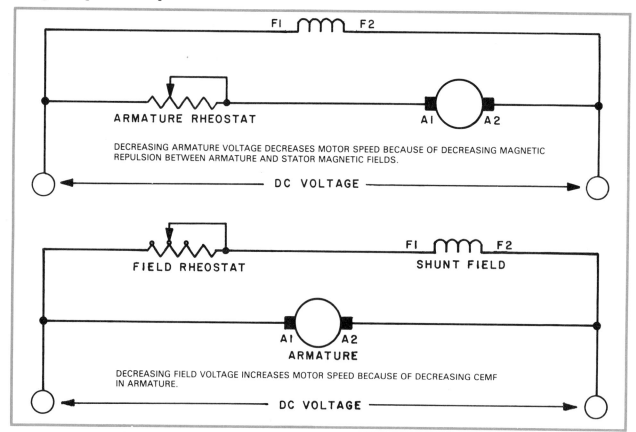

Figure 16-22 A field rheostat or armature rheostat could be used to adjust the speed of a DC shunt motor.

Solid State Speed Control of DC Motors

As previously illustrated in Unit 15, the SCR can perform the same function as resistance in controlling the voltage applied to a load. Since an SCR can be adjusted throughout its range, it is electrically similar to a rheostat (variable resistor). The SCR, however, does not have some of the disadvantages of the rheostat. The SCR is smaller in size for the same rating, energy-efficient in not wasting power within itself as does the rheostat, and less expensive. For these reasons most DC speed controls made today use SCRs instead of rheostats.

Controlling Base Speed of DC Motor

DC motor control is one of the best industrial applications of the SCR. In this application the SCR can be used to control the speed of the DC motor below the base speed by changing the amount of current that flows through the armature circuit or above the base speed by changing the voltage applied to the field.

The speed of a DC motor is controlled by varying the applied voltage across the armature, the field, or both. When armature voltage is controlled, the motor will deliver a constant torque characteristic. When field voltage is controlled, the motor will deliver a constant horsepower characteristic. Figure 16-24 illustrates the effect of reducing the armature or field voltage in relationship to the base speed. The base speed of a motor is the speed at which the motor will run with full line voltage applied to the armature and field.

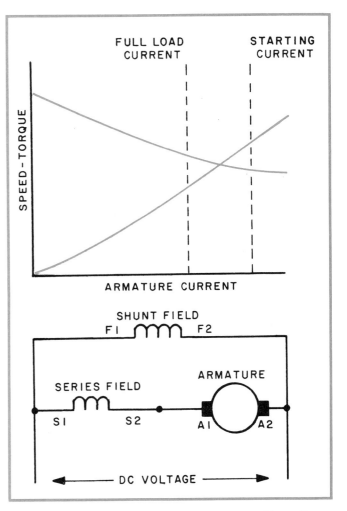

Figure 16-23 A DC compound motor combines the higher starting torque and better speed control characteristics of the DC series and DC shunt motors.

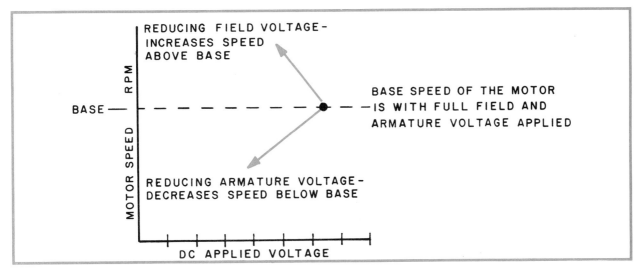

Figure 16-24 Field voltage and armature voltage have an effect on changing the base speed of a DC motor.

Figure 16-25 illustrates how an SCR is used to control the speed of a DC motor. In this circuit the speed can be controlled from zero to the base speed using an SCR. The SCR is controlled by the setting of the potentiometer, which varies the "on" time of the SCR per cycle—and thus varies the amount of average current flow to the armature. The voltage applied to the SCR is AC since the SCR will rectify (as well as control) the AC voltage. A rectifier circuit, as illustrated, is required for the field circuit, since the field circuit must be supplied with DC. If speed control above the base is required, the rectifier circuit in the field can also be changed to an SCR control.

SPEED CONTROL OF AC MOTORS

The AC motor is basically a constant speed motor. This is because the synchronous speed of an induction motor is expressed by the following formula:

$$\text{Synchronous (RPM)} = \frac{120 \times F}{P}$$

F = Supply frequency in cycles/second
P = Number of poles in motor winding.

From this formula we can see that motors designed for 60 Hertz use have synchronous speeds as follows:

Poles:	2	4	6	8	10	12	14	16
RPM:	3600	1800	1200	900	720	600	514	450

All induction motors will have a full load speed somewhat below the synchronous speed. This percentage of reduction in speed is known as the "percent slip." Most motors run from 2 to 10% behind the synchronous speed at no load; as the load is increased, the percentage of slip is increased.

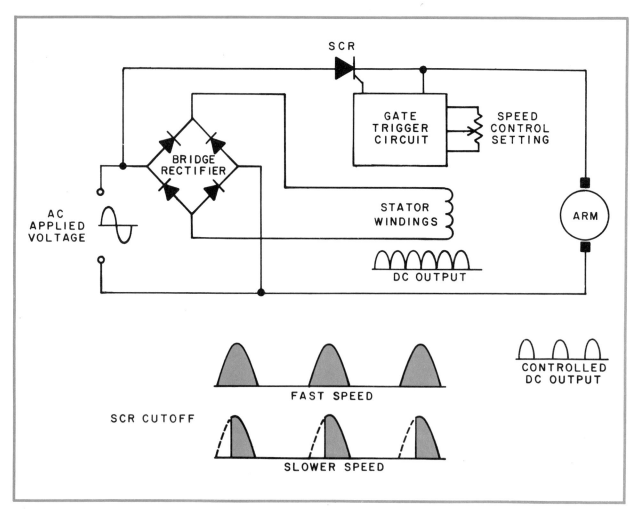

Figure 16-25 An SCR is used to control the speed of a DC motor.

From the formula used to determine the speed of an AC motor, it is evident that the supply frequency and number of poles are the only variables that determine the speed of the motor. Unlike the speed of a DC motor, the speed of an AC motor should not be changed by varying the applied voltage. In fact, if the supply voltage is varied to an AC motor more than 10% above or below rated nameplate voltage, damage may be done to the motor. This is because, in an induction motor, the starting torque and breakdown torque all vary as the square of the applied voltage. For example, with 90% of rated voltage, these torques become $(0.9)^2 = 0.81$, or 81% of what they would be at rated voltage. This is also important to remember in applications where a motor is located at the end of a long line. The line drop at the motor may be great enough to keep the motor from starting the load or developing the required torque to operate satisfactorily.

Since the frequency or number of poles must be changed to change the speed of an AC motor, two methods of speed control are available. These are: (1) changing the frequency applied to the motor, or (2) using a motor with windings that may be reconnected to form different numbers of poles.

Multispeed AC motors, designed to be operated at constant frequency, are provided with stator windings that can be reconnected to provide a change in the number of poles and thus a change in the speed. These multispeed motors are available in two or more fixed speeds which are determined by the connections made to the motor. Two-speed motors usually have one winding that may be connected to provide two speeds, one of which is half the other. Motors with more than two speeds usually include many windings that are connected and reconnected to provide different speeds by changing the number of poles.

In multispeed motors the different speeds are determined by connecting the external winding leads to a multispeed starter. Although this starter can be manual, the magnetic motor starter is the most common means used for speed control. One starter is required for each speed of the motor. It is important that each starter be interlocked (mechanical, auxiliary contact, or pushbutton) to prevent more than one starter from being on at the same time. (The motor can run at only one speed at a time.) For two-speed motors, a standard forward and reversing starter is usually used, since it provides mechanical interlocking.

Basic Speed Control Circuits

Several control circuits can be developed to control a multispeed motor, depending upon the requirements of the circuit. Figure 16-26 illustrates a basic control circuit used with a two-speed motor using a high-low pushbutton station. In this circuit the operator may start the motor from rest at either speed. The stop pushbutton must be operated before changing from low to high speed

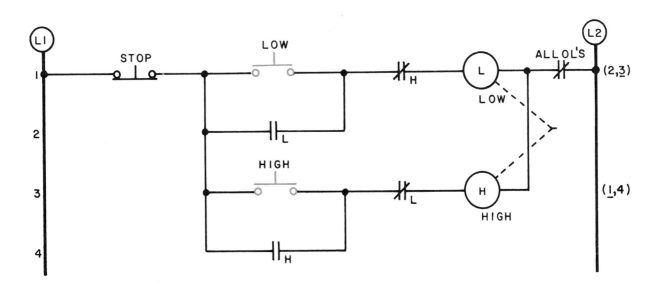

Figure 16-26 In a two-speed motor control circuit, the motor can be started in the low or high speed. The stop pushbutton must be pressed before changing speeds.

or from high to low speed. Figure 16-27 illustrates the wiring diagram for a two-speed motor using a high-low pushbutton station.

Figure 16-28 illustrates another high-low pushbutton control for a two-speed motor. This circuit allows the operator to start the motor from rest at either speed or change from the low speed to the high speed without pressing the stop pushbutton. The stop pushbutton must, however, be operated before it is possible to change from high to low speed. This high-to-low arrangement is intended to prevent both excessive line current and shock to the motor. In addition, the machinery which is driven by the motor is protected from shock that could result from connecting a motor at high speed to a lower speed.

Compelling Circuit Logic. In many applications of speed control, the motor must always be started at the low speed before it can be changed

to the high speed. Figure 16-29 illustrates a circuit that allows the operator to start the motor in one speed before changing to another. This circuit logic is called *compelling circuit logic*.

This circuit does not allow the operator to start the motor at high speed. Instead, the circuit compels the operator to first start the motor at low speed before changing to high speed. This arrangement is intended to prevent the motor and driven machinery from starting at a high speed. The motor and driven machinery are first allowed to accelerate at a low speed before accelerating to a high speed. The circuit also compels the operator to press the stop pushbutton before changing speed from high to low.

Accelerating Circuit Logic. In many applications a motor must be automatically accelerated from the low to the high speed even if the high pushbutton is pressed first. This type of circuit

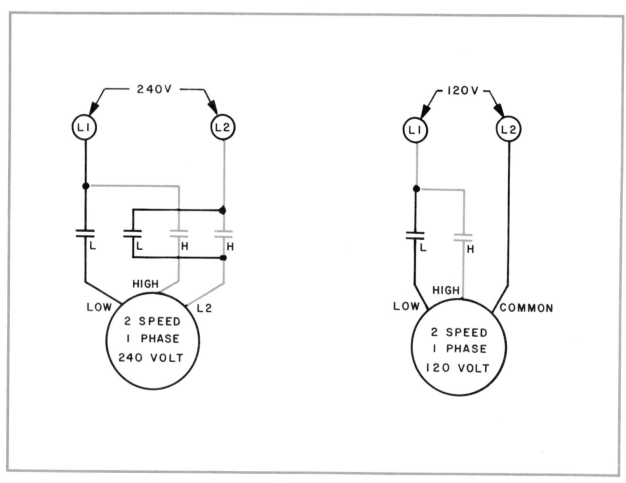

Figure 16-27 The wiring diagram for a two-speed motor used with the high-low control circuit of Figure 16-26.

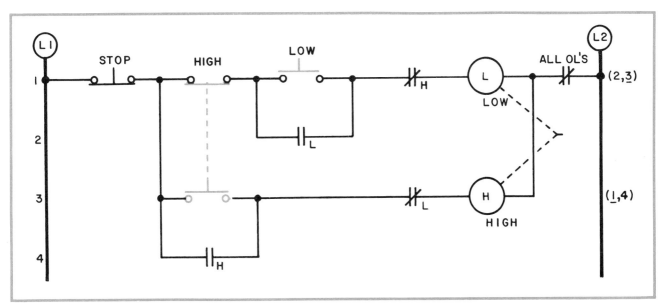

Figure 16-28 In a modified control circuit the motor can be changed from low speed to high speed without first stopping the motor.

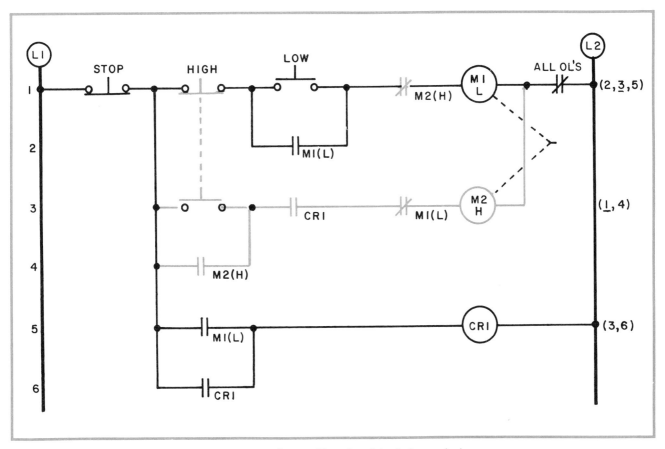

Figure 16-29 Compelling circuit logic is used where the motor must always be started at the low speed before it can be changed to high speed.

logic is called *accelerating circuit logic*. Figure 16-30 illustrates a circuit using accelerating circuit logic. Remember that in any of these circuits the pushbutton may be replaced with any control device—pressure, photoelectric, etc.—without changing the circuit logic.

This circuit allows the operator to select the desired speed by pressing either the low or high pushbutton. If the operator presses the low pushbutton, the motor will start and run in the low speed. If the operator presses the high pushbutton, the motor will start in the low speed and then run in the high speed only after the predetermined time that is set on the timer in the control

circuit. This arrangement is intended to give the motor and driven machinery a definite time period to accelerate from the low to the high speed. The circuit also requires that the operator press the stop pushbutton before changing speed from high to low.

Decelerating Circuit Logic. In some applications a motor or load can not take the stress of changing from a high to a low speed without damage. In these applications the motor must be allowed to decelerate by coasting or braking before being changed to a lower speed. This type of

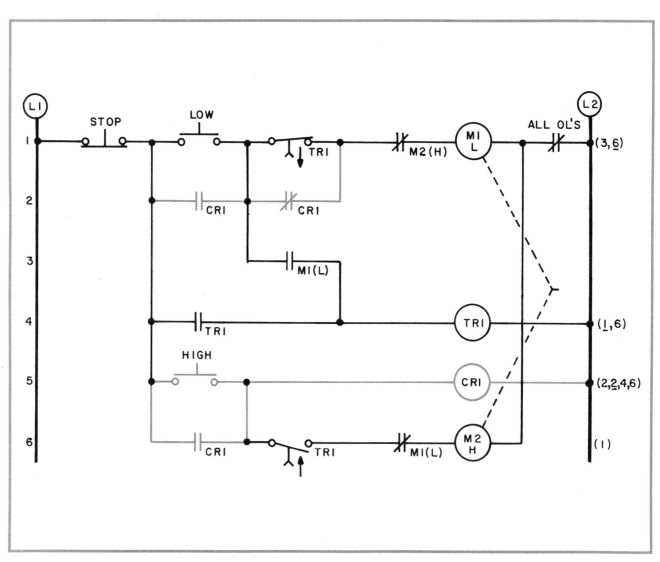

Figure 16-30 With accelerating circuit logic, a motor must be automatically accelerated from the low to the high speed even if the high pushbutton is pressed first.

logic is called *decelerating circuit logic* and is illustrated in Figure 16-31.

This circuit allows the operator to select the desired speed by pressing either the low or high pushbutton. If the low pushbutton is pressed, the motor will start and run at low speed. If the high pushbutton is pressed, the motor will start and run at high speed. If, however, the operator changed from the high to the low speed, the motor will change to the low speed only after a predetermined time. This arrangement is intended to give the motor and driven machinery a different time period to decelerate from a high speed to a low speed.

Changing the Frequency

Speed control of AC squirrel-cage motors can be accomplished if the frequency of the voltage applied to the stator is varied to change the synchronous speed. The change in synchronous speed then causes the motor speed to change. There are two basic methods used to vary the frequency of the AC voltage applied to the motor: using an inverter, or using a converter. The inverter changes DC voltage into AC and can be designed to produce variable frequency. The converter changes the standard 60 Hertz AC power into almost any desired frequency. Both inverter and converter use solid state SCRs for control.

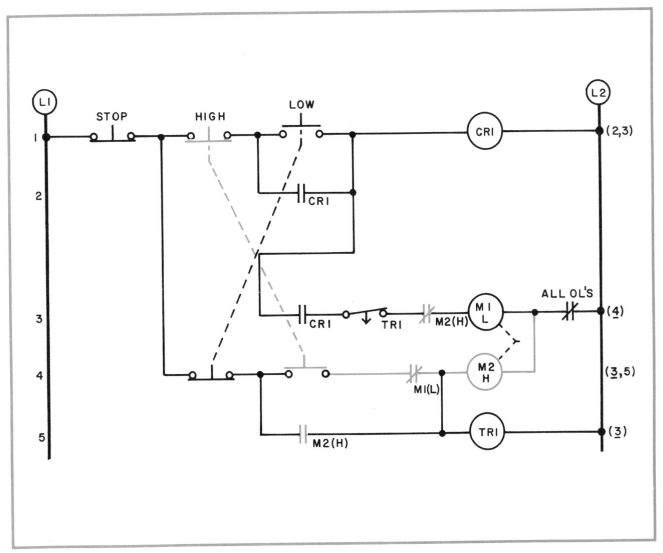

Figure 16-31 With decelerating circuit logic, the motor must be allowed to decelerate by coasting or braking before being changed to a lower speed.

Figure 16-32 illustrates the basic principle of an inverter. In this circuit the adjustable speed AC drive rectifies the three-phase incoming power and delivers the power to the inverter circuit. The inverter circuit changes the DC power back to an adjustable frequency AC output that controls the speed of the motor. Frequency is controlled by the setting of the rheostat. The rheostat controls the firing rate of the SCRs in the inverter, and thus the frequency of the AC output. Each SCR is fired in sequence, one immediately following the other. The rate of firing and length of time the SCR is allowed to remain on determines the frequency. Inverters are usually manufactured for the 5 to 150 HP range.

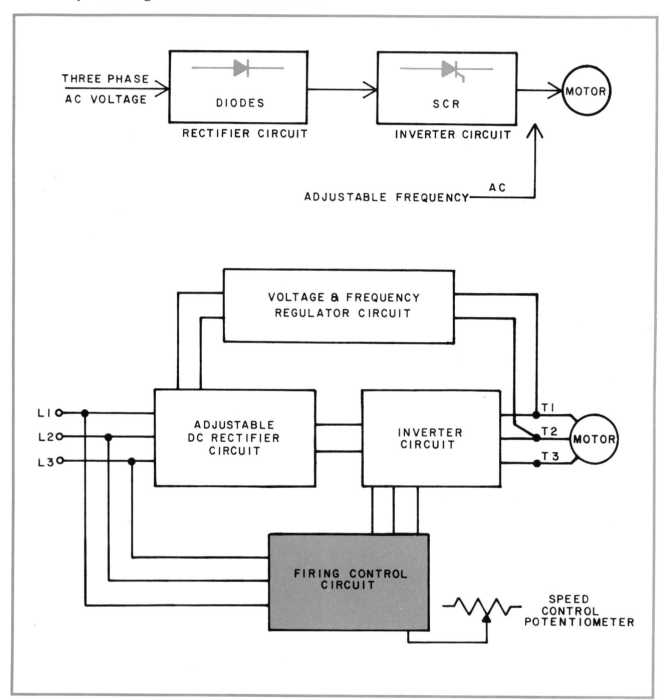

Figure 16-32 An inverter circuit changes DC power back to an adjustable frequency AC output that controls the speed of the motor.

Figure 16-33 illustrates the basic principle of a converter. The operation of a variable frequency converter is basically the same as the inverter. The frequency of the output voltage is determined by the length of time the SCRs are allowed to remain on. Full wave converters require a minimum of 36 SCRs with a complicated control circuit. For this reason their cost is considered high. However, in applications that require several AC motors to have precise speed synchronization, the cost becomes less important. Converters are usually manufactured up to the 10,000 HP range and are often used to control large groups of motors simultaneously.

Changing the Applied Voltage

Although the speed of the standard AC squirrel-cage motor is primarily varied by changing the number of poles or the applied frequency, there is another method that is sometimes used. This method is not a standard method of speed control and caution must be taken in applying it.

Although the speed of the AC motor is determined by the number of poles and applied frequency, it is possible in some applications, when the motor is connected to a load, to control the speed of the load by varying the applied voltage to the motor. By varying the voltage applied to the motor, the torque that the motor can deliver to the load is varied. The torque of a squirrel-cage motor varies as the square of the applied voltage.

We have already seen that the greater the torque, the faster the acceleration time. This was illustrated in the example of using a small and large motor to lift a 50 pound load 100 feet. The large motor produced more torque, so it moved the load faster. If we would reduce the torque of the larger motor to that of the smaller motor, the speed at which the larger motor performed the work would also be reduced.

Although it is possible to reduce the speed of a larger motor by reducing the applied voltage, this method could damage the motor. This damage may come from excess heat buildup in the motor. Most AC motors are not designed to have their voltage varied more than 10% of nameplate rating. However, in applications where the manufacturers can determine the load requirements and motor type and size in advance, they may install this type of speed control. The advantage of less cost for control with a larger motor is the determining factor. This type of speed control is limited to applications of soft-start light loads. Fan motors are sometimes controlled this way. This application is usually used with the shaded-pole or permanent-split capacitor motor. Except in applications that are specifically designed for this type of speed control, it should not be considered a standard method.

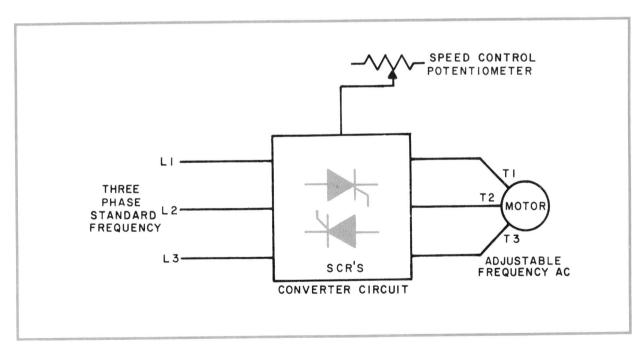

Figure 16-33 In a converter circuit, the frequency of the output voltage is determined by the length of time the SCRs are allowed to remain on.

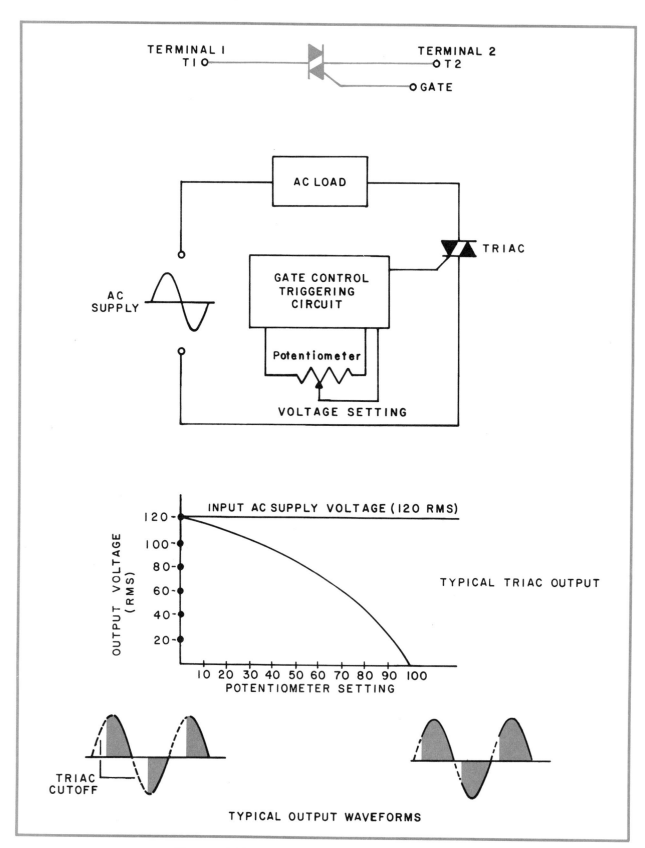

Figure 16-34 Block diagram of typical triac circuit with output wave forms and graph.

Most variable voltage outputs of this type use a full wave triac output to vary the voltage. Figure 16-34 illustrates a typical triac circuit with output wave forms and graph. The triac can vary the voltage by adjusting the point on the AC wave at which the triac is turned on. The triac in this application is very much like the SCR used to control the speed of the DC motor. In fact, the triac is two reverse-parallel connected SCRs connected to allow the AC sine wave to pass in both directions at a controlled level. As with an SCR, the triac's output is controlled by varying a potentiometer in the gate circuitry. This same basic triac circuit is also used to control the heating or light output of heating elements and incandescent lights. If you have a light dimmer at home, it is probably a triac circuit of this same design.

Motor Drives

As we have seen, the standard squirrel-cage induction motor runs at a constant speed for a given frequency and number of poles. The most common and economical running speed of a squirrel-cage motor is about 1800 rpm (lower speeds would mean the addition of more poles to reduce the speed). Because there are many applications that require some speed other than 1800 rpm, but not a variable speed control, some means must be provided to match the motor output speed to the lower or higher speed required by the load without changing the running speed of the motor. This can easily be accomplished with pulleys connected to belts, chains, or gear drives. These belts, chains, or gears are used for smooth speed changes between the motor drive and machine drive.

For any application, a pulley can be used to change the output speed of the motor, provided the manufacturer's limits are not exceeded. To determine the pulley size needed, the speed of the motor (drive rpm), the speed of the machine that is to be driven (driven rpm), and the diameter of the pulley on both the drive motor and driven machine must be considered. The following equation can be used to determine any unknown if three knowns are given.

$$\frac{\text{Drive rpm}}{\text{Driven rpm}} = \frac{\text{Driven Pulley Diameter}}{\text{Drive Pulley Diameter}}$$

If, for example, in Figure 16-35, the drive motor ran at 1800 rpm with a 6″ pulley, and the driven

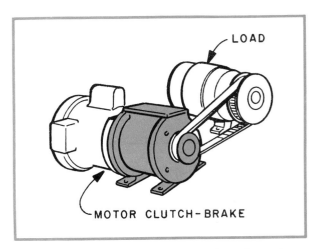

Figure 16-35 The rpm required for the driven machine is obtained by the correct motor rpm and pulley sizes.

machine was to run at 900 rpm, the driven pulley size could be found as follows:

$$\frac{\text{Drive rpm (1800)}}{\text{Driven rpm (900)}} = \frac{\text{Driven Pulley Diameter (X)}}{\text{Drive Pulley Diameter (6″)}}$$

$$\frac{1800}{900} = \frac{\text{X}}{6″} \quad \text{or} \quad \frac{18\cancel{00}}{9\cancel{00}} = \frac{\text{X}}{6″}$$

By cross multiplying and then dividing, the required pulley size is found to be 12:

$$\text{X} = 12″.$$

If the drive motor or driven machine does not have a pulley, a reasonable pulley size can be selected for one or the other and the equation used to solve for the unknown size.

For very low speeds that are required in some applications, a gear-motor or motor connected to a gear drive can be used. Gearmotors are designed with output speeds as low as 1 rpm. Gearmotors and geardrives work on the same simple gear reduction principles of most clocks and watches.

In selecting and using gearmotors, consideration must be given to the application of the motor. This is because mechanical overloads are more likely to cause failure of gears than failure of the motor. Gears are sized for maximum peak and not average loads; they simply cannot take overloads, even for short periods of time. To solve this problem, it is usually a good idea to size the gear motor one size larger and/or use a motor that is designed to withstand the overload shock conditions.

1. What is the basic operating principle of friction brakes?
2. In friction braking, why are large brake shoe linings required?
3. How is braking torque determined?
4. What is the basic operating principle of plugging?
5. Why is interlocking important in plugging circuits?
6. Name two reasons that some types of motors can not be plugged.
7. When a timing relay is used for plugging, what type of timer is used?
8. What is the basic operating principle of electrical braking?
9. Why is interlocking important in electrical braking?
10. What is the basic operating principle of dynamic braking?
11. What is the relationship of force and work?
12. What is the relationship of power and horsepower?
13. What is the relationship between torque, speed, and horsepower?
14. What does an NEMA design letter indicate?
15. What is the relationship of speed and torque in a DC series motor?
16. What is the relationship of speed and torque in a DC shunt motor?
17. What is the relationship of speed and torque in a DC compound motor?
18. How is the speed of a DC motor controlled below the base speed of the motor?
19. How is the speed of a DC motor controlled above the base speed of the motor?
20. What determines the speed of an AC motor?
21. What is the synchronous speed of an AC motor with eight poles?
22. What is meant by compelling circuit logic?
23. What is meant by accelerating circuit logic?
24. What is meant by decelerating circuit logic?
25. What is the main difference between an inverter and a converter?
26. If a drive motor ran at 1200 rpm with a four-inch pulley, and a 10-inch pulley was connected to the driven machine, what is the speed of the driven machine?

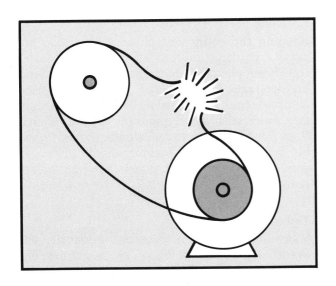

17 Preventive Maintenance and Troubleshooting

Preventive maintenance is an important part of a complete maintenance program. Preventive maintenance consists of inspecting, cleaning, and testing components before failure. When troubleshooting the technician is inspecting, cleaning, testing, and replacing components after failure. Although many of the techniques of preventive maintenance are similar, troubleshooting usually involves the cost of down time and loss of production.

PREVENTIVE MAINTENANCE

In modern industrial plants, preventive maintenance has become an absolute must. Preventive maintenance can no longer be thought of as just another good idea if time and personnel permit. Today's industrial plants and assembly lines are turning out products faster and more economically than in any time in the past. This means that shutdowns of even short periods of time can be extremely costly. Because of this, preventive maintenance has come to such a level that, without it, there is always lower productivity with less and less profit. In today's fast-moving, highly competitive industrial world, this deadly combination is enough to make many companies close.

Preventive maintenance, in theory, is very simple: keep the machines, assembly lines, production, and plant operating with as little (or no)

shutdown as possible by preventing trouble before it happens.

As recently as a few years ago, that simple idea was just about impossible to maintain. The reason was simple—it cost far more in time and personnel than it seemed to be worth. The job of the maintenance department was almost always one of repairing equipment that was "down" and installing new equipment. Unfortunately, equipment was always "down" (with production slowed or stopped) and installation of new equipment was always weeks or months behind schedule. This combination usually meant that preventive maintenance was totally lacking.

Even today, with an increased need to keep production moving, it is difficult or impossible to have the personnel trained for much of a preventive maintenance program. Fortunately, preventive maintenance can today take on a totally new meaning and not take up the time and personnel it once did. This new meaning is made possible by the introduction of a number of different, very inexpensive monitors that can be the eyes and ears of the maintenance department. These monitors can detect and react to almost any problem before it becomes major.

For example, they can monitor for voltage or phase unbalances, voltage losses, phase reversals, over or under voltages, currents or temperatures, loss of a pump's prime and other conditions that may be signs of trouble to come.

Although these monitors are for the most part new to the electrician and their internal principles of operation somewhat complicated, they are very easy to install and operate. They can be installed on new or old equipment and are designed to signal the maintenance department of trouble and take preventive measures until the maintenance personnel arrive. They should become as much a part of your troubleshooting as your volt, amp, and ohm meters are because they will be taking these measurements for you on a 24-hour-a-day basis.

The importance of a good preventive maintenance program cannot be overemphasized. The purpose of a preventive maintenance program is to:

1. Maintain the equipment in such condition as to insure uninterrupted operations for as long as possible.

2. Maintain the equipment in such condition that it will always operate at the highest possible efficiency.

3. Protect the equipment from dirt, dust, moisture, corrosion, electrical and mechanical overloads.

4. Maintain good records of all maintenance work to establish maintenance needs and priorities.

The preventive maintenance program includes the following things.

Inspection

Inspection is one of the most important parts of a good preventive maintenance program. Careful observation of all equipment will usually uncover evidence of a problem before it causes down time. In most cases, time can be saved if problems are corrected before they lead to major breakdowns. Inspection consists of observation for signs of overheating, dirt, loose parts, noise and any other signs of abnormalities.

Cleaning

Keeping the equipment clean, both inside and outside, is important for good operation. Keeping equipment clean will help eliminate overheating, high-voltage leakage, and breakdowns. Most equipment can be cleaned by blowing the dust and dirt away with a *low* pressure dry air stream. Care must be taken to remove the power if possible before cleaning. Cleaning should always be done on any equipment that is serviced for any reason.

Tightening

Vibration will result in loose connections that will eventually cause problems. The importance of a firm mounting and tight connections cannot be overemphasized. All connections should be tightened firmly, but it is equally important not to tighten beyond the pressure for which the connection is intended. Using correct tools of the proper size is usually the answer to most tightening problems.

Adjusting and Lubricating

Routine maintenance such as adjusting and lubricating equipment is part of a good preventive maintenance program. Lubrication of bearings in motors and other rotating equipment will help to eliminate wear and heat. Adjustments in equipment that has been in operation for some time will assure that the equipment will continue to operate properly. In making adjustments and adding lubrication, the electrician must follow the manufacturers' manual and recommendations.

Keeping the Equipment Dry

Electrical equipment operates best in a dry atmosphere. Moisture on copper and other metal surfaces used in electrical equipment can cause corrosion and rust, which lead to higher resistance and heating. Quite often the moisture itself causes leakage currents or short circuits in the equipment. The important thing to remember in keeping equipment dry is to use the correct enclosure for the application.

Electronically Monitoring the Power Circuit

Electronic power monitors are available and easily installed to monitor each of the following conditions.

Phase Unbalance. Unbalance of a three-phase power system will occur when single phase loads are applied, causing one or two of the lines to carry more or less of the load. On new installations of three-phase power systems, normally careful attention is given to balancing loads. However, as more and more single phase loads are added to the system, an unbalance begins to occur.

This phase voltage unbalance causes three-phase motors to run at temperatures beyond their published ratings. These high temperatures soon result in insulation breakdown and other such devices will usually not detect this gradual unbalance and therefore do not afford proper protection.

Phase Loss. The total loss of one of the three phases is an extreme case of a phase unbalance. This condition is generally known as "single-phasing". The most serious result of this condition is that it can go undetected on most systems for a long enough time to burn out a motor. A three-phase motor running on a single phase will continue to run, drawing all its current from two of the lines. In most all cases, this condition will be undetectable by measuring the voltage at the motor terminals because the open winding in the motor is generating a voltage almost equal to the phase voltage that was lost. In this case, the phase angles will be displaced sufficiently to be detected by the electronic power monitor.

Phase loss can occur when a single-phase overload condition causes a fuse to blow, when a three-phase circuit is struck by lightning, or when a mechanical failure within the switching equipment on machinery has occurred. Attempting to start a three-phase motor on a single phase will cause the motor to draw locked-rotor current. Thermal overloads are not always capable of preventing damage to the motor under these conditions.

Phase Reversal. Reversing any two of the three phases may cause damage to driven machinery or injury to personnel. This can occur when modifications are made to power distribution systems, or when maintenance is performed on cabling and switching equipment. The *National Electrical Code®* requires phase reversal protection on all equipment transporting people, such as moving walkways and escalators.

Figure 17-1, top, illustrates a typical three-phase power monitor that is designed to continu-

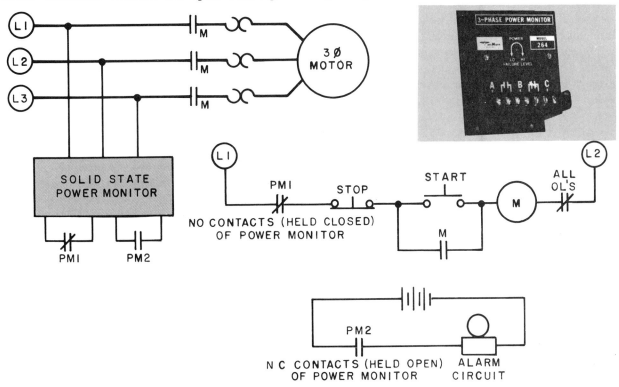

Figure 17-1 Typical three-phase power monitor that is designed to continuously monitor three-phase power lines for abnormal conditions such as loss of any phase, low voltage on any or all phases, and phase reversal. How this power monitor is connected into the power and control circuit is also illustrated. (Time Mark Corporation)

ously monitor three-phase power lines for abnormal conditions such as loss of any phase, low voltage on any or all phases, and phase reversal. Figure 17-1, bottom, illustrates how this power monitor is connected into the power and control circuit. As illustrated, the power monitor has three terminal connections that are used to connect the device to the power lines. If the electronic circuit senses any abnormal conditions from these three lines, the monitor de-energizes a set of NO and NC contacts. The NO contact (held closed during normal operation) is connected in series with the stop pushbutton in the control circuit. The contact, when opened (abnormal condition), will turn off the circuit in the same way that the stop on overload contact would. The NC contact (held open) is connected to an alarm circuit to signal the problem.

TROUBLESHOOTING

Troubleshooting is one of the most rewarding experiences an electrician can have. Unlike any other part of electrical work, troubleshooting is hardly ever routine. When called upon to service a machine or problem, one never knows what to expect. The experience will always be rewarding and educational. With each repair job, the electrician will always gain more experience and knowledge. Always look forward to any troubleshooting that can be done; it is never as hard as it seems. In almost all cases, troubleshooting is straight forward and usually simple. This is because in most cases only one problem exists. To help troubleshoot problems quickly, good tools and test equipment are essential. Refer back to Unit 1, "Electrical Tools, Instruments and Safety," to review hand tool safety. For test equipment review the list of rules provided in this part of the unit.

Basic Rules in Using Test Instruments

1. Always read the manufacturer's instructions first. Save the instructions in a file or safe place for future reference.

2. In using any instrument that has several scales, always start with the highest scale available in order to prevent overloading due to unknown values.

3. Always remove the component to be tested or disconnect the line voltage from the circuit before making any resistance measurements.

4. Never try to use a test instrument beyond its rated capacity.

5. When using a clamp-on instrument, close the jaws tightly. All clamp-on instruments are designed to be clamped around one conductor only. Clamping around two conductors neutralizes the fields and no reading can be taken.

6. All leads must be insulated.

7. For accurate reading, all connections must be tight.

8. Check to make sure that any instrument fuses or batteries are in working condition. If the needle cannot be zeroed on the ohm scale, a new battery is needed.

9. In a clamp-on instrument, the needle should read in the upper half of the scale for greatest accuracy.

10. Always apply basic rules of electrical theory when testing any circuit.

Basic Troubleshooting Procedure

Often times in electrical circuits it takes more time to find the trouble than it does to correct it. For this reason, it is important to learn how to troubleshoot a circuit as quickly as possible. Troubleshooting consists of two steps: (1) finding the problem, and (2) correcting the problem.

It is the first step, finding the problem, that all maintenance personnel must master. The second, correcting the problem, can usually be accomplished by replacing the defective part. If the expertise is available in the maintenance department, the defective part can be serviced and repaired.

Finding the problem will not take much time if the electrician will follow a few basic troubleshooting techniques. These techniques will allow the electrician to approach the problem in a logical manner that will help to isolate the trouble.

Basic Troubleshooting Techniques

The first step in troubleshooting any circuit is to have a clear understanding of the circuit and its function before starting. If you do not understand how a circuit functions when there is no problem, it is almost impossible to troubleshoot the circuit because you may not know what you are looking for. This does not mean you must have a total understanding of the circuit, but it does mean that you must have a general overall knowledge of the circuit's function.

The next step is to eliminate the obvious, no matter how simple it may seem. This includes first checking the fuses, circuit breakers, and overload resets. Careful observation will often yield recognition of out of the ordinary operation. This includes any overheating or warm areas, signs of leakage, smells and recently made changes. It is also important to obtain information from the operator who uses the equipment on a continual basis. Often he can help isolate problems more quickly or possibly forewarn you.

The next step is to isolate the problem to the control circuit, power circuit, load, or incoming power. Each of these areas, although connected and related, can be isolated one from the other to make troubleshooting easier.

Troubleshooting the Control Circuit

Troubleshooting the control circuit is not as hard as it might seem. Figure 17-2 illustrates a typical control circuit that can be used to illustrate the steps to take in troubleshooting any control circuit. Following is a step-by-step procedure.

1. Eliminate the incoming power supply as the source of problems by connecting a voltmeter or test light across L1 and L2, as illustrated in Step Number 1.

2. Since the sole function of the control circuit is to deliver power to the control coil in a prede-

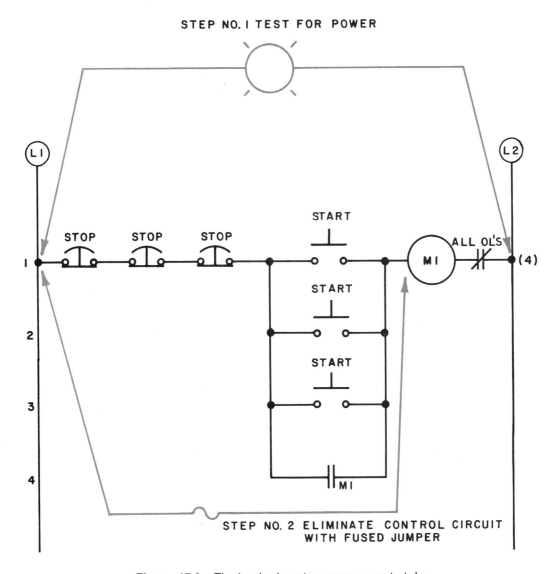

Figure 17-2 The beginning steps necessary to take in troubleshooting any control circuit.

termined manner, it can be eliminated to isolate any problems. This is accomplished by connecting a fused jumper, as illustrated in Step Number 2. (The fuse in the jumper should be larger than the current drawn by the starting coil, but small enough to protect from a short circuit). With the jumper in place, the starter should energize. If it does, the control circuit has been found to contain the problem. To further isolate the problem, remove the jumper wire.

3. Once the control circuit has been determined to be the source of the problem, finding the problem is fairly straight forward. Continue to jump each control device until the problem is found, as illustrated in Figure 17-3. A jumper is the best device to use in troubleshooting the controls because there is only one line (L1) connected to and from each control device. A test light or voltmeter will not indicate a voltage unless it can also be connected to L2, which is not part of the control circuit which is external to the enclosure containing the starter. Since the only purpose of a control switch such as a pushbutton, limit switch, float switch, etc., is to allow the electricity to pass through it when closed, the jumper wire can serve to test any switch.

This test procedure can be used to test the control circuit from the coil to the L2 side of the power source. Here the only component under test is the overload contact.

Troubleshooting the Power Circuit

In troubleshooting the power circuit, the following steps should be taken.

1. Test each of the incoming power lines as illustrated in Figure 17-4, Step Number 1. The voltage must be present and at the correct power level.

2. Eliminate the control circuit by "jumping" out the control circuit as illustrated under Step Number 2. This is not necessary on starters that provide a manual control closing of the power contacts. On this type of starter, a place is provided for a screwdriver to close the contact.

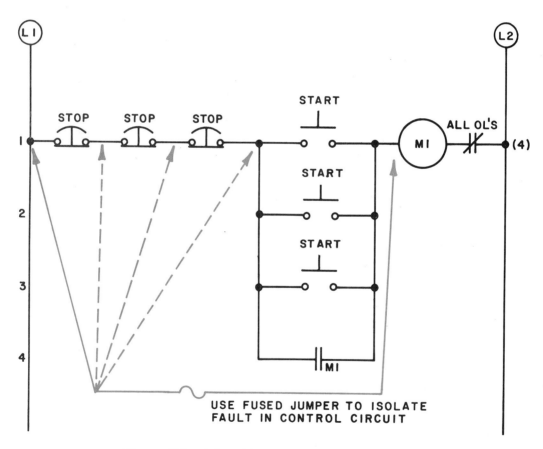

Figure 17-3 A fused isolation jumper can be used to quickly isolate faults in the control circuit.

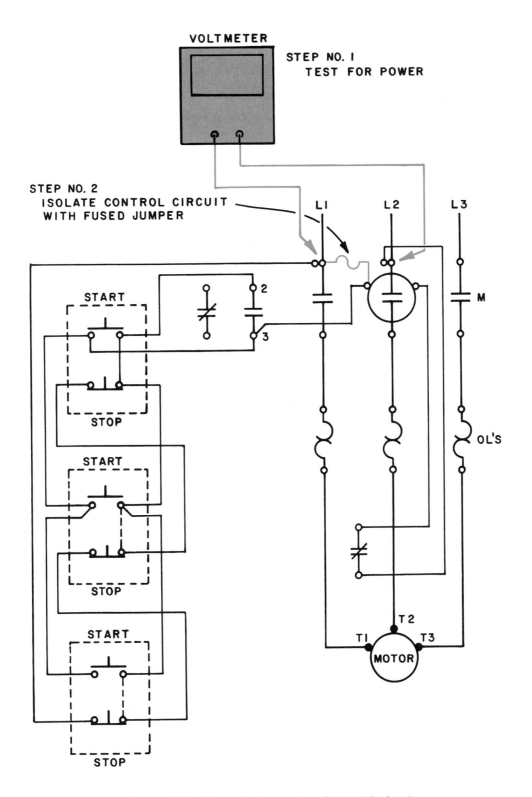

Figure 17-4 The preliminary steps for troubleshooting the power circuit.

3. With the power contacts closed as illustrated in Figure 17-5, check the outgoing power at the T1, T2 and T3 terminals. The power should be present and at the proper level. If the power is not present at the terminals, test the power coming directly out of the power contacts. If the power is present, the overload heater must be checked.

Checking the Incoming Power Supply, Fuses and Circuit Breakers

One of the most common troubles found in all electrical circuits is a blown fuse or tripped circuit breaker. This is because the fuse or breaker is intentionally sized to open or trip in case of trouble. The fuse or breaker that has opened or tripped has done so because of some fault on the line. This fault may be in any load connected to the line, combination of loads, the wiring, or even the fuse or breaker itself. In any case, the fuse or breaker that has opened and removed the power will have to be found and replaced or reset.

Although circuit breakers usually have an indicator to tell when they are tripped, the following test can still be used for testing the incoming

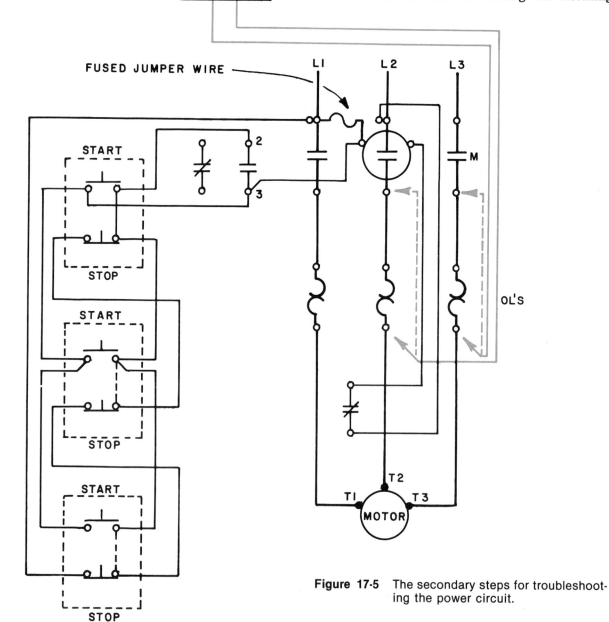

Figure 17-5 The secondary steps for troubleshooting the power circuit.

power to both breakers and fuses that are faulty.

Before a suspected fuse or breaker is tested, the incoming power supply must be checked to make sure power is coming to the fuse or breaker. A mistake often made is to assume that the power is present on the incoming lines. Remember that all incoming power lines are fed by other fuses and breakers through the distribution system and that the power may be broken at any point on the distribution system.

Figure 17-6 illustrates how the incoming power can be tested. Although a test light can be used in most applications, a voltmeter should be used. The reason a voltmeter should be used is that a routine check of the power should be made at this time. This includes checking the voltage between each pair of power leads. The incoming supply voltage should be within 10% of the voltage rating which appears on the nameplate of the loads that are connected to the line. If there is a high or low voltage, a definite problem has been found even if it is not part of the original problem. If voltage is present and there is a correct reading on each line, then the enclosure should be tested to make sure it is grounded. To test for ground, connect one side of the voltmeter or test light to a metal (unpainted) part of the enclosure and touch the other side to each of the line terminals. A voltage difference should be indicated on the voltmeter or the test light should light.

If the incoming power is present, the fuse can be tested for an open circuit. Figure 17-7 illustrates how to test for an open circuit fuse. In this test, the power must be applied to the fuses. To prevent any possible feedback through the connected loads, the loads on the load side of the fuses may have to be disconnected, although this is usually not required. After this is done, one side of the voltmeter or test light should be connected to one line side and another (never the same) load side as illustrated. A voltage reading indicates a good fuse; no voltage reading indicates a bad fuse. This procedure must be repeated for each fuse in the line.

If an ohmmeter is available, a fuse can be easily checked for an open. Figure 17-8 illustrates how an ohmmeter is used to check for a bad fuse. After the fuse is removed from the circuit, the ohmmeter is connected to the fuse. The ohmmeter should be placed on the lowest (R1) scale and a reading taken. An infinite reading indicates that the fuse is open and bad. A zero reading indicates that the fuse has continuity and is good.

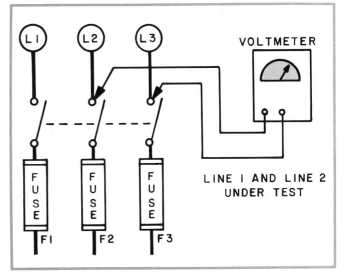

Figure 17-6 How to test for incoming power using a voltmeter.

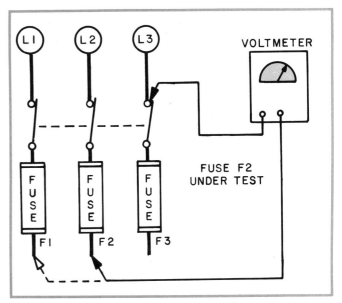

Figure 17-7 How to test for an open fuse using a voltmeter.

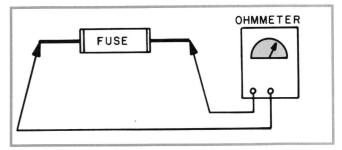

Figure 17-8 How to test for an open fuse using an ohmmeter.

Testing the Control Transformer

The control transformer is used in the circuit to step down the supply voltage in order to provide a safer voltage level for the control circuit. If there is a problem in the control circuit that appears to be related to the power supply, the control transformer should be checked. Following is a procedure for testing the control transformer.

1. With the power supply energized, check the input and output voltages of the transformer as illustrated in Figure 17-9. Both the input and output voltages should equal the transformer's nameplate rating within five percent (10% at the very maximum). If the voltage is within the rating or proportionally low, the transformer is good as far as the voltage is concerned.

2. Each transformer is capable of delivering a limited current output at a given voltage. The amount of current times the voltage will be the power limit of any transformer. This power limit is listed on the nameplate of the transformer under KVA rating. This rating indicates the apparent power the transformer can deliver. If this limit is exceeded, the transformer will overheat and the control circuit will not function properly. It is important to remember this every time a new control load such as a relay, counter, or timer is added to the circuit. With each additional load the transformer comes closer to reaching its limit. Figure 17-10 illustrates how an ammeter can be used to determine the current drawn by the control circuit. When this current reading is multiplied times the voltage reading, the apparent power drawn by the control circuit is determined. If the VA drawn is greater than the rating of the

transformer, a larger size transformer is required.

3. On new transformer installations, or if a ground problem is suspected, a ground test should be made on the transformer. Figure 17-11 illustrates how a test circuit can be used to check for a ground. Connect one lead of the test circuit to the transformer. (Do not connect to a painted or varnished surface). Then connect the second lead of the test circuit to each lead of the transformer on both the primary and secondary. If the light bulb lights on any lead, the transformer is defective and must be replaced. (*Note:* This test can also be used to check for a ground on motors.)

Troubleshooting Electric Motors

Electric motors are essentially reliable machines and require little maintenance in comparison to the rest of the circuit. It is not uncommon for a typical motor to run satisfactorily for 20 or more years without repair. Examples of how dependable motors are can be found around your home as well as in the plant. For example, refrigerators, air conditioners, and heating systems often last more than 20 years without any work to the motors at all. Any repair that is required almost always involves the related components. However, when a motor fails, the electrician must be able to identify and correct the problem.

Today it is far more common to replace a motor that has failed than it is to repair it. If any repair is required, the motor is almost always replaced with a new motor and the bad motor sent out to be repaired or repaired in the maintenance shop by trained personnel. Smaller motors simply cost

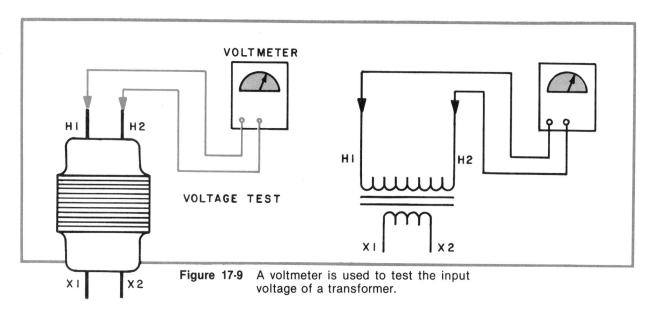

Figure 17-9 A voltmeter is used to test the input voltage of a transformer.

more to repair than to replace, and most larger motors can be replaced with a more energy efficient motor that more than justifies the extra expenditure.

The cost involved with a motor that has failed is almost always in downtime of the operation and maintenance time involved. For this reason, two major factors are important in motor maintenance. They are:

1. Locating the motor as the problem as quickly as possible.

2. Determining the reason the motor failed, to eliminate the cause and thereby prevent the problem from returning.

Locating the Motor as the Problem as Quickly as Possible

When a motor has failed to start, it is important to determine if the fault is with the motor, incoming power, or load connected to the motor. There are four basic categories that most motor problems will fall under. They are:

1. The motor fails to start.

2. The motor runs very hot.

3. The motor runs slower than the normal speed.

4. The motor runs noisily.

Regardless of which category the motor problem falls under, there are some basic preliminary

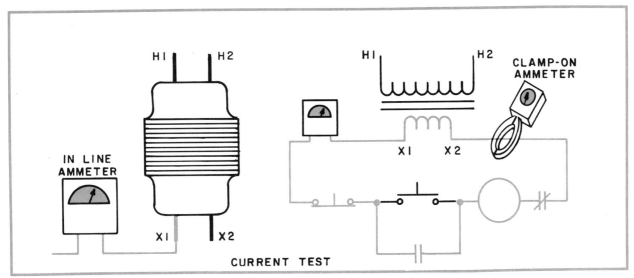

Figure 17-10 An in line ammeter or clamp-on ammeter can be used to determine current drawn by the control circuit.

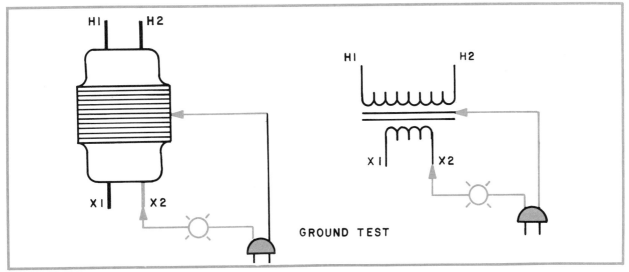

Figure 17-11 A lamp test circuit can be used to test for a ground condition.

steps that should be taken to determine the problem. These steps are listed below.

1. As illustrated in Figure 17-12, check the incoming voltage to make sure that it is the correct voltage level (that it matches the motor nameplate). If the voltage is within 10% of the nameplate voltage, the incoming power can temporarily be eliminated as the source of trouble. If the voltage is not within 10% of the nameplate voltage, the motor can temporarily be eliminated as the source of trouble, and the incoming power must be investigated. This is true even if the motor is burned out because a fault is located somewhere ahead of the motor and replacing the motor will not eliminate the fault.

2. If the power connecting the motor is correct, the next step is to turn off and lock out the incoming power to the motor. With the power off, try to rotate the shaft of the motor. If the motor is connected to a load that makes rotation impossible, the load must be disconnected from the motor. If the motor and the load that the motor is connected to rotate freely, the load can temporarily be eliminated as the source of trouble and the motor should be replaced and the probable cause of failure investigated. If the load had to be disconnected from the motor, the motor can temporarily be turned back on. If the motor runs with no problems with the power applied and load disconnected, the motor can temporarily be eliminated as the source of trouble. In this case, the load must be investigated for possible jam-ups.

3. In each application in which a motor is replaced with a new one, careful observation is required. If the original problem exists, turn off the power immediately. Disconnect the new motor and proceed with locating the problem. If the new motor appears to work in the system, start to determine the reason the first motor failed so that the same problem will not damage the new motor.

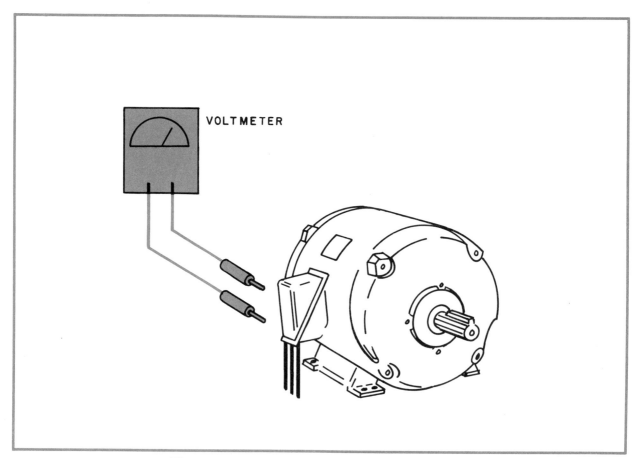

Figure 17-12 How to use a voltmeter to check the incoming voltage to make sure it is the correct voltage level (that it matches the motor nameplate).

Determining the Reason the Motor Failed

It is important to realize that it is normal for a motor to fail after many years of good service. Every motor, no matter how well it is protected, will eventually fail. It is just a matter of time. However, any motor that fails before its time must be investigated to determine why it failed.

By far the major cause of motor failure is the deterioration of the insulation. Over the past years, better motor insulation has been manufactured; but the insulation is still by far the weakest link in any motor. The deterioration and failure of the insulation is almost always caused by excessive temperatures. A rule of thumb commonly applied to motor insulation life is that for every 10° C rise above the rated temperature of the motor, the winding life of the motor will be reduced by one-half. Following are some of the common causes of overheating in electric motors.

Excessive Mechanical Load. When the motor is called upon to do more work than it is designed to do, the motor attempts to do the work by drawing more current. The draw of more current will result in increased motor temperature. The heat build up in the motor is proportional to the square of the current drawn. Should the overload persist, the overload relay should disconnect the motor from the line. Although the overload relay has disconnected the motor from the line, this should be the exception and not a continual event. In the case of an overload relay with an automatic reset, the motor can cycle with the overload until enough heat builds up to destroy the motor windings.

Whenever you replace a burned out motor with a new one, you can test the amount of work the motor is required to deliver by using a clamp-on ammeter as illustrated in Figure 17-13. Compare the amount of current drawn by the motor to that of the current rating on the nameplate of the motor. The closer the reading on the ammeter is to the rated current, the more the motor is working. The rating listed on the motor is the maximum current that should be drawn by the motor to deliver full power and should not be considered the normal current draw of the motor. If the new motor load current is almost equal to or greater than the listed current rating, a larger motor may have to be installed. The larger motor may be needed to handle an increased load resulting from any change from the original installation (change

in pulley or gear sizes, wear of driven equipment, etc.).

A Low or High Line Voltage. Motors are usually designed for operation with a 10% range above or below their rated voltage. Motor life will be shortened if a motor is operated at a voltage above or below the limits. The effect of low voltage on motors is a reduction in starting torque and an increased full load temperature rise. The effect of high voltage on motors is an increased torque and starting current. The increased current will cause a saturation of the iron and a consequent steep rise in current and heating. In troubleshooting or replacing a motor, always check the applied voltage against the motor's nameplate voltage.

Unbalanced Voltages. Unbalanced voltage applied to a motor is a source of motor heating that is not generally recognized. An unbalanced voltage may be more damaging than a low or high voltage. An unbalanced voltage can result in a current unbalance in the motor winding equal to several times the voltage unbalance. The result of the unbalance will be increased heat, a decrease in torque, noise and vibration. Always check to make sure the voltages applied to the motor are at

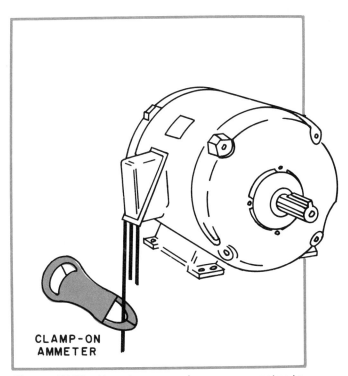

CLAMP-ON
AMMETER

Figure 17-13 How to use a clamp-on ammeter to check current drawn by motor against that which is stated on the motor nameplate as normal.

the same voltage level. Figure 17-14 illustrates a simple way to test a three-phase line. Each phase is measured to ground, and the three voltage readings are compared. The three readings should be about equal.

Single-Phasing. Single-phasing is caused when one of the lines on the three-phase power is opened. The line may open from a mechanical breakage of a conductor, burned or open switch contact, or blown fuse. If the single-phasing

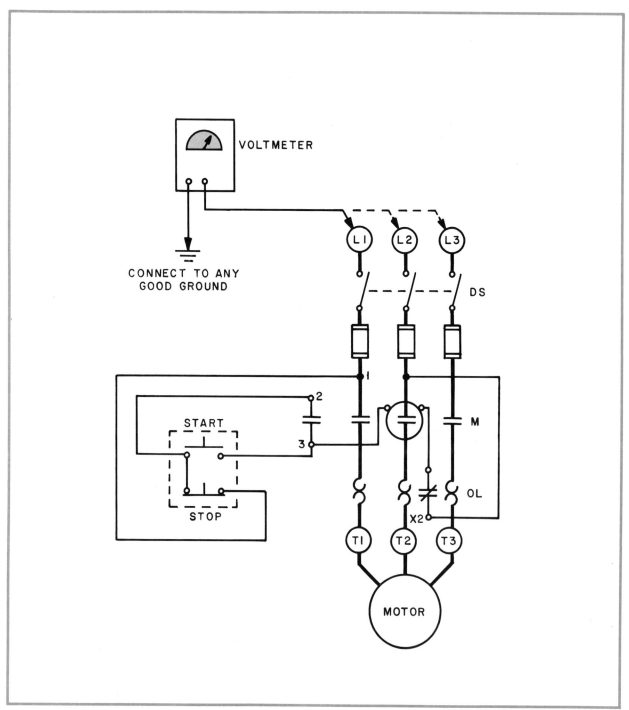

Figure 17-14 This shows a simple way to test a three-phase line using a voltmeter. Each phase is measured to ground and the three voltage readings are compared. The readings of all three lines should be about equal.

occurs while the motor is running, the motor slip will increase and the motor will continue to run as long as the load torque is not greater than the motor torque. This will cause severe damage to the motor—due to increased heat and vibration—if the motor is not disconnected.

If single-phasing occurs while the motor is at standstill, the motor will hum and probably not start the load. It is important for the electrician to check for single-phasing on the power line when a motor has a fault. If the electrician assumes that a humming motor which does not start is bad and replaces it, he will find out that the new motor will have the same problem because of single-phasing. Single-phasing is also very damaging for one other reason. With the loss of one phase, a three-phase motor can start in either direction. This may cause severe damage to the drive unit the motor is connected to.

Excessive Duty Cycle. Since motors draw a much higher starting current than running current, any motor that is repeatedly plugged, jogged, or reversed can be damaged. Heat build-up in the motor is also caused because much of the self-cooling effect of the revolving motor is lost. If a motor is to have an excessive duty cycle, it will be necessary to use a motor of a higher horse-power rating than is simply required to drive the load.

High Ambient Temperature. A high ambient temperature will produce the same heating effect in the motor as a high current would. This means that motor life will be shortened in high ambient environments unless the motor is cooled or de-rated. High ambient temperature can occur when the motor is in a relatively enclosed environment or when the motor is located in an area subject to external heat. Since motor overloads will not detect the high ambient temperature because they are monitoring line current, the high temperature can burn the motor out without being detected.

Insufficient Ventilation. When air passages within the motor become clogged due to dirt, lint, or other material, the natural cooling effect of the motor is lost. In this situation it is possible for the motor to reach excessive temperatures without drawing excessive current. This means that, if gone undetected, the motor is sure to burn out in a short time. Always clean the motor whenever the passages appear to need it. This is why any good service person that works on any motor-run appliance in your home always vacuums the motor as a matter of routine maintenance.

Remarking the Connections of the Three-Phase Induction Motor

The three-phase induction motor is by far the most common motor used in industrial applications. Part of the reason it is used in so many applications is that it will operate for many years with little or no maintenance required. It is not uncommon to find three-phase induction motors that have been in operation for 10 to 20 years in certain applications. One of the problems the electrician may find when troubleshooting a motor that has been in operation for a long period of time is that the markings of the external leads are sometimes defaced or missing. This may also happen to a new or rebuilt motor that has been in the maintenance shop for some time. To insure proper operation the electrician must remark each motor lead before troubleshooting and reconnecting the motor to a power source.

There are two basic types of motors that are the most common three-phase types. They are the three-phase, three-lead single voltage motor and the three-phase, nine-lead dual voltage motor. Either the three-lead or nine-lead motors may be internally connected for a Wye or Delta connection.

The three-lead single voltage motor is of no particular problem in remarking. This is because the three leads of the motor can be marked as T1, T2 and T3 in any order without mismarking the motor leads. The motor can then be connected to the rated voltage and allowed to rotate. If the rotation is in the wrong direction, any two leads may be interchanged (interchanging T1 and T3 is the industrial standard to follow).

Test for Wye or Delta Connections

The problem of correctly remarking the motor leads comes with the dual voltage motor. The standard dual voltage motor will have nine leads extending from the motor and be internally connected as a Wye or Delta motor. The first step is to determine if the nine-lead motor is internally connected as a Wye or Delta motor. To do this, an ohmmeter is used to measure resistance or a test light circuit is used.

It is not difficult to determine if a dual voltage motor is a Wye or Delta motor because the Wye motor will have four separate circuits and the

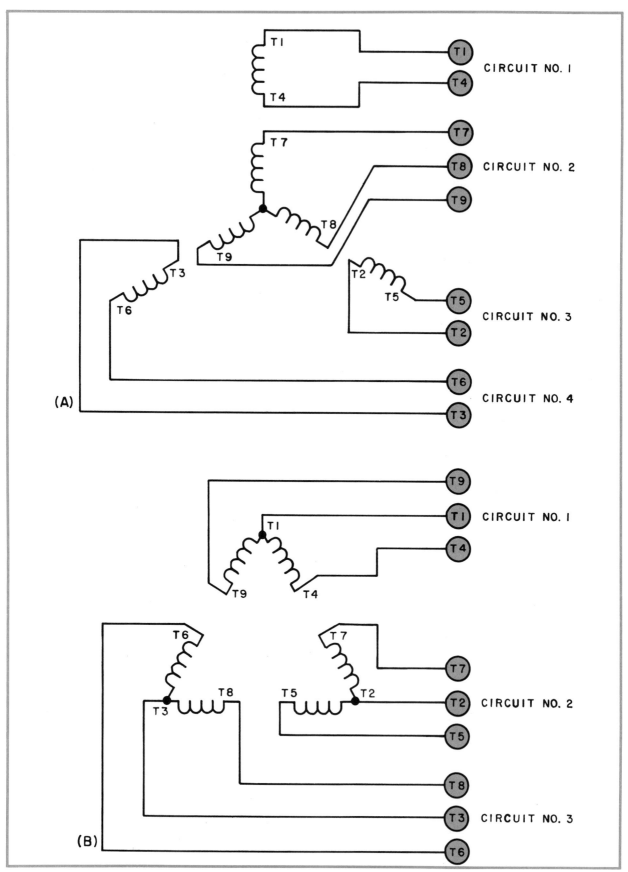

Figure 17-15 The separate circuits found in the Wye motor and the Delta motor.

Delta motor will have three separate circuits. Figure 17-15 illustrates the four separate circuits found in the Wye motor and the three separate circuits found in the Delta motor. As illustrated, the Wye motor will have three circuits of two leads each (T1-T4, T2-T5, T3-T6) and one circuit of three leads (T7-T8-T9). The Delta motor will have three circuits of three leads each (T1-T4-T9, T2-T5-T7, T3-T6-T8).

To determine a circuit on an unmarked motor using an ohmmeter, refer to Figure 17-16. With no power connected to the motor, connect one lead of the ohmmeter to any motor lead and temporarily connect the other lead to each of the remaining motor leads. A reading of resistance other than infinity will indicate a circuit which is represented by the motor's coil.

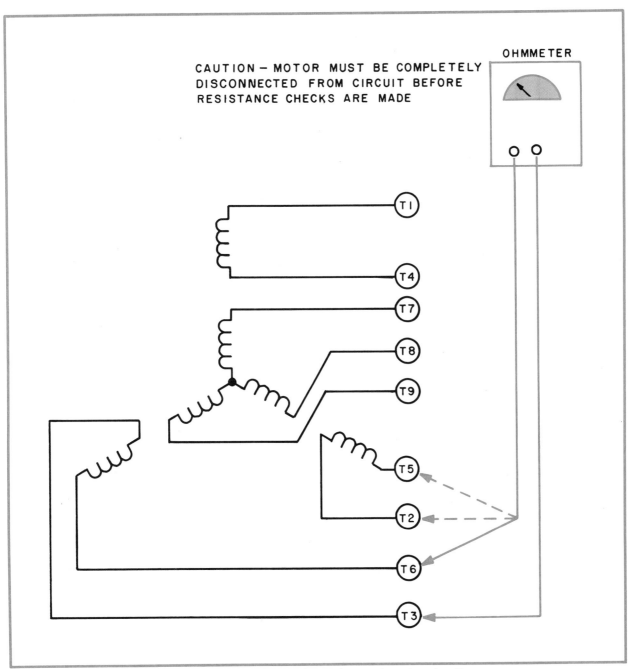

Figure 17-16 How to determine each motor circuit on an unmarked motor using an ohmmeter.

To determine a circuit on an unmarked motor using a test light circuit, refer to Figure 17-17. With power applied to the test light circuit as illustrated, connect one side of the test circuit to any motor lead and temporarily connect the other lead to each of the remaining motor leads. When the light glows, a circuit which is represented by the motor coil has been found. For each connection that indicates a circuit, mark that circuit by tapping or pairing the leads together. It is important that once a pair of leads is found they be checked with all the remaining motor leads to determine if the circuit is a two- or three-lead circuit. If three circuits of two leads and one circuit of three leads is found, the motor is a Wye motor. If three circuits of three leads is found, the motor is a Delta motor.

Remarking the Dual Voltage Wye Motor

The first step in remarking the leads of a Wye connected motor is to take the one three-lead circuit and mark the leads T7, T8 and T9 in any order. Separate the other motor leads in pairs, making sure none of the wires are touching each other. Next connect the motor to the correct disconnected supply voltage as illustrated in Figure 17-18. This means to connect T7 to L1, T8 to L2, T9 to L3. The correct supply voltage is the lowest voltage rating of the dual voltage rating given on

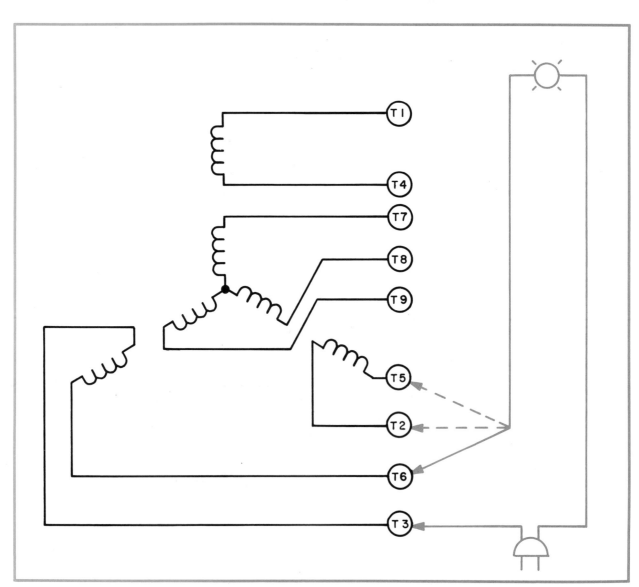

Figure 17-17 How to determine each motor circuit on an unmarked motor using a test light.

the nameplate of the motor. This lower voltage will usually be 220 volts, since the standard dual voltage motor operates on 220–440 volts. For any other voltages, all the following test voltages should be changed in proportion to the motor rating.

Next, turn on the supply voltage and let the motor run. The motor should run with no apparent noise or problem. If the motor is too large to be started by connecting it directly to the line, the starting voltage should be reduced through a reduced voltage starter as in the regular operation of the motor.

The next step is to take a voltmeter that is set on at least a 440 AC voltage scale and measure the voltage across each of the three open circuits while the motor is still running. Care must be taken in measuring the voltage because you are working with high voltage and a motor that is running. For this reason, insulated test leads must be used and you must connect only one test lead at a time. The voltage measured should be about 127 volts (or slightly less) and should be the same on all three circuits.

The reason a voltage will be read even though the two-wire circuits are not connected to the power lines is that the voltage applied to the three-lead circuit will induce a voltage through transformer action into the two-wire circuits.

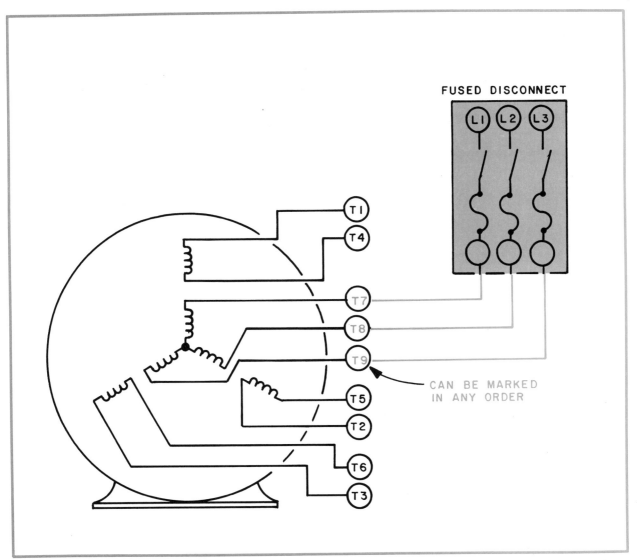

Figure 17-18 The three common leads should be marked T7, T8 and T9 in any order and then connected to the supply voltage.

Make a drawing of the three-phase Wye motor and mark your voltage readings on that drawing as illustrated in Figure 17-19.

Next, take one lead of any two-wire circuit and connect it to T7 and the other lead to one side of the AC voltmeter as illustrated in Figure 17-20. Temporarily mark the lead connected to T7 as T4 and the lead connected to the voltmeter as T1. Take the other lead of the AC voltmeter and connect it first to T8 and then to T9. If the two voltages are both the same value and equal to about 335 volts, mark T1 and T4 permanently. If the voltages are unequal, you have the wrong two-wire circuits and must try another two-wire circuit until the correct measurements are found.

After the first test is completed, you will have found T1, T7 and T4. Next, take one of the two remaining unmarked two-wire circuits and connect one lead to T8 and the other lead to one side of the AC voltmeter. Temporarily mark the lead connected to T8 as T5 and the lead connected to the voltmeter as T2. Take the other side of the voltmeter and connect it to T7 and T9, measuring the voltage. As in the first test, measurements

and changes should be made until a position is found at which both voltages are equal and of a value of about 335 volts.

After the second test is completed, you will have found T2, T5 and T8. Next, check the third circuit in the same way until a position is found at which both voltages are equal and about 335 volts. After the third test is completed, you will have found T3, T6 and T9.

After each motor lead is found, turn off the motor and reconnect the motor to the supply voltage as follows. Connect L1 to T1 & T7, L2 to T2 & T8, L3 to T3 & T9, and connect T4, T5, and T6 together. After the motor is connected, start the motor and let it run. Next, check the current on each power line (L1, L2 & L3) with a clamp-on ammeter. If the current is about even and normal, you can assume the markings are correct and mark each lead permanently.

Remarking the Dual Voltage Delta Motor
Figure 17-21 illustrates the nine lead Delta motor with three separate circuits. Each separate circuit

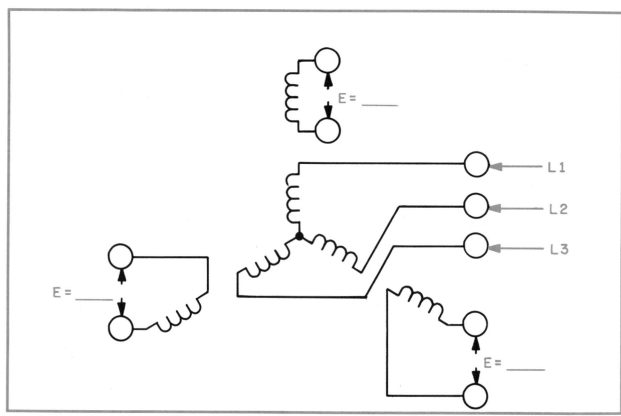

Figure 17-19 How to make a drawing of the unknown motor leads so that voltage measurements can be taken.

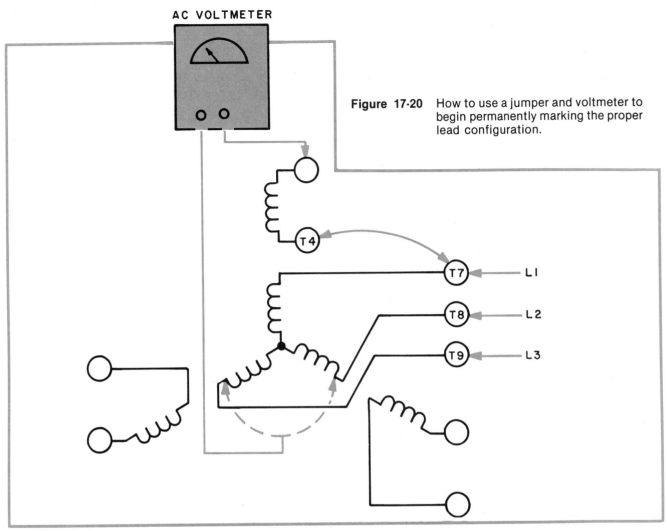

AC VOLTMETER

Figure 17-20 How to use a jumper and voltmeter to begin permanently marking the proper lead configuration.

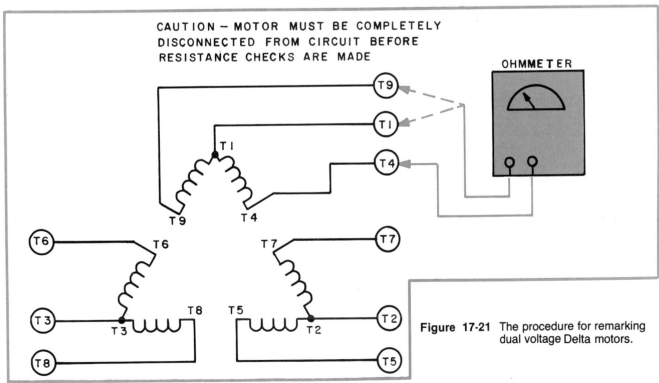

CAUTION — MOTOR MUST BE COMPLETELY
DISCONNECTED FROM CIRCUIT BEFORE
RESISTANCE CHECKS ARE MADE

OHMMETER

Figure 17-21 The procedure for remarking dual voltage Delta motors.

has three motor leads connected to make that circuit (T1-T4-T9, T2-T5-T7, and T3-T6-T8). As with the Wye motor, the first step is to find each of these three groups. Although this can be done with a test light circuit, an ohmmeter should be used with the Delta circuit. The reason for using an ohmmeter is that the ohmmeter can isolate each separate circuit and also find the center taps of each separate circuit which are leads T1, T2 and T3.

To find the center of each separate circuit, measure the resistance of each circuit. The center lead can be found by the approximate measurement of resistance, since the resistance from say T4 to T9 will be twice the resistance from T1 to T4 or T1 to T9. Using this resistance check, separate the three circuits and mark the center terminal for each circuit.

The next step is to take each of the center terminals and mark them permanently as T1, T2 and T3. Also temporarily mark the two leads with the T1 group as T4 and T9, and the two leads with the T2 group as T5 and T7, and the two leads with the T3 group as T6 and T8. Disconnect the ohmmeter at this time and remove it from the area since no further resistance measurements will be taken. This precaution is taken since the next measurements to be taken will be voltage, and any ohmmeter that is accidently connected to a voltage will be damaged.

After each terminal is permanently (T1, T2, T3) or temporarily (T4, T5, T6, T7, T8, T9) marked, separate the leads back into their original three-wire groups. Take the group marked T1, T4 and T9 and connect them to L1, L2 and L3 of a 220 volt power supply. (This should be the lower voltage rating on the nameplate of the motor with 220/440 being the standard). Each of the other six leads should be left disconnected and must not be touching, since a voltage will be transformed through the other coils into these leads even though these leads are not connected to power.

With this power applied to T1, T4 and T9, the motor can be turned on and let run. The same care should now be taken as with the Wye motor: use insulated test leads, work one-handed, and be very careful.

Set your voltmeter on at least a 460 volt AC range. Now connect T4 (which is also connected to L2) to T7. Using the voltmeter, read and record the voltage between T1 and T2. If the measured voltage is equal to about 440 volts, the lead markings for T4, T9, and T7 and T5 are correct. If the measured voltage is equal to about 380 volts,

interchange T5 with T7 or T4 with T9, measuring the voltage again. If the measured voltage is equal to about 440 volts, the lead markings are correct. However, if the new measured voltage is equal to only approximately 220 volts, interchange *both* T5 with T7 and T4 with T9 and measure the voltage again. The voltage should now equal about 440 volts and T4, T9, T7 and T5 may be permanently marked.

You should now have all leads permanently marked except T6 and T8. To correctly identify them, connect T6 and T8 and measure the voltage from T1 and T3. If the measured voltage is equal to about 440 volts, T6 and T7 may be permanently marked. If the voltage does not equal 440 volts, interchange leads T6 and T8 and measure the voltage again at T1 and T3. The measured voltage should now equal about 440 volts and T6 and T8 may be permanently marked.

The motor should now be turned off and reconnected to a second set of motor leads. This means to connect L1 to T2, L2 to T5 and L3 to T7. Restart the motor and observe the direction of rotation. The motor should rotate in the same direction as with the previous connection. After the motor has run and the direction been determined, turn off the motor and reconnect the motor to the third set of motor leads. Connect L1 to T3, L2 to T6, and L3 to T8. Restart the motor and again observe the direction of rotation. The motor should rotate in the same direction as the first two connections. If the motor does not rotate in the same direction for any set of leads, you will have to start over carefully, remarking each lead.

The motor should now be turned off and reconnected for a low voltage connection. That is L1 should be connected to T1-T6-T7, L2 to T2-T4-T8 and L3 to T3-T5-T9. Then the motor should be restarted and a current reading taken on L1, L2 and L3 with a clamp-on ammeter. If the motor current is about equal on each line, you can assume the markings are correct.

Remarking the Connections of DC Motors

Figure 17-22 illustrates the three basic types of DC motors. They are the series, shunt, and compound wound motors. All three types may have the same armature and frame, differing only in the way the field coils and armature are connected. For all DC motors, terminal markings A-1 and A-2 always indicate the armature leads. S-1 and S-2 always indicate the series-field leads. F-1 and F-2 always indicate the shunt-field leads.

If any or all of the terminal markings are missing from a DC motor, they are not hard to remark. The motor terminals can be remarked by using an ohmmeter. Take a resistance reading for each pair of wires. A pair of wires must have a resistance reading or they are not a pair. Since each DC motor type must have an armature, the field readings can be compared to the armature reading. The series-field will usually have a reading less than the armature, and the shunt-field will have a reading considerably larger than the armature. The armature can easily be identified by rotating the shaft of the motor when taking the readings. The armature will move the ohmmeter needle as it makes and breaks different windings. One final check can be made by lifting one of the brushes or placing a piece of paper under the brush. The ohmmeter will move to the infinity reading.

From this information, identification of the type of motor and terminal markings is easy. For example, if a motor has two pairs of leads (four wires) coming out, it is either a series or shunt motor. If the reading of the coil that is not the armature coil is less than the armature resistance, the coil is the series-field. If the reading is considerably larger than the armature resistance, the coil is the shunt-field.

Terminal Connections

Terminal connections are a source of some of the more troublesome problems found in the control and power circuits. This is because a loose terminal connection can cause a load to burn out from under voltage. When the burned out load is located and replaced, the trouble is thought to be taken care of, only to turn up again and again until the real problem is found.

A loose connection in a lug can cause trouble in more than one way. Any loose connection means an increase in resistance. When current passes through resistance of any kind, there is a voltage drop at the resistance point and heat develops. The voltage drop and heat produced by a loose connection can cause many problems.

If the loose terminal is on a circuit breaker, the heat at the terminal can be carried by the wire to the thermal overload inside the breaker. The heat from the loose contact, added to the current in the overload, may cause the breaker to trip on a current far below the rating. This condition may lead the electrician to incorrectly suspect an overloaded circuit or faulty breaker.

If the loose terminal develops a high enough voltage drop across it, the loads that are connected to the circuit will develop all kinds of problems. The simplest problems will be burn out of the loads; the more difficult problems will be the drop out of coils (solenoids, starters, etc.) and the resetting of timers and counters.

The heat developed at a loose connection may also cause other problems. If left long enough, the insulation around the terminal may be destroyed, leaving the possibility of a short circuit. This heat may also destroy any device that is connected to or near the loose connection.

To avoid loose connection, be careful how the wires are placed in the lugs. If the wire is solid wire, be sure the lug will clamp the wire tightly. This is especially true with aluminum wire because the aluminum is softer than copper and will not hold its shape as well. Aluminum will also expand and contract more than copper, which may cause a loose connection. If two wires are used in the same lug, make sure that both wires fit tightly. It is also a good practice to always keep an eye open for loose terminals and check possible problem areas.

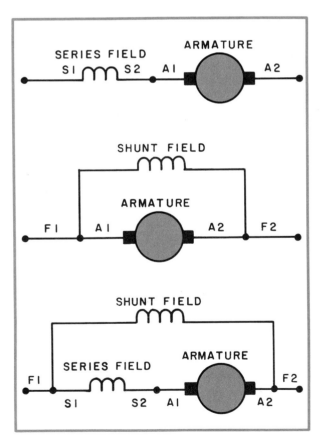

Figure 17-22 The three basic types of DC motors are the series, shunt, and compound (series and shunt) wound motors.

Locating a Circuit in a Switchboard, Panelboard or Load Center

Quite often the electrician must locate one circuit in a switchboard, panelboard or load center in order to turn off the power before troubleshooting or working on the circuit. This may not be as simple as one might think, because switchboards, panelboards and load centers are often very crowded with wires that are not marked or that are mismarked. The industrial electrician simply cannot start turning off each circuit individually until the correct circuit is found. As you turn off a circuit, you disconnect all loads connected to that circuit. This means resetting timers, counters, clocks, starters and other control devices or stopping critical equipment such as alarms and safety circuits.

A simple and safe way of locating a circuit is illustrated in Figure 17-23. As illustrated, a flashing incandescent lamp and a clamp-on ammeter are all that is required. The flashing lamp is plugged into any receptacle on the circuit that is to be disconnected. As the lamp is flashing on and off, a clamp-on ammeter is used to check each circuit. Each circuit will display a constant current reading except one. The circuit with the flashing light will display a swinging motion on the ammeter equal to the time of flashing on the lamp. This circuit may then be turned off and worked on.

Checking Capacitors

Capacitors are used on capacitor start and capacitor run motors as well as on other equipment found in the plant. Capacitors have a limited life and should be checked whenever a problem is suspected. The following procedure is used to check the type of capacitor used most often in electrical circuits.

1. Visually check the capacitor for any sign of leakage, cracks, or bulges. Replace if any of these signs are present.

2. Check the capacitor using an ohmmeter as illustrated in Figure 17-24. Remove the capacitor from the circuit and set the ohmmeter on the Rx10 scale. Take an insulated screwdriver and short the two leads of the capacitor. (A spark should be expected, and is normal.) Next, place the ohmmeter probes on the capacitor terminals and observe the movement of the needle. The following results can be expected:

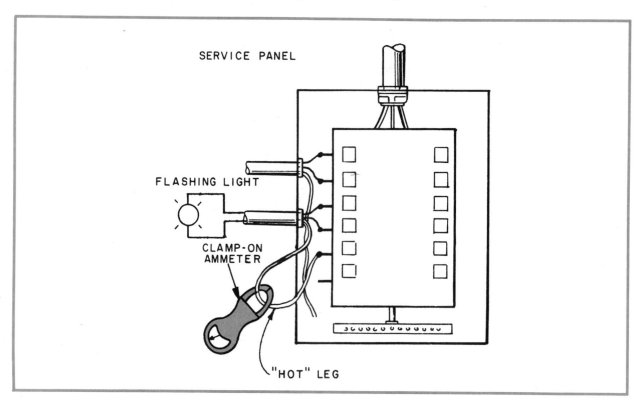

Figure 17-23 A flashing light and a clamp-on ammeter can be used to isolate a particular circuit with tracing wires.

Good Capacitor. The needle will swing to zero resistance and slowly move one fourth to one third to one half of the way across the scale to infinity. When the needle reaches the half-way point, remove one of the leads and wait 30 seconds. When the lead is reconnected again, the needle should swing back to the half-way point and continue to infinity. If the needle swings back to zero, the capacitor is not holding a charge and should be replaced.

Shorted Capacitor. The needle will swing to zero and not move. This capacitor is bad and must be replaced.

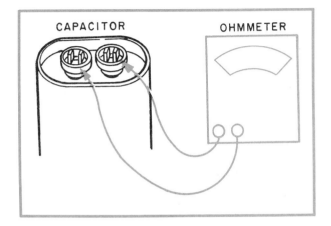

Figure 17-24 How to check a capacitor using an ohmmeter.

Open Capacitor. The needle will not move off from infinity. This capacitor is bad and must be replaced.

Testing Diodes for Opens and Shorts
Diodes are very reliable; however, they are not indestructible. High voltages, improper connections, and overheating can damage a diode. The electrician or technician may be responsible for determining the condition of the diode. Opens and shorts are the two most common diode problems. The ohmmeter is very helpful in determining the condition of the diode.

The procedure used when testing a diode is illustrated in Figure 17-25. A diode is usually tested twice. If the diode is good, there is a low resistance reading in one direction and a high resistance reading in the opposite direction. If the diode is

shorted, a low resistance reading registers in both directions. If the diode is open, there is a high resistance reading in both directions. If a diode is open or shorted, it must be replaced.

Testing a Zener Diode
A zener diode either provides voltage regulation, or it fails. If it fails, the zener diode must be replaced in order to return the circuit to proper operation. Occasionally, the zener diode may appear to fail only in certain situations. These types of failures are called *intermittents*. To check for intermittents, the zener diode must be tested while in operation. An oscilloscope is used for testing the characteristics of a zener diode in an operating situation.

An oscilloscope displays the dynamic operating characteristics of the zener diode. Figure 17-26 illustrates the test display if the zener diode is good.

Testing Thermistors
The proper connection of a thermistor to any electronic circuit is very important. Loose or corroded connections create a high resistance in series with the thermistor resistance. The control circuit may sense the additional resistance as a false temperature reading.

The hot and cold resistance of a thermistor can be checked with an ohmmeter, but one end of the thermistor must be disconnected from the circuit as shown in Figure 17-27. The hot and cold resistance of a thermistor can then be tested as follows:

1. Connect the ohmmeter leads to the thermistor leads and place the thermistor and a thermometer in a mixture of ice and water.

2. Record the temperature and resistance readings.

3. Place the thermistor and thermometer in hot water (not boiling).

4. Record the temperature and resistance readings.

5. Compare these hot and cold readings with the manufacturer's specification sheet or a similar thermistor that is known to be good.

Testing Solid State Pressure Sensors
The condition of a solid state pressure sensor can be tested as follows:

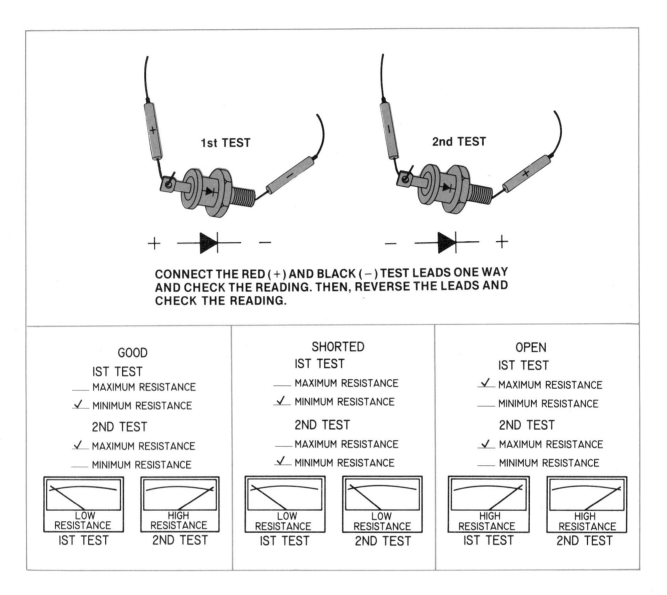

Figure 17-25 The diode is tested twice to determine if the diode is shorted, open, or good.

1. Disconnect the pressure sensor from the circuitry.

2. Connect the ohmmeter leads to the pressure sensor.

3. Activate the device being monitored (compressor, air tank, etc.) until pressure builds up. Record the resistance of the pressure sensor at the high pressure setting.

4. Open the relief or exhaust valve and reduce the pressure on the sensor. Record the resistance of the pressure sensor at the low pressure setting.

5. Compare these high and low resistance readings with manufacturer specification sheets.

When specification sheets are not available, use a replacement pressure sensor that is known to be good.

6. As an additional check, the voltage supplied to the PC board and to the switching device should be checked while the circuit is in operation.

Testing Photoconductive Cells

Humidity and contamination are the primary causes of photoconductive cell failure. The use of quality components that are hermetically sealed is essential for long life and proper operation. Some plastic units are less rugged and more susceptible

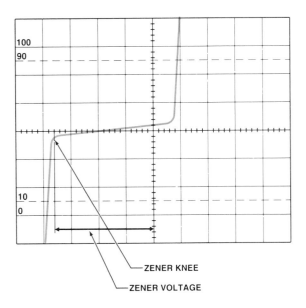

ZENER KNEE

ZENER VOLTAGE

Figure 17-26 An oscilloscope test display will indicate if the zener diode is good.

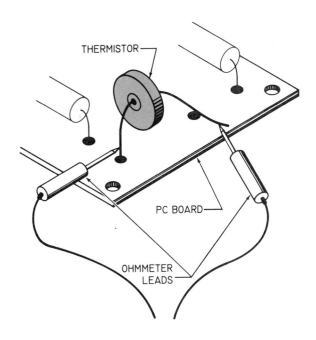

THERMISTOR

PC BOARD

OHMMETER LEADS

Figure 17-27 One terminal lead is disconnected from the circuit when testing a thermistor.

to temperature changes than glass units.

The resistance of a photoconductive cell is tested with an ohmmeter, but the photoconductive cell must have one lead disconnected from the circuitry:

1. Connect the ohmmeter leads to the photoconductive cell.

2. Cover the photoconductive cell and record its dark resistance.

3. Shine a light on the photoconductive cell and record its light resistance.

4. Compare these resistance readings with the manufacturer's specification sheets. When specification sheets are not available, use a similar photoconductive cell that is known to be good.

As with other resistance-type sensors, the connection to the control circuit is very important. All connections should be tight and corrosion-free.

Testing Transistors for Opens and Shorts

High voltages, improper connections, and overheating can damage a transistor. The technician is responsible for determining the condition of the transistor. Although many types of transistor testers are available, the technician may not have one when it is needed. Therefore, an ohmmeter can be substituted for a transistor tester when finding opens and shorts. The transistor may be considered back-to-back diodes when conducting tests for opens and shorts as illustrated in 17-28.

NOTE: A diode passes current when it is forward biased. It blocks current when it is reverse biased. When making resistance measurements, the ohmmeter should indicate a low resistance one way and a high resistance the opposite way.

The first diode junction to test is the emitter-to-base junction as illustrated in Figure 17-29. Placing the leads one way and then reversing the leads

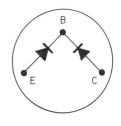

PNP TRANSISTOR

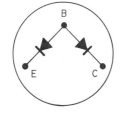

NPN TRANSISTOR

Figure 17-28 The PN junctions of a transistor can be considered as back-to-back diodes when testing for opens and shorts.

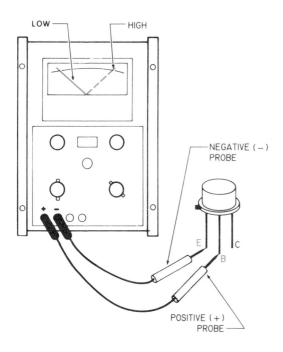

Figure 17-29 Placing the ohmmeter leads on the emitter-base one way then reversing the leads should result in a high forward-to-reverse resistance ratio.

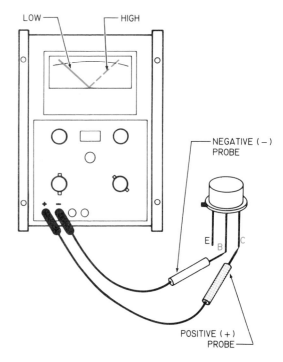

Figure 17-30 Placing ohmmeter leads on the base-collector one way then reversing the leads should result in a high forward-to-reverse resistance ratio.

should result in a high forward-to-reverse resistance ratio. The second diode junction of the base-collector should show a similar high forward-to-reverse resistance ratio as illustrated in Figure 17-30. The last check is from emitter to collector as illustrated in Figure 17-31. It should show a high resistance in both directions for the transistor to be considered good.

A comparison should be made of the resistance measurements of the emitter-base section in forward and reverse. The measurements should indicate a high resistance in one direction and a low resistance in the opposite direction if the transistor is good. If both emitter-base resistances are low, the transistor is shorted. If both resistances are high, the transistor is open.

The measurements of the base-collector terminals should indicate a high resistance in one direction and a low resistance in the opposite direction if the transistor is good. If both base-collector resistances are low, the transistor is shorted. If both resistances are high, the transistor is open. If the resistance is low in either direction, the transistor is shorted.

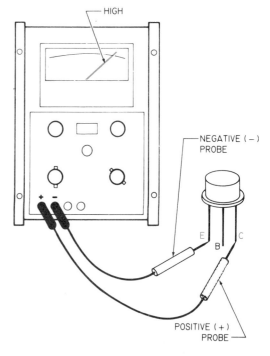

Figure 17-31 When comparing the measurements of the emitter-collector terminals, the measurements should show a high resistance in both directions.

Testing an SCR

Special test equipment, such as an oscilloscope, is needed to properly test an SCR under operating conditions. However, a rough test, using a test circuit, can be made with an ohmmeter as illustrated in Figure 17-32. The procedure for a rough test is:

1. Place the ohmmeter on the "R × 100" scale.

2. Connect the negative lead of the ohmmeter to the cathode.

3. Connect the positive lead of the ohmmeter to the anode. The resistance reading should be so high that the ohmmeter indicates infinity. (The resistance will in fact be over 250,000 ohms.)

4. Short-circuit the gate to the anode by closing switch S.

5. The resistance reading should be so low that the ohmmeter will indicate almost zero. (The resistance will be about 10 to 50 ohms, but this will not register on the "R × 100" scale.)

6. When switch S is opened, the zero resistance reading should remain.

7. Remove and reconnect the ohmmeter leads with the positive lead to the cathode and the negative lead to the anode. The resistance reading should be so high that the ohmmeter indicates infinity. (The resistance will in fact be over 250,000 ohms.)

8. Short-circuit the gate to the anode by closing switch S. The resistance should remain high since the SCR is reverse-biased and cannot conduct.

9. Open switch S. The high resistance should remain since the SCR is reverse-biased and has no gate current.

NOTE: If the SCR does not respond as indicated for each of these steps, it is defective and must be replaced.

Testing a Diac

The ohmmeter can be used to test a diac for a short circuit. The procedure used to test for a short circuit is:

1. Place the ohmmeter on the "R × 100" scale.

2. Connect the ohmmeter leads as shown in A of Figure 17-33, and record the resistance reading.

3. Reconnect the ohmmeter leads as shown in B of Figure 17-33, and record the resistance reading.

NOTE: Since the diac is essentially two zener diodes connected in series, both readings should show high resistance. Also, testing a diac in this manner will only show that the component is shunted. If it is suspected that the diac is open, a second test, using an oscilloscope, should be performed.

The procedure for testing for an open diac, using an oscilloscope is shown in Figure 17-34.

1. Set up the test circuit.

2. Apply power to the circuit.

3. Adjust the oscilloscope.

NOTE: If the diac is good, a trace similar to that shown in Figure 17-34 will appear.

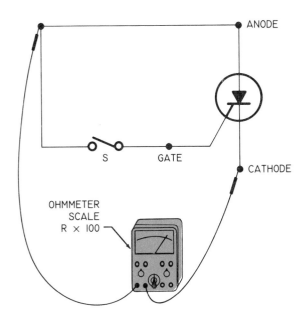

Figure 17-32 An SCR can be tested using a test circuit with an ohmmeter. The ohmmeter should show a high resistance when switch S is open and a low resistance when switch S is closed.

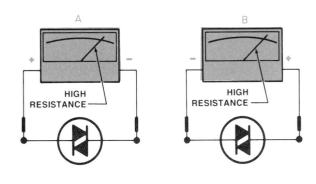

Figure 17-33 When testing a diac for a short circuit with an ohmmeter, the resistance reading will be high in both directions if the diac is good.

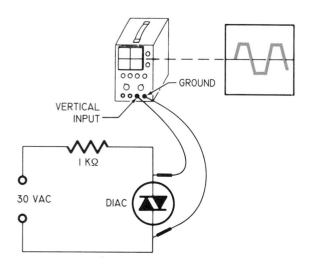

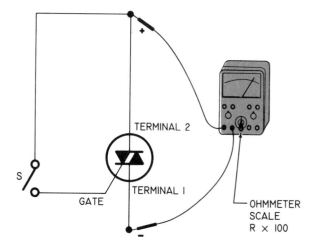

Figure 17-34 When testing a diac with an oscilloscope and test circuit, the waveform shown will be displayed if the diac is good.

Figure 17-35 An ohmmeter can be used to make a rough test of the operating condition of a triac.

Testing a Triac

Special test equipment, such as an oscilloscope, is needed to test a triac under operating conditions. However, a rough test can be made, using a test circuit, with an ohmmeter as shown in Figure 17-35. To make a rough test:

1. Place the ohmmeter on the "R × 100" scale.
2. Connect the negative lead of the ohmmeter to Terminal 1.
3. Connect the positive lead of the ohmmeter to Terminal 2. The resistance will be so high that the ohmmeter indicates infinity. (The resistance will, in fact, be over 250,000 ohms.)
4. Short circuit the gate by closing switch S. The resistance reading should be so low that the ohmmeter will indicate almost zero. (The resistance will be about 10-50 ohms, but will not register on the "R × 100" scale.)
5. When the gate switch is opened, the zero resistance reading should remain.
6. Remove and reconnect the ohmmeter leads with the positive lead to Terminal 1 and the negative lead to Terminal 2. The resistance will be so high that the ohmmeter indicates infinity. (The resistance will, in fact, be over 250,000 ohms.)

7. Short circuit the gate by closing switch S. The resistance reading should be so low that the ohmmeter will indicate almost zero. (The resistance will be about 10-50 ohms, but will not register on the "R × 100" scale.)

8. When the gate switch is opened, the zero resistance reading should remain.

NOTE: If the triac does not respond as indicated in each of these steps, it is probably defective and must be replaced.

TROUBLESHOOTING HINTS

On the job, you will run across many problems. Often the cause and remedy is fairly routine. Figure 17-36 illustrates some of the common troubles, causes and remedies with standard industrial controls. Learning these will give you a good start in the field of electrical control.

Motor Control Trouble-Remedy Table

TROUBLE	CAUSE	REMEDY
Magnetic Contactors and Starters		
CONTACTS		
Contact Chatter	1. Broken shading coil.	1. Replace magnet and armature.
	2. Poor contact in control circuit.	2. Replace the contact device or use holding circuit interlock (3 wire control).
	3. Low voltage.	3. Correct voltage condition. Check momentary voltage dip during starting.
Welding or Freezing	1. Abnormal inrush of current.	1. Check for grounds, shorts or excessive motor load current or use larger contactor.
	2. Rapid jogging.	2. Install larger device rated for jogging service.
	3. Insufficient tip pressure.	3. Replace contacts and springs, check contact carrier for deformation or damage.
	4. Low voltage preventing magnet from sealing.	4. Correct voltage condition. Check momentary voltage dip during starting.
	5. Foreign matter preventing contacts from closing.	5. Clean contacts with Freon. Contactors, starters, and control accessories used with very small current or low voltage, should be cleaned with Freon.
	6. Short circuit.	6. Remove short or fault and check to be sure fuse or breaker size is correct.
Short Tip Life or Overheating of Tips	1. Filing or dressing.	1. Do not file silver tips. Rough spots or discoloration will not harm tips or impair their efficiency.
	2. Interrupting excessively high currents.	2. Install larger device or check for grounds, shorts or excessive motor currents.
	3. Excessive jogging.	3. Install larger device rated for jogging.
	4. Weak tip pressure.	4. Replace contacts and springs, check contact carrier for deformation or damage.
	5. Dirt or foreign matter on contact surface.	5. Clean contacts with Freon.
	6. Short circuits.	6. Remove short or fault and check to be sure fuse or breaker size is correct.
	7. Loose connection.	7. Clean and tighten.
	8. Sustained overload.	8. Check for excessive motor load current or install larger device.
COILS		
Open Circuit	1. Mechanical damage.	1. Handle and store coils carefully.
Roasted Coil	1. Over voltage or high ambient temperature.	1. Check application, circuit, and correct.
	2. Incorrect coil.	2. Install correct coil.

Figure 17-36 Typical troubles, causes, and remedies for standard industrial controls. (Square D Company)

Motor Control Trouble-Remedy Table (continued)

TROUBLE	CAUSE	REMEDY
	Magnetic Contactors and Starters	
Roasted Coil (continued)	3. Shorted turns caused by mechanical damage or corrosion.	3. Replace coil.
	4. Under voltage, failure of magnet to seal in.	4. Correct system voltage.
	5. Dirt or rust on pole faces increasing air gap.	5. Clean pole faces.
OVERLOAD RELAYS		
Tripping	1. Sustained overload.	1. Check for grounds, shorts, or excessive motor currents and correct cause.
	2. Loose connection on load wires.	2. Clean and tighten.
	3. Incorrect heater.	3. Heater should be replaced with correct size.
MAGNETIC & MECHANICAL PARTS		
Noisy Magnet	1. Broken shading coil.	1. Replace magnet and armature.
	2. Magnet faces not mating.	2. Replace magnet and armature.
	3. Dirt or rust on magnet faces.	3. Clean.
	4. Low voltage.	4. Check system voltage and voltage dips during starting.
Failure to Pick-up and Seal	1. Low voltage.	1. Check system voltage and voltage dips during starting.
	2. Coil open or shorted.	2. Replace.
	3. Wrong coil.	3. Replace.
	4. Mechanical obstruciton.	4. WITH POWER OFF check for free movement of contact and armature assembly.
Failure to Drop-Out	1. Gummy substance on pole faces.	1. Clean pole faces.
	2. Voltage not removed.	2. Check coil circuit.
	3. Worn or rusted parts causing binding.	3. Replace parts.
	4. Residual magnetism due to lack of air gap in magnet path.	4. Replace magnet and armature.
PNEUMATIC TIMERS		
Erratic Timing	1. Foreign matter in valve.	1. Replace timing head complete or return timer to factory for repair and adjustment.
Contacts Do Not Operate	1. Maladjustment of actuating screw.	1. Adjust as per instruction in service bulletin.
	2. Worn or broken parts in snap switch.	2. Replace snap switch.
LIMIT SWITCHES		
Broken Parts	1. Overtravel of actuator.	1. Use resilient actuator or operate within tolerances of the device.
MANUAL STARTERS		
Failure to Reset	1. Latching mechanism worn or broken.	1. Replace starter.

COMPENSATORS (MANUAL)

Welding of Contacts on Starting Side	1. Inching, jogging and operating handle slowly.	1. Excessive inching and jogging not recommended (caution operator). Move handle swiftly and surely to start position.
Welding of contacts Running Side	1. Moving handle slowly to run position.	1. Move handle swiftly and surely to run position as motor approaches full speed.
	2. Lack of sufficient spring pressure.	2. Replace contacts and contact springs.
Damaged or Burned Transformer	1. Repeated inching and jogging.	1. Excessive inching and jogging not recommended (caution operator).
	2. Holding handle in start position for long periods.	2. Hold handle in start position only until motor approaches full speed.

QUESTIONS ON UNIT 17

1. Why is preventive maintenance so important in today's industrial plant?
2. What are some of the basic requirements of a good preventive maintenance program?
3. What effect does an unbalanced power supply have on a three-phase motor?
4. When is phase reversal protection required?
5. In using an instrument with several scales, what scale should the instrument be set for first?
6. What is the first step in troubleshooting any circuit?
7. Why does a jumper work better than a test light or voltmeter when you are troubleshooting control devices such as pushbuttons and pressure switches?
8. Before troubleshooting the power circuit, how can the control circuit be eliminated quickly?
9. What step should be taken before testing a fuse or breaker?
10. When using an ohmmeter to test a fuse, what type of resistance reading would indicate a good fuse?
11. In testing a transformer for a proper voltage level, what is the maximum percent of variation from the rated voltage that should exist?
12. What are the four basic categories that most motor problems fall under?
13. Why is it important to lock out the incoming power before rotating the shaft of a motor?
14. What is the major cause of motor failure?
15. What are some of the common causes of overheating in electric motors?
16. How can you test the amount of work a motor is performing?
17. How does a high ambient temperature affect a motor?
18. In testing a motor to determine the type, how many individual circuits will a Wye motor have? A Delta motor?
19. How are the terminals marked on the armature, series-field, and shunt-field of a DC motor?
20. What effect will a loose terminal connection on a circuit breaker have?
21. What does a reading of zero ohms indicate when checking a capacitor?
22. What consideration must the electrician be aware of when testing an SCR with a clamp-on ammeter?

APPENDIX

DESIGN AND LAYOUT DATA

DC CIRCUIT CHARACTERISTICS

Ohm's Law:

$$E = IR \qquad I = \frac{E}{R} \qquad R = \frac{E}{I}$$

E = voltage impressed on circuit (volts)
I = current flowing in circuit (amperes)
R = circuit resistance (ohms)

In direct current circuits, electrical power is equal to the product of the voltage and current:

$$P = EI = I^2R = \frac{E^2}{R}$$

P = power (watts)
E = voltage (volts)
I = current (amperes)
R = resistance (ohms)

AC CIRCUIT CHARACTERISTICS

The instantaneous values of an alternating current or voltage vary from zero to a maximum value each half cycle. In the practical formulae which follow, the "effective value" of current and voltage is used, defined as follows:

Effective value = 0.707 × maximum instantaneous value

Impedance:

Impedance is the total opposition to the flow of alternating current. It is a function of resistance, capacitive reactance and inductive reactance. The following formulae relate these circuit properties:

$$X_L = 2\pi fL \qquad X_C = \frac{1}{2\pi fC} \qquad Z = \sqrt{R^2 + (X_L - X_C)^2}$$

X_L = inductive reactance (ohms)
X_C = capacitive reactance (ohms)
Z = impedance (ohms)
f = frequency (cyles per second)
C = capacitance (farads)
L = inductance (henrys)
R = resistance (ohms)
π = 3.14

Ohm's Law for AC Circuits:

$$E = I \times Z \qquad I = \frac{E}{Z} \qquad Z = \frac{E}{I}$$

POWER FACTOR

Power factor of a circuit or system is the ratio of actual power (watts) to apparent power (volt-amperes), and is equal to the cosine of the phase angle of the circuit:

$$PF = \frac{actual\ power}{apparent\ power} = \frac{watts}{volts \times amperes} = \frac{KW}{KVA} = \frac{R}{Z}$$

KW = kilowatts
KVA = kilovolt-amperes = volt-amperes × 1,000
PF = power factor (expressed as decimal)

SINGLE-PHASE CIRCUITS

$$KVA = \frac{EI}{1,000} = \frac{KW}{PF} \qquad KW = KVA \times PF$$

$$I = \frac{P}{E \times PF} \qquad E = \frac{P}{I \times PF} \qquad PF = \frac{P}{E \times I}$$

$$P = E \times I \times PF$$
P = power (watts)

THREE-PHASE CIRCUITS, BALANCED STAR OR WYE

$$I_N = O \qquad I = I_p \qquad E = \sqrt{3}\,E_p = 1.73\,E_p$$

$$E_p = \frac{E}{\sqrt{3}} = \frac{E}{1.73} = 0.577E$$

I_N = current in neutral (amperes)
I = line current per phase (amperes)
I_p = current in each phase winding (amperes)
E = voltage, phase to phase (volts)
E_p = voltage, phase to neutral (volts)

THREE-PHASE CIRCUITS, BALANCED DELTA

$$I = 1.732 \times I_p \qquad I_p = \frac{I}{\sqrt{3}} = 0.577 \times I$$

$$E = E_p$$

POWER: BALANCED 3-WIRE, 3-PHASE CIRCUIT, DELTA OR WYE

For unity power factor (PF = 1.0):

$$P = 1.732 \times E \times I$$

$$I = \frac{P}{\sqrt{3}\,E} = \frac{0.577P}{E} \qquad E = \frac{P}{\sqrt{3} \times I} = \frac{0.577P}{I}$$

P = total power (watts)

For any load:

$$P = 1.732 \times E \times I \times PF \qquad VA = 1.732 \times E \times I$$

$$E = \frac{P}{PF \times 1.73 \times I} = \frac{0.577 \times P}{PF \times I}$$

$$I = \frac{P}{PF \times 1.73 \times E} = \frac{0.577 \times P}{PF \times E}$$

$$PF = \frac{P}{1.73 \times I \times E} = \frac{0.577 \times P}{I \times E}$$

VA = apparent power (volt-amperes)
P = actual power (watts)
E = line voltage (volts) phase to phase
I = line current (amperes)

POWER LOSS: ANY AC OR DC CIRCUIT

$$P = I^2R \qquad I = \sqrt{\frac{P}{R}} \qquad R = \frac{P}{I^2}$$

P = power heat loss in circuit (watts)
I = effective current in conductor (amperes)
R = conductor resistance (ohms)

Source: Midco Components

Installation and Maintenance of Motors

1. SAFETY PRECAUTIONS

Use safe practices when handling, lifting, installing, operating, and maintaining motors and motor-operated equipment.

Install motors and electrical equipment in accordance with the National Electrical Code (NEC) or sound local electrical and safety codes and practices, and, when applicable, the Occupational Safety and Health Act (OSHA).

Ground motors securely. Make sure that "grounding" wires and devices are, in fact, properly grounded. FAILURE TO GROUND MOTOR PROPERLY MAY CAUSE SERIOUS INJURY TO PERSONNEL.

Before servicing or working on or near motors or motor driven equipment, disconnect power source from motor and accessories.

SAFETY CAUTION

Motors subjected to overload, locked rotor, current surge or inadequate ventilation conditions may experience rapid heat build-up, presenting risks of motor damage or fire. To minimize such risks, use of motors with proper overload protectors is advisable for most applications.

Consult the manufacturer for application assistance.

DO NOT use motors with automatic-reset protectors where automatic restarting might be hazardous to personnel or to equipment. Use motors with manual-reset protectors where such hazards exist. Such applications include conveyors, compressors, tools, most farm equipment, etc.

Remove shaft key from keyways of uninstalled motors before energizing motor. Be sure keys, pulleys, fans, etc. are fully secured on installed motors before energizing motor.

Make sure fans, pulleys, belts, etc. are properly guarded if they are in a location that could be hazardous to personnel.

Provide proper safeguards against failure of motor-mounted brakes, particularly on applications involving overhauling loads.

Provide proper safeguards on applications where a motor is mounted on or through a gear reducer to a holding or overhauling application. Do not depend on gear friction to hold the load.

Do not lift motor-driven equipment with motor lifting means. If eyebolts are used for lifting motors, they must be securely tightened and the direction of the lift must not exceed a 15-degree angle with the shank of the eyebolt.

2. LOCATION/MOTOR ENCLOSURES

a. **Open, Drip-proof Motors** are designed for use in areas that are reasonably dry, clean, and well ventilated—usually indoors. If installed outdoors, it is recommended that the motor be protected with a cover that does not restrict flow of air to the motor.

b. **Totally Enclosed Motors** are suitable for use where exposed to dirt, moisture, and most outdoor locations, but not for very moist or for hazardous (explosive vapor or dust-filled atmosphere) locations.

c. **Severe-duty Enclosed Motors** are suitable for use in corrosive or excessively moist locations.

d. **Explosion-proof Motors** are made to meet Underwriters Laboratories Standards for use in hazardous (explosive) locations shown by the UL label on the motor.

Certain locations are hazardous because the atmosphere does or may contain gas, vapor or dust in explosive quantities. The National Electrical Code (NEC) divides these locations into Classes and Groups according to the type of explosive agent which may be present. Listed below are some of the agents in each classification. For complete list, see Article 500 of the NEC.

CLASS I (Gases, Vapors)

Group A—Acetylene (NOTE: MOTORS NOT AVAILABLE FOR THIS GROUP.)

Group B—Butadiene, ethylene oxide, hydrogen, propylene oxide (NOTE: MOTORS NOT AVAILABLE FOR THIS GROUP.)

Group C—Acetaldehyde, cyclopropane, diethyl ether, ethylene, isoprene.

Group D—Acetone, acrylonitrile, ammonia, benzene, butane, ethylene dichloride, gasoline, hexane, methane, methanol, naphtha, propane, propylene, styrene, toluene, vinyl acetate, vinyl chloride, xylene.

CLASS II (Combustible Dusts)

Group E—Aluminum, magnesium and other metal dusts with similar characteristics.

Group F—Carbon black, coke or coal dust.

Group G—Flour, starch or grain dust.

e. **Ambient temperature** around motors should not exceed 40 C unless motor nameplate specifically permits a higher value.

3. MOUNTING

a. Unless specified otherwise, listed motors can be mounted in any position or any angle. However, drip-proof motors must be mounted in the normal horizontal position to meet the enclosure definition.

b. Mount motor securely to the mounting base of equipment or to a rigid, flat surface, preferably metallic.

c. For direct-coupled applications, align shaft and coupling carefully, using shims as required, under motor base. Use a flexible coupling, if possible, but not as a substitute for good alignment practices.

d. For belted applications, align pulleys and adjust belt tension so approximately 1/2″ of belt deflection occurs when thumb force is applied midway between pulleys. With sleeve-bearing motors, position motor so belt pull is away from the oil hole in the bearing (approximately under the oiler of the motor).

4. POWER SUPPLY/CONNECTIONS

a. Connect motor for desired voltage and rotation according to the connection diagram on the nameplate or in the terminal box.

b. Voltage, frequency, and phase of power supply should be consistent with the motor nameplate rating. Motor will operate satisfactorily on voltage within 10% of nameplate value or frequency within

continued

Installation and Maintenance of Motors

5%; combined variation not to exceed 10%.

c. OPERATION OF 230-VOLT MOTORS ON 208-VOLT SYSTEMS

Motors rated 230 volts will operate satisfactorily on 208-volt systems on most applications requiring nominal starting torques. Starting and maximum running torques of a 230-volt motor will be reduced approximately 25% when operated on 208-volt systems. Fans, blowers, centrifugal pumps, and similar loads will normally operate satisfactorily at these reduced torques. Where the application torque requirements are high, it is recommended that the next higher horsepower 230-volt motor or a 200-volt motor can be used. External motor controls for 230-volt motors on 208-volt systems should be selected from 230-volt nameplate data.

5. WIRING

a. All wiring and electrical connections should comply with the National Electrical Code (NEC) and with local codes and sound practices.

b. Undersize wire between the motor and the power source will adversely limit the starting and load carrying abilities of the motor. Minimum wire sizes for motor branch circuits are recommended in the tables following.

Individual Branch Circuit Wiring
For Single-phase Induction Motors

Motor Data		Copper Wire Size Minimum AWG No.				
		Branch Circuit Length				
Hp	Volts	0-25 Ft	50 Ft	100 Ft	150 Ft	200 Ft
1/6	115	14	14	14	12	10
	230	14	14	14	14	14
1/4	115	14	14	12	10	8
	230	14	14	14	14	14
1/3	115	14	12	10	8	6
	230	14	14	14	14	12
1/2	115	14	12	10	8	6
	230	14	14	14	14	12
3/4	115	12	10	8	6	4
	230	14	14	14	12	10
1	115	12	10	8	6	4
	230	14	14	14	12	10
1-1/2	115	10	10	6	4	4
	230	14	14	12	10	8
2	115	10	8	6	4	
	230	14	12	12	10	8
3	115	6	6	4		
	230	10	10	10	8	8
5	230	8	8	8	6	4

Individual Branch Circuit Wiring
For Three-phase Squirrel-cage
Induction Motors

Motor Data		Copper Wire Size Minimum AWG No.				
		Branch Circuit Length				
Hp	Volts	0-25 Ft	50 Ft	100 Ft	150 Ft	200 Ft
1/2	230	14	14	14	14	14
	460	14	14	14	14	14
3/4	230	14	14	14	14	14
	460	14	14	14	14	14
1	230	14	14	14	14	14
	460	14	14	14	14	14
1-1/2	230	14	14	14	14	12
	460	14	14	14	14	14
2	230	14	14	14	12	12
	460	14	14	14	14	14
3	230	14	14	12	10	10
	460	14	14	14	14	14
5	230	12	12	10	10	8
	460	14	14	14	14	14
7-1/2	230	10	10	10	8	6
	460	14	14	14	14	12
10	230	8	8	8	6	4
	460	12	12	12	12	12
15	230	6	6	6	4	4
	460	10	10	10	10	10
20	230	4	4	4	4	3
	460	8	8	8	8	8
25	230	4	4	4	3	2
	460	8	8	8	8	8
30	230	3	3	3	2	1
	460	6	6	6	6	6
40	230	1	1	1	1	0
	460	6	6	6	6	6
50	230	00	00	00	00	00
	460	4	4	4	4	4

See NEC for exact sizing

Source: General Electric Company

Examples of Overcurrent Protection for Control Circuits

(Some of these schemes may be required by applicable electrical codes)

1

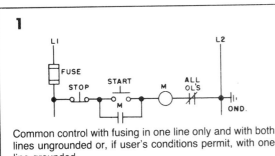

Common control with fusing in one line only and with both lines ungrounded or, if user's conditions permit, with one line grounded.

2

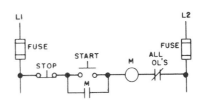

Common control with fusing in both lines and with both lines ungrounded.

3

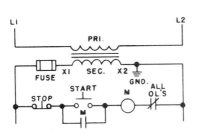

Control circuit transformer with fusing in one secondary line and with both secondary lines ungrounded or, if user's conditions permit, with one line grounded.

4

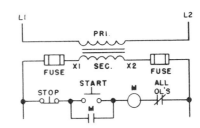

Control circuit transformer with fusing in both secondary lines and with both secondary lines ungrounded.

5

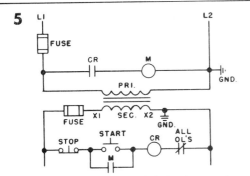

Control circuit transformer with fusing in one primary and one secondary line, and with all lines ungrounded, or, if user's conditions permit, with one primary and one secondary line grounded.

6

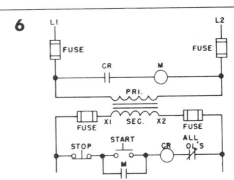

Control circuit transformer with fusing in both primary lines and both secondary lines and with all lines ungrounded.

7

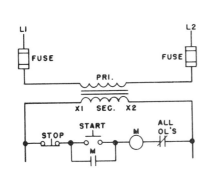

Control circuit transformer with fusing in both primary lines, with no secondary fusing and with all lines ungrounded.

Source: Square D Company

PROGRAMMABLE LOGIC RELAY

SL 100

SL 100

❋ Logic relay »or« - »nor« - »and« - »nand«.
❋ Negative logic.
❋ 32 possibilities of combination on 5 inputs.
❋ 10 A SPDT output relay.
❋ LED-indication for relay on.
❋ AC- or DC supply voltage.

TECHNICAL DATA

Common technical data and ordering key
Pages 10–12.

Inputs
3 signal inputs:
Pins 5-6 and 7.

2 function inputs:
Pins 8 and 9.

Common negative:
Pin 11.

Data on inputs
The signal- and the function inputs have equal technical data.

Control voltage
Internal voltage:
10 VDC.

Negative logic
Logical »0« corresponds to open input (10 VDC).
Logical »1« corresponds to short-circuited input (0 VDC).

Input voltage for positive »0«-function
Min. 6 V.

Input voltage for positive »1«-function
Max. 3 V.

Input resistance
10 KΩ.

Short-circuit current
1 mA.

Pulse duration
10 ms - ∞

Contact resistance
Max. 8 KΩ.

Accessories
Bases.
Hold down spring.
Mounting rack.
Base cover.
Front mounting bezel.

WIRING DIAGRAMS

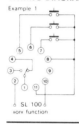

Example 1

SL 100
»or« function

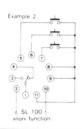

Example 2

SL 100
»nor« function

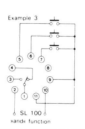

Example 3

SL 100
»and« function

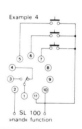

Example 4

SL 100
»nand« function

MODE OF OPERATION

This logic relay, being equipped with 3 signal inputs and 2 inputs controlling the function, is universally applicable for »or«, »nor«, »and«, »nand« functions. The function is at any time determined by the way in which signals on the function terminals are combined. See examples 1 to 4.
The signal voltage (10 VDC) is supplied by the logic relay.
Signals can, on all 5 inputs, be handled by ordinary metallic switches and/or by NPN-transistors with open collector. The emitters are connected to pin 11.

Example 1 The relay operates when one or more of the three signal inputs are connected to pin 11.
(»or«-function).

Example 2 The relay releases when one or more of the three signal inputs are connected to pin 11.
(»nor«-function).

Example 3 The relay operates when all three signal inputs are connected to pin 11.
(»and«-function).
Unused inputs to be connected directly to pin 11.

Example 4 The relay releases when all three signal inputs are connected to pin 11.
(»nand«-function).
Unused inputs to be connected directly to pin 11.

OPERATION DIAGRAM

	Example 1 »or«	Example 2 »nor«	Example 3 »and«	Example 4 »nand«
Supply voltage				
Activation pin 5				
Activation pin 6				
Activation pin 7				
Activation pin 8				
Activation pin 9				
Relay on				

Source: Electronic Components, Ltd.

GLOSSARY

Absolute pressure. Pressure measured relative to a perfect vacuum. Usually expressed as pounds per square inch absolute (psia).

Accuracy. The limits in which the stated value of a measurement might vary relative to its true value. Generally expressed as a percent of full scale, such as ± 1%.

Accuracy (range). The minimum and maximum limits in which the stated value of a measurement might vary relative to the total range setting. Expressed as "range accuracy," such as + 5% on maximum setting and − 10% on minimum setting. As with most measured values, accuracy is the best when using the middle of the range. For example, a typical delay on operate timer is available in the following time ranges:

A) .15 to 3 seconds
B) .8 to 18 seconds
C) 3 to 60 seconds
D) 8 to 180 seconds
E) 30 to 600 seconds

Range Accuracy

+ 5% on max.
− 10% on min.

If a known time setting of 3 seconds is required for a given application any one of three times can be used: timer A, B or C. However, timer B (.8 to 18 seconds) will give the best accuracy of time when set for 3 seconds.

Address. A reference number assigned to a unique memory location. Each memory location has an address and each address has a memory location.

Ambient temperature. The average temperature of the surrounding medium such as air, water, or earth into which the equipment comes in contact. The ambient temperature range for example, may be listed as: 20°C to +50°C (−4°F to +122°F) when the equipment is operating and −50°C to +85°C (−58°F to 185°F) when the equipment is in storage.

Amplification. The production of an output larger than the input.

Amplifier. A device whose output is a larger reproduction of the input signal.

Analog device. Apparatus that measures continuous information (e.g., voltage current). The measured analog signal has an infinite number of possible values. The only limitation on resolution is the accuracy of the measuring device.

Analog input interface. An input circuit that employs an analog-to-digital converter to convert an analog value to a digital value that can be used for processing.

Analog output interface. An output circuit that employs a digital-to-analog converter to convert a digital value to an analog value that will control a connected analog device.

Analog signal. Signal having the characteristic of being continuous and changing smoothly over a given range, rather than switching suddenly between certain levels as with discrete signals.

AND logic. Operation yielding logical "1" if and only if all inputs are "1."

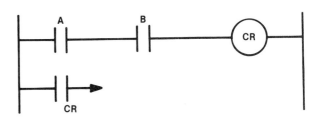

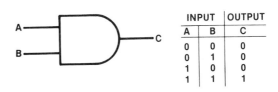

INPUT		OUTPUT
A	B	C
0	0	0
0	1	0
1	0	0
1	1	1

AND logic

Asymmetrical. Used in timer terminology to describe a recycle timer in which the on time and off time can be two different settings. For example, the off time (P_T) can be set anywhere within this time range and the on time (A_T) can also be set anywhere within this time range.

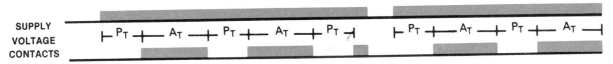

| SUPPLY VOLTAGE CONTACTS | ⊢ P_T ⊣ ⊢ A_T ⊣ ⊢ P_T ⊣ ⊢ A_T ⊣ ⊢ P_T ⊣ ⊢ P_T ⊣ ⊢ A_T ⊣ ⊢ P_T ⊣ ⊢ A_T ⊣ |

asymmetrical

Bimetallic thermometer. A strip of two metals having different coefficients of expansion, bonded together usually in the form of a spiral or strip. Movement of the bonded metals caused by a temperature change becomes a measure of temperature.

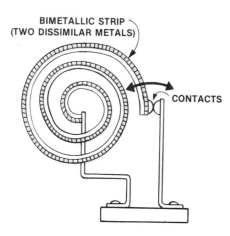

bimetallic thermometer

Binary coded decimal (BCD). A binary number system in which each decimal digit from 0 to 9 is represented by four binary digits (bits). The four positions have a weighted value of 1, 2, 4, and 8, respectively, starting with the least significant bit.

BINARY	BCD	DECIMAL
0	0000	0
1	0001	1
10	0010	2
11	0011	3
100	0100	4
101	0101	5
110	0110	6
111	0111	7
1000	1000	8
1001	1001	9

binary coded decimal (BCD)

Binary number system. A number system that uses two numerals (binary digits), 0 and 1. Each digit position for a binary number has a place value of 1, 2, 4, 8, 16, 32, 64, 128, and so on, beginning with the least significant (right-most) digit. Example: 1101 = 1 (1) + 0 (2) + 1 (4) + 1 (8) = 13.

Binary word. A related grouping of 1's and 0's having coded meaning assigned by position, or as a group, has some numerical value. 10010010 is an eight-bit binary word in which each bit could have coded significance, or as a group represent the number 146 in decimal.

Bit. One binary digit. The smallest unit of binary information. A bit can have a value of 1 or 0.

Boolean algebra. Shorthand notation for expressing logic functions. Used to understand the logic of a circuit to simplify the circuit when working with programmable controllers.

Boolean operators. Logical operators such as AND, OR, NAND, NOR, NOT, and Exclusive-OR, that can be used singly or in combination to form logical statements or circuits.

Burden current. The operating current in a line powered (3 wire) sensor. This current does not pass through the load.

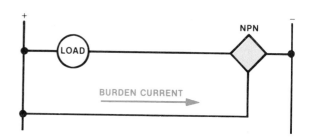

burden current

Capacitance detector. A device with single or multiple probes that, with an object coming within proximity of the detector, will cause a change in probe capacitance. This change in probe capacitance will allow the detector to turn the load on or off.

Central processing unit. That part of the programmable controller that governs system activities, including the interpretation and execution of programmed instruction. In general the CPU consists of the arithmetic-logic unit, timing circuitry, counting circuitry, scratch pad memory, program counter and address stack, and an instruction register. The central processing unit is sometimes called the processor of the CPU.

Clear. To remove data from a single memory location or all memory locations, and return to a nonprogrammed state or some initial condition (normally "O" in a programmable controller).

Clock signal. A clock pulse that is periodically generated and used throughout the system to synchronize equipment operation.

CMOS (Complementary metal oxide semiconductor). A low power IC in which almost no current flows when a gate is not switching.

Conductive level detector. A device with single or multiple probes. A change in level completes an electrical circuit between the container and/or probes.

Control point. The level at which a system is to be maintained (usually temperature).

Control voltage and/or current, nominal. The normal control voltage and/or current intended to be applied to the input of an SSR.

Control voltage and/or current, maximum. The maximum control voltage and/or current intended to be applied to the input of an SSR.

Current, leakage off-state. The effective current that flows in the load circuit when an SSR is in the off-state. This current is usually about 2 mA to 10 mA. This leakage current can be high enough to turn low current loads on (such as an input to a P.C.). In this case, a load resistor will be required.

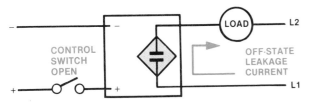

current, leakage off-state

Current loop. A two wire communication link in which the presence of a 20 mA current level indicates a binary "1" (mark), and its absence indicates no data, a binary "0" (space).

Current, inrush (also called Surge Currents). The current flowing in a load circuit immediately following turn-on. For capacitive, transformer, motor and tungsten lamp loads the level of the inrush usually will exceed the steady state current for some period of time following turn-on. This surge of current may damage a controller.

Current, load. The current that flows in the load circuit when the control switch is on. This current must not exceed the current limit of the control switch.

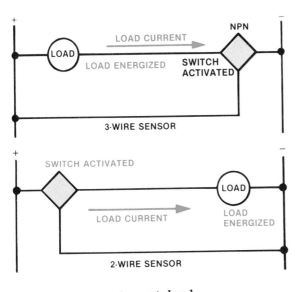

current, load

Current, nonrepetitive surge. The maximum nonrepetitive peak surge current that may be conducted by the control switch for a specific duration. Control may be lost during and following the surge due to excessive heating.

Current, on-state holding, minimum. The minimum current required to maintain a solid state switch in the on-state.

Current sinking sensor (NPN or N-type). The negative terminal of DC system is called the sink because conventional current flows into it. A current sinking sensor "sinks" the current from the load.

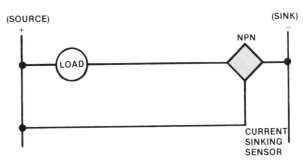

current sinking sensor (NPN or N-type)

Current sourcing sensor (PNP or P-type). The positive terminal of a DC system is called the source because conventional current flows out of it. A current sourcing sensor "sources" the current to the load.

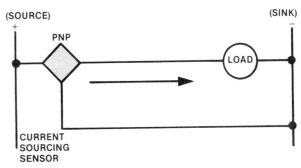

current sourcing sensor (PNP or P-type)

Dark operated (D.O.). Control operating mode in which the output is energized when the light is blocked, the photosensor is dark.

Dead band. The range through which the input can be varied without initiating response. Dead band is usually expressed in a percent of the span or engineering units.

Debouncing. The act of removing intermediate noise states from a mechanical switch.

Decimal number system. A number system that uses ten numeral digits (decimal digits), 0, 1, 2, 3, 4, 5, 6, 7, 8, 9. Each digit position has a place value of 1, 10, 100, 1000, and so on, beginning with the least significant (right-most) digit.

Diac. A three-layer bidirectional device used primarily as a triggering device.

Dielectric strength. The maximum allowable AC rms voltage which may be applied between input and output, input to case and output to case.

dv/dt, off-state, maximum. The maximum rate of rise off-state voltage to be applied to the output. Higher dv/dt may result in turn-on of the SSR.

Differential (also called hysteresis). For an on/off controller, it refers to the temperature difference between the temperature at which the controller turns heat off and the temperature at which the heat is turned back on. It is usually expressed in degrees. Also used with other control devices, such as pressure switches.

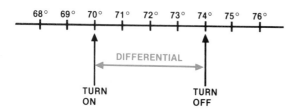

differential (hysteresis)

Diffuse scan. A reflective scan technique in which reflection from a nearby non-shiny surface illuminates the photosensor. Sometimes called "proximity scan" because of the required nearness of light source and photosensor to the reflecting surface.

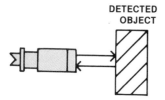

diffuse scan

Digital. The representation of data in the form of pieces (bits or digits). It is possible to express in binary digital form all information stored, transferred, or processed.

Direct memory access (DMA). A process in which a direct transfer of data to or from the memory of a processor-based system can take place without involving the central processing unit.

Direct scan. A scan technique in which the light source is aimed directly at the photoreceiver. The object to be detected must pass between the transmitter and receiver.

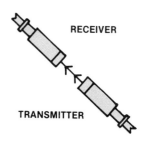

direct scan

EAROM. Electrically alterable read only memory.
EEPROM. Electrically erasable programmable read only memory.
EPROM. Erasable programmable read only memory. A ROM that can be erased with ultraviolet light and then reprogrammed.
Filter. Electrical device used to suppress undesirable electrical noise.
Flip-flop. An electronic circuit having two stable states, or conditions, usually designated "set" and "reset." When in either state, application of an input signal or pulse causes the circuit to change rapidly to the other state.

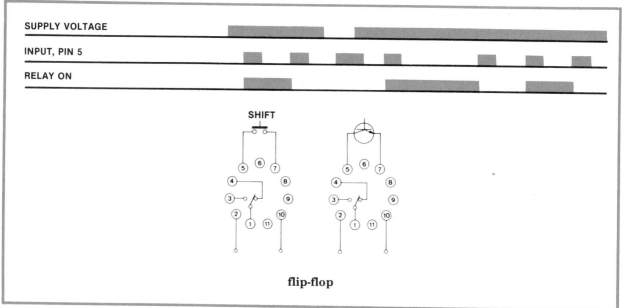

flip-flop

Frequency. Occurrence of a periodic function (with time as the independent variable), generally specified as a certain number of cycles per unit time.

Half cycling. A false turn-on of an SSR for a portion of one-half cycle. It is usually caused by voltage transients appearing across the output that exceed off-state dv/dt or breakover voltage capabilities of the SSR.

Handshaking. Two-way communication between two devices in order to ensure successful data transfer. Based on a data ready/data received scheme, one device alerts the other that it is ready to send. The other device signals when ready to receive and acknowledges when the data is received. Handshaking can be accomplished through hardware, using special lines, or through software, using special codes.

Hard copy. A printed document of what is stored in memory.

Heat sink. A piece of metal used to dissipate the heat of solid state components mounted on it.

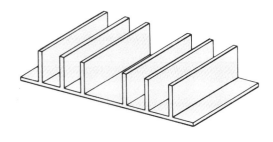

heat sink

Hysteresis. (See Differential.)

Inductive detector. A level measuring system incorporating an oscillator and electromagnetic field.

Infrared (IR). The invisible radiation (as opposed to visible light) that certain LEDs emit. Infrared light is used with photoelectric controls.

Interface. A circuit that permits communication between the central processing unit and a field input or output device. Different devices require different interfaces. Typical interfaces are AC inputs, AC outputs, DC inputs, DC outputs, analog inputs, analog outputs.

Interference, electrical. Any stray voltage or current arising from external sources and appearing in the circuits of a device.

Interference, magnetic field. A form of interference induced into a circuit due to the presence of a magnetic field.

Inversion. The inversion function on a relay allows for the relay contacts to be changed from one state, such as normally open to closed contacts. The inversion function can then be used to set for light operated or dark operated in photoelectric switches and turn on or turn off in temperature type switches.

Isolation. The value of insulation resistance; measured between the input and output, input to case, and output to case.

kVA. Kilovolt-amperes (1000 volt-amps).

Lag. A delay in output change following a change in input.

LASCR (light-activated silicon controlled rectifier). A type of SCR that is triggered by an impulse of light.

Latch. An instruction or component that retains its state even though the conditions that caused it to latch ON may go OFF. A latched output must be unlatched. A latched output will retain its last state (on or off) if power is removed.

Leakage current (residual). A small amount of current that flows through a non-conducting (open) load-powered sensor.

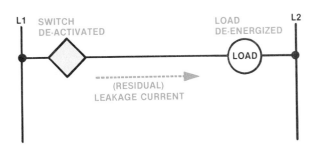

leakage current (residual)

Light emitting diode (LED). A PN junction that emits light when forward biased.

Light operated (L.O.). Control operating mode in which the output is energized when the light beam is not blocked when the photosensor is illuminated (light).

Limiting. A boundary imposed on the upper or lower range of a controller.

Line diagram (also called ladder diagram). An industry standard for representing control systems.

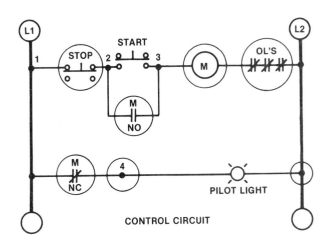

line diagram (ladder diagram)

Line-powered sensor, 3 wire. A sensor that draws its operating current (burden current) directly from the line. The operating current does not flow through the load. Three connections are required.

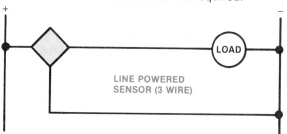

line-powered sensor, 3 wire

Liquid crystal display (LCD). A display device consisting basically of a liquid crystal hermetically sealed between two glass plates. The readout is either dark characters on a dull white background or white characters on a dull black background. LCDs are used on many handheld programmers.

Load-powered sensor, 2 wire. A sensor that draws its operating current (residual current) through the load. Load-powered sensors require only two connections.

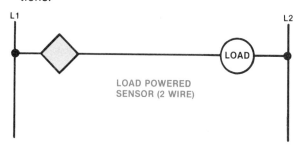

load-powered sensor, 2 wire

Machine language. A program written exclusively in binary form.

Memory. That part of the programmable controller where data and instructions are stored.

Minimum load current. The minimum on-state load current that will ensure proper operation of a load and sensor.

Modulated light source (MLS) control. A photoelectric control that operates on modulated (pulsed) infrared light and responds to modulating frequency rather than light intensity. Modulated LED controls should always be used outdoors.

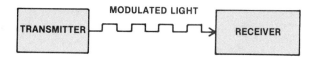

modulated light source (MLS) control

MOV. Metal oxide varistor.

Multiplex. The act of channeling two or more signals to one source.

NAND logic. Operation yielding logical "0" if and only if all inputs are "1."

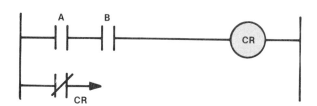

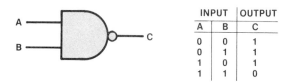

INPUT		OUTPUT
A	B	C
0	0	1
0	1	1
1	0	1
1	1	0

NAND logic

Negative logic. The use of binary logic in such a way that "0" represents the voltage level normally associated with logic 1 (e.g., 0 = +5V, 1 = 0V) Positive logic is more conventional (e.g., 1 = 5V, 0 = 0V).

Noise. Any condition that interferes with the desired signal in a circuit. Noise can produce false information or erratic operations. Noise can be picked up on lines in many ways. If two wires are run together for some distance, signals from one wire can be transmitted into the other.

NOR logic. Operation yielding logical "1" if and only if all inputs are "0."

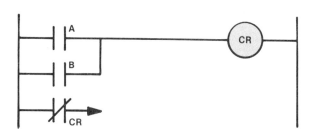

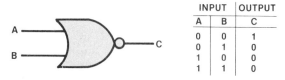

INPUT		OUTPUT
A	B	C
0	0	1
0	1	0
1	0	0
1	1	0

NOR logic

NOT logic. Operation yielding logical "1" if the input is "0" and yielding "0" if the input is "1."

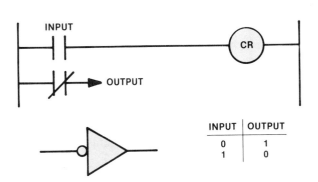

INPUT	OUTPUT
0	1
1	0

NOT logic

OR logic. Operation yielding logical "1" if one or any number of inputs is "1."

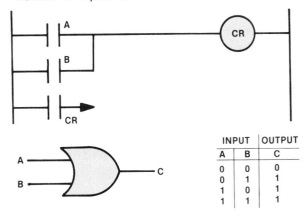

INPUT		OUTPUT
A	B	C
0	0	0
0	1	1
1	0	1
1	1	1

OR logic

Opacity. The characteristic of an object that prevents light from passing through it. Opposite of translucent.

Operational diagram. The operational diagram is a standard way of illustrating the function of a circuit or control module. It illustrates the relationship between input (usually supply voltage, contact closure or set valve) and output (usually relay turn-on). (Illustrated are typical operational diagrams for timers.)

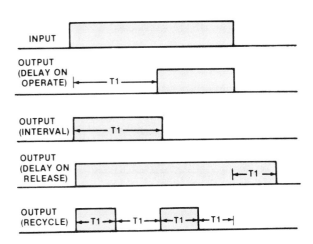

operational diagram

Optical isolation. Two circuits which are connected only through an LED transmitter and photoelectric receiver with no electrical continuity between the two circuits.

Phase asymmetry. When three phase power is generated and transmitted, each of the three power lines are 120 electrical degrees apart. It is important to good motor operation that the incoming power be kept at the balanced 120° spacing. If an unbalance does accrue, the motor will run at temperatures beyond the normal rating. This higher temperature can result in insulation breakdown and shortened motor life.

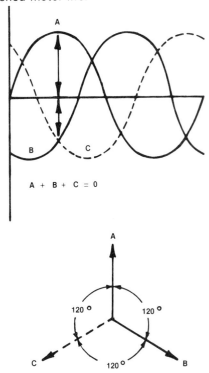

phase asymmetry

Phase breaking (also called single-phasing). When one of the three-phase power lines is lost, usually due to a fuse or contact opening, an extreme case of phase unbalance will exist. A three-phase motor running on single phase will continue to run, drawing all its current from two of the three lines. This condition is usually undetectable by measuring the voltage at the motor terminals because the open winding in the motor is generating a voltage almost equal to the phase voltage that is lost. This condition can exist long enough to burn out the motor.

Phase sequence. Reversing any two of the three phase power lines will reverse the direction of rotation of a three-phase motor. This can cause damage to machinery and injury to personnel. The National Electrical Code® requires protection against incorrect phase sequencing on all equipment transporting people (such as elevators).

Positive logic. The use of binary logic in such a way that "1" represents a positive logic level (e.g., 1 = 5V, 0 = 0V). This is the conventional use of binary logic.

PROM. Programmable read-only memory. A digital storage device that can be written into only once, but can be read continually.

Pulse duration. The time that an input pulse (single) must be present to correctly be registered by the control module. This duration is given as a minimum time, such as 15ms, or as a time range such as 15ms to infinity.

Radio frequency interference (RFI). Electromagnetic interference in the radio frequency range.

Random turn-on. Initial turn-on may occur at any point on the AC line voltage cycle.

Reflective scan. A scan technique in which the light source is aimed at a reflecting surface to illuminate the photoreceiver. Retroreflective, specular, and diffuse scan are all reflective scan techniques.

Relay, mechanical. An electromechanical device that completes or interrupts a circuit by physically moving electrical contacts into contact with each other.

Relay, solid state. A solid state switching device that completes or interrupts a circuit electrically, with no moving parts.

Resistive load. An electrical load characterized by its not having any significant inrush current. When a resistive load is energized, the current rises instantly to its steady state value, without first rising to a surge value.

Retroreflective scan. The reflective scan technique that uses a special reflector (retroreflector) to return light along the same path it is sent.

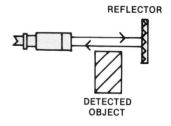

retroreflective scan

Reverse polarity protection. Internal protective circuitry that prevents damage to the sensor in case of accidental reverse polarity connection (plus to minus, minus to plus).

ROM (read-only memory). A digital storage device specified for a single function. Data is loaded permanently into the ROM when it is manufactured. This data is available whenever the ROM address lines are scanned.

RS232. An EIA standard, originally introduced by the Bell system, for the transmission of data over a twisted-wire pair. It defines pin assignments, signal levels, etc. for receiving and transmitting devices.

RS422. An EIA standard for the electrical characteristics of balanced voltage digital interface.

Scan time. The time required to read all inputs, execute the control program, and update local and remote I/O. This is effectively the time required to activate an output that is controlled by programmed logic.

SCR (silicon controlled rectifier). A three-junction semiconductor device that is normally an open circuit until a signal is applied to the gate terminal, at which time it rapidly switches to the conducting state.

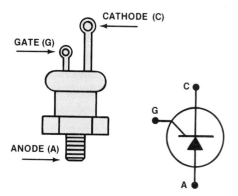

SCR (silicon controlled rectifier)

Single phasing. The condition that occurs when one phase of a three phase system opens.

Snubber circuit. Circuits that suppress transient spikes, preventing a false turn-on.

Specular scan. A reflective scan technique in which reflection from a shiny surface illuminates the photoreceiver.

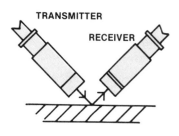

specular scan

Surge current. A current of short duration that occurs when power is first applied to loads.

Symmetrical. Used in timer terminology to describe a recycle timer in which the on time and off time are the same and cannot be set differently.

Translucent. The quality of allowing light to pass through.

SUPPLY VOLTAGE CONTACTS

symmetrical

Thermistor. An electrical resistor whose resistance varies sharply with temperature.

Thermocouple. A device constructed of two dissimilar metals that generates a small voltage as a function of the temperature difference between a measuring and reference junction.

Thyristor. A bistable semiconductor device that can be switched from the off-state to the on-state, or vice versa.

Transducer. A device used to convert physical parameters, such as temperature, pressure, weight, into electrical signals.

Transient. A temporary current or voltage that occurs randomly and rides the AC voltage sine wave.

TTL. Abbreviation for transistor/transistor logic. A family of integrated circuit logic. (Usually 5 volts is high or "1" and 0 volts is low, or "0"; 5V = 1, 0V = 0.) This family of devices is characterized by high speed and medium power dissipation.

Upper range value. The highest quantity that a device can adjust to or measure.

Voltage, breakover. The voltage across the output of a solid state switch in the off-state at which a breakover turn-on occurs.

Voltage, nonrepetitive blocking. The maximum voltage to be applied to the output of an SSR in the off-state. Higher voltages may result in breakover turn-on.

Wiring diagram. The wiring diagram is intended to show as closely as possible the actual connection and placement of all component parts in a device or circuit, including the power circuit wiring.

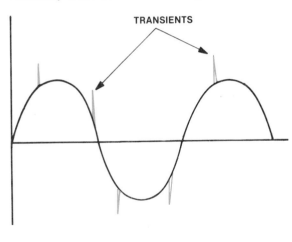

TRANSIENTS

transient

Triac. A solid state switching device used to switch alternating current.

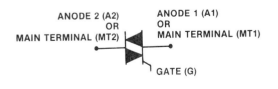

ANODE 2 (A2) OR MAIN TERMINAL (MT2)

ANODE 1 (A1) OR MAIN TERMINAL (MT1)

GATE (G)

triac

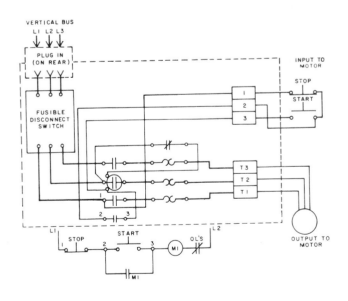

wiring diagram

Yawing. To deviate from the straight course. Term used in windmill control. In windmill control "forced yawing" is sometimes required to deviate the windmill angle from its set course by the wind force.

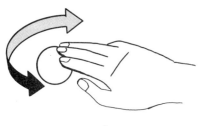

yawing

Zero current turn-off. Turn-off at essentially the zero crossing of the load current that flows through an SSR. A thyristor will turn off only when the current falls below the minimum holding current. If input control is removed when the current is a higher value, turn-off will be delayed until the next zero current crossing.

Zero voltage turn-on. Initial turn-on occurs at a point near zero crossing of the AC line voltage. If input control is applied when the line voltage is at a higher value, initial turn-on will be delayed until the next zero crossing.

INDEX

Illustrations are listed in italic.